Springer Series in Vision Research

More information about this series at http://www.springer.com/series/10633

About the Series

The Springer Series in Vision Research is a comprehensive update and overview of cutting-edge vision research, exploring, in depth, current breakthroughs at a conceptual level. It details the whole visual system, from molecular processes to anatomy, physiology and behaviour and covers both invertebrate and vertebrate organisms from terrestrial and aquatic habitats. Each book in the series is aimed at all individuals with interests in vision including advanced graduate students, post-doctoral researchers, established vision scientists and clinical investigators. The series editors are N. Justin Marshall, Queensland Brain Institute, The University of Queensland, Australia, and Shaun P. Collin, Neuroecology Group within the School of Animal Biology and the Oceans Institute at the University of Western Australia. This volume on Human Colour Vision covers many recent developments in the field and provides descriptions of new methods and emerging hypotheses. Although relatively colour blind, or at least compromised, compared to some other animals, humans are particularly concerned with colour. We fill our world with it and in common with other species use colour for object detection and discrimination of certain features, basing many of our day-to-day judgements on colour differences. We therefore hope that this book will be of interest to anyone with interests in the biology of colour vision, the medical aspects of what happens when it fails, other areas of colour science and within the world of art and design.

Jan Kremers • Rigmor C. Baraas
N. Justin Marshall
Editors

Human Color Vision

Springer

Editors
Jan Kremers
University of Erlangen-Nürnberg
Erlangen, Germany

N. Justin Marshall
Queensland Brain Institute
The University of Queensland
Brisbane, Australia

Rigmor C. Baraas
Department of Optometry and Visual Science
University College of Southeast Norway
Kongsberg, Norway

Springer Series in Vision Research
ISBN 978-3-319-44976-0 ISBN 978-3-319-44978-4 (eBook)
DOI 10.1007/978-3-319-44978-4

Library of Congress Control Number: 2016955085

Printed on acid-free paper

This Springer imprint is published by Springer Nature
The registered company is Springer International Publishing AG
The registered company address is: Gewerbestrasse 11, 6330 Cham, Switzerland

Preface

Color vision is considered to be a visual sensation that is intimately related to emotions. If the world is "colorful," then it is positively full of wonder and surprises. When it is "bleak" or "gray," then the prospects are sad and pessimistic. Red is warm but can mean danger. Blue is cold. Yellow is a color of warning.

Color vision is a marvelous subdiscipline in vision research, embraced by those who study it and sometimes carefully avoided by those for whom it is only of indirect interest.

In the recent years, our understanding of human color vision has been tremendously advanced by many new developments, ideas, and achievements. It has been and still is an exciting time for color vision scientists. We therefore think that it is timely that these new developments are brought together in a book, particularly if it is one in a series on Vision Research. In this book, many new developments have been assembled, covering many different levels from genetics to perception, and studied with state-of-the-art methods such as genetics, morphology, imaging techniques, electrophysiology, psychophysics, and computational neuroscience. The genetics of cone photopigments is discussed in Chap. 1. Further new exiting developments have been obtained in the study of cone mosaics (Chap. 3), in the physiology of color vision in retinal (Chaps. 2 and 4) and cortical circuitries (Chap. 7), and in color psychophysics and perception (Chaps. 5, 6 and 8). Going beyond the questions about the processes leading to visual perception within an individual, the book also considers the latest computational models (Chap. 9), clinical implications and the question how retinal disorders can compromise color vision (Chap. 10), and finally the evolution of color vision (Chap. 11). We hope that the reader will find the chapters inspiring and helpful in defining scientific topics that will be of interest in the future. We think that there will be many interesting challenges. To name but a few, the following topics may emerge: the molecular basis of color vision; the study of single cells and pathways and their visual responses in the living retina; the responses of cells in their intact circuitries; the mathematical description of color processing; the improved use of color vision in diagnosing and monitoring inherited and acquired disorders of the retina; a better understanding of the many perceptual aspects of color vision.

We were supported by world experts who contributed to the book and wrote chapters on the new developments in their field of interest. We encouraged them to seek contact and collaborate with other experts. The result often was an interesting discussion amongst the authors and with the editors. We are extremely glad and proud that all authors have put so much effort in writing their chapters. We asked the authors to keep the text as simple and understandable as they could (without compromising on the scientific content), so that it also would entice and interest nonexpert scientists and students. The result is a book of which we think highly of and we are confident that it brings the latest developments in color vision research for a broader scientific audience. We hope you, as a reader, will agree.

We would like to thank the authors for their brilliant efforts. We appreciate it enormously. The collaboration between the series editors and with Springer was also extremely positive and inspiring.

It remains to thank those who continuously supported us. More particularly:

Rigmor C. Baraas: Finn Erik, Rasmus, the rest of the family, and all present and former members of the lab.

Jan Kremers: Andrea, Leon, Finy, the rest of the family, and all present and former members of the lab.

Justin Marshall: Sue for endless support and patience

Erlangen, Germany — Jan Kremers
Kongsberg, Norway — Rigmor C. Baraas
Brisbane, Australia — N. Justin Marshall

Contents

Contributors

Rigmor C. Baraas National Centre for Optics, Vision and Eye Care, Department of Optometry and Visual Science, Faculty of Health Sciences, University College of Southeast Norway (USN), Kongsberg, Norway

John Barbur Applied Vision Research Centre, School of Health Sciences, City University London, London, UK

Livia S. Carvalho Lions Eye Institute, University of Western Australia, Perth, WA, Australia

Vebjørn Ekroll Laboratory of Experimental Psychology, University of Leuven (KU Leuven), Leuven, Belgium

Wolf M. Harmening Department of Ophthalmology, University of Bonn, Bonn, Germany

Cord Huchzermeyer Department of Ophthalmology, University Hospital Erlangen, Erlangen, Germany

David M. Hunt, Ph.D., F.R.S.B. Lions Eye Institute, University of Western Australia, WA, Australia

School of Animal Biology, University of Western Australia, Stirling Highway, Perth, WA, Australia

Gerald H. Jacobs Department of Psychological and Brain Sciences, University of California, Santa Barbara, CA, USA

Elizabeth N. Johnson Department of Neurobiology and the Duke Institute for Brain Sciences, Duke University School of Medicine and Duke University, Durham, NC, USA

Almut Kelber Department of Biology, Lund University, Lund, Sweden

Frederick A.A. Kingdom Department of Ophthalmology, McGill Vision Research, McGill University, Montreal, QC, Canada

Jan Kremers Department of Ophthalmology, University Hospital Erlangen, Erlangen, Germany

School of Optometry and Vision Science, University of Bradford, Bradford, West Yorkshire, UK

Barry B. Lee School of Optometry, State University of New York, New York, NY, USA

Max-Planck Institute for Biophysical Chemistry, Göttingen, Germany

Declan J. McKeefry School of Optometry and Vision Science, University of Bradford, Bradford, West Yorkshire, UK

Kathy T. Mullen Department of Ophthalmology, McGill Vision Research, Montreal, QC, Canada

Maria Olkkonen Department of Psychology, Science Laboratories, Durham University, Durham, UK

Institute of Behavioural Sciences, University of Helsinki, Helsinki, Finland

Neil R.A. Parry School of Optometry and Vision Science, University of Bradford, Bradford, West Yorkshire, UK

Centre for Hearing and Vision Research, Institute of Human Development, University of Manchester, Manchester, UK

Vision Science Centre, Manchester Royal Eye Hospital, Central Manchester University Hospitals NHS Foundation Trust, Manchester Academic Health Science Centre, Manchester, UK

Austin Roorda School of Optometry, University of California, Berkeley, CA, USA

Ramkumar Sabesan School of Optometry, University of California, Berkeley, CA, USA

Lawrence C. Sincich Department of Optometry and Vision Science, University of Alabama at Birmingham, Birmingham, AL, USA

Luiz Carlos L. Silveira Tropical Medicine Nucleus and Biological Sciences Institute, Federal University of Pará, Belém, Pará, Brazil

Ceuma University, São Luís, Maranhão, Brazil

William S. Tuten School of Optometry, University of California, Berkeley, CA, USA

Thomas Wachtler Computational Neuroscience, Department Biologie II, Ludwig-Maximilians-Universität München, Planegg-Martinsried, Germany

Christian Wehrhahn Vision Center Laboratory, The Salk Institute, La Jolla, CA, USA

Andrew J. Zele School of Optometry and Vision Science and Institute of Health and Biomedical Innovation, Queensland University of Technology (QUT), Brisbane, QLD, Australia

Chapter 1
The Genetics of Color Vision and Congenital Color Deficiencies

David M. Hunt and Livia S. Carvalho

Abstract Primates are unique among mammals in possessing trichromacy. In Old World primates, it is based on three cone classes in the retina, each expressing a different class of visual pigment. These pigment classes are each orthologues of pigments present throughout the vertebrate kingdom, the short wavelength-sensitive (SWS1, SWS2, LWS and MWS) pigment and two representatives of the long wavelength-sensitive (LWS) pigment, L cone opsin and M cone opsin. The latter two pigments arose from a duplication of the *LWS* gene that occurred at the base of the Old World primate lineage to give an array of two closely adjacent opsin genes on the X chromosome. This close proximity and the extensive sequence identity of the *L* and *M* genes promotes mispairing of the genes and thereby underlies the high frequency of red-green color blindness seen in humans. The consequences of this mispairing are the loss of either the *L* or *M* gene to give full dichromacy, or the generation of hybrid genes to give anomalous trichromacy. Generally, red-green color blindness is not associated with loss of acuity, although this is present in a rare form of dichromacy called Bornholm eye disease where cone dysfunction and myopia is also present. Other forms of color blindness include the X-linked disorder of blue cone monochromatism where L and M cones are absent, the dominant disorder of tritanopia where S cone are severely reduced or absent, and the recessive disorder of achromatopsia where all cone classes may be absent.

Keywords Color vision • Evolution • Color blindness • Trichromacy • Dichromacy • Achromatopsia • Visual pigments • Visual opsins

D.M. Hunt, B.Sc., Ph.D., FRSB (✉)
Lions Eye Institute, University of Western Australia,
Verdun Street, Perth, WA 6009, Australia

School of Animal Biology, University of Western Australia,
Stirling Highway, Perth, WA 6009, Australia
e-mail: david.hunt@uwa.edu.au

L.S. Carvalho, B.Sc., M.Sc., PhD
Lions Eye Institute, University of Western Australia,
Verdun Street, Perth, WA 6009, Australia
e-mail: liviacarvalho@lei.org.au

J. Kremers et al. (eds.), *Human Color Vision*, Springer Series in Vision Research, DOI 10.1007/978-3-319-44978-4_1

1.1 Introduction

The vertebrate retina contains photoreceptors cells that are specialized for the capture of light. These cells are subdivided into two classes, the rods and cones. Rods are responsible for monochromatic vision in dim light and cones for vision at normal light levels and for color vision. In rod and cone photoreceptors, the outer segments are composed of a stack of membranous disks in which the key molecules for photon capture, the photosensitive visual pigments, are embedded.

The mechanism of color vision depends critically on a comparison of the amount of light (photon capture) falling on spectrally different types of cone photoreceptors that are maximally sensitive to different wavelengths (For further details see Chap. 4). This is the process of cone opponency whereby each photoreceptor type is stimulated to a different extent by light of differing spectral content. Comparison of these signals by the brain provides the sensation of color. From this it follows that color vision requires a minimum of two spectrally different types of cone photoreceptors to be present.

In primates, trichromatic color vision is provided by the presence of three classes of cone photoreceptors with wavelengths of maximal sensitivity (λ_{max}) in the yellow-green (around 560 nm, longwave-sensitive, L), green (around 530 nm, middlewave-sensitive, M), and blue (around 430 nm, shortwave-sensitive, S) regions of the spectrum [1–3]. The light-sensitive components of photoreceptors, the visual pigments, are members of the GPCR family of proteins. They comprise a seven transmembrane (TM) opsin protein that is covalently linked to a Lys residue via a Schiff base (SB) to the chromophore. In mammals, this is invariably 11-*cis*-retinal derived from vitamin A1, so the peak spectral absorption (λ_{max}) of a visual pigment is determined not by the chromophore, but by the amino acid sequence of the opsin protein, with certain residues tuning the pigment to particular spectral locations.

1.2 Phototransduction Cascade

The ability to see in dim and bright light arises in vertebrates from differences in the light sensitivity of rods and cones. In general, rod photoreceptors are more sensitive than cones, have less "dark noise', adapt over a much narrower range of light intensities, but have slower response kinetics [4–7]. In contrast, cones have a higher frequency of spontaneous thermal isomerizations of the chromophore [8], and a substantial proportion of cone opsin exists in an apo form lacking bound chromophore. This apo-pigment is able to activate phototransduction and may account, in part, for the faster and larger response of cones [9, 10]. The presence of either a rod or cone pigment however makes only a minor contribution to the kinetics of the photoresponse, as demonstrated by the photoreceptors present in the retina of the nocturnal Tokay gecko, *Gecko gecko*, which have a rod-like morphology [11] and rod-like photokinetics [12], but contain only cone visual pigments [13]. The fundamental difference between rods and cones with respect to sensitivity to light is not therefore simply determined by the presence of either a rod or cone visual pigment.

The visual process involves the conversion of the signal from the light-activated visual pigment to an electrical impulse and this is achieved by the process of phototransduction within the photoreceptor; this process also results in a substantial amplification of the original signal. A striking feature of vertebrate photoreceptors is the number of rod- and cone-specific isoforms that form the phototransduction cascade (Fig. 1.1). Absorption of light causes the isomerisation of the chromophore, 11-*cis*-retinal, to the all-*trans* form in a photobleaching sequence with consequent conformational changes in the opsin protein, leading to the activation of the G protein transducin by the activated form of the visual pigment, metarhodopsin II (meta II) (Reviewed in Ref. [14]). Meta II activates the heterotrimeric GTP-binding protein transducin, which is composed of α, β, and γ subunits. GDP bound to transducin is replaced by GTP and the GTP-α-subunit conjugate dissociates from the $\beta\gamma$ component. Different isoforms for all three of these subunits are present in rods and cones. Phosphodiesterase (PDE) in rods is composed of catalytic α and β-subunits and two inhibitory γ subunits, whereas the cone form is composed of two identical α' subunits and two inhibitory γ subunits. Activation of PDE involves the interaction with GTP-α-transducin and the dissociation of the inhibitory γ subunits. This activation results in the breakdown of cGMP and the closing of the cGMP-gated (CNG) channels, which also differ between rod and cones, leading to reduced levels of intracellular Ca^{2+}. Cone channels are generally more permeable to Ca^{2+} than rod channels, and this may underlie, in part, the more rapid and larger light dependent changes in Ca^{2+} concentration in cone cells [15].

The restoration of the cGMP and Ca^{2+} resting states is achieved by the activation of retinal-specific guanylate cyclases (GCs) by the Ca^{2+}-binding guanylate cyclase activating proteins (GCAPs) [16]. At the low levels of Ca^{2+} that exist after the closure of the CNG channels, GCAPs activate the cyclase function of GC, resulting in the production of cGMP, the reopening of the CNG channels and an increase in Ca^{2+} concentration to the resting state level. Two isoforms of GC, GC1 and GC2, have been identified, but GC1 is clearly the more important enzyme for phototransduction since null mutations in the corresponding *GUCY2D* gene cause the severe blinding disease of Leber Congenital Amaurosis [17], and altered photoreceptor survival in gene knockout mutant mice [18]. Missense mutations are also a major cause of dominant cone dystrophy [19, 20] (see Chap. 10). *GUCY2D* is expressed in both rods and cones, although the level of expression is higher in the latter [21]. Multiple isoforms of retinal GCAPs have also been identified [22–25], but GCAP1 encoded by *GUCA1A* appears to play the more important role. Mutations in GCAP1 are known to cause dominant cone dystrophy [26–30].

The inactivation of the cascade is necessary for recovery from a photoresponse. This occurs as a two-step process, involving phosphorylation of the activated pigment (meta II) by rhodopsin kinase targeted to serine residues in the carboxy terminus of the opsin protein, followed by binding of the inhibitory protein arrestin [31]. Two retinal-specific kinases, GRK1 [32, 33] and GRK7 [34], are present in the

a

Gene	Subunit	Isoforms: Rod	Isoforms: Cone
Transducin	Alpha	*GNAT1*	*GNAT2*
	Beta	*GNB1*	*GNB3*
	Gamma	*GNGT1*	*GNGT2*
Phosphodiesterase	Catalytic	*PDE6A*, *PDE6B*	*PDE6C*
	Inhibitory	*PDE6G*	*PDE6H*
Cyclic nucleotide gated channel	Alpha	*CNGA1*	*CNGA3*
	Beta	*CNGB1*	*CNGB3*
Rhodopsin kinase		*GRK1*	*GRK7*
Arrestin		*SAG*	*ARR3*

b

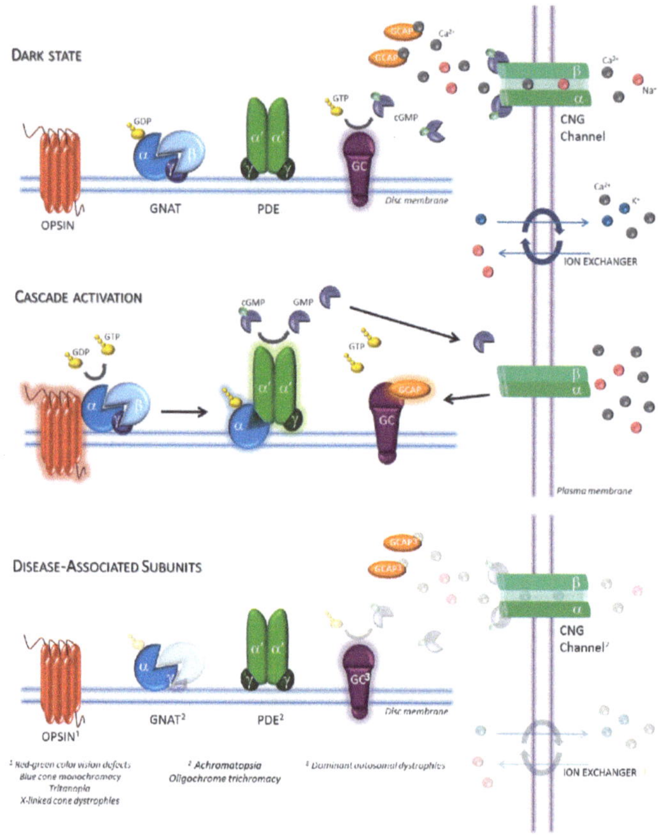

Fig. 1.1 Phototransduction cascade. (**a**) Gene isoforms of component processes that are expressed in either rods or cones. (**b**) Schematic diagram showing role of component processes in the activation of the cascade. D*ark state* shows the component processes with the cGMP-gated CNG channels open through the production of cGMP by activated retGC. *Cascade activation* results from the conversion of GDP to GTP and the release of the α-transducin subunit. This in turn activates PDE by removal of the inhibitory γ-subunit, leading to the breakdown of cGMP, the closure of the CNG channels and a fall in Ca^{2+} levels. Ca^{2+}-free GCAPs bind to retGC, enabling the activation of the cyclase activity and the production of cGMP. The restored levels of cGMP lead to a reopening of the CNG channels and the restoration of Ca^{2+} levels. *Disease-associated subunits* identify the component processes that are involved in defect in color vision

human retina; *GRK7* is preferentially expressed in human cones [35]. Inactivation is then completed by the binding of the inhibitory protein arrestin [31] to the phosphorylated terminus of the pigment; two isoforms of arrestin are present, S-antigen (SAG) in rods, and C- or X-arrestin (ARR3) in cones [36].

1.3 Evolution of Visual Pigments in Vertebrates

1.3.1 Ancestral Vertebrate Complement

The vertebrate blueprint for color vision consists of four spectrally distinct visual opsins, each encoded by a different gene, that first arose in the agnathans, as demonstrated by their retention by the southern hemisphere lamprey, *Geotria australis* [37, 38]. Orthologues of these four classes survived the split between the agnathans and gnathostomes which occurred over 540 MYA [million years ago] [39, 40] and have been retained by most vertebrate classes. They consist of a longwave-sensitive (LWS) pigment with λ_{max} 500–570 nm, a middlewave-sensitive (MWS or RH2) pigment with λ_{max} 480–530 nm, and two shortwave-sensitive pigment classes, SWS2 with λ_{max} 400–470 nm and SWS1 with λ_{max} 355–445 nm. In vertebrate visual pigments with λ_{max} values >385 nm, the SB is protonated, with a negatively charged residue at site 113 (usually Glu113) acting as a counterion to stabilize the proton of the SB [41]. In a subset of SWS1 pigments that show ultraviolet-sensitivity with λ_{max} values around 360 nm, the SB is unprotonated in the resting state [1, 42] (For an extended account of evolution of color vision see Chap. 11).

1.3.2 Loss of Cone Pigment Classes in the Early Evolution of Mammals

Not all of these cone pigment classes are found in mammals. Cone pigment loss is thought to have arisen during a nocturnal phase that marks the early evolution of the mammals around 150–200 million years ago (MYA) [43]. In marsupials and eutherian mammals, the *LWS* gene is paired with the *SWS1* gene, with the loss of the *SWS2* and *RH2* genes. The egg-laying protherian mammals, the platypus, *Ornithorhynchus anatinus*, and echidna, *Tachyglossus aculeatus*, belonging to the Order Monotremata that diverged from the marsupial/placental mammal lineage around 200 MYA, have also retained the *LWS* gene but this is paired the *SWS2* gene [44, 45]. In contrast therefore to marsupial and eutherian mammals, protherians have lost the *SWS1* and *RH2*. The presence of the *SWS2* gene in monotremes also means that ancestral mammals prior to the protherian–therian split must have retained both *SWS* genes which, in combination with the *LWS* gene, would have provided the basis for trichromacy.

1.4 Evolution of Trichromacy in Primates

1.4.1 Gene Duplication and Gene Conversion in Old World Primates

Amongst the mammals, only primates show true trichromatic color vision. The evolutionary drive behind the acquisition of trichromacy is thought to be improved color discrimination in the red/green region of the spectrum for the detection and evaluation of ripe fruits [46–49] and young nutritious leaves [50] against the green foliage of the rainforest. This has been achieved in Old World primates by a ~40 kb duplication of the X-linked *LWS* gene which occurred at the base of the Old World primate lineage [3, 51, 52] to give rise to an array of two adjacent genes (Fig. 1.2). The duplication generated a copy of the entire coding regions of the *LWS* opsin gene plus an almost complete copy of the *TEX28* gene [52, 53], a gene that is expressed in testis and not thought to be involved in vision. The duplicated *LWS* genes have subsequently diverged to give an upstream copy (*OPN1LW*) encoding a long wavelength-sensitive (L) pigment with λ_{max} around 560 nm and a downstream copy (*OPN1MW*) encoding a middle wavelength-sensitive (M) pigment with λ_{max} around 535 nm. In humans, intron 1 of the *L* opsin gene is generally longer by ~2.9 kb than intron 1 in the *M* opsin gene. This long intron is rarely present (1 %) in the *M* opsin gene of Caucasians but is polymorphic in African-Americans at a frequency of 35 % [54]. Individual cones express just one copy to give L and M cones, which together with S cones expressing the autosomal *SWS1* gene (*OPN1SW*), give rise to full trichromacy.

In females with two X chromosomes, the process of X chromosome inactivation will ensure that only one opsin gene array is active in any given photoreceptor, but another mechanism must exist that ensures that only one gene, either *L* or *M*, is switched on. This may be the role of the locus control region (LCR), a highly conserved segment of DNA located in humans between 3.1 and 3.7 kb upstream of the opsin gene array [55]. The LCR acts as an enhancer and is present in species where there is only a single X-linked opsin gene [52]; activation of this gene is nevertheless thought to require interaction between the LCR and the promoter region immediately

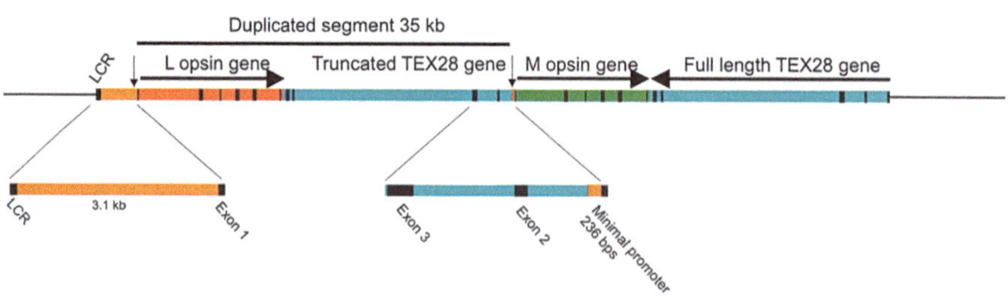

Fig. 1.2 Origin of opsin gene duplication on the X chromosome of Old World primates. The exons for the *L* and *M* opsin genes and for the *TEX28* gene are shown as *black bars*. *Arrows* indicate the direction of transcription. The positions of the upstream LCR and the minimal promoter for the *M* opsin gene are shown

upstream of the coding region [56]. Since the cone-specific phototransduction genes described above are all expressed in both L and M cones, the selection of either an *L* or *M* opsin gene for expression [57] would appear to be the single determinant for the production of either an L or M cone. This process may depend on the stochastic selection of one of two stable and mutually exclusive states. The *L* and *M* genes each possess a minimal promoter immediately upstream of exon 1, so this region could interact in a gene-specific manner with the LCR to activate only the adjacent gene. This was tested with transgenic mice that carried the human LCR and the L and M promoters driving different reporter genes within a transgene [58]. As shown in Fig. 1.3, the activity of the gene driven by the more proximal promoter to the LCR

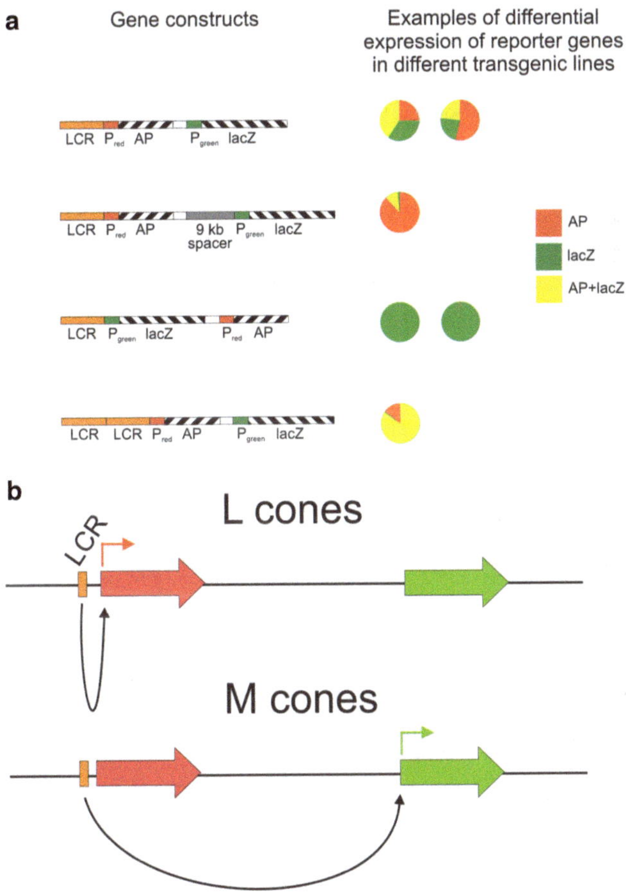

Fig. 1.3 The role of the LCR in directing reporter gene expression from *L* and *M* opsin gene promoters. (**a**) The relative activity of the *L* and *M* promoters in four different transgene arrays assessed in a number of transgenic mouse lines by scoring individual cone cells for the expression of the reporter genes. P_L human *L* opsin gene promoter, P_M human *M* opsin gene promoter, *AP* human placental alkaline phosphatase, *lacZ E. coli* β-galactosidase. Redrawn from Ref. [58]. (**b**) Potential mechanism for the stochastic expression of *L* or *M* opsin genes via LCR-opsin gene promoter interaction

was higher, while a 9 kb spacer interposed between the L and M reporter genes reduced the expression of the downstream gene even further, indicating that physical distance from the LCR may reduce the frequency of promoter activation. In this way therefore, only a single opsin gene within the array is activated, with the probability of interaction with either the L or M promoter dependent on the distance from the LCR and on certain key regions of the promoter, with different promoters showing stronger or weaker affinities [56]. This may also in part explain the differences in the relative number of L and M cones in different individuals (For more on L:M ratios see Chaps. 2, 3 and 5). Most subjects show a fourfold range [59] but this can extend in rare cases to a 30-fold range. It should be emphasized however that this variation in L and M cone ratios is not associated with color vision defects.

1.4.2 Spectral Tuning of Primate Visual Pigments

The spectral shifts between primate L and M pigments are largely due to substitutions at three sites, 180 encoded by exon 3 and 277 and 285 encoded by exon 5 [60], although smaller changes arise from substitution at 116 in exon 2, and 230 and 233 in exon 4 [61]. The residues present at the three former sites are polar Ser, Tyr, and Thr, respectively, in the L pigment and nonpolar Ala, Phe, and Ala, respectively, in the M pigment [51, 62]. As shown by site-directed mutagenesis and in vitro expression studies, the spectral shifts achieved by substitution at these sites are approximately additive [63, 64]. Site 180 is polymorphic in humans; Ser180 is the more common residue in the L pigment, but Ala180 is present at a significant frequency to give a short wavelength-shifted L pigment in some individuals [65].

UV-sensitive (UVS) SWS1 pigments are found throughout the vertebrate kingdom and almost certainly represent the ancestral form of the pigment [2, 66]. Within the mammalia, UVS pigments are relatively common in marsupials but are restricted in eutherians to just a subset of species from the Orders Rodentia, Chiroptera, and Insectivora [1]. All species of primates possess violet-sensitive (VS) pigments [67, 68] but the tuning method to change from UV to violet sensitivity remains uncertain. It has been proposed from a comparison with the mouse UVS pigment that tuning from UV to violet requires the simultaneous replacement of residues at three sites, Phe86Leu, Thr93Pro, and Ser118Thr [69, 70]. However, other residue changes at these sites are found in members of the Haplorrhini (New World and Old World monkeys and tarsiers), and the Strepsirrhini (prosimians excluding tarsiers), including Phe86 in one species, the aye-aye, a lemur endemic to Madagascar [71]. Phe86 is generally associated with a UVS pigment [66] and it is the replacement of this residue by either Tyr or Ser or Val [72–74] that is responsible for the loss of UV-sensitivity in a number of non-primate species [75]. In vitro expression of the aye-aye pigment however gave a peak sensitivity at 409 nm [71], so Phe86 does not in this case result in a UVS pigment. The only residue that is consistently present in primate VS pigments is Pro93. Pro93 is also found in the VS pigment of the clawed frog [76] so a Thr93Pro substitution may underlie primate VS pigments. Consistent with this is the observation that a Pro93Thr substi-

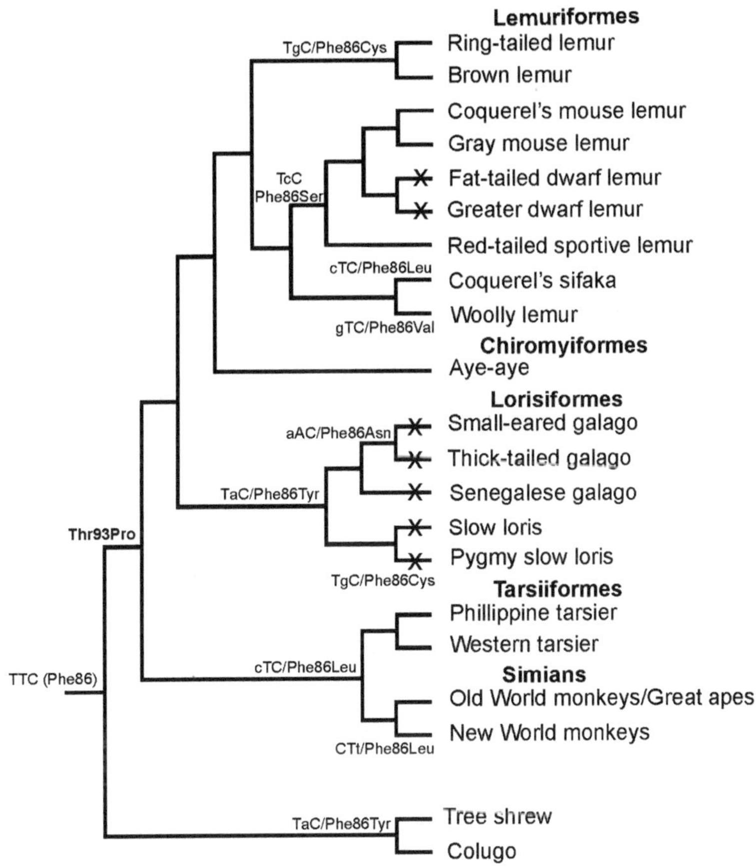

Fig. 1.4 Phylogenetic relationships of different primate species showing the different residues found at site 86 in S cone opsins, together with the corresponding codon sequences. Under the scenario presented, the Thr93Pro substitution occurred at the base of the primate lineage. Phe86 was present in the ancestral primate and retained by the aye-aye. Replacement of Phe by either Leu, Cys, Asn, Val, Ser, or Cys then occurred by single nucleotide changes at each step to generate the amino acid changes. Crosses on lineages indicate species where *SWS1* pseudogenes are present. Based on data from Ref. [78]

tution into the aye-aye pigment by site-directed mutagenesis causes a shift to 371 nm in the UV [71]. This does not however rule out a role for site 86 in the initial shift to violet sensitivity, especially since Tyr86 is found in the S cone opsins of the tree shrew and colugo [77], close relatives of the primates. As shown in Fig. 1.4, the most parsimonious scenario is that Phe86 was present at the base of the primate lineage but underwent a series of substitutions in the various primate lineages. A subset of these changes (Ser, Tyr, and Val) have been shown in other species to generate a shift into the violet [75], so these changes may have been responsible for the origin of the VS pigment in primates. However, if substitution at site 93 also occurred at the base of the primates, then this

Table 1.1 Prevalence of red-green color deficiency in European populations

Color vision deficiency	Cone(s) affected	Male prevalence (%)
Anomalous trichromacy		
Protanomaly	L	1.1
Deuteranomaly	M	4.6
Dichromacy		
Protanopia	L	1.0
Deuteranopia	M	1.3
Combined		8.0

Data from Ref. [80]

would have removed constraints on the residue at site 86, thereby allowing different substitutions at this site (which include Leu, Cys, Ser, Val, Tyr and Asn in different primate species) to fine tune the λ_{max} of the S pigment [71].

1.5 Color Vision Deficiencies and Color Blindness

1.5.1 L and M Opsins in Red-Green Color Vision Defects

The most common forms of color vision deficiencies in humans affect color discrimination in the red/green region of the spectrum and are associated with changes in the X-linked *L* and *M* opsin genes. Red-green color blindness is also referred to as Daltonism after John Dalton, the famous chemist, who was the first to describe the condition and who himself was red-green color blind [79]. The high frequency of these defects amongst males is a direct consequence of hemizygosity of X-linked genes (Table 1.1).

The original duplication in Old World primates generated a tandem opsin gene array on the X-chromosome, with the two copies diverging at key tuning sites to encode the spectrally distinct L and M opsin pigments. The coding regions of these genes share 98 % identity compared to only 40 % with S opsin, and the introns also show a high level of identity. A direct consequence of this close proximity and high level of sequence homology of the duplicate copies of the gene is mispairing during meiosis (Fig. 1.5). When this is followed by intergenic crossover within the array, extra downstream copies of the genes are placed into the array, with a consequent expansion of the number of copies of the *M* opsin gene. Copy number for *M* opsin is known to range from one to five with a mode of two [51, 54, 81, 82]. Mispairing within the array followed by intergenic crossover will also result in the deletion of gene copies and, where the crossover is intragenic between mispaired genes, this will give rise to hybrid genes. It is the latter two events that are responsible for the high frequency of red-green color vision deficiencies in humans [83, 84]. Dichromacy arises when either the *L* opsin gene (protanopia) or *M* opsin gene (deuteranopia) is missing; the prevalence among Caucasian males of these two forms of red-green color blindness is 1.01 % and 1.27 % respectively [80, 85]. Anomalous trichromacy arises from the production of hybrid *L*/*M* genes. The precise position of the intra-

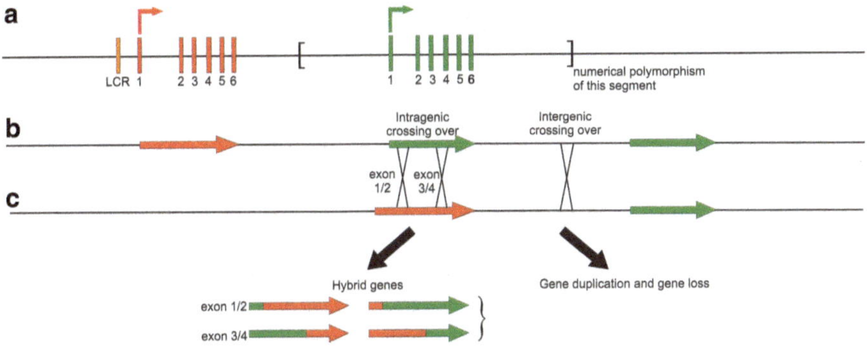

Fig. 1.5 Mispairing and crossover within the L and M opsin gene array. (**a**) Genomic organization of the L and M opsin genes. (**b**) Mispairing of the L and M genes, followed by either intragenic or intergenic crossover. (**c**) Gene duplication/loss and example products of intragenic and intergenic crossover between mispaired genes

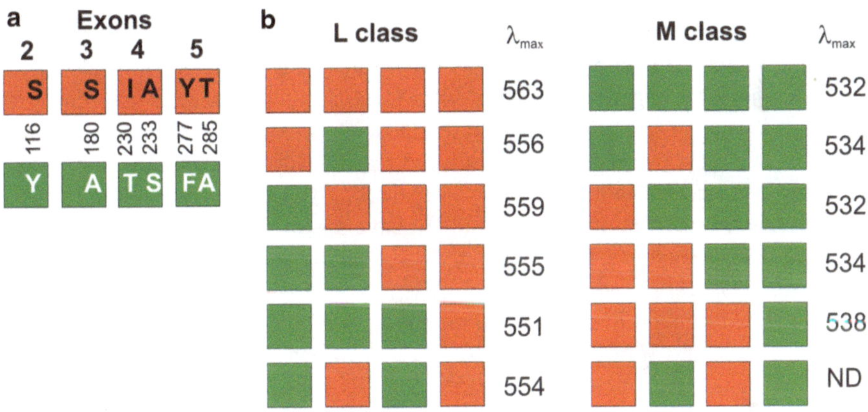

Fig. 1.6 Spectral tuning of L and M pigments. (**a**) Major spectral tuning sites in exons 2–5. (**b**) *L*, *M*, and hybrid genes and λ_{max} values for corresponding pigments. Data from Ref. [63]

genic crossover will determine the relative contribution of *L* or *M* sequence to the hybrid gene. Since the tuning sites 277 and 285 encoded by exon 5 make the major contribution to the spectral shift between L and M pigments, the location of the spectral peak will be largely determined by the origin of exon 5 from an *L* or *M* opsin gene. The most common form of anomalous trichromacy is deuteranomaly, where M cones express a hybrid pigment; the prevalence of this disorder in Caucasian males is 4.63 % [80, 85]. Protanomaly, where L cones express a hybrid pigment, is less common at 1.08 % in Caucasian males. With a few rare exceptions, only the first two genes in the array are expressed [86, 87] so L cone photoreceptors will express a normal L pigment whereas the M cone photoreceptors will express the hybrid pigment. Depending on the particular combination of *L* and *M* sequences in the hybrid gene (Fig. 1.6), there will be a reduction in the spectral separation between the two pigments, thereby reducing color discrimination in the red-green region of the spectrum.

The presence of multiple copies of genes within the opsin gene array and ongoing mispairing and nonhomologous crossover, has resulted in extensive intermixing of *L* and *M* opsin gene sequences. Only the first two genes in the array are generally expressed, so anomalous trichromacy will arise even when intact *L* and *M* opsin genes are present, if the first or second position in the array is occupied by a hybrid gene [84]. Protanomaly arises when a normal *L* opsin gene is absent, so the two pigments produced from the array come from a normal *M* and an *M*-like hybrid gene. It is however not possible to derive such a combination of genes from mispairing between arrays that have the ancestral organization of an upstream L and downstream M opsin genes, so such an arrangement must derive from multiple events [88]. This process of nonhomologous exchanges between genes and the generation of hybrids by mispairing and crossover has been referred to as gene conversion, a process whereby sequences of related genes become homogenized; in the present case, conversion has resulted in the coding and noncoding regions of the *L* and *M* opsin genes becoming more like each other within species than the same regions of either the *L* or *M* orthologues between species [62, 89–91]. In humans, it is only exons 2–5 that remain distinct, presumably due to natural selection acting on the tuning residues encoded by these exons [92]. Surprisingly, conversion is also seen in the X-linked opsin gene arrays of Old World monkeys, even though gene changes associated with red-green color vision deficiencies is extremely rare in nonhuman primates, with just two reports of genetic defects in the opsin gene array [93]. In the first study carried out in a colony of 744 male long-tailed macaques, *Macaca fascicularis*, three male animals were found to have a hybrid *L/M* gene [94], giving a frequency of anomalous trichromacy of 0.4 %, and in the second study involving 58 male chimpanzees, *Pan troglodytes*, an *L/M* hybrid was identified in just a single male [95]. Conversion is also seen in gibbons, but in this case, it is located in the introns of the *L* and *M* opsin genes [96]. Multiple *M* opsin genes is also seen in gibbons, with an average incidence of 23.5 % across all species studied. The corollary of this, that is the loss of genes from the array, was not seen, nor were any hybrid genes identified, in 152 individual animals genotyped. Overall therefore, this would appear to be strong evidence that red-green color vision deficiencies are highly detrimental to survival within nonhuman primate communities and are removed by natural selection.

1.5.2 Blue-Cone Monochromacy

Blue (S) cone monochromatism (BCM) is a rare X-linked congenital disorder that affects approximately 1 in 100,000 individuals. Affected males lack L and M cones so vision is dependent on S cones and rod photoreceptors alone [83, 97]. BCM typically presents with reduced visual acuity (6/24 to 6/60), pendular nystagmus and photophobia. Patients often have myopia, but the S cone ERG is well preserved [98].

The different genetic mechanisms underlying BCM all involve changes in the *L/M* opsin gene array, resulting in nonfunctional L and M photopigments and thus inactive L and M cones [63, 99]. These changes fall into three classes. The first class that accounts for approximately 60 % of cases involves a two-step process whereby nonhomologous recombination between the *L* and *M* opsin genes has reduced the number of genes in the opsin array to one, followed by a mutation in the remaining gene that inactivates the encoded pigment. Two missense mutations (Cys203Arg and Pro307Leu) and a nonsense mutation (Arg247Stop) have been reported [99–101]. Where a Cys203Arg mutation is the cause of BCM, it most commonly occurs in an *L/M* hybrid gene [102]. Imaging of the retina by adaptive optics in such individuals shows a relatively undisrupted cone mosaic even though cone density is reduced. This implies that cones expressing the mutant opsin undergo degeneration prior to final packing [103]. The second class of changes that accounts for approximately 40 % of cases involves a deletion upstream of the *L/M* opsin gene array that includes the LCR; the loss of the LCR eliminates expression from both the *L* and *M* opsin genes even though both may be fully intact and unaltered [101, 104–106]. Imaging of individuals with an LCR deletion showed that cone density was reduced and the organization of the mosaic was also disrupted [106]. Finally, exon deletion has been observed, either in an *L* opsin gene [107] or in a hybrid L/M gene [102]; in both cases, the array had been reduced to this single nonfunctional gene. BCM is generally considered to be a stationary disorder although longitudinal assessments have found some evidence for progression [100, 108].

1.5.3 Tritanopia

The loss of functional S cones, a condition called tritanopia, arises from mutations in the *S* opsin gene on chromosome 7. It is very rare compared to red-green color blindness with an estimated frequency of between 1 in 13,000 and 1 in 65,000 [109], although a later study puts the frequency much higher at 1 in 500 [110]. The absence of the blue-sensitive S pigment limits blue-yellow color discrimination. Tritanopia is inherited as an autosomal dominant disorder arising from missense mutations in the S opsin gene, which means that the presence of a single mutant copy of the S gene is sufficient to cause the visual defect. Reported amino acid substitutions caused by missense mutations are the replacement of a conserved Leu at site 56 with Pro (Leu56Pro) [111], Gly79Arg, Thr190Ile [112], Ser214Pro, Pro264Ser [113, 114], and Arg283Gln [115]. In all four cases, the substitutions are within the helical transmembrane domains of the opsin. Three of the substitutions involve either the gain or loss of Pro residues; since Pro residues introduce kinks, it would be expected to result in significant changes in the conformation of the pigment. The other changes involve either the gain (Gly79Arg) or loss (Arg283Gln) of a charged residue; in both cases, the change would be expected to impact adversely on pigment function. What is perhaps surprising is that the production of mutant opsin protein from a single mutant *S* opsin allele is sufficient to cause the loss of S cone function, even though a normal allele encoding wild type S opsin is also present. In this regards, tritanopia is similar to the dominant forms of retinitis pigmentosa that arise from mutations in the rod opsin (*RH1*) gene [116–119]. Further evidence of the impact of these mutations comes from a study of two subjects (father and daughter) with the same Arg283Gln mutation but showing different degrees of color vision deficiency [115]. In vivo imaging shows that S cones are present at a near normal density in the daughter but totally absent in the father at an eccentricity in the retina where the peak S cone density normally occurs. It would appear from this family that tritanopia is a progressive disorder that shows incomplete penetrance in younger subjects.

1.5.4 Enhanced S-Cone Syndrome

Enhanced S-cone syndrome (ESCS), also known as Goldmann–Favre syndrome, is a recessive form of retinal degeneration in humans that is characterized by severely reduced rod sensitivity, loss of visual acuity, and atypical ERGs that show little or no responses to dim light (scotopic) stimuli, but have large, slow responses to brighter (photopic) stimuli [120–122]. This photopic ERG originates from an excess of S-cones in the retina, with reduced contributions from L and M cones. However, despite the reduction in L and M cone sensitivity, color vision in ESCS individuals is generally assessed to be normal. The genetic basis for ESCS has been shown by [123] to be due to mutations in the *NR2E3* gene. *NR2E3*, together with *NRL* [124], act as regulators of rod photoreceptor cell fate [125]. Both gene products are required for rod photoreceptor development, with *NR2E3* also required to suppress cone development [126]. In the absence of *NR2E3* expression, normal rod development is blocked [127], resulting in a severe reduction in rods and an increase in S cones.

The thyroid hormone system has been shown to play an important role in the development of cone photoreceptors. Thyroid hormone (TH) and the cone-specific TH receptor THRβ2, a nuclear transcription factor encoded by the *thrb* gene, have been shown to control the relative production of S and L opsin during cone development and maturation; in mice with a deletion for *thrb*, there is a selective loss of L cones with a concomitant increase in S cones [128]. In a recent study of an infant with a cone photoreceptor disorder associated with severe thyroid hormone resistance [129], a severely reduced photopic response was reported characterized by severely reduced L and M cone responses and an increased S cone response. The affected child was found to be a compound heterozygote for two missense mutations in the *THRβ2* gene, Arg338Trp and Arg429Trp. In both cases, the substitutions are in the ligand-binding domain, and it is suggested that the severity of the phenotype may be due to a selective disruption of homodimer formation, but not heterodimer formation, with retinoid X receptor (RXR) [130]. The increased S cone response in this disorder is similar therefore to ESCS. However, in ESCS, the increase in S cones is at the expense of rod photoreceptors whereas in this disorder, the enhanced S cone phenotype derives from a conversion of L and M cones to S cones.

1.5.5 Achromatopsia

Complete or incomplete achromatopsia (ACHM), also referred to as rod monochromacy, is characterized by reduced or complete absence of color vision caused by nonfunctional cone photoreceptors. It has been estimated to affect 1 in 30,000–50,000 people worldwide and has an autosomal recessive inheritance pattern [131, 132]. So far, mutations in five genes have been shown to cause achromatopsia, accounting for 93 % of patients [131, 133–139]. In all cases, the genes encode components of the phototransduction cascade (Fig. 1.1), with disease-causing mutations reported in genes encoding the α-subunit (*CNGA3*) and β-subunit (*CNGB3*) of the CNG channel, the α-subunit (*PDE6C*) and inhibitory γ-subunit (*PDE6H*) of cone PDE6, and the α-subunit of cone transducin (*GNAT2*). The mutational spectrum reported so far includes missense and nonsense, deletions and insertions, and splice mutations. However the vast majority of ACHM cases are due to mutations in the *CNGA3* and *CNGB3* genes, with *CNGB3* defects accounting for 50 % of total cases in the Caucasian population [140–142]. Of the remaining cases, mutations in *GNAT2* [137, 138, 143–146] and *PDE6C* [139, 147, 148] each have a prevalence of <2 %, while *PDE6H* disease variants have a prevalence of 0.3 % [136, 149].

Despite the small percentage of *GNAT2* and *PDE6C* mutation-related ACHM cases, the number of disease-causing variants reported for these genes has not been negligible, with 11 mutant variants reported for *GNAT2* [137, 138, 143–146] and 20 for *PDE6C* [139, 147, 148, 150, 151]. Interestingly, the majority of *GNAT2* mutations are nonsense changes while *PDE6C* has a much higher incidence of missense changes. Since both these α-subunits are highly conserved proteins that are integral for the correct activation of different components of the phototransduction cascade, the extremely low number of missense mutations (1/11) [137] in the *GNAT2* gene is surprising. Different subunit interactions between cone transducin and cone phosphodiesterase could account for different selective pressures and mutational tolerance between residues amongst these two genes.

The majority of ACHM cases are due to defects in the *CNGB3* gene, with the most common mutation being a 1 base pair frameshift deletion c1148delC (p.Thr383Ile fs*13); this accounts for around 70 % of *CNGB3* disease-causing alleles [140, 142, 152, 153]. Besides this frameshift deletion, more than 40 different disease-causing variants have so far been reported, with a high proportion of nonsense mutations [135, 142, 152, 154–158]. In contrast, disease-causing mutations in the *CNGA3* gene show greater variability with more than 80 variants described so far and a majority of missense mutations [153]. The CNG channel is a heterotetrameric complex of the α- and β-subunits [159] and this divergence in the type of variants found between the *CNGA3* and *CNGB3* genes aligns well with the different primary and modulatory functions of the α- and β-subunits, respectively [160], indicating a lower tolerance for genetic variance in the *CNGA3* gene.

Most of the reported ACHM studies looking at *CNGA3* and *CNGB3* mutations in large cohorts have been with patients of European or American origin or descent [161], which may give a skewed prevalence of disease variants. Recently, a large study of 130 ACHM patients distributed across 49 different families within the

Israeli and Palestinian populations determined the prevalence and spectrum of *CNGA3* and *CNGB3* mutations [157]. In this study, the most common cause of ACHM in western populations (*CNGB3*, c.1148delC) was only found in 8 % of families in their cohort while in 84 % of families, the disease was due to mutations in the *CNGA3* gene. The high incidence of *CNGA3* disease-variants in this population can be explained by the presence of two founder mutations in a common ancestor of Arab-Muslims and Oriental Jews. Another interesting finding of this study was the increased prevalence of ACHM in this population; extrapolation from these data indicated that ACHM prevalence could be as high as 1 in 5000 within the local Arab Muslim community, making it the second highest ACHM prevalence after the 4–10 % incidence in the population on the Pingelap Island in the Pacific Ocean [132]. Other studies have reported a different prevalence of *CNGB3* and *CNGA3* mutations between Northern European and Middle Eastern descent populations [162–164] but more studies using large cohorts of ACHM patients from differing genetic/populational backgrounds are needed to gain a more complete understanding of ACHM prevalence and disease-causing variants.

Clinical symptoms of ACHM start at birth/early infancy and include congenital pendular nystagmus, poor visual acuity, severe photophobia/hemeralopia, and severe defects in the protan, deutan, and tritan color axes [131, 140]. Hypermetropic refractive errors are common and affected subjects can also present with small central scotoma, eccentric fixation, and paradoxical pupillary responses [131, 149, 151, 165]. Fundus appearance is often normal, but macula changes have been reported, ranging from subtle retinal pigment epithelium (RPE) changes to atrophy and foveal autofluorescence [166]. Clinical symptoms can be highly variable from patient to patient and most studies have been unable to find a correlation between phenotype and genotype and/or specific disease variants, although one study comparing *CNGB3* and *CNGA3* patients reported a slightly more affected macular appearance and reduced visual acuity in the *CNGA3* patients [140].

Analysis of electroretinogram (ERG) and psychophysical color testing responses has led to the division of the disease into complete and incomplete ACHM subtypes [140]. Patients with the complete subtype have no recordable ERG cone function and completely lack of color discrimination, while those with incomplete ACHM retain some residual cone function on ERG, varying degrees of color discrimination and higher visual acuity [139, 143, 145, 162, 167, 168]. None of these studies however has been able to correlate the complete and incomplete subtypes with genotype, age or mutation, so it is possible that differences between ERG parameters and apparatus, psychophysical color vision tests and other traditionally used measures of cone vision amongst the studies could account for the discrepancy in the subtypes. An alternative explanation for residual cone function in some patients is that it is mediated by rods which in ACMH patients show a reduced desensitization under light-adapted conditions [156]. Historically, rod function in ACHM patients has been described as normal, although several recent studies, including some longitudinal recordings, have reported a reduction and/or deterioration of rod-driven ERG [140, 146, 157, 168–170] and rod-derived dark adaptation [170]. Rod dysfunction has been reported in some but not all patients [140, 157, 168] and does not seem to correlate directly with genotype, with mutations in *CNGA3*, *CNGB3*, and *GNAT2* implicated. Surprisingly, abnormal rod function has not been reported in ACHM patients with *PDE6C* or *PDE6H* defects but this could be due to the relatively low number of cases. What is still unclear is why and how rod function is affected by nonfunctional cones. Hypotheses include structural deficit due to loss of cones, shortening of rods inner and outer segments, secondary apoptosis induced by dying cones and anomalies in the rod signaling pathway.

ACHM is usually considered a stationary disease with congenital symptoms and vision assessments remaining stable throughout a patient's life. However several cross-sectional and longitudinal studies have reported disease progression over time, mainly in *CNGA3* and *CNGB3* patients [140, 141, 158, 166, 168, 171, 172] but also in *GNAT2* [143, 172] and *PDE6C* [173] patients. Some of these studies indicate that there might be a higher rate of deterioration within the first years of life which then becomes stationary in older patients [171, 173]. Interestingly, assessments of ACHM mouse models have shown an initial peak in cone photoreceptor cell death which subsequently stabilizes [174–176], but the *CNGB3* deficient dog model was reported to have a continuous, but slow loss of cones [177]. If ACHM is indeed a progressive disorder, this will have important implications for future clinical trial design as treatment during the early stages of the disease may provide the best outcome, as shown by the presence of an optimal window of treatment for gene therapy in the *CNGB3* mouse [178] and dog models [177]. However, two recent studies have questioned the effect of age in disease progression [172, 179]; Sundaram et al. [179] failed to find a correlation between cone structure loss, deterioration in best-corrected visual acuity (BCVA), contrast sensitivity, or reading acuity with advancing age but did find a significant decline in microperimetry-based retinal sensitivity with age. This discrepancy with previous studies has been attributed to the cross-sectional nature of this study, genetic diversity of patients and lack of standardization in cone structure measurements between studies [180, 181]. However, in a follow-up study, the same group came to a similar conclusion that phenotype did not correlate with age or genotype and although ACHM may be progressive, it is slow and subtle in most patients [172]. This is the largest and longest prospective longitudinal study of ACHM to date [172] and it remains to be seen whether similar findings will be reported in other patient cohorts. Initial studies were unable to determine whether viable cone cells are still present in the retina of ACHM patients but several recent studies have taken advantage of the advancing technology in in vivo imaging to address this issue. Spectral-domain optical coherence tomography (SD-OCT) and adaptive optics scanning light ophthalmoscopy (AOSLO) have demonstrated a highly variable range of photoreceptor integrity and structure between patients and that although cone structure is most likely disrupted, cones are still present [179, 182–188]. These studies have not been able to identify an altered cone structure and cone prevalence in *CNGA3* and *CNGB3* genotypes, but one study has indicated that better cone structure preservation might be present in the *GNAT2*-based disease [185].

1.5.6 Oligocone Trichromacy

Oligocone trichromacy (OT) is a rare cone dysfunction syndrome [189], which is characterized by reduced visual acuity, mild photophobia, normal fundi, and a reduced amplitude of the cone ERG. However, color vision is within normal limits. It has been proposed that these patients might have a reduced number of normal functioning cones, with each of the three cone types affected equally, thereby maintaining normal proportions and permitting trichromacy [190]. Recent studies have indicated that OT, at least in some instances, may be caused by mutations in the same genes that cause achromatopsia, namely *CNGA3*, *CNGB3*, and *PDE6C* [191, 192]. Some forms of OT may therefore represent a mild form of achromatopsia.

1.6 Cone and Cone–Rod Dystrophies

1.6.1 Dominant Autosomal Dystrophies

Altered color vision is associated with a number of cone (COD) and cone–rod dystrophies (CORD) (see also Chap. 10 for clinical features), most likely arising from the progressive loss of cones in these disorders. One of the best characterized forms of COD arises from dominant mutations in the *GUCA1A* gene which encodes the retinal guanylate cyclase activating protein (GCAP1) [193]. GCAPs bind calcium and act as activators of retinal guanylate cyclase (retGC) in the restoration of cGMP levels in photoreceptors after light activation (Fig. 1.1). Disease causing GCAP1 mutations are located in the EF-hands of the protein which alter its ability to bind calcium and thereby downregulate retGC activity [16]. Detailed clinical assessment of patients carrying a Tyr99Cys substitution in EF2 of the GCAP1 protein [29] has shown that the disorder presents as a mild photophobia with reduced central vision and defective color vision, with all three color axes affected. As the disorder progresses, all color vision may be lost. Confirmation that cones are progressively lost comes from two studies, firstly in a mouse line carrying a *GUCA1A* transgene with a Tyr99Cys mutation [194] and secondly in a gene-targeted mouse line carrying a knockin Glu155Gly mutation into EF4 which also causes COD [195, 196]. In both cases, cone photoreceptors are progressively lost.

Dominant mutations in the gene for retGC type 1 encoded by the autosomal *GUCY2D* gene account for a significant proportion of cases of CORD [19, 20], with the majority of disease-causing changes involving replacement of Arg at site 837. Site 837 is in a coiled-coil region of the protein that is important for GCAP binding [197]; as for the *GUCA1A* mutations described above, mutation in this region of the retGC1 protein results in constitutive activation of cyclase activity [198]. Clinical assessment of a patient with a *GUCY2D* mutation revealed the absence of color discrimination on a tritan axis, with a significant retention of red–green discrimination [199], implying that S cones may be preferentially lost in this disorder.

1.6.2 X-Linked Cone Dystrophies

An X linked disorder named Bornholm eye disease (BED) where red-green color deficiency is associated with myopia was originally reported in a large five generation Danish family that had its origins on the Danish island of Bornholm [200–202]. The disorder presents with moderate to high myopia combined with astigmatism and impaired visual acuity. Affected members in this original family are deuteranopes [201], but other families showing this combination of myopia and cone dysfunction are protanopes [200], so it would appear that the disorder can be associated with both types of dichromacy. Linkage analysis performed in the original BED family mapped the locus to Xq28, in the vicinity of the *L/M* opsin gene array, and a detailed study of several families with BED identified two mechanisms. Firstly, the deleterious Cys203Arg mutation, a change that is known to disrupt opsin folding [203] and is responsible for the loss of gene function in BCM [99, 100, 104], was found in one family [200, 204]. All other families lack this mutation but uniquely possess an L opsin gene in which exon 3 encodes an unusual combination of amino acids that is not found elsewhere [204]. The combination (haplotype) is across sites 153, 171, 174, 178, and 180 with either Leu, Ile, Ala, Val, Ala (LIAVA), or Leu, Val, Ala, Val, Ala (LVAVA) present compared to the wild type (WT) haplotype at these sites of LVAIS. Neither of these variant haplotypes is seen in normal trichromats or in dichromats that otherwise possess normal vision [205, 206], but were previously identified as the only change in a BCM family [99] and in a dichromatic individual showing a loss of cone photoreceptors [207]. Functional in vitro analysis of the variant pigments demonstrated a consistent reduction in yield compared with WT, indicating that the variant haplotypes may interfere with retinal chromophore handling [204]. An alternative and more likely explanation is however that the variant nucleotide sequence of exon 3 rather than the amino acid substitutions is responsible for the disorder. The variant nucleotide sequence causes aberrant transcript splicing [208] such that exon 2 is joined to exon 4 with the elimination of exon 3; the resulting protein would be truncated and therefore incapable of forming a functional pigment and the subset of cone phototreceptors expressing the variant gene would be nonfunctional, thereby giving rise to dichromacy. The individuals studied [208] both have the LIAVA haplotype but a more recent study has shown that exon 3 skipping is also present with the LVAVA haplotype and with another rare opsin haplotype of MIAVA [209]. In all three cases, the key nucleotide change that disrupts splicing may be an A>G transition in codon 532 [209]. Finally, an in vivo imaging study of retinal structure [210] showed that subjects expressing these variant opsins have a thinner retina that extended to both the inner retina and the outer nuclear layer, suggesting that degenerative changes in L/M cone photoreceptors are present that also damage neighboring cells (i.e., S cones and rods). The regularity of the cone mosaic is also disrupted (Fig. 1.7). These changes would appear to account therefore for the cone dysfunction seen in BED subjects.

Aberrant splicing of L or M opsin transcripts is therefore the underlying cause of BED and, as myopia is a constant component of the BED phenotype, this must also

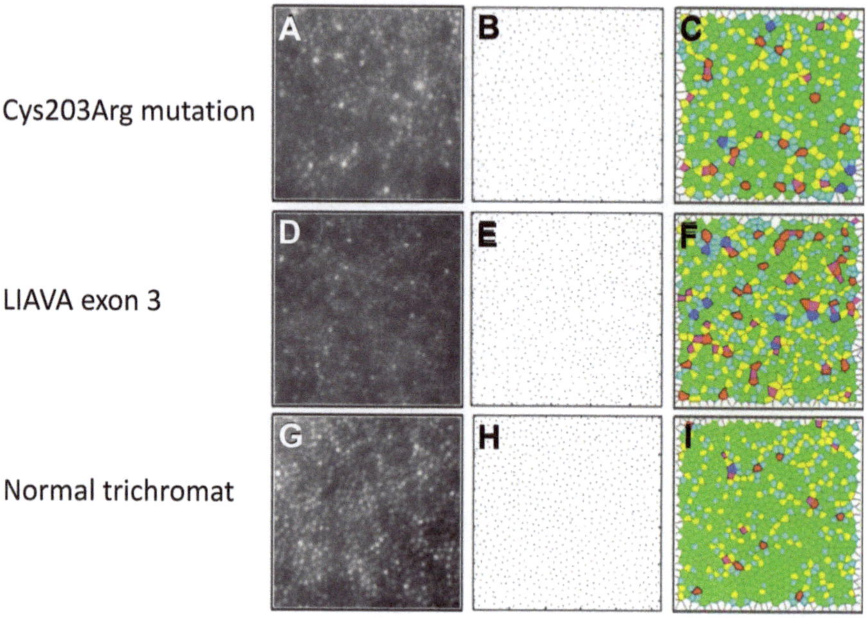

Fig. 1.7 Assessment of cone packing in two subjects with altered L opsins. *Left column* are foveal images obtained by adaptive optics imaging [103] from the (**a**) a subject with a Cys203Arg mutation, (**d**) a subject with an LIAVA variant L opsin, and (**g**) a normal trichromat. *Middle column* shows the corresponding plots of cone locations. *Right column* is color-coded to indicate the number of neighbors to each cone (*magenta* = 4, *cyan* = 5, *green* = 6, *yellow* = 7, *red* = 8, and *purple* = 9) and thereby an indication of the regularity of the photoreceptor mosaic. Large regions of *green polygons* indicate a regular triangular mosaic whereas other colors indicate disruptions in hexagonal packing. Note the extensive areas of *green polygons* in the normal trichromat and the disrupted patterns as shown by the substantial presence of polygons of other colors in the two subjects with variant L opsins. Redrawn from Ref. [103] with permission

arise from the opsin gene dysfunction. Myopia is caused by a defect in the process of emmetropization whereby the growth of eye length is regulated by visual experience to match the eye's optics and to compensate for variation in corneal/lens curvature and power. The signals that guide this process are initiated largely by light absorption of the photopigments found in L and M cones. Changes in the pattern of light and dark in the retinal image that characterize blurred versus sharply focused images are monitored to stop eye growth when the correct length for coordinated plano (neutral) optics is achieved. In BED patients, the emmetropization malfunctions as the result of the opsin pigment variation and a change in the organization of the cone mosaic [207].

In addition to the deleterious Cys203Arg mutation described above, other mutations in *L* and *M* opsin genes have been reported to cause cone dysfunction and color vision deficiencies, although the number of such mutations is relatively few compared to mutations in the rod opsin gene that cause dominant and recessive forms of retinitis pigmentosa [211]. A cone dysfunction with a protan vision deficiency [212] is caused by a novel Glu41Lys mutation in a hybrid *L*/*M* opsin gene. Retinal imaging reveals extensive cone disruption and spectral analysis of in vitro generated mutant opsin shows that it fails to form a pigment and fails to be trafficked to the cell membrane. However, although this is the only opsin gene in the *L*/*M* opsin gene array of the affected subject, the reported protanopia indicates that the mutant opsin must retain some activity in vivo. A similar cone dystrophy disorder showing deterioration of visual acuity and color vision with linkage to the *L*/*M* opsin gene array at Xq26.1-qter is caused by a Trp177Arg substitution encoded by exon 3 that is present in both the *L* and *M* opsin genes in the affected family members [213]. Interestingly, the sequence of exon 3 carrying the mutation, although *M* opsin-like, was identical in both *L* and *M* opsin genes indicating that the mutation most likely occurred in an *M* opsin gene but was subsequently incorporated into *L* opsin by gene conversion. In the two brothers with the Trp177Arg substitution assessed by [208], both had good S cone function but lacked *L*/*M* cone function. Imaging of the retina revealed a relatively healthy retina. Finally, retinal imaging of a male subject with a normal *L* opsin gene and an *M* opsin gene encoding a novel Trp149X premature termination codon revealed a significantly reduced cone density [214], indicating that cones expressing the mutant opsin show reduced viability.

1.7 Conclusion

Although not life threatening, loss of one or several aspects of normal color vision creates a major impairment and decline in the life quality of patients. Currently there are no proven treatments but the increase in molecular and genetic testing has aided in providing better diagnosis. This has enabled clinicians to offer improved counseling and educational support. The prognosis for future treatments is however looking promising. Recent advances in retinal gene delivery platforms have generated several studies that have resulted in successful visual restoration in several animal models of retinal disease. Achromatopsia is a particularly strong candidate for gene replacement therapy; around 90 % of cases have already been attributed to mutations in a limited number of genes and the successful restoration of cone function via gene replacement therapy has been achieved in several genetically distinct animal models [78, 177, 178, 215, 216]. Understanding the genetics and molecular mechanisms behind human color vision and its disorders will not only lead to advances in gene replacement therapy but will also aid in the discovery and validation of other modes of pharmacological and neuroprotective interventions.

References

1. Hunt DM, Peichl L. S cones: evolution, retinal distribution, development and spectral sensitivity. Vis Neurosci. 2014;31(2):115–38.
2. Hunt DM, Carvalho LS, Cowing JA, Davies WL. Evolution and spectral tuning of visual pigments in birds and mammals. Philos Trans R Soc Lond B Biol Sci. 2009;364(1531):2941–55.

3. Hunt DM, Dulai KS, Cowing JA, Julliot C, Mollon JD, Bowmaker JK, et al. Molecular evolution of trichromacy in primates. Vision Res. 1998;38(21):3299–306.
4. Ebrey T, Koutalos Y. Vertebrate photoreceptors. Prog Retin Eye Res. 2001;20(1):49–94.
5. Miller JL, Picones A, Korenbrot JI. Differences in transduction between rod and cone photoreceptors: an exploration of the role of calcium homeostasis. Curr Opin Neurobiol. 1994;4(4):488–95.
6. Pugh Jr EN, Nikonov S, Lamb TD. Molecular mechanisms of vertebrate photoreceptor light adaptation. Curr Opin Neurobiol. 1999;9(4):410–8.
7. Yau KW. Phototransduction mechanism in retinal rods and cones. The Friedenwald Lecture. Invest Ophthalmol Vis Sci. 1994;35(1):9–32.
8. Baylor DA, Matthews G, Yau KW. Two components of electrical dark noise in toad retinal rod outer segments. J Physiol. 1980;309:591–621.
9. Kefalov VJ, Estevez ME, Kono M, Goletz PW, Crouch RK, Cornwall MC, et al. Breaking the covalent bond—a pigment property that contributes to desensitization in cones. Neuron. 2005;46(6):879–90.
10. Travis GH. DISCO! Dissociation of cone opsins: the fast and noisy life of cones explained. Neuron. 2005;46(6):840–2.
11. Underwood G. Reptilian retinas. Nature. 1951;167(4240):183–5.
12. Kleinschmidt J, Dowling JE. Intracellular recordings from gecko photoreceptors during light and dark adaptation. J Gen Physiol. 1975;66(5):617–48.
13. Kojima D, Okano T, Fukada Y, Shichida Y, Yoshizawa T, Ebrey TG. Cone visual pigments are present in gecko rod cells. Proc Natl Acad Sci U S A. 1992;89(15):6841–5.
14. Hunt DM, Collin SP. The evolution of photoreceptors and visual photopigments in vertebrates. In: Hunt DM, Hankins MW, Collin SP, Marshall NJ, editors. Evolution of visual and non-visual pigments. New York: Springer; 2014. p. 163–218.
15. Frings S, Seifert R, Godde M, Kaupp UB. Profoundly different calcium permeation and blockage determine the specific function of distinct cyclic nucleotide-gated channels. Neuron. 1995;15(1):169–79.
16. Hunt DM, Buch P, Michaelides M. Guanylate cyclases and associated activator proteins in retinal disease. Mol Cell Biochem. 2009;334(1–2):157–68.
17. Perrault I, Rozet JM, Calvas P, Gerber S, Camuzat A, Dollfus H, et al. Retinal-specific guanylate cyclase gene mutations in Leber's congenital amaurosis. Nat Genet. 1996;14(4):461–4.
18. Yang RB, Robinson SW, Xiong WH, Yau KW, Birch DG, Garbers DL. Disruption of a retinal guanylyl cyclase gene leads to cone-specific dystrophy and paradoxical rod behavior. J Neurosci. 1999;19(14):5889–97.
19. Kelsell RE, Gregory-Evans K, Payne AM, Perrault I, Kaplan J, Yang RB, et al. Mutations in the retinal guanylate cyclase (RETGC-1) gene in dominant cone-rod dystrophy. Hum Mol Genet. 1998;7(7):1179–84.
20. Payne AM, Morris AG, Downes SM, Johnson S, Bird AC, Moore AT, et al. Clustering and frequency of mutations in the retinal guanylate cyclase (GUCY2D) gene in patients with dominant cone-rod dystrophies. J Med Genet. 2001;38(9):611–4.
21. Dizhoor AM, Lowe DG, Olshevskaya EV, Laura RP, Hurley JB. The human photoreceptor membrane guanylyl cyclase, RetGC, is present in outer segments and is regulated by calcium and a soluble activator. Neuron. 1994;12(6):1345–52.
22. Imanishi Y, Yang L, Sokal I, Filipek S, Palczewski K, Baehr W. Diversity of guanylate cyclase-activating proteins (GCAPs) in teleost fish: characterization of three novel GCAPs (GCAP4, GCAP5, GCAP7) from zebrafish (Danio rerio) and prediction of eight GCAPs (GCAP1–8) in pufferfish (Fugu rubripes). J Mol Evol. 2004;59(2):204–17.
23. Haeseleer F, Sokal I, Li N, Pettenati M, Rao N, Bronson D, et al. Molecular characterization of a third member of the guanylyl cyclase-activating protein subfamily. J Biol Chem. 1999;274(10):6526–35.
24. Palczewski K, Subbaraya I, Gorczyca WA, Helekar BS, Ruiz CC, Ohguro H, et al. Molecular cloning and characterization of retinal photoreceptor guanylyl cyclase-activating protein. Neuron. 1994;13(2):395–404.

25. Dizhoor AM, Olshevskaya EV, Henzel WJ, Wong SC, Stults JT, Ankoudinova I, et al. Cloning, sequencing, and expression of a 24-kDa Ca(2+)-binding protein activating photoreceptor guanylyl cyclase. J Biol Chem. 1995;270(42):25200–6.
26. Hunt DM, Wilkie SE, Newbold R, Deery E, Warren MJ, Bhattacharya SS, et al. Dominant cone and cone-rod dystrophies: functional analysis of mutations in retGC1 and GCAP1. Novartis Found Symp. 2004;255:37–49. discussion 49–50, 177–8.
27. Jiang L, Wheaton D, Bereta G, Zhang K, Palczewski K, Birch DG, et al. A novel GCAP1(N104K) mutation in EF-hand 3 (EF3) linked to autosomal dominant cone dystrophy. Vision Res. 2008;48(23–24):2425–32.
28. Kitiratschky VB, Behnen P, Kellner U, Heckenlively JR, Zrenner E, Jagle H, et al. Mutations in the GUCA1A gene involved in hereditary cone dystrophies impair calcium-mediated regulation of guanylate cyclase. Hum Mutat. 2009;30:E782–96.
29. Michaelides M, Wilkie SE, Jenkins S, Holder GE, Hunt DM, Moore AT, et al. Mutation in the gene GUCA1A, encoding guanylate cyclase-activating protein 1, causes cone, cone-rod, and macular dystrophy. Ophthalmology. 2005;112(8):1442–7.
30. Payne AM, Downes SM, Bessant DA, Taylor R, Holder GE, Warren MJ, et al. A mutation in guanylate cyclase activator 1A (GUCA1A) in an autosomal dominant cone dystrophy pedigree mapping to a new locus on chromosome 6p21.1. Hum Mol Genet. 1998;7(2):273–7.
31. Brannock MT, Weng K, Robinson PR. Rhodopsin's carboxyl-terminal threonines are required for wild-type arrestin-mediated quench of transducin activation in vitro. Biochemistry. 1999;38(12):3770–7.
32. Palczewski K, Buczylko J, Lebioda L, Crabb JW, Polans AS. Identification of the N-terminal region in rhodopsin kinase involved in its interaction with rhodopsin. J Biol Chem. 1993;268(8):6004–13.
33. Zhao X, Haeseleer F, Fariss RN, Huang J, Baehr W, Milam AH, et al. Molecular cloning and localization of rhodopsin kinase in the mammalian pineal. Vis Neurosci. 1997;14(2): 225–32.
34. Weiss ER, Raman D, Shirakawa S, Ducceschi MH, Bertram PT, Wong F, et al. The cloning of GRK7, a candidate cone opsin kinase, from cone- and rod-dominant mammalian retinas. Mol Vis. 1998;4:27.
35. Tachibanaki S, Tsushima S, Kawamura S. Low amplification and fast visual pigment phosphorylation as mechanisms characterizing cone photoresponses. Proc Natl Acad Sci U S A. 2001;98(24):14044–9.
36. Craft CM, Whitmore DH. The arrestin superfamily: cone arrestins are a fourth family. FEBS Lett. 1995;362(2):247–55.
37. Collin SP, Knight MA, Davies WL, Potter IC, Hunt DM, Trezise AE. Ancient colour vision: multiple opsin genes in the ancestral vertebrates. Curr Biol. 2003;13(22):R864–5.
38. Davies WL, Cowing JA, Carvalho LS, Potter IC, Trezise AE, Hunt DM, et al. Functional characterisation and regulation of visual pigment gene expression in an anadromous lamprey. FASEB J. 2007;21:2713–24.
39. Shu DG, Morris SC, Han J, Zhang ZF, Yasui K, Janvier P, et al. Head and backbone of the Early Cambrian vertebrate Haikouichthys. Nature. 2003;421(6922):526–9.
40. Xian-guang H, Aldridge RJ, Siveter DJ, Siveter DJ, Xiang-hong F. New evidence on the anatomy and phylogeny of the earliest vertebrates. Proc Biol Sci. 2002;269(1503):1865–9.
41. Nathans J. Determinants of visual pigment absorbance: identification of the retinylidene Schiff's base counterion in bovine rhodopsin. Biochemistry. 1990;29(41):9746–52.
42. Fasick JI, Applebury ML, Oprian DD. Spectral tuning in the mammalian short-wavelength sensitive cone pigments. Biochemistry. 2002;41(21):6860–5.
43. Davies WI, Collin SP, Hunt DM. Molecular ecology and adaptation of visual photopigments in craniates. Mol Ecol. 2012;21(13):3121–58.
44. Davies WL, Carvalho LS, Cowing JA, Beazley LD, Hunt DM, Arrese CA. Visual pigments of the platypus: a novel route to mammalian colour vision. Curr Biol. 2007;17(5): R161–3.

45. Wakefield MJ, Anderson M, Chang E, Wei KJ, Kaul R, Graves JA, et al. Cone visual pigments of monotremes: filling the phylogenetic gap. Vis Neurosci. 2008;25(3):257–64.
46. Mollon JD. "Tho' she kneel'd in that place where they grew…" The uses and origins of primate colour vision. J Exp Biol. 1989;146:21–38.
47. Osorio D, Vorobyev M. Colour vision as an adaptation to frugivory in primates. Proc R Soc Lond B Biol Sci. 1996;263(1370):593–9.
48. Regan BC, Julliot C, Simmen B, Vienot F, Charles-Dominique P, Mollon JD. Fruits, foliage and the evolution of primate colour vision. Philos Trans R Soc Lond B Biol Sci. 2001;356(1407):229–83.
49. Sumner P, Mollon JD. Catarrhine photopigments are optimized for detecting targets against a foliage background. J Exp Biol. 2000;203(Pt 13):1963–86.
50. Dominy NJ, Lucas PW. Ecological importance of trichromatic vision to primates. Nature. 2001;410(6826):363–6.
51. Nathans J, Thomas D, Hogness DS. Molecular genetics of human color vision: the genes encoding blue, green, and red pigments. Science. 1986;232(4747):193–202.
52. Dulai KS, von Dornum M, Mollon JD, Hunt DM. The evolution of trichromatic color vision by opsin gene duplication in New World and Old World primates. Genome Res. 1999;9(7): 629–38.
53. Hanna MC, Platts JT, Kirkness EF. Identification of a gene within the tandem array of red and green color pigment genes. Genomics. 1997;43(3):384–6.
54. Jorgensen AL, Deeb SS, Motulsky AG. Molecular genetics of X chromosome-linked color vision among populations of African and Japanese ancestry: high frequency of a shortened red pigment gene among Afro-Americans. Proc Natl Acad Sci U S A. 1990;87(17):6512–6.
55. Wang Y, Macke JP, Merbs SL, Zack DJ, Klaunberg B, Bennett J, et al. A locus control region adjacent to the human red and green visual pigment genes. Neuron. 1992;9(3):429–40.
56. Shaaban SA, Deeb SS. Functional analysis of the promoters of the human red and green visual pigment genes. Invest Ophthalmol Vis Sci. 1998;39(6):885–96.
57. Winderickx J, Battisti L, Motulsky AG, Deeb SS. Selective expression of human X chromosome-linked green opsin genes. Proc Natl Acad Sci U S A. 1992;89(20):9710–4.
58. Smallwood PM, Wang Y, Nathans J. Role of a locus control region in the mutually exclusive expression of human red and green cone pigment genes. Proc Natl Acad Sci U S A. 2002;99(2):1008–11.
59. Carroll J, Neitz J, Neitz M. Estimates of L:M cone ratio from ERG flicker photometry and genetics. J Vis. 2002;2(8):531–42.
60. Neitz M, Neitz J, Jacobs GH. Spectral tuning of pigments underlying red-green color vision. Science. 1991;252(5008):971–4.
61. Williams AJ, Hunt DM, Bowmaker JK, Mollon JD. The polymorphic photopigments of the marmoset: spectral tuning and genetic basis. EMBO J. 1992;11(6):2039–45.
62. Ibbotson RE, Hunt DM, Bowmaker JK, Mollon JD. Sequence divergence and copy number of the middle- and long-wave photopigment genes in Old World monkeys. Proc R Soc Lond B. 1992;247(1319):145–54.
63. Asenjo AB, Rim J, Oprian DD. Molecular determinants of human red/green color discrimination. Neuron. 1994;12(5):1131–8.
64. Merbs SL, Nathans J. Role of hydroxyl-bearing amino acids in differentially tuning the absorption spectra of the human red and green cone pigments. Photochem Photobiol. 1993;58(5):706–10.
65. Sanocki E, Lindsey DT, Winderickx J, Teller DY, Deeb SS, Motulsky AG. Serine/alanine amino acid polymorphism of the L and M cone pigments: effects on Rayleigh matches among deuteranopes, protanopes and color normal observers. Vision Res. 1993;33(15):2139–52.
66. Hunt DM, Cowing JA, Wilkie SE, Parry J, Poopalasundaram S, Bowmaker JK. Divergent mechanisms for the tuning of shortwave sensitive visual pigments in vertebrates. Photochem Photobiol Sci. 2004;3:713–20.
67. Bowmaker JK, Astell S, Hunt DM, Mollon JD. Photosensitive and photostable pigments in the retinae of Old World monkeys. J Exp Biol. 1991;156:1–19.

68. Hunt DM, Cowing JA, Patel R, Appukuttan B, Bowmaker JK, Mollon JD. Sequence and evolution of the blue cone pigment gene in Old and New World primates. Genomics. 1995;27(3):535–8.
69. Shi Y, Radlwimmer FB, Yokoyama S. Molecular genetics and the evolution of ultraviolet vision in vertebrates. Proc Natl Acad Sci U S A. 2001;98(20):11731–6.
70. Yokoyama S, Shi Y. Genetics and evolution of ultraviolet vision in vertebrates. FEBS Lett. 2000;486(2):167–72.
71. Carvalho LS, Davies WL, Robinson PR, Hunt DM. Spectral tuning and evolution of primate short-wavelength-sensitive visual pigments. Proc Biol Sci. 2012;279(1727):387–93.
72. Cowing JA, Poopalasundaram S, Wilkie SE, Robinson PR, Bowmaker JK, Hunt DM. The molecular mechanism for the spectral shifts between vertebrate ultraviolet- and violet-sensitive cone visual pigments. Biochem J. 2002;367(Pt 1):129–35.
73. Parry JW, Poopalasundaram S, Bowmaker JK, Hunt DM. A novel amino acid substitution is responsible for spectral tuning in a rodent violet-sensitive visual pigment. Biochemistry. 2004;43(25):8014–20.
74. Yokoyama S, Takenaka N, Agnew DW, Shoshani J. Elephants and human color-blind deuteranopes have identical sets of visual pigments. Genetics. 2005;170(1):335–44.
75. Hunt DM, Carvalho LS, Cowing JA, Parry JW, Wilkie SE, Davies WL, et al. Spectral tuning of shortwave-sensitive visual pigments in vertebrates. Photochem Photobiol. 2007;83(2):303–10.
76. Starace DM, Knox BE. Cloning and expression of a Xenopus short wavelength cone pigment. Exp Eye Res. 1998;67(2):209–20.
77. Moritz GL, Lim NT-L, Neitz M, Peichl L, Dominy NJ. Expression and evolution of short wavelength sensitive opsins in colugos: a nocturnal lineage that informs debate on primate origins. Evol Biol. 2013;40:542–53.
78. Carvalho LS, Davies WL, Robinson PR, Hunt DM. Spectral tuning and evolution of primate short-wavelength-sensitive visual pigments. Proc Biol Sci. 2012;279(1727):387–93.
79. Hunt DM, Dulai KS, Bowmaker JK, Mollon JD. The chemistry of John Dalton's color blindness. Science. 1995;267(5200):984–8.
80. Sharpe LT, Stockman A, Jagle H, Nathans J. Opsin genes, cone photopigments, color vision and color blindness. In: Gegenfurtner KR, Sharpe LT, editors. Color vision: from genes to perception. Cambridge: Cambridge University Press; 1999. p. 3–51.
81. Drummond-Borg M, Deeb SS, Motulsky AG. Molecular patterns of X chromosome-linked color vision genes among 134 men of European ancestry. Proc Natl Acad Sci U S A. 1989;86(3):983–7.
82. Feil R, Aubourg P, Heilig R, Mandel JL. A 195-kb cosmid walk encompassing the human Xq28 color vision pigment genes. Genomics. 1990;6(2):367–73.
83. Nathans J, Piantanida TP, Eddy RL, Shows TB, Hogness DS. Molecular genetics of inherited variation in human color vision. Science. 1986;232(4747):203–10.
84. Neitz M, Neitz J. Molecular genetics of color vision and color vision defects. Arch Ophthalmol. 2000;118(5):691–700.
85. Simunovic MP. Colour vision deficiency. Eye (Lond). 2010;24(5):747–55.
86. McMahon C, Neitz J, Neitz M. Evaluating the human X-chromosome pigment gene promoter sequences as predictors of L:M cone ratio variation. J Vis. 2004;4(3):203–8.
87. Hagstrom SA, Neitz M, Neitz J. Cone pigment gene expression in individual photoreceptors and the chromatic topography of the retina. J Opt Soc Am A Opt Image Sci Vis. 2000;17(3):527–37.
88. Neitz J, Neitz M. The genetics of normal and defective color vision. Vision Res. 2011. doi:10.1016/j.visres.2010.12.002.
89. Balding DJ, Nichols RA, Hunt DM. Detecting gene conversion: primate visual pigment genes. Proc Biol Sci. 1992;249(1326):275–80.
90. Shyue SK, Li L, Chang BH, Li WH. Intronic gene conversion in the evolution of human X-linked color vision genes. Mol Biol Evol. 1994;11(3):548–51.

91. Shyue SK, Hewett-Emmett D, Sperling HG, Hunt DM, Bowmaker JK, Mollon JD, et al. Adaptive evolution of color vision genes in higher primates. Science. 1995;269(5228):1265–7.
92. Zhao Z, Hewett-Emmett D, Li WH. Frequent gene conversion between human red and green opsin genes. J Mol Evol. 1998;46(4):494–6.
93. Verrelli BC, Lewis Jr CM, Stone AC, Perry GH. Different selective pressures shape the molecular evolution of color vision in chimpanzee and human populations. Mol Biol Evol. 2008;25(12):2735–43.
94. Onishi A, Koike S, Ida M, Schichida Y, Takemaka O, Hanazawa A, et al. Dichromaticsm in macaque monkeys. Nature. 1999;402(6758):139–40.
95. Terao K, Mikami A, Saito A, Itoh S, Ogawa H, Takenaka O, et al. Identification of a protoanomalous chimpanzee by molecular genetic and electroretinogram analyses. Vis Res. 2005;45(10):1225–35.
96. Hiwatashi T, Mikami A, Katsumura T, Suryobroto B, Perwitasari-Farajallah D, Malaivijitnond S, et al. Gene conversion and purifying selection shape nucleotide variation in gibbon L/M opsin genes. BMC Evol Biol. 2011;11:312.
97. Michaelides M, Hunt DM, Moore AT. The cone dysfunction syndromes. Br J Ophthalmol. 2004;88(2):291–7.
98. Gouras P, MacKay CJ. Electroretinographic responses of the short-wavelength-sensitive cones. Invest Ophthalmol Vis Sci. 1990;31(7):1203–9.
99. Nathans J, Davenport CM, Maumenee IH, Lewis RA, Hejtmancik JF, Litt M, et al. Molecular genetics of human blue cone monochromacy. Science. 1989;245(4920):831–8.
100. Michaelides M, Johnson S, Simunovic MP, Bradshaw K, Holder G, Mollon JD, et al. Blue cone monochromatism: a phenotype and genotype assessment with evidence of progressive loss of cone function in older individuals. Eye. 2005;19:2–10.
101. Nathans J, Maumenee IH, Zrenner E, Sadowski B, Sharpe LT, Lewis RA, et al. Genetic heterogeneity among blue-cone monochromats. Am J Hum Genet. 1993;53(5):987–1000.
102. Gardner JC, Michaelides M, Holder GE, Kanuga N, Webb TR, Mollon JD, et al. Blue cone monochromacy: causative mutations and associated phenotypes. Mol Vis. 2009;15:876–84.
103. Carroll J, Baraas RC, Wagner-Schuman M, Rha J, Siebe CA, Sloan C, et al. Cone photoreceptor mosaic disruption associated with Cys203Arg mutation in the M-cone opsin. Proc Natl Acad Sci U S A. 2009;106(49):20948–53.
104. Ayyagari R, Kakuk LE, Bingham EL, Szczesny JJ, Kemp J, Toda Y, et al. Spectrum of color gene deletions and phenotype in patients with blue cone monochromacy. Hum Genet. 2000;107(1):75–82.
105. Ayyagari R, Kakuk LE, Coats CL, Bingham EL, Toda Y, Felius J, et al. Bilateral macular atrophy in blue cone monochromacy (BCM) with loss of the locus control region (LCR) and part of the red pigment gene. Mol Vis. 1999;5:13.
106. Carroll J, Rossi EA, Porter J, Neitz J, Roorda A, Williams DR, et al. Deletion of the X-linked opsin gene array locus control region (LCR) results in disruption of the cone mosaic. Vision Res. 2010;50(19):1989–99.
107. Ladekjaer-Mikkelsen AS, Rosenberg T, Jorgensen AL. A new mechanism in blue cone monochromatism. Hum Genet. 1996;98(4):403–8.
108. Fleischman JA, O'Donnell Jr FE. Congenital X-linked incomplete achromatopsia. Evidence for slow progression, carrier fundus findings, and possible genetic linkage with glucose-6-phosphate dehydrogenase locus. Arch Ophthalmol. 1981;99(3):468–72.
109. Wright WD. The characteristics of tritanopia. J Opt Soc Am. 1952;42(8):509–21.
110. Went LN, Pronk N. The genetics of tritan disturbances. Hum Genet. 1985;69(3):255–62.
111. Gunther KL, Neitz J, Neitz M. A novel mutation in the short-wavelength-sensitive cone pigment gene associated with a tritan color vision defect. Vis Neurosci. 2006;23:403–9.
112. Baraas RC, Hagen LA, Dees EW, Neitz M. Substitution of isoleucine for threonine at position 190 of S-opsin causes S-cone-function abnormalities. Vision Res. 2012;73:1–9. doi:10.1016/j.visres.2012.09.007.

113. Weitz CJ, Miyake Y, Shinzato K, Montag E, Zrenner E, Went LN, et al. Human tritanopia associated with two amino acid substitutions in the blue-sensitive opsin. Am J Hum Genet. 1992;50(3):498–507.
114. Weitz CJ, Went LN, Nathans J. Human tritanopia associated with a third amino acid substitution in the blue-sensitive visual pigment. Am J Hum Genet. 1992;51(2):444–6.
115. Baraas RC, Carroll J, Gunther KL, Chung M, Williams DR, Foster DH, et al. Adaptive optics retinal imaging reveals S-cone dystrophy in tritan color-vision deficiency. J Opt Soc Am A Opt Image Sci Vis. 2007;24(5):1438–47.
116. al-Maghtheh M, Gregory C, Inglehearn C, Hardcastle A, Bhattacharya S. Rhodopsin mutations in autosomal dominant retinitis pigmentosa. Hum Mutat. 1993;2(4):249–55.
117. Berson EL, Sandberg MA, Dryja TP. Autosomal dominant retinitis pigmentosa with rhodopsin, valine-345-methionine. Trans Am Ophthalmol Soc. 1991;89:117–28. discussion 28–30.
118. Sung CH, Davenport CM, Nathans J. Rhodopsin mutations responsible for autosomal dominant retinitis pigmentosa. Clustering of functional classes along the polypeptide chain. J Biol Chem. 1993;268(35):26645–9.
119. Dryja TP, McGee TL, Reichel E, Hahn LB, Cowley GS, Yandell DW, et al. A point mutation of the rhodopsin gene in one form of retinitis pigmentosa. Nature. 1990;343(6256):364–6.
120. Jacobson SG, Marmor MF, Kemp CM, Knighton RW. SWS (blue) cone hypersensitivity in a newly identified retinal degeneration. Invest Ophthalmol Vis Sci. 1990;31(5):827–38.
121. Marmor MF, Jacobson SG, Foerster MH, Kellner U, Weleber RG. Diagnostic clinical findings of a new syndrome with night blindness, maculopathy, and enhanced S cone sensitivity. Am J Ophthalmol. 1990;110(2):124–34.
122. Marmor MF, Tan F, Sutter EE, Bearse Jr MA. Topography of cone electrophysiology in the enhanced S cone syndrome. Invest Ophthalmol Vis Sci. 1999;40(8):1866–73.
123. Kobayashi M, Takezawa S, Hara K, Yu RT, Umesono Y, Agata K, et al. Identification of a photoreceptor cell-specific nuclear receptor. Proc Natl Acad Sci U S A. 1999; 96(9):4814–9.
124. Swaroop A, Xu JZ, Pawar H, Jackson A, Skolnick C, Agarwal N. A conserved retina-specific gene encodes a basic motif/leucine zipper domain. Proc Natl Acad Sci U S A. 1992;89(1):266–70.
125. Mears AJ, Kondo M, Swain PK, Takada Y, Bush RA, Saunders TL, et al. Nrl is required for rod photoreceptor development. Nat Genet. 2001;29(4):447–52.
126. Haider NB, Demarco P, Nystuen AM, Huang X, Smith RS, McCall MA, et al. The transcription factor Nr2e3 functions in retinal progenitors to suppress cone cell generation. Vis Neurosci. 2006;23(6):917–29.
127. Swaroop A, Kim D, Forrest D. Transcriptional regulation of photoreceptor development and homeostasis in the mammalian retina. Nat Rev Neurosci. 2010;11(8):563–76.
128. Ng L, Hurley JB, Dierks B, Srinivas M, Salto C, Vennstrom B, et al. A thyroid hormone receptor that is required for the development of green cone photoreceptors. Nat Genet. 2001;27(1):94–8.
129. Weiss AH, Kelly JP, Bisset D, Deeb SS. Reduced L- and M- and increased S-cone functions in an infant with thyroid hormone resistance due to mutations in the THRbeta2 gene. Ophthalmic Genet. 2012;33(4):187–95.
130. Machado DS, Sabet A, Santiago LA, Sidhaye AR, Chiamolera MI, Ortiga-Carvalho TM, et al. A thyroid hormone receptor mutation that dissociates thyroid hormone regulation of gene expression in vivo. Proc Natl Acad Sci U S A. 2009;106(23):9441–6.
131. Kohl S, Hamel C. Clinical utility gene card for: achromatopsia—update 2013. Eur J Hum Genet. 2013;21(11).
132. Sundin OH, Yang JM, Li Y, Zhu D, Hurd JN, Mitchell TN, et al. Genetic basis of total colour blindness among the Pingelapese islanders. Nat Genet. 2000;25(3):289–93.
133. Roosing S, Thiadens AA, Hoyng CB, Klaver CC, den Hollander AI, Cremers FP. Causes and consequences of inherited cone disorders. Prog Retin Eye Res. 2014;42:1–26.

134. Kohl S, Marx T, Giddings I, Jagle H, Jacobson SG, Apfelstedt-Sylla E, et al. Total colour blindness is caused by mutations in the gene encoding the alpha-subunit of the cone photoreceptor cGMP-gated cation channel. Nat Genet. 1998;19(3):257–9.
135. Kohl S, Baumann B, Broghammer M, Jagle H, Sieving P, Kellner U, et al. Mutations in the CNGB3 gene encoding the beta-subunit of the cone photoreceptor cGMP-gated channel are responsible for achromatopsia (ACHM3) linked to chromosome 8q21. Hum Mol Genet. 2000;9(14):2107–16.
136. Kohl S, Coppieters F, Meire F, Schaich S, Roosing S, Brennenstuhl C, et al. A nonsense mutation in PDE6H causes autosomal-recessive incomplete achromatopsia. Am J Hum Genet. 2012;91(3):527–32.
137. Kohl S, Baumann B, Rosenberg T, Kellner U, Lorenz B, Vadala M, et al. Mutations in the cone photoreceptor G-protein alpha-subunit gene GNAT2 in patients with achromatopsia. Am J Hum Genet. 2002;71(2):422–5.
138. Aligianis IA, Forshew T, Johnson S, Michaelides M, Johnson CA, Trembath RC, et al. Mapping of a novel locus for achromatopsia (ACHM4) to 1p and identification of a germline mutation in the alpha subunit of cone transducin (GNAT2). J Med Genet. 2002;39(9):656–60.
139. Thiadens AA, den Hollander AI, Roosing S, Nabuurs SB, Zekveld-Vroon RC, Collin RW, et al. Homozygosity mapping reveals PDE6C mutations in patients with early-onset cone photoreceptor disorders. Am J Hum Genet. 2009;85(2):240–7.
140. Thiadens AA, Slingerland NW, Roosing S, van Schooneveld MJ, van Lith-Verhoeven JJ, van Moll-Ramirez N, et al. Genetic etiology and clinical consequences of complete and incomplete achromatopsia. Ophthalmology. 2009;116(10):1984–9. e1.
141. Thiadens AA, Roosing S, Collin RW, van Moll-Ramirez N, van Lith-Verhoeven JJ, van Schooneveld MJ, et al. Comprehensive analysis of the achromatopsia genes CNGA3 and CNGB3 in progressive cone dystrophy. Ophthalmology. 2010;117(4):825–30. e1.
142. Kohl S, Varsanyi B, Antunes GA, Baumann B, Hoyng CB, Jagle H, et al. CNGB3 mutations account for 50% of all cases with autosomal recessive achromatopsia. Eur J Hum Genet. 2005;13(3):302–8.
143. Michaelides M, Aligianis IA, Holder GE, Simunovic M, Mollon JD, Maher ER, et al. Cone dystrophy phenotype associated with a frameshift mutation (M280fsX291) in the alpha-subunit of cone specific transducin (GNAT2). Br J Ophthalmol. 2003;87(11):1317–20.
144. Ouechtati F, Merdassi A, Bouyacoub Y, Largueche L, Derouiche K, Ouragini H, et al. Clinical and genetic investigation of a large Tunisian family with complete achromatopsia: identification of a new nonsense mutation in GNAT2 gene. J Hum Genet. 2011;56(1):22–8.
145. Rosenberg T, Baumann B, Kohl S, Zrenner E, Jorgensen AL, Wissinger B. Variant phenotypes of incomplete achromatopsia in two cousins with GNAT2 gene mutations. Invest Ophthalmol Vis Sci. 2004;45(12):4256–62.
146. Pina AL, Baumert U, Loyer M, Koenekoop RK. A three base pair deletion encoding the amino acid (lysine-270) in the alpha-cone transducin gene. Mol Vis. 2004;10:265–71.
147. Katagiri S, Hayashi T, Yoshitake K, Sergeev Y, Akahori M, Furuno M, et al. Congenital achromatopsia and macular atrophy caused by a novel recessive PDE6C mutation (p.E591K). Ophthalmic Genet. 2015;36(2):137–44.
148. Grau T, Artemyev NO, Rosenberg T, Dollfus H, Haugen OH, Cumhur Sener E, et al. Decreased catalytic activity and altered activation properties of PDE6C mutants associated with autosomal recessive achromatopsia. Hum Mol Genet. 2011;20(4):719–30.
149. Kohl S, Jagle H, Wissinger B. Achromatopsia. In: Pagon RA, Adam MP, Ardinger HH, Wallace SE, Amemiya A, Bean LJH, editors. GeneReviews(R). Seattle: University of Washington; 1993.
150. Huang L, Zhang Q, Li S, Guan L, Xiao X, Zhang J, et al. Exome sequencing of 47 Chinese families with cone-rod dystrophy: mutations in 25 known causative genes. PLoS One. 2013;8(6):e65546.
151. Chang B, Grau T, Dangel S, Hurd R, Jurklies B, Sener EC, et al. A homologous genetic basis of the murine cpfl1 mutant and human achromatopsia linked to mutations in the PDE6C gene. Proc Natl Acad Sci U S A. 2009;106(46):19581–6.

152. Johnson S, Michaelides M, Aligianis IA, Ainsworth JR, Mollon JD, Maher ER, et al. Achromatopsia caused by novel mutations in both CNGA3 and CNGB3. J Med Genet. 2004;41(2):e20.
153. Aboshiha J, Dubis AM, Carroll J, Hardcastle AJ, Michaelides M. The cone dysfunction syndromes. Br J Ophthalmol. 2016;100(1):115–21.
154. Wawrocka A, Kohl S, Baumann B, Walczak-Sztulpa J, Wicher K, Skorczyk-Werner A, et al. Five novel CNGB3 gene mutations in Polish patients with achromatopsia. Mol Vis. 2014;20:1732–9.
155. Corton M, Nishiguchi KM, Avila-Fernandez A, Nikopoulos K, Riveiro-Alvarez R, Tatu SD, et al. Exome sequencing of index patients with retinal dystrophies as a tool for molecular diagnosis. PLoS One. 2013;8(6):e65574.
156. Nishiguchi KM, Sandberg MA, Gorji N, Berson EL, Dryja TP. Cone cGMP-gated channel mutations and clinical findings in patients with achromatopsia, macular degeneration, and other hereditary cone diseases. Hum Mutat. 2005;25(3):248–58.
157. Zelinger L, Cideciyan AV, Kohl S, Schwartz SB, Rosenmann A, Eli D, et al. Genetics and disease expression in the CNGA3 form of achromatopsia: steps on the path to gene therapy. Ophthalmology. 2015;122:997–1007.
158. Michaelides M, Aligianis IA, Ainsworth JR, Good P, Mollon JD, Maher ER, et al. Progressive cone dystrophy associated with mutation in CNGB3. Invest Ophthalmol Vis Sci. 2004;45(6):1975–82.
159. Matveev AV, Quiambao AB, Browning Fitzgerald J, Ding XQ. Native cone photoreceptor cyclic nucleotide-gated channel is a heterotetrameric complex comprising both CNGA3 and CNGB3: a study using the cone-dominant retina of Nrl–/– mice. J Neurochem. 2008;106(5):2042–55.
160. Kaupp UB, Seifert R. Cyclic nucleotide-gated ion channels. Physiol Rev. 2002;82(3): 769–824.
161. Kohl S, Varsanyi B, Antunes GA, Baumann B, Hoyng CB, Jagle H, et al. CNGB3 mutations account for 50% of all cases with autosomal recessive achromatopsia. Eur J Hum Genet. 2005;13(3):302–8.
162. Wissinger B, Gamer D, Jagle H, Giorda R, Marx T, Mayer S, et al. CNGA3 mutations in hereditary cone photoreceptor disorders. Am J Hum Genet. 2001;69(4):722–37.
163. Zelinger L, Greenberg A, Kohl S, Banin E, Sharon D. An ancient autosomal haplotype bearing a rare achromatopsia-causing founder mutation is shared among Arab Muslims and Oriental Jews. Hum Genet. 2010;128(3):261–7.
164. Ahuja Y, Kohl S, Traboulsi EI. CNGA3 mutations in two United Arab Emirates families with achromatopsia. Mol Vis. 2008;14:1293–7.
165. Ben Simon GJ, Abraham FA, Melamed S. Pingelapese achromatopsia: correlation between paradoxical pupillary response and clinical features. Br J Ophthalmol. 2004;88(2):223–5.
166. Fahim AT, Khan NW, Zahid S, Schachar IH, Branham K, Kohl S, et al. Diagnostic fundus autofluorescence patterns in achromatopsia. Am J Ophthalmol. 2013;156(6):1211–9. e2.
167. Stockman A, Smithson HE, Michaelides M, Moore AT, Webster AR, Sharpe LT. Residual cone vision without alpha-transducin. J Vis. 2007;7(4):8.
168. Khan NW, Wissinger B, Kohl S, Sieving PA. CNGB3 achromatopsia with progressive loss of residual cone function and impaired rod-mediated function. Invest Ophthalmol Vis Sci. 2007;48(8):3864–71.
169. Moskowitz A, Hansen RM, Akula JD, Eklund SE, Fulton AB. Rod and rod-driven function in achromatopsia and blue cone monochromatism. Invest Ophthalmol Vis Sci. 2009;50(2):950–8.
170. Aboshiha J, Luong V, Cowing J, Dubis AM, Bainbridge JW, Ali RR, et al. Dark-adaptation functions in molecularly confirmed achromatopsia and the implications for assessment in retinal therapy trials. Invest Ophthalmol Vis Sci. 2014;55(10):6340–9.
171. Thomas MG, McLean RJ, Kohl S, Sheth V, Gottlob I. Early signs of longitudinal progressive cone photoreceptor degeneration in achromatopsia. Br J Ophthalmol. 2012;96(9):1232–6.

172. Aboshiha J, Dubis AM, Cowing J, Fahy RT, Sundaram V, Bainbridge JW, et al. A prospective longitudinal study of retinal structure and function in achromatopsia. Invest Ophthalmol Vis Sci. 2014;55(9):5733–43.
173. Thiadens AA, Somervuo V, van den Born LI, Roosing S, van Schooneveld MJ, Kuijpers RW, et al. Progressive loss of cones in achromatopsia: an imaging study using spectral-domain optical coherence tomography. Invest Ophthalmol Vis Sci. 2010;51(11):5952–7.
174. Ding XQ, Harry CS, Umino Y, Matveev AV, Fliesler SJ, Barlow RB. Impaired cone function and cone degeneration resulting from CNGB3 deficiency: down-regulation of CNGA3 biosynthesis as a potential mechanism. Hum Mol Genet. 2009;18(24):4770–80.
175. Trifunovic D, Dengler K, Michalakis S, Zrenner E, Wissinger B, Paquet-Durand F. cGMP-dependent cone photoreceptor degeneration in the cpfl1 mouse retina. J Comp Neurol. 2010;518(17):3604–17.
176. Biel M, Seeliger M, Pfeifer A, Kohler K, Gerstner A, Ludwig A, et al. Selective loss of cone function in mice lacking the cyclic nucleotide-gated channel CNG3. Proc Natl Acad Sci U S A. 1999;96(13):7553–7.
177. Komaromy AM, Alexander JJ, Rowlan JS, Garcia MM, Chiodo VA, Kaya A, et al. Gene therapy rescues cone function in congenital achromatopsia. Hum Mol Genet. 2010;19(13):2581–93.
178. Carvalho LS, Xu J, Pearson RA, Smith AJ, Bainbridge JW, Morris LM, et al. Long-term and age-dependent restoration of visual function in a mouse model of CNGB3-associated achromatopsia following gene therapy. Hum Mol Genet. 2011;20(16):3161–75.
179. Sundaram V, Wilde C, Aboshiha J, Cowing J, Han C, Langlo CS, et al. Retinal structure and function in achromatopsia: implications for gene therapy. Ophthalmology. 2014;121(1):234–45.
180. Thomas MG, Gottlob I. Re: Sundaram et al.: Retinal structure and function in achromatopsia: implications for gene therapy (Ophthalmology 2014;121:234–45). Ophthalmology. 2014;121(9):e46–7.
181. Thiadens AA, Klaver CC. Re: Sundaram et al.: Retinal structure and function in achromatopsia: implications for gene therapy (Ophthalmology 2014;121:234–45). Ophthalmology. 2014;121(8):e40–1.
182. Carroll J, Choi SS, Williams DR. In vivo imaging of the photoreceptor mosaic of a rod monochromat. Vision Res. 2008;48(26):2564–8.
183. Thomas MG, Kumar A, Kohl S, Proudlock FA, Gottlob I. High-resolution in vivo imaging in achromatopsia. Ophthalmology. 2011;118(5):882–7.
184. Genead MA, Fishman GA, Rha J, Dubis AM, Bonci DM, Dubra A, et al. Photoreceptor structure and function in patients with congenital achromatopsia. Invest Ophthalmol Vis Sci. 2011;52(10):7298–308.
185. Dubis AM, Cooper RF, Aboshiha J, Langlo CS, Sundaram V, Liu B, et al. Genotype-dependent variability in residual cone structure in achromatopsia: toward developing metrics for assessing cone health. Invest Ophthalmol Vis Sci. 2014;55(11):7303–11.
186. Greenberg JP, Sherman J, Zweifel SA, Chen RW, Duncker T, Kohl S, et al. Spectral-domain optical coherence tomography staging and autofluorescence imaging in achromatopsia. JAMA Ophthalmol. 2014;132(4):437–45.
187. Scoles D, Sulai YN, Langlo CS, Fishman GA, Curcio CA, Carroll J, et al. In vivo imaging of human cone photoreceptor inner segments. Invest Ophthalmol Vis Sci. 2014;55(7):4244–51.
188. Yang P, Michaels KV, Courtney RJ, Wen Y, Greninger DA, Reznick L, et al. Retinal morphology of patients with achromatopsia during early childhood: implications for gene therapy. JAMA Ophthalmol. 2014;132(7):823–31.
189. Michaelides M, Holder GE, Bradshaw K, Hunt DM, Mollon JD, Moore AT. Oligocone trichromacy: a rare and unusual cone dysfunction syndrome. Br J Ophthalmol. 2004;88(4):497–500.
190. Michaelides M, Rha J, Dees EW, Baraas RC, Wagner-Schuman ML, Mollon JD, et al. Integrity of the cone photoreceptor mosaic in Oligocone Trichromacy. Invest Ophthalmol Vis Sci. 2011;52(7):4757–64.

191. Andersen MK, Christoffersen NL, Sander B, Edmund C, Larsen M, Grau T, et al. Oligocone trichromacy: clinical and molecular genetic investigations. Invest Ophthalmol Vis Sci. 2010;51(1):89–95.
192. Vincent A, Wright T, Billingsley G, Westall C, Heon E. Oligocone trichromacy is part of the spectrum of CNGA3-related cone system disorders. Ophthalmic Genet. 2011;32(2):107–13.
193. Jiang L, Baehr W. GCAP1 mutations associated with autosomal dominant cone dystrophy. Adv Exp Med Biol. 2010;664:273–82.
194. Olshevskaya EV, Calvert PD, Woodruff ML, Peshenko IV, Savchenko AB, Makino CL, et al. The Y99C mutation in guanylyl cyclase-activating protein 1 increases intracellular Ca^{2+} and causes photoreceptor degeneration in transgenic mice. J Neurosci. 2004;24(27):6078–85.
195. Buch PK, Mihelec M, Cottrill P, Wilkie SE, Pearson RA, Durran Y, et al. Dominant cone-rod dystrophy: a mouse model generated by gene targeting of the GCAP1/*guca1a* gene. PLoS One. 2011;6(3):e18089.
196. Wilkie SE, Li Y, Deery EC, Newbold RJ, Garibaldi D, Bateman JB, et al. Identification and functional consequences of a new mutation (E155G) in the gene for GCAP1 that causes autosomal dominant cone dystrophy. Am J Hum Genet. 2001;69(3):471–80.
197. Ramamurthy V, Tucker C, Wilkie SE, Daggett V, Hunt DM, Hurley JB. Interactions within the coiled-coil domain of RetGC-1 guanylyl cyclase are optimized for regulation rather than for high affinity. J Biol Chem. 2001;276(28):26218–29.
198. Wilkie SE, Newbold RJ, Deery E, Walker CE, Stinton I, Ramamurthy V, et al. Functional characterization of missense mutations at codon 838 in retinal guanylate cyclase correlates with disease severity in patients with autosomal dominant cone-rod dystrophy. Hum Mol Genet. 2000;9(20):3065–73.
199. Gregory-Evans K, Kelsell RE, Gregory-Evans CY, Downes SM, Fitzke FW, Holder GE, et al. Autosomal dominant cone-rod retinal dystrophy (CORD6) from heterozygous mutation of GUCY2D, which encodes retinal guanylate cyclase. Ophthalmology. 2000;107(1):55–61.
200. Michaelides M, Johnson S, Bradshaw K, Holder GE, Simunovic MP, Mollon JD, et al. X-linked cone dysfunction syndrome with myopia and protanopia. Ophthalmology. 2005;112(8):1448–54.
201. Schwartz M, Haim M, Skarsholm D. X-linked myopia: Bornholm eye disease. Linkage to DNA markers on the distal part of Xq. Clin Genet. 1990;38(4):281–6.
202. Young TL, Deeb SS, Ronan SM, Dewan AT, Alvear AB, Scavello GS, et al. X-linked high myopia associated with cone dysfunction. Arch Ophthalmol. 2004;122(6):897–908.
203. Kazmi MA, Sakmar TP, Ostrer H. Mutation of a conserved cysteine in the X-linked cone opsins causes color vision deficiencies by disrupting protein folding and stability. Invest Ophthalmol Vis Sci. 1997;38(6):1074–81.
204. McClements M, Davies WI, Michaelides M, Young T, Neitz M, Maclaren RE, et al. Variations in opsin coding sequences cause x-linked cone dysfunction syndrome with myopia and dichromacy. Invest Ophthalmol Vis Sci. 2013;54(2):1361–9.
205. Sharpe LT, Stockman A, Jagle H, Knau H, Klausen G, Reitner A, et al. Red, green, and red-green hybrid pigments in the human retina: correlations between deduced protein sequences and psychophysically measured spectral sensitivities. J Neurosci. 1998;18(23):10053–69.
206. Ueyama H, Kuwayama S, Imai H, Oda S, Nishida Y, Tanabe S, et al. Analysis of L-cone/M--cone visual pigment gene arrays in Japanese males with protan color-vision deficiency. Vision Res. 2004;44(19):2241–52.
207. Carroll J, Neitz M, Hofer H, Neitz J, Williams DR. Functional photoreceptor loss revealed with adaptive optics: an alternate cause of color blindness. Proc Natl Acad Sci U S A. 2004;101(22):8461–6.
208. Ueyama H, Muraki-Oda S, Yamade S, Tanabe S, Yamashita T, Shichida Y, et al. Unique haplotype in exon 3 of cone opsin mRNA affects splicing of its precursor, leading to congenital color vision defect. Biochem Biophys Res Commun. 2012;424:152–7.

209. Gardner JC, Liew G, Quan YH, Ermetal B, Ueyama H, Davidson AE, et al. Three different cone opsin gene array mutational mechanisms with genotype-phenotype correlation and functional investigation of cone opsin variants. Hum Mutat. 2014;35(11):1354–62.
210. Carroll J, Dubra A, Gardner JC, Mizrahi-Meissonnier L, Cooper RF, Dubis AM, et al. The effect of cone opsin mutations on retinal structure and the integrity of the photoreceptor mosaic. Invest Ophthalmol Vis Sci. 2012;53(13):8006–15.
211. Malanson KM, Lem J. Rhodopsin-mediated retinitis pigmentosa. Prog Mol Biol Transl Sci. 2009;88:1–31.
212. McClements M, Davies WI, Michaelides M, Carroll J, Rha J, Mollon JD, et al. X-linked cone dystrophy and colour vision deficiency arising from a missense mutation in a hybrid L/M cone opsin gene. Vision Res. 2013;80:41–50.
213. Gardner JC, Webb TR, Kanuga N, Robson AG, Holder GE, Stockman A, et al. X-linked cone dystrophy caused by mutation of the red and green cone opsins. Am J Hum Genet. 2010;87(1):26–39.
214. Wagner-Schuman M, Neitz J, Rha J, Williams DR, Neitz M, Carroll J. Color-deficient cone mosaics associated with Xq28 opsin mutations: a stop codon versus gene deletions. Vision Res. 2010;50(23):2396–402.
215. Alexander JJ, Umino Y, Everhart D, Chang B, Min SH, Li Q, et al. Restoration of cone vision in a mouse model of achromatopsia. Nat Med. 2007;13(6):685–7.
216. Michalakis S, Muhlfriedel R, Tanimoto N, Krishnamoorthy V, Koch S, Fischer MD, et al. Restoration of cone vision in the CNGA3−/− mouse model of congenital complete lack of cone photoreceptor function. Mol Ther. 2010;18(12):2057–63.

Chapter 2
The Retinal Processing of Photoreceptor Signals

Jan Kremers, Luiz Carlos L. Silveira, Neil R.A. Parry, and Declan J. McKeefry

Abstract Color vision is the ability to perceive differences in the wavelength content of a light source, a process which starts with absorption of photons of different wavelengths and energies by the photopigments. In this chapter, the photopigments and the efficiency with which light of different wavelengths are absorbed by a photopigment are discussed. In addition, the translation of a photoisomerization to a photoreceptor excitation is considered as well as the signal transmission from the photoreceptors to post-receptoral cells and the post-receptoral processing of this signal in the retina. A large part of the chapter provides an overview of recent evidence that retinal processes in the major retino-geniculate pathways that are

J. Kremers, Ph.D. (✉)
Department of Ophthalmology, University Hospital Erlangen, Schwabachanlage 6, 91054 Erlangen, Germany

School of Optometry and Vision Science, University of Bradford, Richmond Road, Bradford, West Yorkshire BD7 1DP, UK
e-mail: Jan.kremers@uk-erlangen.de

L.C.L. Silveira, Ph.D. (deceased)

N.R.A. Parry, Ph.D.
School of Optometry and Vision Science, University of Bradford, Richmond Road, Bradford, West Yorkshire BD7 1DP, UK

Centre for Hearing and Vision Research, Institute of Human Development, University of Manchester, Manchester, UK

Vision Science Centre, Manchester Royal Eye Hospital, Central Manchester University Hospitals NHS Foundation Trust, Manchester Academic Health Science Centre, Manchester, UK
e-mail: neil.parry@manchester.ac.uk

D.J. McKeefry, Ph.D.
School of Optometry and Vision Science, University of Bradford, Richmond Road, Bradford, West Yorkshire BD7 1DP, UK
e-mail: d.mckeefry@bradford.ac.uk

J. Kremers et al. (eds.), *Human Color Vision*, Springer Series in Vision Research, DOI 10.1007/978-3-319-44978-4_2

relevant for luminance and color vision, can be studied in the intact visual system by electroretinography (ERG), thus providing the possibility of direct study of human retinal physiology.

Keywords Photopigments • Fundamentals • Photoreceptors • Retinal pathways • Electroretinography • Silent substitution • Horizontal cells • Bipolar cells • Retinal ganglion cells • Post-receptoral circuitry • Cone opponency

2.1 Photopigments and Excitation of Photoreceptors

2.1.1 Spectral Sensitivities of Photopigments and Fundamentals

Photoreceptors are neurons that are specialized for the absorption of light and transforming light information into electrical signals. Signals that are further processed in the retina and transmitted to the brain for visual perception and non-conscious behaviors such as eye movements, pupillary reflexes, and circadian rhythms. The chromophore, an aldehyde of vitamin A called retinal, and the opsin protein, which is an integral protein of the plasma membrane, together form the photopigment. The visual response starts with absorption of photons by the photopigments resulting in an isomerization of the retinal chromophore. Retinal isomerization triggers a cascade of internal changes that finally leads to photoreceptor hyperpolarization, as a result of changes in ion transfer across the cell membrane thanks to opening and closing of the ion pores controlled by the chromophore. Unlike most other neurons, photoreceptors (and the majority of other retinal neurons except retinal ganglion cells) do not propagate electrical signals through action potentials, but do so instead via graded potential changes.

The absorption of a photon leads to deformation or photo-isomerization of the retinal chromophore from the 11-*cis*- to the all-*trans*-configuration. The photopigments are mainly present in the photoreceptors but some retinal ganglion cells contain melanopsin and as a result are also photoresponsive [1–6]. These recently discovered intrinsically photosensitive ganglion cells (ipRGCs) are thought to have little or no role in vision (but see Ecker et al. [7] for a challenging idea) and are probably not involved in color vision. They are therefore not considered in depth in this chapter.

2.1.2 Variability of Pigment Spectra and Its Consequences for Psychophysics

The Principle of Univariance [8] states that the effect of every photon that is absorbed is the same, independent of the wavelength or energy content of the photon. Thus, a photoisomerization always leads to the same configuration change of the retinal chromophore as mentioned above and to the same response (i.e., hyperpolarization) of the photoreceptor.

Importantly, the chance that a photon is absorbed and that a photoisomerization takes place is not always the same and depends on the wavelength of the light. The probability of a photoisomerization as a function of wavelength is described by the absorption spectrum. The absorption spectrum is determined by the properties of the retinal and the amino-acid sequence of the opsin protein in which the retinal is embedded. In humans, there are three types of cone opsin, the configurations of which are genetically determined. The genes for the L- and M-cone opsins are located on the X-chromosome. The gene for S-cone opsins is located on chromosome 7 (see Chap. 1). In situ, the absorption spectrum is also influenced by other parameters such as pigment concentration and length of the photoreceptor outer segments. These factors influence the absorption spectrum through self-screening effects. The photopigment is ordered in layers in the stacks of membranes in the photoreceptor outer segments. The first layer absorbs more light with wavelengths close to the maximum of its spectral sensitivity. As a result, the light reaching the second layer contains less photons with wavelengths close to this maximum (the first layers is thus screening the second layer) and therefore will also absorb less light close to maximum and more of the "submaximal" wavelengths in comparison with the first layer. This self-screening effect is propagated through all layers. The overall absorption spectrum thus will become broader the more layers the light will have to pass.

In order to be able to appreciate the effect of light with a certain wavelength content, the absorption spectrum alone is generally not sufficient to provide all the necessary information. For many cases and experimental situations (basically those in which measurements are performed with intact eyes), pre-retinal absorption (mainly by the cornea and the lens) alters the wavelength content of light reaching the retina. In these cases, not the rod and cone absorption spectra but the rod and cone fundamentals, that include pre-retinal absorption, should be taken into account. The fundamentals therefore describe the efficiency of light at the level of the pupil entrance of the eye to stimulate the photoreceptors. Figure 2.1 shows the fundamentals of the human photoreceptors. The most often used fundamentals are those of Smith and Pokorny [9, 10] and of Stockman and colleagues [11–14] (see Chap. 5).

The absorption spectra are more appropriate than the fundamentals when considering the results of in vitro or ex vivo experiments (in which the retina has been excised from the eye or pre-retinal tissue has been removed).

2.1.3 *Responses of Photoreceptors to Stimuli*

The efficiency with which a stimulus is able to excite a photopigment is calculated by the multiplication of the fundamentals with the emission spectrum of the stimulus integrated over the wavelength range. Following the Principle of Univariance, the photopigment excitation is proportional to the response (hyperpolarization) of the photoreceptor and thus describes the sensitivity of the photoreceptor to the stimulus.

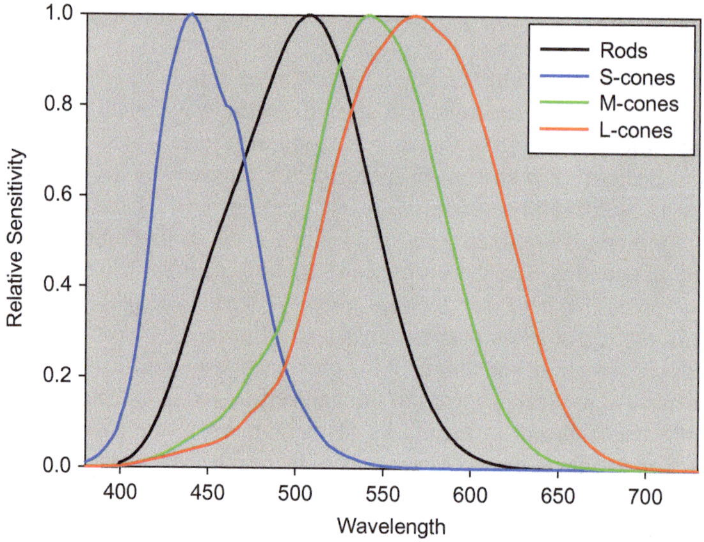

Fig. 2.1 Normalized cone [11, 14] and rod fundamentals. The fundamentals describe the sensitivity of the different photoreceptor types for stimulation at the pupil. The sensitivities are normalized to unity at their maxima. The rod fundamental is identical to the scotopic spectral luminosity function [V'_λ; [15]]. The cutoff at short wavelengths is caused by pre-retinal absorption mainly by the lens and the cornea

Sensitivity is defined as the inverse of the stimulus strength required to evoke a criterion response or a psychophysical threshold. Alternatively, particularly in physiological experiments, sensitivity can be quantified by a gain that is defined as the change in response amplitude caused by a change in stimulus strength (i.e., the slope of the response amplitude vs. stimulus strength curve). Crucially, however, the photoreceptors are dynamic, changing their sensitivity via adaptation when the stimulus is strong, or presented over an extended time.

The strength of a repetitive stimulus around a mean level is quantified by Michelson contrast (*MC*):

$$\mathrm{MC} = \frac{E_{\mathrm{max}} - E_{\mathrm{min}}}{E_{\mathrm{max}} + E_{\mathrm{min}}}$$

where E is the cone or rod excitation (or photoreceptor sensitivity), E_{max} and E_{min} are the maximal and minimal excitations, respectively. Observe that the same formula is also used to calculate luminance contrast where E is replaced by the luminance. The luminance is calculated in a similar manner as E with the difference that instead of the fundamentals, the emission spectra are multiplied by the photopic spectral luminosity function, V_λ, This definition of stimulus strength is useful when the modulation of the excitation is evenly distributed around the mean level (such as in sine-wave or square-wave stimuli). In such instances the addition of the two (i.e., $E_{\mathrm{max}} + E_{\mathrm{min}}$) equals twice the mean excitation. The main advantage in using this metric is that stimulus strength (cone and rod contrast; similarly for luminance contrast) can be changed without changing of the mean level and vice versa.

For pulsed stimuli, Weber contrast (*WC*) is more useful to describe stimulus strength:

$$\mathrm{WC} = \frac{E_{\max} - E_{\min}}{E_{\min}}$$

Here, $E_{\min}$ equals the excitation caused by the background and, for short-lived weak pulses (when $E_{\max} - E_{\min}$ is small in comparison with $E_{\min}$), also the mean excitation. The problem with pulses is that the mean excitation changes with pulse strength and duration. This may cause the visual system to adapt and change its sensitivity.

There are two ways to isolate the responses from a particular photoreceptor type. In the early days of psychophysical studies, when there was no precise knowledge about the fundamentals or the physiological properties of photoreceptors and when the technical possibilities for stimulus presentation were limited, this was achieved by selective desensitization of the photoreceptors that were not of interest by a background and selective isolation of the photoreceptor under study with a flash stimulus [12, 16, 17] (See also Ch. 5).

Although a well-established technique, this method of selective desensitization has its disadvantages. First, complete isolation is difficult, if not impossible, to achieve, particularly when flash and background wavelengths are not very different. Second, the abovementioned disadvantages of pulsed stimuli also count here. Third, to study different photoreceptor types, different backgrounds need to be used which induce different states of adaptation in the retina and the visual system. This may affect the outcome of the measurements and the results from different photoreceptor isolation conditions cannot therefore be easily compared.

An alternative is the silent substitution method using periodic stimuli around a mean luminance and chromaticity (Fig. 2.2). This method was first introduced by Donner and Rushton [18] and further developed by Estévez and Spekreijse [19, 20]. With the technical advance of cost-effective stimulus devices that can deliver stimuli with well-defined spectral, temporal, and spatial properties, this method has become extremely attractive and has now generally replaced the method of selective adaptation. The silent substitution method can be applied to flashed and to continuously modulating stimuli. In the latter case, as mentioned before, the state of adaptation can be kept constant even when different mechanisms are studied. Furthermore, isolation of the responses of different photoreceptor types can, at least theoretically, be more complete than with the selective adaptation method [21]. A silent substitution is the replacement of one stimulus by another of different spectral composition. It is possible to choose the luminance (intensity) of the two stimuli so that the number of photoisomerizations in one photoreceptor type, and thus its excitation, does not change (and cone or rod contrast, as defined above, equals zero). The principle is illustrated in Fig. 2.2 for two monochromatic stimuli and two hypothetical photopigments. The number of photoisomerizations depends on the fundamentals of the photoreceptors and on the intensity of the stimuli. In Fig. 2.2 the number of photoisomerizations of the two hypothetical photoreceptor types, containing

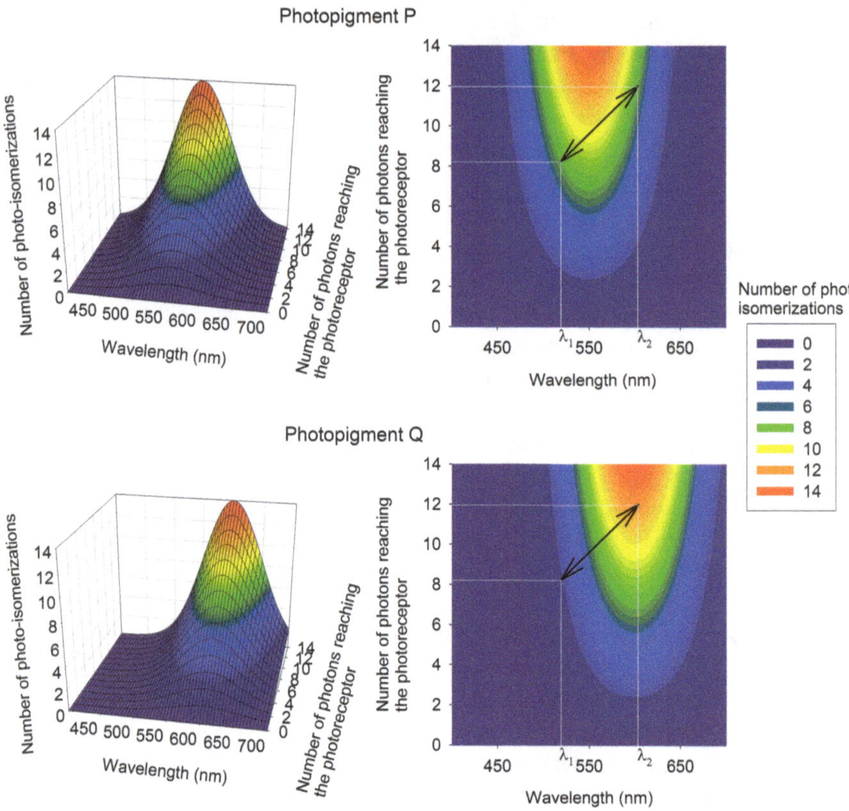

Fig. 2.2 Explanation of the silent substitution method. The numbers of photoisomerizations in two photoreceptor types with different photopigments (P and Q) are given as function of wavelength and intensity of the stimuli. The effect of wavelength is described by the fundamental. Stimulus intensity is assumed to have linear effect: if the intensity is doubled, the number of isomerizations is also doubled. In the *right plot* an example of a stimulus (an exchange between two monochromatic lights, λ_1 and λ_2) is shown. The intensity of the two lights is chosen to counteract the difference in probability of an absorption of a photon in photopigment P as described by the fundamentals. As a result, the number of isomerizations does not change by the wavelength exchange. This stimulus results in a silent substitution for photoreceptor P and the isolation of the response of photoreceptor Q. Reprinted with permission from Kremers [21]

photopigments P and Q, are given as a function of wavelength and of intensity. Three dimensional plots are given on the left. On the right, the same is displayed in a color coded manner. The double headed arrows signify a modulation between two monochromatic stimuli (λ_1 and λ_2). The intensities of these two stimuli are chosen such that the number of photoisomerizations in photopigment P (and thus the excitation of the photoreceptor containing photopigment P) is not altered. The stimulus is a silent substitution of that photoreceptor. The number of photoisomerizations in photopigment Q is strongly altered. Only the photoreceptor containing photopigment Q responds to the stimulus and thus is isolated.

The silent substitution method can be extended to stimuli with broadband and complex emission spectra, such as LEDs or CRT monitors. The method is not only suitable for silencing one or more photoreceptor types, but also allows the choice of any stimulus strength for each photoreceptor type, provided the stimuli are within the gamut of the stimulator. The number of photopigments that can be independently stimulated can be increased when more light sources are used. Theoretically, the number of independent light sources should at least equal the number of photopigments present. Since humans have normally four different photoreceptor types, a four primary stimulator is necessary [22]. If the melanopsin pigment also has to be taken into account, a five primary stimulator is desirable [23].

In practice, the silent substitution method may be contaminated by errors, miscalculations, misinterpretations, and oversimplifications. For instance macular pigment, which only covers the central retina, influences pre-retinal absorption and as a consequence will affect the wavelength composition of light reaching the photoreceptors differently in the central compared to the peripheral retina. Furthermore, individual differences in the optical density of the macular pigments [24, 25] result in individual differences in central cone fundamentals. It is often difficult to take this spatial and individual variability into account in psychophysical and physiological measurements. However, as work with dichromats and other control experiments have often shown [26–29], the silent substitution method is generally very effective. For instance, the isolation of L-cone driven responses can be tested in protanopes who lack the L-cones. Stimuli that isolate L-cones elicit electrophysiological responses and psychophysically measured perception in trichromats and deuteranopes but not in protanopes. Similarly, M-cone isolating stimuli elicit much smaller responses in deuteranopes (they are not always completely abolished; we return to this issue later). Rod isolating stimuli elicit small ERG responses when full field stimuli and high luminances are used. At low luminances and with smaller stimuli, the rod responses are increased in amplitude through stimulation of the dark adapted surrounding retina through stray light (McKeefry, Maguire, Parry, Murray, Kommanapalli, Aher, Kremers unpublished data; see also Park et al. [30]). In psychophysical experiments, the degree of rod isolation can be tested by bleaching the rods prior to the measurements. If only rods are stimulated, the stimuli should not be perceivable directly after bleaching. Perception of the stimuli should return only after several minutes of dark adaptation. Cone-driven perception returns to normal much more quickly after bleaching [31].

2.2 Physiological Basis of Color Vision

As mentioned above, color vision is the ability to perceive, within limits, the wavelength content of light emitted by a luminous source or reflected by an illuminated surface. Physiologically, there are two prerequisites for color vision. First, photoreceptor types with different absorption spectra are necessary. In humans and under photopic conditions, three different cone types provide the basis for color vision. Depending on the wavelength content of the light reaching the cones, their relative excitations will be different which will influence the percept of color. Because humans normally have three cone types, color vision is trichromatic: all perceived colors can be obtained by mixing the outputs of three independent primaries (independent in this context means that the output of one of the three primaries cannot be obtained by mixture of the output of the other two). With these cone types, wavelengths between approximately 380 and 780 nm can be distinguished (see Fig. 2.1).

The second prerequisite for color vision is a mechanism that can determine the relative cone excitations. These mechanisms are thus post-receptoral. Probably, multiple comparisons are performed at different stages in the visual system.

2.3 Post-receptoral Processing

The signals coming from the photoreceptors are processed in a parallel manner directly at the first synapse, where connections are made with different types of horizontal cells and bipolar cells that have different anatomical and physiological properties.

Color vision can only occur when the relative excitations of the three cone types are determined by post-receptoral processing of cone signals. In the retina, the signals are processed in parallel in different post-receptoral channels or pathways. The main retinal pathways are already separated at the level of the bipolar cells. The properties of the different pathways and signal processing therein are subject of Chap. 4. Here, we summarize the properties with an emphasis on how the signals, originating in the different cone types, are distributed and processed.

Many studies on primate color vision are performed in Old and New World monkeys. New World monkeys are closely related to humans and Old World monkeys but show different photoreceptor phenotypes. Therefore they make it possible to study the relationship between cone phenotypes and retinal wiring and the effect this has on their color vision [32]. Most New World monkeys have only one gene coding for a cone photopigment on the X-chromosome, but there are three or even more alleles of this gene in the population. As a result, color vision is polymorphic with all males and the homozygous females are dichromats whereas the heterozygous females are trichromats. One species, the howler monkey, displays full trichromacy similar to Old World monkeys [33, 34]. However, trichromacy in these species has probably evolved independently [35, 36]. Finally, the nocturnal owl monkeys lack S cones and are therefore obligatory monochromats [37, 38]. These species form natural experiments in which the signals of different photoreceptor types can be studied.

2.3.1 Horizontal Cells and Their Connectivity

Horizontal cells and bipolar cells form the first steps of the neuronal processing of photoreceptor outputs and bipolar cells transmit information from the inner to the outer plexiform layers [32]. Amacrine cells and ganglion cells further process the photoreceptor signals, whereas inter-plexiform cells transmit information from the inner plexiform layer back to the outer plexiform layer. Ganglion cells send the results of retinal processing to higher visual centers located in the mesencephalon and diencephalon [32]. We refer to Chap. 4 for a more detailed discussion of the responses of retinal ganglion cells and neurons of the lateral geniculate nucleus with respect to cone opponent processing and generation of the information necessary for color perception.

The retinae of all primates so far studied (including humans) have post-receptoral neurons, comprising horizontal cells, bipolar cells, amacrine cells, inter-plexiform cells, and ganglion cells. Their morphologies and functional roles are similar across species. The morphology of these cells, the connections that they make in the inner and outer plexiform layers, and the two-dimensional mosaics they form in the retinal layers have been studied using techniques that stain individual cells [33–35] or cell populations [36–41]. Intracellular injection of neurotracers is a powerful method to study both the morphology of individual cells and the properties of cell populations [42–46].

Primates have two classes of horizontal cells, H1 [33] and H2 [47], whose morphology, cone connectivity, and distribution have been extensively described in humans [47–50] and several diurnal monkeys [45, 51–55] (see Fig. 2.3). H1 and putative H2 horizontal cells have also been described in the single extant nocturnal monkey, the owl monkey, although H2 cells seem to be very rare and exhibit a simple morphology [56]. H1 horizontal cells have large cell bodies and radiate stout primary dendrites bearing conspicuous clusters of terminals which can be traced to axon pedicles of M or L cones. In the majority (85 %) of H1 horizontal cells, the dendritic terminals make synapses exclusively with M and L cones. In the other 15 %, a few synapses contact S cones [54]. H1 horizontal cells typically have 4 dendritic clusters in the foveal region, increasing to 8 at about 1 mm of eccentricity, and 30 or more at higher eccentricities [49, 55]. From the fovea to the retinal periphery, the inter-cluster distance strictly follows the inter-cone distance, as measured in cone mosaics visualized in retinal flat-mounts [56]. H1 horizontal cells have axons that are thick, run away from the cell body in straight paths for a long distance and then exhibit fan-shaped terminals with compact branches and tiny knobs, which can be traced to rod spherules.

H2 horizontal cells have medium-sized cell bodies and thin, curved, intertwined, profusely branched dendrites with dendritic terminals that, unlike H1 cells, are not clearly organized to form clusters. The dendrites of H2 horizontal cells are directed to contact all cone pedicles—S, M, or L—overlying their dendritic fields. However, S cone pedicles make more synapses with H2 dendrites than M or L cone pedicles do [45] (Fig. 2.3). In an anatomical study of the marmoset retina, about 11 % of synapses to a single H2 horizontal cell were formed by a single S cone pedicle which represented only 4 % of all cones under the cell dendritic tree [54]. The H2 axons are short and convoluted, branch one or more times, and bear a small number of terminals along their paths that also contact S cone pedicles.

H1 and H2 dendritic trees and H1 axon terminals form three cell networks in the outer plexiform layer extensively coupled by specific types of gap junctions [58–60]. Thus, the response to light of a single horizontal cell occurs in a retinal region much

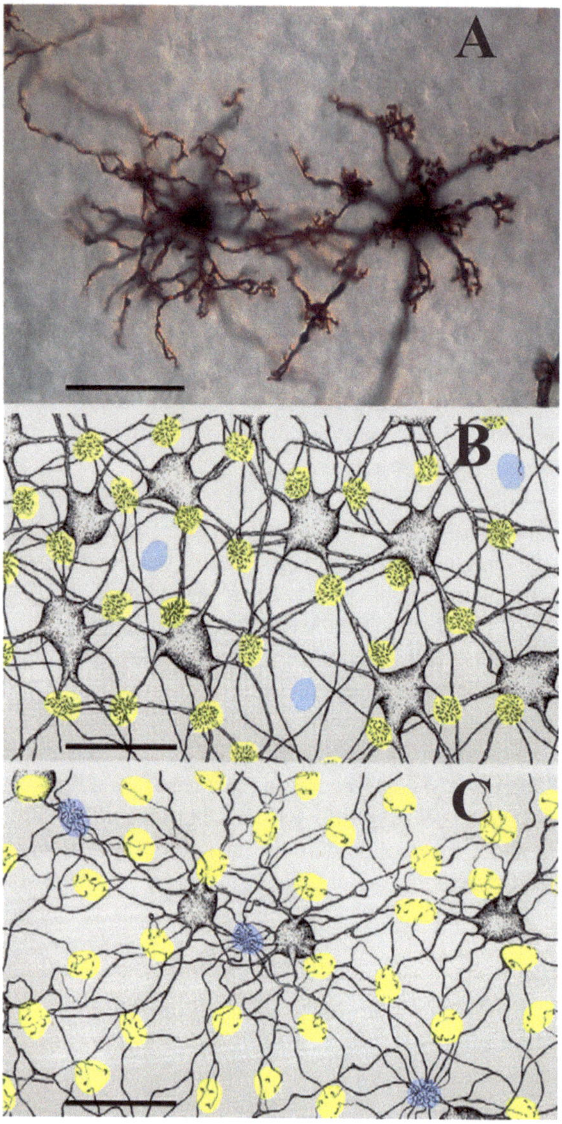

Fig. 2.3 H1 and H2 horizontal cells and their cone contacts. The morphology, connectivity, and retinal distribution of retinal cells can be studied by combining different techniques to stain individual cells, to label a particular cell population, and to double label the overlying cone mosaic. (**a**) Photomicrograph of macaque H1 and H2 cells stained with the method of Golgi, which uses heavy metals such as silver and mercury compounds to impregnate neurons. Cells are shown in flat view focused on the dendritic trees and terminal dendritic clusters. (**b**) Drawings of macaque H1 cells stained by neurobiotin injection in one of the cells. The tracer diffused to the neighboring H1 cells through gap junctions between them. H1 cells form dense terminal dendritic clusters at all M- and L-cone pedicles (indicated by *yellow patches*), but almost completely miss three S-cones (*blue patches*). (**c**) Drawing of macaque H2 cells similarly labeled with neurobiotin. They form dense terminal dendritic clusters at the three S-cones (*blue*), but also contact the M- and L-cones. Scale bars = 25 μm. Images kindly provided by Dennis Dacey to Leo Peichl. Reproduced with permission from: Peichl, L. [57]

larger than that covered by its processes. The strength of coupling in the cell network is highly dynamic and is modulated by light-dependent changes in the levels of several neuromodulators including dopamine, nitric oxide and retinoic acid [60].This allows horizontal cells to play an important role in the spatial extent of the interactions between photoreceptors and post synaptic interneurons [61]. Horizontal cells send feed-back and feed-forward inhibitory signals to cones and rods and bipolar cells, thereby mediating lateral inhibition. Thus they serve to integrate and control photoreceptor output, allowing the retina to adapt to a broad range of light intensities and contributing to center-surround organization of bipolar and ganglion cells [61].

The difference between the two types of horizontal cells suggests that H2 cells may have a role in the blue-yellow pathway that receives S cone signals. The exact function of horizontal cells in color vision is, however, unclear. Considering the H1 morphologies of trichromatic, dichromatic, and monochromatic Old- and New-World monkeys, there are no major inter-species differences that could be related to their cone phenotypes. This implies that horizontal cells play only a minor role in primate color vision [54–56, 62, 63]. The sparseness and simple morphology of H2 horizontal cells in the retina of the owl monkey is probably related to the absence of S cones [56]. Thus, horizontal cells in primates (and other mammals) seem to exhibit cone specificity without cone opponency [45] (Fig. 2.4). This is in contrast with fish and turtles where horizontal cells seem to be the basis of color vision [64, 65].

2.3.2 *Bipolar Cells and Their Connectivity*

Primate bipolar cells comprise several morphological classes with specific connections in the outer and inner plexiform layers, and with putatively diverse roles in vision [33, 34, 66–69]. The work of Boycott and Wässle [35] established the current classification of primate bipolar cells. Specific antibodies are now available that selectively label several bipolar cell classes. As a result there is increasing interest in a detailed characterization of primate bipolar cell morphology, spatial distribution, cone connections, and synaptic targets in the inner retina. In spite of the amount of information about individual cells provided by the Golgi method [35] or DiI labeling [70], immunocytochemistry has been particularly critical for the advancement of knowledge on the populations of these small, very numerous, and diversified retinal interneurons [38, 70–72].

The cone signals diverge to several cone bipolar cell classes, suggesting that, at the very first synapse of the visual pathway, visual information is distributed into parallel pathways for further processing. Thus, the identification and characterization of the bipolar cell classes is paramount to understand how parallel processing in the visual system starts and is organized through the retinal layers. Boycott and Wässle [35] described ten different bipolar cell classes in the retina of rhesus macaque, characterizing them by the number of cones or rods they contact with their dendrites, the depth of their cell body in the inner nuclear layer and the shape and level of branching of their axon terminals in the inner plexiform layer. Further

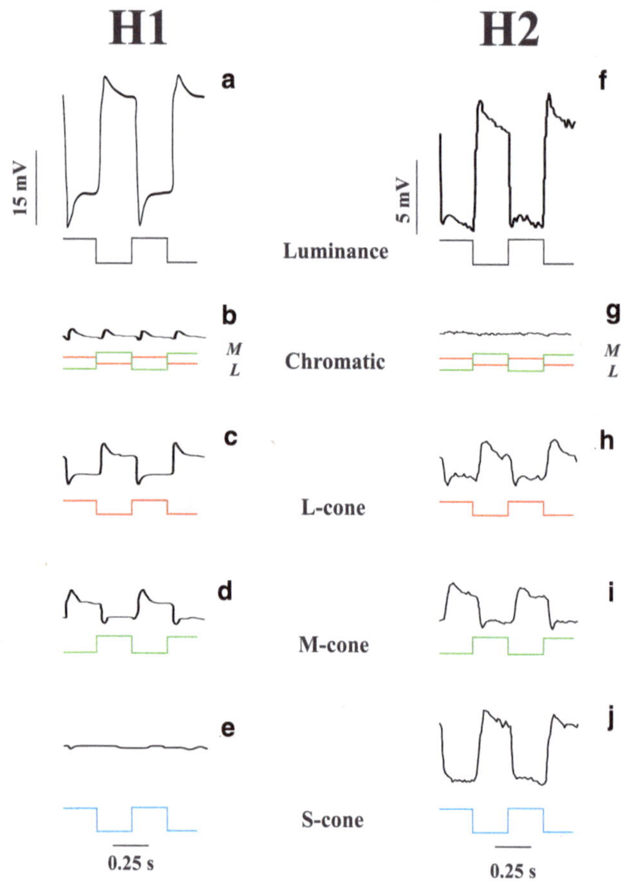

Fig. 2.4 Electrophysiology of cone inputs to H1 and H2 horizontal cells. Responses of horizontal cells to different stimuli are depicted below each trace. The five rows display responses to luminance (in phase modulation of S-, M-, and L-cones), chromatic (counter-phase modulation of M- and L-cones), and to L-, M-, and S-cone isolating stimuli respectively. H1 cells (*left column*) display sustained hyperpolarizing responses to luminance increments (**a**) and small responses at twice the stimulus frequency to chromatic modulation (**b**). Selective modulation of L- or M-cones also elicits a hyperpolarizing response (**c** and **d**). Selective S-cone stimulation elicits no response in H1 cells (**e**). H2 cells respond (*right column*) to luminance modulation with a sustained hyperpolarization (**f**). There is no response to chromatic modulation (**g**). H2 cells hyperpolarize in response to L- and M-cone excitation (**h** and **i**) but also to S-cone excitation (**j**). Modified with permission from: Dacey et al. [45]

details of macaque bipolar cell classes were provided by a series of immunocytochemistry studies [38, 39, 41, 71, 73]. More recently, another series of studies revealed that marmosets and other New World monkeys with abovementioned polymorphic color vision possess the same classes of bipolar cells as macaques, indicating that color processing in bipolar cells is not based on its morphology [70, 72, 74, 75]. However, there is evidence that the cone inputs to centers and

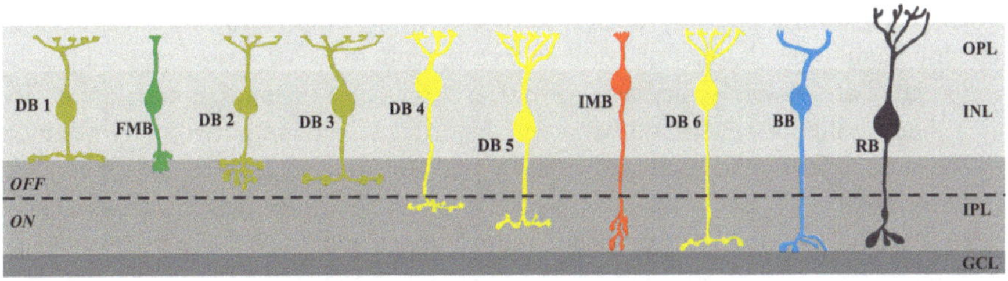

Fig. 2.5 Schematic diagram of bipolar cell classes found in the primate retina. Cells were studied in retinal flat mounts or in transverse sections impregnated with the method of Golgi [35]. They were classified using morphological criteria: shape of dendritic trees, number of dendritic clusters, cell body position in the inner nuclear layer, shape and position of their axon terminals in the inner plexiform layer. *OPL* outer plexiform layer, *INL* inner nuclear layer, *IPL* inner plexiform layer, *GCL* ganglion cell layer, *OFF* superficial strata of the inner plexiform layer where the axon terminals of OFF bipolar cells branch, *ON* deep strata of the inner plexiform layer where the axon terminals of ON bipolar cells branch. Modified with permission from Boycott and Wässle [35]

surrounds of parvocellular retinal ganglion and LGN cells are more selective than expected on the basis of random wiring, indicating that in course of evolution the functional wiring of bipolar cells may have changed such that responses to color stimuli were increased [76] (see also Chap. 4). Below, we argue that electroretinograms (ERGs), that presumably reflect bipolar cell activity, can display responses that are reminiscent of those of the parvocellular pathway, suggesting that some bipolar cells (probably the midget bipolar cells) indeed may be involved in chromatic processing.

Bipolar cell classes comprise a single class of rod bipolar cells (RB cells) and at least nine different classes of cone bipolar cells: two midget bipolars (FMB cells and IMB cells), six diffuse bipolars (DB1–6 cells), and a single class of S-cone bipolars (BB cells) [32] (Fig. 2.5). In addition, giant cone bipolar cells (GB cells) have been reported in several studies [33, 35, 48, 68, 77, 78].

Midget bipolar (MB) cells make very distinct connections and form the origin of the M/L cone opponent parvocellular pathway which is considered to be the basis for the red-green channel of color vision. They connect small patches of the photoreceptor mosaic with small dendritic trees of midget ganglion cells (also called parvocellular or PC ganglion cells) in the inner plexiform layer [35, 41]. In the central region, MB cells have a single primary dendrite and a single dendritic cluster that contact the pedicles of one L- or M-cone. Single-cone MB cells are found up to 45 deg of eccentricity, but at greater eccentricities there is an increasing proportion of MB cells that have two to four dendritic clusters each contacting a distinct cone [41]. MB cells comprise two separate populations with distinctive morphology and connectivity: FMB cells make flat synapses with cone pedicles and send axons to the upper half of the inner plexiform layer while IMB cells make invaginating synapses with cone pedicles and send axons to the lower half of the inner plexiform layer [35, 41, 66]. Every M or L cone pedicle contacts an FMB and an IMB cell [79]. Below, we discuss the functional implication of these morphological findings for red-green color vision and for recent ERG findings.

BB cells form a pathway dedicated to conveying information from S cones to specific ganglion cells and thus provide S cone input to S-/ML-cone opponent pathways. They are probably the basis for blue-yellow color vision. BB cells are easily identified by their long, smoothly curved dendrites which are horizontally oriented in the outer plexiform layer and contact between one and three cone pedicles. Double labeling of BB cells and S cones showed that their dendrites are clearly cone-selective and reach toward one or more S cones, making invaginating synapses with them. Their axon terminals are relatively large and stratify in the deeper stratum of the inner plexiform layer, close to the ganglion cell layer.

DB cells comprise several different classes, some of them providing M or L cone input to S-/ML-cone opponent pathways, for which they provide the "yellow" input of the blue-yellow color channel. DB cells can be divided into two groups using similar criteria of flat and invaginating synapses as used for distinguishing MB cells: DB1, DB2, and DB3 make flat synapses with cone pedicles and send axons to the upper half of the inner plexiform layer while DB4, DB5, and DB6 cells make invaginating or flat synapses with cone pedicles and send axons to the lower half of the inner plexiform layer; the different classes have been sequentially numbered following the depth of their axon terminals in the inner plexiform layer [35, 38, 71]. A subdivision of DB3 cells into DB3a and DB3b has recently been proposed [72, 80, 81]. DB cells contact all cones in their dendritic trees by the means of multiple dendritic clusters (4–10 clusters in the retinal periphery) [35, 38, 71, 73, 82, 83], but DB4 and DB6 are biased against S cones and make proportionally more dendritic contacts with M and L cones [84] while there is no evidence for a bias in favor or against S cones in the other DB classes [82].

2.3.3 *The Transmission of Photoreceptor Signals to Bipolar Cells*

Photoreceptor signals are transmitted to a series of post-receptoral elements that are described in the previous section which process these signals and send them to second order retinal neurons. The result is coded in the train of neural impulses sent by about 1,100,000 retinal ganglion cell axons to higher visual centers of the human visual system (described in Chap. 4). There are important limits to vision imposed by the properties of photoreceptors and post-receptoral neurons and, consequently, several aspects of the information content present in the retinal image are considerably transformed by retinal neural circuits. The existence of ON and OFF pathways, the temporal responses of different class of neurons, the spatial pooling by retinal elements of the same class, and the coding of spectral reflectance are relevant aspects of visual information processing that take place in the retina. In this section we describe those more directly relevant for color vision.

Cones and rods transmit their signals directly to horizontal cells and bipolar cells through complex synapses located at their axon terminals—cone pedicles and rod spherules, respectively. Cones of all classes and rods respond to light with changes of electrical potential across their plasma membrane, always with the same polarity.

When the light level is increased, cones and rods hyperpolarize and decrease the amount of glutamate released by the cone pedicles or rod spherules and, conversely, when the light level is decreased, cones and rods depolarize and increase the amount of glutamate released by their axon terminals. However, in the subsequent steps of the visual pathway there are not only neurons that respond to light in the same manner as photoreceptors, but also others that respond in the opposite way. Neurons that conserve the photoreceptor response polarity are called OFF cells while those that invert the photoreceptor response polarity are called ON cells [85, 86]. This is possible because there are complex mechanisms of photoreceptor signal transmission that conserve or invert the sign of photoreceptor responses to light based on different kinds of glutamate receptors that are found in the photoreceptor synapses [87]. Both classes of horizontal cells—H1 and H2 cells—and about half of the bipolar cell classes—DB1, FMB, DB2, DB3a, and DB3b cells, with axon terminals branching in the upper half of the inner plexiform layer—have AMPA and/or KA glutamate receptors (iGluRs) in their post-synaptic membranes facing the cones and rods axon terminals [88]. The ionotropic nature of these receptors means that they act directly by opening or closing ion pores in the plasma membrane, in this case responding to glutamate by increasing conductance to cations. A decrease of glutamate release by the photoreceptors in response to light results in a decrease of the iGluR activation and in a hyperpolarization of the post-receptoral horizontal and bipolar cells. These are consequently of the OFF type. The sign conserving synapses are located in horizontal cell dendrites (contacting cone pedicles) or axons (contacting rod spherules) that form the lateral elements of the triads in invaginating synapses. Flat synapses of cone pedicles with bipolar cell dendrites are also sign conserving [79]. Therefore, these bipolar cells are also of the OFF type. Their axon terminals are located in the outer half of the plexiform layer. BB and RB cells do not make flat synapse contacts with the S cones and rods, respectively, and they are exclusively of the ON-type.

DB4, DB5, IMB, DB6, BB, and RB bipolar cells, which have axon terminals progressively deeper in the inner half of the inner plexiform layer, have metabotropic receptors, i.e., they do not act directly via ion pores (although this can be the end effect of the cascade) but modify a G-protein molecule which in turn activates a secondary messenger molecule. These bipolars possess the glutamate receptor mGluR6 in their postsynaptic membranes [89]. The mGluR6 is a G-coupled receptor that activates a membrane phosphodiesterase, decreasing intracellular cGMP levels and, consequently, decreasing cGMP-dependent membrane conductance for cations. Thus, a decrease of glutamate release by the photoreceptors in response to increased light results in increased intracellular cGMP levels leading to an increased membrane cation conductance, which depolarizes the postsynaptic bipolar cells. Thus, these bipolar cells are ON cells and their dendrites constitute the central elements of the triads of the invaginating synapses with rod spherules or cone pedicles [79].

The retinal image is sampled at a spatial resolution that is initially determined by the photoreceptor mosaic. However, convergence of signals along the visual pathway may progressively degrade the spatial resolution. Convergence starts at the first synapses between photoreceptors and bipolar cells, and further increases in synapses between bipolar cells and ganglion cells. In macaque, convergence of cones onto cone bipolar cells differs according to the bipolar cell class and varies only slightly as a function of retinal eccentricity: MB cells contact a single cone (1–4 in the retinal periphery), BB cells 1–3 cones, and DB cells 4–10 cones (with slight differences for different DB classes) [35, 39, 41]. The one-to-one connectivity between cones, MB cells, and PC-ganglion cells is preserved in the foveal region, but the convergence of MB cells onto PC-ganglion cells increases steeply with increasing retinal eccentricity [41]. In other visual channels, such as the one formed by cones connected to DB cells and thence to magnocellular (MC-) ganglion cells, the convergence of DB cells onto MC-ganglion cells also increases towards retinal periphery. Thus, the spatial resolution of PC-, MC-, and KC-(koniocellular) channels is set by their respective ganglion cell mosaics.

The ON/OFF dichotomy of bipolar cells and the convergence of single cones onto MB bipolar cells have important consequences for the M/L cone opponent mechanism of color vision. They result in the existence of four kinds of MB cells [32]. Two of these receive sign conserving synapses from a single M or L cone. These are the M-OFF FMB and L-OFF FMB cells. A further two receive sign inverting synapses from a single M or L cone. These are the M-ON IMB and L-ON IMB cells. In the central retina, each MB bipolar cell makes synapses with a single PC ganglion cell. Thus, PC ganglion cells have center-surround receptive fields with four different mechanisms driving the receptive field center, provided by the MB cells described above, and receptive field surrounds of opposite polarity whose origin is not well understood: M-OFF/L-ON, L-OFF/M-ON, M-ON/L-OFF, and L-ON/M-OFF cells. It is believed that these four ganglion cell subclasses underpin red-green color vision (reviewed in Chap. 4).

2.3.4 Midget, Parasol and Bi-stratified Ganglion Cells

There are three major retinal pathways that process cone signals for conscious visual perception including color vision. The magnocellular (MC-) pathway consists of diffuse bipolar cells and parasol ganglion cells that project to the two ventral layers of the lateral geniculate nucleus (LGN). MC-ganglion cells receive additive input from the L- and M-cones, which makes them luminance sensitive. They are therefore the physiological basis of luminance vision [62, 63, 90–92] (see Chap. 4). They can respond to high temporal frequencies and have relatively large receptive fields and this is probably why their responses are used for motion processing [93]. Their large receptive fields make them probably less important for form perception, although their high responsivity for luminance stimuli allows them to code for spatial position on a hyperacuity level [94–96]. The receptive fields can be subdivided in centers and surrounds which are antagonistic. As described above for bipolar cells, stimuli leading to a luminance increase in the center (or a luminance decrease in the surround) lead to a depolarization in the ON-center cells and to a hyperpolarization in the OFF-center cells. Conversely, ON-center cells are hyperpolarized by luminance decreases in the centers and luminance increases in the surrounds. These stimuli lead to depolarizations in the OFF-center cells.

The PC-pathway contains midget bipolar cells and midget ganglion cells, which project to the four dorsal layers of the LGN. Like the MC-pathway, the PC-pathway processes exclusively L- and M-cone signals. However, in contrast to the MC-pathway, the cone signals provide opponent inputs. Furthermore, there are strong indications that the cone strengths are much more balanced compared to the MC-pathway, since the ratio of L-cone to M-cone responses in the parvocellular pathway is close to unity [97]. Again they have receptive fields with centers and surrounds. Foveal parvocellular cells have centers that receive input from only one cone. It is not clear yet if the surrounds receive inputs from the two cone types or whether there is a bias towards the cone type that is not represented in the center [98, 99]. With a random cone input, cone opponency would only occur because of the obligatory cone selective center. It is not clear in that case how an L:M cone ratio of about unity can arise. The centers of peripheral PC-cells are larger and again may receive mixed cone input (see above). However, under such circumstances the cells would lose their cone opponency.

Again there are ON- and OFF-center cells so that in the end there are four types of parvocellular ganglion cells (+L−M, +M−L as the ON-center cells; OFF-center cells are −L+M and −M+L) The L-/M-cone opponency in the parvocellular ganglion cells is the physiological basis for red-green color vision. The small receptive field sizes are probably used for the perception of forms of objects [100, 101].

The KC-pathway processes input from all three cone types, in which the S-cone signals are antagonistic to those from L- and M-cones (i.e., the cone input can be described as +S-[L+M]). As described above, two types of bipolar cells are involved: DB cells provide the L- and M-cone inhibitory signals whereas the S-cone excitatory signal are transmitted by BB cells [102]. The ganglion cells belonging to this pathway are of the small-field bi-stratified type [44] and project to the interlaminar regions and the areas surrounding the parvocellular and magnocellular layers of the LGN. The antagonistic input is not organized in center and surround substructures of the receptive fields; instead they are spatially co-extensive and constitute Hubel and Wiesel's [103] type 2 cells. Therefore these cells are thought to be unimportant for spatial vision but provide the basis for blue-yellow color vision.

The physiological properties of retinal ganglion cells and neurons in the LGN are well described. Those of primate bipolar cells are less well known. From recent ERG data, it can be inferred that the properties of the diffuse and midget bipolar cells are very similar to those of the MC- and PC-ganglion cells respectively (see below).

2.4 Photoreceptor and Post-receptoral Processes Leading to the Electroretinogram

The electroretinogram (ERG) is an electrical signal of retinal origin that is elicited by the same excitation of photopigments that leads to a visual response. The ERG has important clinical value because it is a non-invasive electrophysiological, and thus objective, method which can be used to monitor the functional integrity of the

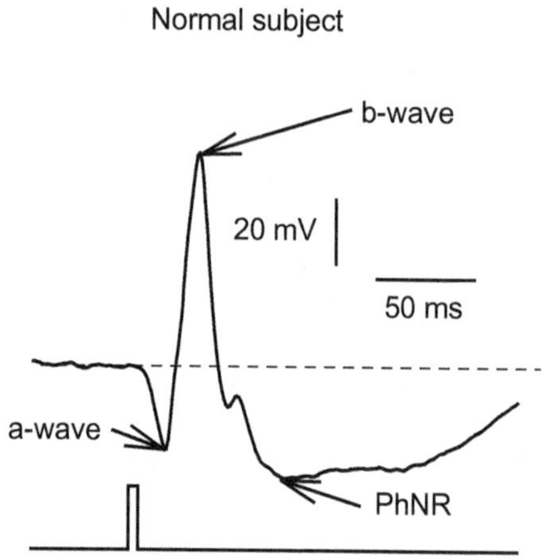

Fig. 2.6 Example of a full field flash ERG measured in a normal subject. The main components are identified

retina. In addition, the ERG has emerged as an important tool for the measurement of retinal function in animals in vivo. This has been particularly evident following the development of transgenic mouse lines, where the retinal organization has been genetically altered, and the use of mouse models of retinal diseases.

Originally, the ERG was measured to short light pulses [104]. These flash ERGs are still measured. However, increasing numbers of other stimuli and resulting ERG types are measured nowadays. Below, we describe those that are important for revealing cone opponent processes. In the flash ERG, three components can be distinguished that have different cellular origins (Fig. 2.6). The early negative a-wave originates in activity of the photoreceptors and off-bipolar cells The next wave is positive b-wave that is mainly determined by activity of on-bipolar cell [105]. The b-wave of the flash ERG is in fact an addition of a positivity as a response to stimulus onset and a slightly delayed positive response to stimulus offset known as the d-wave [106]. The d-wave has identical implicit times to stimulus offset as the a-wave after stimulus onset indicating that they have the same cellular origins [107]. The b-wave is followed by a photopic negative response (PhNR) that reflects activity of retinal ganglion cells [108–110].

The value of the ERG in studying vision, and more particularly color vision, is, however, not straightforward. Armington [111] in his book "The Electroretinogram" emphasized the potential of the ERG in the study of visual processing in the retina. Yet at the same time he was also aware of the difficulties in establishing correlations between the ERG and psychophysical data.

2.4.1 Early Research on Chromatic Processes in the ERG

Armington [111] dedicated a whole chapter to the ERG related to human spectral sensitivity and color vision. As discussed above, different cone spectral sensitivities are necessary but not sufficient for color vision. Cone opponent processing is a second prerequisite for color vision. At the time when Armington's book appeared, the notion of cone opponency was still quite novel [112]. Riggs and colleagues [113, 114] showed, using equiluminant alternating gratings (5.35 Hz; 10.7 reversals per second), that the wavelength separation for a criterion ERG response is similar to psychophysically derived measures of wavelength discrimination with minima around 500 and 600 nm, indicating that repetitive stimuli may elicit ERGs that reflect cone-opponent processing.

Beginning in the early 1970s, data on the spectral sensitivity of the b-wave in flash ERGs emerged [115–119]. The b-wave spectral sensitivity as a function of wavelength of flashed stimuli upon a background displayed similarities with psychophysically determined detection sensitivities using similar stimuli [118, 120]. Such spectral sensitivities are an indication that detection is mediated by cone opponent mechanisms. Nowadays, the b-wave is thought to originate from bipolar cell activity [121]. Armington [122] found spectral sensitivities that also indicated spectral opponent processing as early as 1959, but at that time the data were difficult to interpret. The similarity with the psychophysical data indicates that the same post-receptoral mechanisms may underlie the psychophysical and ERG data.

On the other hand, the a-wave exhibits different spectral sensitivity which does not reflect cone opponent processing, indicating that it has a different cellular origin. Indeed, it is now thought that the a-wave mainly reflects the activity of the photoreceptors and the OFF-bipolar cells. The main origin of the b-wave resides in the ON-bipolar cells [105].

Single cell recordings in the 1960s showed that retinal ganglion cells and LGN cells could display cone opponent processing [103, 123], indicating that the subtractive interaction between cone signals, necessary for color vision, occurs already in the retina. The exact site was, and still is, not exactly known. Data from fish and turtle indicated that cone opponency occurs in the outer retina in these species and Baron [124] proposed that, similarly, opponency might also occur there in primate and would therefore be discernable in the primate ERG. He described a component that possibly reflects cone opponency in the foveal ERG of monkeys using intra-retinal electrodes. This component could subsequently also be identified in the ERG as measured at the cornea [125]. However, the data are not easy to interpret as being related to cone opponency. Nowadays it is clear that cone opponent processing in "lower" vertebrates and in primates is quite different. Horizontal cells in turtle and fish retinae show distinct cone opponency [65] that is absent in horizontal cells of the primate [45].

2.4.2 Recent Developments

One possible reason for the limited correlation between early ERG and psychophysical data is that ERGs were mainly measured to short flashes of light. Possibly, the circuitries leading to an ERG and those leading to a visual percept have substantially different response properties when using very short stimuli. In the following section, we provide evidence that the use of repetitive stimuli, leading to steady state responses, may reveal more similarities between ERG and psychophysical data. One example of the use of repetitive stimuli is the heterochromatic flicker photometry (HFP) method that is also used in psychophysical measurements to determine the luminance of a light source. Briefly, two lights, a reference and a test light, are modulated in counter-phase at relatively high temporal frequencies (≥16 Hz) at equal and fixed contrast. The mean luminance of the test light is altered by the subject to minimize the perception of flicker, where, by definition, the two lights are equiluminant. Using the same stimuli, it proved to be possible to also minimize the ERG response. The spectral sensitivities that could be measured in this way were also identical in psychophysical and ERG experiments (resulting into the photopic spectral luminosity function, V_λ [126–131]). Furthermore, the HFP method leads to psychophysical results that differ between different individuals. This inter-individual variability was also present in the ERG data [129] and is the result of a variability in the ratio of L- to M-cone signals in the luminance channel. The L/M cone ratio could also be measured directly in psychophysical and ERG experiments and it was shown that the amplitude ratio of the high frequency ERG in response to L- and M-cone isolating stimuli can be correlated to the sensitivity ratio for detection of L- and M-cone isolating at high temporal frequencies [132, 133]. At high temporal frequencies, the ratio is on average about 2:1 and reflects the ratio of L- to M-cone packing densities [134]. Data from RNA expression have shown that the far peripheral retina may contain nearly exclusively L-cones [135–137]. This peripheral L-cone dominance was also found with ERG measurements [138, 139]. Finally, changes in L/M cone sensitivity ratio in individual observers, caused by difference states of adaptation, were again also represented in the L/M cone amplitude ratio in the high temporal frequency ERG [28].

The correlations described above between psychophysics and ERGs were found for high temporal frequency stimuli (about 25 Hz and higher) where the psychophysical data were mediated by the luminance channel of which the MC-pathway is the physiological basis. The question is: are there also ERG responses to modulating stimuli that can be linked to chromatic sensitivities? Psychophysically, chromatic sensitivity can be mainly measured at lower temporal frequencies [92, 97, 140, 141]. If stimuli contain both luminance and chromatic modulation, then psychophysical detection thresholds are mediated by the chromatic channel at low temporal frequencies and by the luminance channel at high temporal frequencies [92, 141]. With L- and M-cone isolating stimuli, chromaticity and luminance are simultaneously modulated and detection thresholds for these stimuli are indeed determined by the luminance content at high temporal frequencies and by the chromatic modulation at low temporal frequencies [132].

Recently, it has indeed been found that the ERG responses to intermediate temporal frequency stimuli (around 12 Hz), containing red-green chromatic contrast, can reflect activity of the L-M opponent parvocellular pathway. The evidence for this comes from several experiments. In the first type of experiment, L- and M-cone isolating stimuli are used (see Fig. 2.7). L- and M-cone driven responses are of about equal amplitudes and thus the L:M ratio is close to unity at intermediate temporal frequencies (about

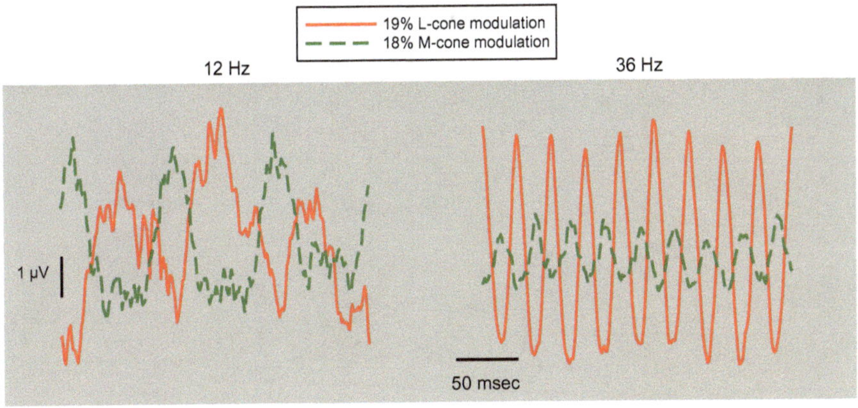

Fig. 2.7 ERG responses in a trichromatic subject to full field L- and M-cone isolating stimuli of similar contrast (18 % M-cone contrast: *dashed green line*; 19 % L-cone contrast: *solid red line*) at two temporal frequencies: 12 Hz (*left*) and 36 Hz (*right*). Triple silent substitution conditions were used (i.e., only the photoreceptor type of interest—either L- or M-cones—were stimulated; the three others were silenced). The 12 Hz responses have similar amplitudes whereas the L-cone driven responses are larger than the M-cone driven responses at 36 Hz [143, 144]

12 Hz). At high temporal frequencies (30 Hz and above), the L-cone driven responses are larger than the responses to M-cone isolating stimuli [142–144]. Furthermore, the phases of the L- and M-cone driven responses can behave differently at low and high temporal frequencies. Generally, the phases of the response to L- and M-cone isolating stimuli at intermediate temporal frequencies are about 180° apart indicating cone opponency. The phase differences at high temporal frequencies seem to be more variable. They can be large as well (as is seen in Fig. 2.7) but there are conditions in which the phase of L- and M-cone driven responses are relatively similar [26].

In the second type of experiment, red and green lights were modulated sinusoidally in counter-phase. The contrasts of the two were varied such that the sum of the red and green contrasts (R + G) was kept constant. Thus, the fraction of red contrast relative to the total contrast (R/(R + G)) was varied (and the fraction of green contrast was complementary; see Fig. 2.8a). The chromatic contrast (Fig. 2.8b left plot; red line) in these stimuli was constant. The luminance contrast (Fig. 2.8b left plot; black line), however, varies linearly as a function of red fraction, reaching zero when R/(R + G) is about 0.5 (depending on the present L:M ratio of the signals entering the luminance signal). Also, the phase has different characteristics (Fig. 2.8b right plot): the phase of the chromatic output is constant whereas the luminance modulation follows the modulation of the green light for small values of R/(R + G) and is determined by the red stimulus when R/(R + G) is larger than 0.5. Using this type of stimulus it was found that the responses to high frequency stimuli (Fig. 2.8c) were very similar to the expected activity of the luminance channel with a clear minimum [145]. In accordance with the notion that the responses reflect luminance activity, the phase changed by 180° at the amplitude minimum. These data are in accordance with the well-known flicker photometric data, described

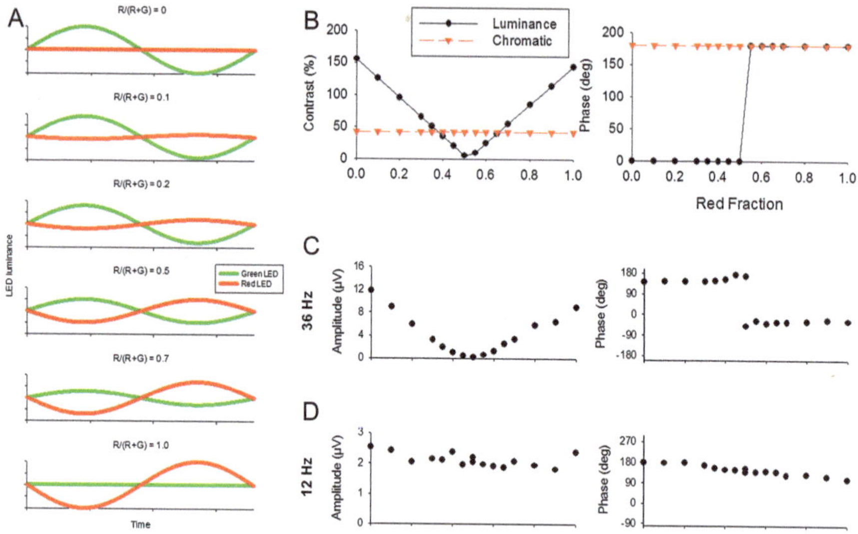

Fig. 2.8 Results of ERG experiments with red-green sinusoidal modulation. (**a**) Six examples of stimuli. One period is shown. The red and green lights (from light emitting diodes; LEDs) were modulated sinusoidally in counterphase. The mean luminance levels were equal for the red and green lights and were not varied. The contrast in the two lights was varied, but the sum of the *red* and *green* contrast was constant. (**b**) Expected output amplitude (*left plots*) and phase (*right plot*) of the luminance channel (*black line*) and red-green chromatic channel (*red line*) as a function of red fraction in the stimulus. (**c**) ERG response amplitude and phase to 36 Hz stimuli as a function of red fraction. Observe the resemblance with the expected response of the luminance channel in **b**. (**d**) Response characteristics measured in the same observer at 12 Hz. These responses are like those of the red-green chromatic channel in **b** (Adapted from Kremers et al. [145].)

above. More importantly, it was found that at an intermediate temporal frequency (12 Hz) neither amplitude nor phase changed appreciably as a function of red fraction (Fig. 2.8d), thereby resembling the properties of the red-green chromatic channel.

An important control experiment was performed with dichromats, who lack a functional red-green chromatic channel because they only possess L or M cones. In these subjects, the responses at all temporal frequencies clearly followed the response properties of the sole cone type (Fig. 2.9), indicating that the mechanism that is responsible for the ERG responses at 12 Hz, is not present in the dichromats, and thus most probably originates in the red-green chromatic channel.

In the third type of experiment, red and green lights were again modulated in counter-phase. But now they were alternately activated. That means that when one light is switched on the other is switched off and vice versa. When activated, the output of the light sources followed a raised cosinusoidal profile. The luminance output of the red and green light were carefully balanced to be equal in luminance. As a result, the chromaticity of the stimulus is modulated at a temporal frequency that is half the luminance frequency (see Fig. 2.10). In this experiment, the temporal frequency of the stimulus was varied. In contrast to the second experiment, where the LEDs were

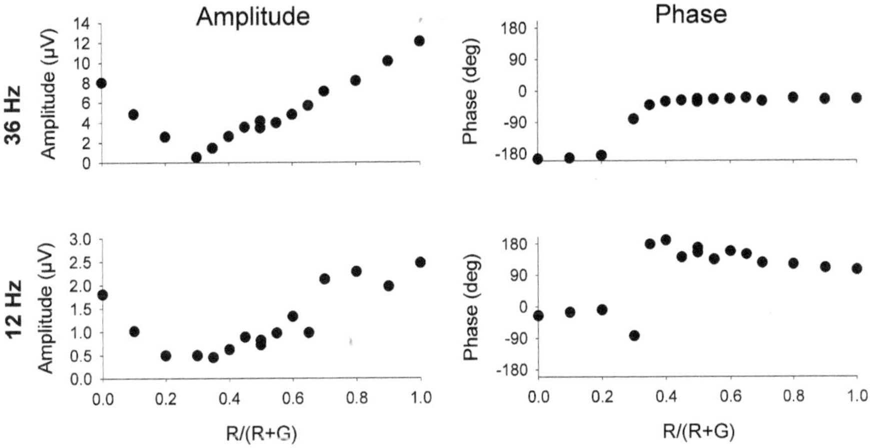

Fig. 2.9 Responses measured in a deuteranope. The same stimuli as displayed in Fig. 2.8 were used. Both the 36 Hz (*upper plots*) and 12 Hz (*lower plots*) responses display the characteristics of a luminance reflecting mechanism. The minimum has shifted towards smaller values of the red fraction in the stimulus. This can be expected for responses that are exclusively driven by L-cones. (Adapted from Kremers et al. [145])

always simultaneously activated, here only one LED is activated at a time. Furthermore, in this experiment the luminance was always modulated with 100 % contrast, whereas the luminance modulation in the second experiment was variable and did not exceed 50 % (for the conditions in which R/(R+G) were 0 and 1). In single cell recordings from the macaque retina it was found that the PC-ganglion cells mainly responded at the frequency of the chromatic modulation whereas MC-cells mainly responded to the luminance modulation (i.e., at twice the response frequency of the parvocellular cells; [146]). These stimuli were also used while recording the ERGs (Fig. 2.10, upper plots). In these experiments, it was found that the high temporal frequency ERGs of normal trichromats (Fig. 2.10, left two columns) contained, almost exclusively, frequency components that were identical to those of the luminance modulation in the stimulus. No response was observed at the chromatic modulation frequency. At low and intermediate temporal frequencies, a clear response component at the frequency of the chromatic modulation was present. In Fig. 2.10, this can be observed because the responses to the red and green stimuli are dissimilar at low temporal frequencies (here particularly for the 6 Hz stimulus conditions). The latter component could not be observed in red-green dichromats (Fig. 2.10, the two right columns) because the responses to the red and green stimuli were very similar in the dichromats also at low temporal frequencies. This result strongly indicates that this component indeed originates in the red-green chromatic channel [147].

In a fourth experiment, transient ERG responses to L- and M-cone isolating stimuli with sawtooth and square-wave temporal profiles were measured [148, 149]. With sawtooth and square wave stimuli, On- and Off-responses (i.e., responses to sudden increments or decrements in cone excitation) can be studied separately. These measurements revealed a response inversion: responses to L-increments

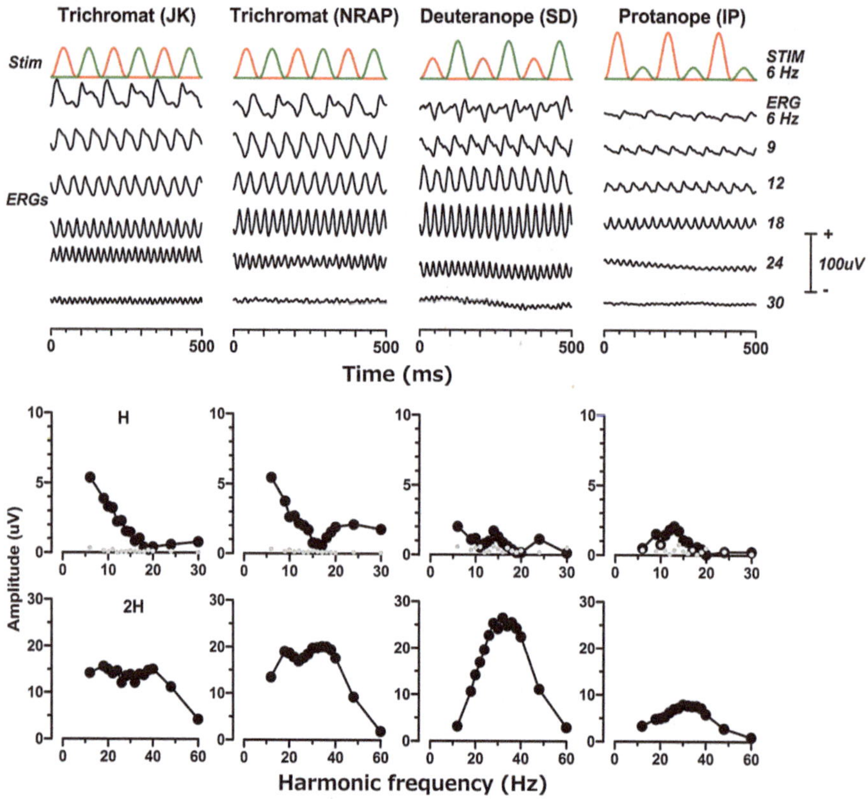

Fig. 2.10 Results of ERG recordings performed with alternating red and green stimuli. *Top rows*: Depiction of the stimuli with Raw ERG responses. In the *left two columns*, recordings in trichromats are displayed, whereas the *right two columns* show the responses measured in dichromats (a deuteranope and a protanope). The temporal frequencies, given on the *right*, indicate the frequency of the luminance modulation in the stimulus. The temporal frequency of the chromatic modulation equals half the luminance frequency. *Two bottom rows*: Amplitude as a function of temporal frequency of the first (chromatic) and second (achromatic) harmonics. Note the marked reduction in the chromatic response in the dichromats, particularly at low temporal frequencies. Adapted from Parry et al. [147]

(L-On) resembled those to M-decrements (M-Off) and the responses to L-decrements (L-Off) and M-increments (M-On) were similar in waveform (see Fig. 2.11 for responses to sawtooth stimuli). The responses to L-cone isolating stimuli had the same polarity as the responses to luminance modulation with the same temporal profile [148, 149]. The response inversion between L- and M-cone driven responses suggests that cone opponent processes are involved. Furthermore, the data show that the textbook explanation of the origins of the different components [105], in which only On- and Off-bipolar cells without a distinction between the responses from for example midget and diffuse bipolar cells, is probably only part of the story. The data strongly suggest that such a distinction between different types of midget bipolar cells should be taken into account. Although the response inversion seems to be

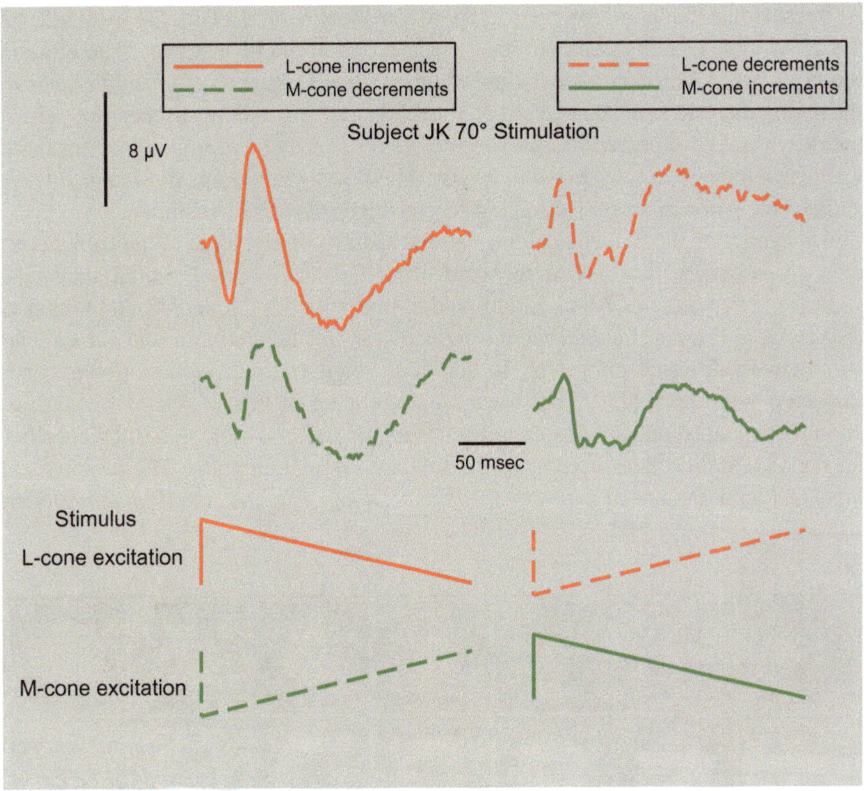

Fig. 2.11 Responses to L- and M-cone isolating sawtooth stimuli (4 Hz temporal frequency). The *left column* displays the responses to L-cone increment stimuli (*solid red line*; the L-cone excitation is rapidly increased and gradually decreased as shown in the stimulus sketch below) and to M-cone decrements (having the opposite profile; *dashed green line*). The *right column* shows the responses to L-cone decrements (*dashed red line*) and M-cone increments (*solid green line*). Observe that the responses to L- and M-cone isolating stimuli of opposite polarities resemble each other [148]. This effect was also found in responses to square wave stimuli [149] but may be less clear for full field stimuli [148]

present for all response components (a-, b-, d-wave-like and other components), the picture is more complicated because the ratios of the L- and M-cone signal strengths is different for the different components. The early components (a- and d-wave) display large L:M ratios and resemble the ratios of cone numbers and of the psychophysical luminance channel. The later components (b-wave and photopic negative response) display an L/M ratio closer to unity, suggesting stronger chromatic input [148]. The responses to M- and L-cone increments in protanopes and deuteranopes respectively had similar waveforms indicating that the inversion of responses is not present in dichromats [148, 149].

From these experiments, a picture emerges that, in trichromatic observers, ERG responses to stimuli that contain luminance and red-green chromatic modulation reflect activity of the luminance channel and the MC-pathway at high temporal frequencies (≥30 Hz). This is also the basis of the ERG results with the HFP paradigm.

At intermediate temporal frequencies (between about 8 and 14 Hz), the ERG reflects the activity of the red-green chromatic channel and of the PC-pathway. Therefore in the years since Armington's book appeared, evidence has been obtained in favor of the notion that ERG signals indeed can reflect the properties of post-receptoral retinal pathways and have perceptual relevance. This result may be of importance for basic vision research because it now will be possible to study properties of these pathways and channels in human subjects using objective non-invasive techniques.

To investigate the influence of the size and spatial configuration of the stimulus on the two pathways, ERGs were recorded using L- and M-cone isolating sinusoidal stimuli at intermediate (8 and 12 Hz) and at high (30, 36, 42, and 48 Hz) temporal frequencies. These responses were measured for full field stimuli and for circular stimuli with different diameters. In addition, annular stimuli with different inner diameters were used [139]. The response amplitudes at high temporal frequencies strongly depended on stimulus area. The responses to 8 and 12 Hz stimuli were similar for all stimulus areas except for the smallest circular stimulus. The responses at high temporal frequencies increase with increasing stimulus size (Fig. 2.12). This

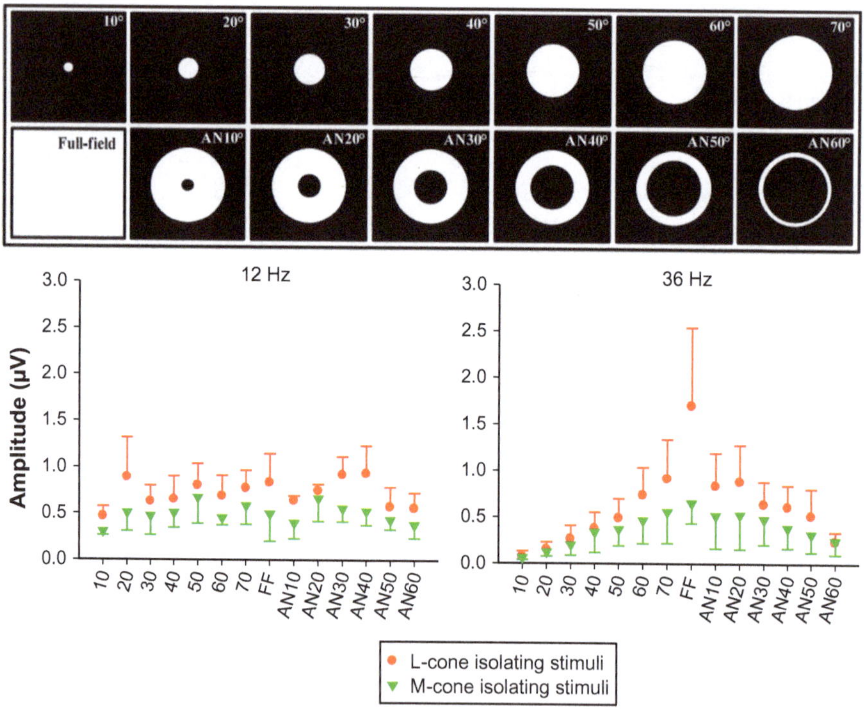

Fig. 2.12 Response amplitudes to sinusoidal modulation of L- and M-cone excitation for different stimulus spatial configurations. Responses at two different temporal frequencies (*left plot*: 12 Hz; *right plot*: 36 Hz) are shown. The sketch above the plots shows the spatial stimulus configurations that were used and corresponds to the labels on the *x*-axes of the plots. Amplitude does not vary strongly with spatial configuration at 12 Hz. The 36 Hz stimuli strongly depend on the spatial arrangement of the stimuli. Redrawn from Ref. [139]

indicates that the PC- and MC pathways may either have different spatial distributions or their response characteristics change differently with retinal eccentricity.

The physiological properties of retinal ganglion cells and neurons in the LGN are well described (see above and Chap. 4). Those of primate bipolar cells are less well known. ERG responses to repetitive stimuli mainly reflect activity of bipolar cells [121]. From the results of the ERG measurements, it can be inferred that the properties of the diffuse and midget bipolar cells are probably very similar to those of the MC- and PC-ganglion cells respectively, indicating that these properties are already present after the first synapse (from photoreceptors to bipolar cells) in the visual pathway. Sperling and colleagues' work, showing that the b-wave may also reflect cone opponent processing [115–118], is in agreement with this proposal because the b-wave is also thought to originate in bipolar cell activity [105]. It should be noted, however, that bipolar cells not only receive feed forward signals from photoreceptors but also from horizontal cells. They also receive feedback signals from horizontal and amacrine cells. These signals influence their response properties. Furthermore, KC-cell responses make use of two types of bipolar cells, one of which (the diffuse bipolar cell) has properties similar to that of the magnocellular channel. Therefore it is possible that the bipolar cells belonging to the +S-(L+M) pathway do not display similar physiological properties as the small field bi-stratified cells. Combined with the fact that S-cones and KC+S-(L+M) ganglion cells are much sparser than the other cone and ganglion cell types, respectively, possibly make the signals of this pathway in the ERG difficult to detect.

2.5 Open Questions

These new developments open up many possibilities to answer basic scientific and clinical questions. However, some important issues arise from this work and in the following section we give an overview of the key questions that we have identified.

2.5.1 *How Can a Chromatic ERG Signal Appear at all with Four PC-Cell Types?*

ERG responses may reflect photoreceptor activity because all photoreceptor types hyperpolarize. However, there are two types of luminance sensitive (DB) bipolar cells: the ON- and the OFF-DB cells. Thus, if one type is hyperpolarized the other will be depolarized and vice versa. Therefore it could be expected that the responses of the two cell groups may cancel each other out (particularly when the two populations contain similar cell numbers) and no luminance ERG would be measurable unless there are physiological differences between ON- and OFF-DB cells. That is indeed the case (such as the abovementioned differences in glutamate receptors). Thus, luminance ERG responses can perhaps be measured because of the physiological differences between ON and OFF DB cells.

Although the differences between ON- and OFF-bipolar cells may explain why a luminance-based ERG can be measured, the situation is more complicated for ERGs that might reflect bipolar cell activity belonging to the PC-pathway. Instead of two there are four cell types present: +L-M, -L+M, +M-L, and -M+L. Again there may be a difference between ON- (+L-M and +M-L) and OFF- (-L+M and -M+L) center cells. But to be able to record an ERG driven by these cells, another asymmetry must be present because the responses to for example the +L-M and the +M-L cells could be expected to cancel each other. One possibility is that the differences in L- and M-cone numbers may also lead to different numbers of cone opponent bipolar cells (so that subjects with a many more L-cones than M-cones also have many more +L-M than +M-L bipolar cells). That would indicate a correlation between that the amplitude of the cone opponent ERG response and the L:M ratio. Although this has never been directly studied, there are no indications that such a correlation exists.

2.5.2 What Are the Effects of Selective Adaptation?

We have observed that small cone-selective adaptations using a CRT screen may strongly influence the amplitudes of L- and M-cone driven ERG responses, resulting in a dramatic change in the L:M ratio. This is paralleled by a similar change in the psychophysical L/M sensitivity ratio [28]. However, Jacob et al. showed that even large changes in state of adaptation do not have a dramatic influence on ERG responses [139]. The latter results were obtained with a four primary LED stimulator. In contrast to this LED stimulator, complete isolation is not possible when a trichromatic observer views a CRT screen. However, it is improbable that the type of stimulator influences the results dramatically [150]. Furthermore, control experiments with the CRT screen showed that rod driven responses probably have little influence on the results. In conclusion, it seems that adaptation can influence ERG and psychophysical results dramatically, but it has not been possible to systematically describe this influence, let alone understand the underlying mechanisms. The fact that the effects of adaptation were also found in psychophysical experiments shows that answering this question may be very important for a better understanding of retinal adaptation processes.

2.5.3 Are There Fundamental Differences Between Protanopes and Deuteranopes?

During the course of several years of studying deuteranopes and protanopes with ERG and psychophysical techniques, some important differences have been revealed. L-cone isolated stimuli normally do not lead to measurable ERG responses in protanopes but M-cone isolating stimuli elicit small but measurable ERGs in deuteranopes [28, 143, 144, 148, 149]. In addition, in psychophysical experiments, protanopes cannot detect L-cone isolating responses even at the highest possible contrast, whereas M-cone isolating stimuli can often be detected by deuteranopes (although the threshold contrasts are substantially higher than for L-cone isolating stimuli [28]). In addition, adapting the L-cones has no effect on the ERG responses and psychophysical thresholds in protanopes. In deuteranopes, M-cone adaptation has effects that are similar to those of trichromats [28]. From control experiments, it was inferred that these differences cannot be caused by rod stimulation and adaptation that were not controlled with the stimuli displayed on a CRT screen. Deuteranopes often display larger responses than protanopes (see the responses of the dichromats displayed in Fig. 2.10), an observation that cannot be explained by differences in L- and M-cone contrast in the stimuli. Finally, as described in Chap. 10 a substantially larger number of deuteranopes pass color vision tests in comparison to protanopes.

The differences between deuteranopes and protanopes were found systematically in several experiments. The cause of these differences is still unclear. One possibility may lie in the fact that that on the X-chromosome there is normally only one copy of the L-opsin gene but that multiple copies of the M-opsin gene may be present [151]. Thus, if the first copy is defective, the subject might be diagnosed to be a deuteranope. If the downstream copies are normal they might lead to a residual activity of M-opsin. However we found no relation between the number of M-opsin genes and the magnitude of residual M-opsin activity [28] although more detailed analyses might be needed. Another explanation might be that an additional pigment might be involved. If the residual responses were present with M-cone stimuli and not with L-cone stimuli it can be expected that this pigment mainly absorbs shorter wavelengths. It is attractive to speculate that melanopsin-driven responses, present in the internally photoreceptive retinal ganglion cells (ipRGCs) may lead to measurable ERG responses and psychophysical flicker detection. Finally, small errors in the assumed density of the macular pigment and variability in photopigment absorption spectra may lead to particularly large for M-cone isolating conditions (that should be abolished in the deuteranopes) in comparison with L-cone isolating stimuli (that should not elicit responses in protanopes). Recently, a simulation of the effects of this variability has indeed shown that variability in L-cone fundamentals may lead to small L-cone stimulations in intended M-cone isolating stimuli. Although the L-cone stimuli are small they may be detectable in the absence of other responses. The residual stimulation of M-cones in L-cone isolating conditions is much smaller (Huchzermeyer and Kremers, Journal of the Optical Society of America A, Accepted for publication). The first two possibilities are highly speculative, the third explanation needs less assumptions. Remains, that there are systematic differences between deuteranopes and protanopes that are very conspicuous.

2.5.4 How Can the Intermediate Temporal Frequency (8–16 Hz) ERG Have Very Similar Response Amplitudes for a Large Range of Different Spatial Configurations of a Stimulus?

It was found that the ERG responses to intermediate temporal frequencies (reflecting activity of the PC-pathway) are relatively similar for different sizes of circular and annular stimuli (see Fig. 2.12) [139]. If this were the case only for circular stimuli, the results could be explained by the possibility that the pathway leading to the ERG responses would be mainly located at the central retina (up to about 10 deg eccentricity). However, the response amplitudes were often very similar with annular stimuli, in which this central area is not stimulated. A possible explanation is that the responses may summate over a retinal area up to a maximum at which it then saturates so that further increase of stimulus area has no effect. Signal processing would be similar at different retinal eccentricities, indicating that the retina would be relatively homogeneous up to an eccentricity of at least 35 deg. This is a surprising idea that certainly needs further experimental testing.

2.6 Future Perspectives

The recently reported correlations between ERG and psychophysical data give rise to the likelihood that the ERG may experience a renaissance in studies of human retinal processes that are relevant for visual perception. Furthermore, the correlation between ERG data and the physiology of the two main retino-geniculate (PC- and MC-) pathways suggests that properties of these pathways can now be studied directly in human observers using electrophysiological techniques. This also opens perspectives for understanding retinal mechanisms underlying color vision.

The abovementioned questions are waiting to be investigated. Answers to them will provide important information about color processing. Cone selective adaptation may for instance shed light upon adaptation mechanisms in the PC-pathway that may then be compared to adaptation mechanisms in the MC-pathway. If the differences between deuteranopes and protanopes are confirmed then this may point at fundamental differences in L- and M-cone driven responses and novel fundamental properties of retinal physiology may become apparent.

Developments in ERG recording techniques may not only have influence on basic visual neuroscience but also on clinical vision science and ophthalmology. The ERG has the potential to allow for better diagnoses of retinal disorders and for clearer identification of their cellular origins. If a disease primarily affects a particular photoreceptor type (as AMD is thought to initially affect the rods) or a particular cell circuitry (as glaucoma was for some time thought to lead to defects mainly in

MC-ganglion cells) then we can expect that ERG recordings that assay the responses of these cell types may be suitable for sensitive disease diagnosis. These techniques may also give an indication about how vision can be affected by disease processes and importantly, may allow the investigation as to how therapeutic interventions can affect visual performance in patients.

Of course other technical developments may bring a huge increase in our understanding of cone opponency and color processing in the primate retina. Recordings with multi-electrode arrays will bring a huge amount of new data. Furthermore, they may provide information about response interactions between different types of retinal neurons. The emerging field of optogenetics also has the potential to allow studying the retina and its constituent neurons in a very precise manner. Using optical techniques, the production of certain proteins can be manipulated and their function then be studied. These developments are extremely exciting and we expect them to bring completely new insights.

Acknowledgements Luiz Carlos da Silva Silveira passed away on 10th July 2016. This work was supported by German Research Council (DFG) grants KR 1317/9-1, KR1317/9-2 and KR1317/13-1.

References

1. Berson DM, Dunn FA, Takao M. Phototransduction by retinal ganglion cells that set the circadian clock. Science. 2002;295(5557):1070–3. doi:10.1126/science.1067262.
2. Hattar S, Liao HW, Takao M, Berson DM, Yau KW. Melanopsin-containing retinal ganglion cells: architecture, projections, and intrinsic photosensitivity. Science. 2002;295(5557):1065–70. doi:10.1126/science.1069609.
3. Dacey DM, Liao HW, Peterson BB, Robinson FR, Smith VC, Pokorny J, et al. Melanopsin-expressing ganglion cells in primate retina signal colour and irradiance and project to the LGN. Nature. 2005;433(7027):749–54. doi:10.1038/nature03387.
4. Brown TM, Gias C, Hatori M, Keding SR, Semo M, Coffey PJ, et al. Melanopsin contributions to irradiance coding in the thalamo-cortical visual system. PLoS Biol. 2010;8(12):e1000558. doi:10.1371/journal.pbio.1000558.
5. Lall GS, Revell VL, Momiji H, Al Enezi J, Altimus CM, Guler AD, et al. Distinct contributions of rod, cone, and melanopsin photoreceptors to encoding irradiance. Neuron. 2010;66(3):417–28. doi:10.1016/j.neuron.2010.04.037.
6. Allen AE, Storchi R, Martial FP, Petersen RS, Montemurro MA, Brown TM, et al. Melanopsin-driven light adaptation in mouse vision. Curr Biol. 2014;24(21):2481–90. doi:10.1016/j.cub.2014.09.015.
7. Ecker JL, Dumitrescu ON, Wong KY, Alam NM, Chen SK, LeGates T, et al. Melanopsin-expressing retinal ganglion-cell photoreceptors: cellular diversity and role in pattern vision. Neuron. 2010;67(1):49–60. doi:10.1016/j.neuron.2010.05.023.
8. Naka KI, Rushton WA. S-potentials from colour units in the retina of fish (Cyprinidae). J Physiol (London). 1966;185:536–55.
9. Smith VC, Pokorny J. Spectral sensitivity of the foveal cone photopigments between 400 and 500 nm. Vision Res. 1975;15:161–71.
10. DeMarco P, Pokorny J, Smith VC. Full-spectrum cone sensitivity functions for X-chromosome-linked anomalous trichromats. J Opt Soc Am A. 1992;9(9):1465–76.

11. Stockman A, MacLeod DIA, Johnson NE. Spectral sensitivities of the human cones. J Opt Soc Am A. 1993;10(12):2491–521.
12. Stockman A, Sharpe LT. Cone spectral sensitivities and color matching. In: Gegenfurtner K, Sharpe LT, editors. Color vision: from genes to perception. Cambridge: Cambridge University Press; 1999. p. 53–88.
13. Sharpe LT, Stockman A, Jägle H, Knau H, Klausen G, Reitner A, et al. Red, green and red-green hybrid pigments in the human retina: correlations between deduced protein sequences and psychophysically-measured spectral sensitivities. J Neurosci. 1998;18:10053–69.
14. Stockman A, Sharpe LT, Fach C. The spectral sensitivity of the human short-wavelength sensitive cones derived from thresholds and color matches. Vision Res. 1999;39(17): 2901–27.
15. Wyszecki G, Stiles W. Color science: concepts and methods, quantitative data and formulas. New York: Wiley; 1982.
16. Stiles WS. Increment thresholds and the mechanisms of colour vision. Doc Ophthalmol. 1949;3:138–63.
17. Stiles WS. Color vision: the approach through increment threshold sensitivity. Proc Natl Acad Sci U S A. 1959;45:100–14.
18. Donner KO, Rushton WAH. Retinal stimulation by light substitution. J Physiol. 1959;149:288–302.
19. Estévez O, Spekreijse H. A spectral compensation method for determining the flicker characteristics of the human colour mechanisms. Vision Res. 1974;14:823–30.
20. Estévez O, Spekreijse H. The "silent substitution" method in visual research. Vision Res. 1982;22:681–91.
21. Kremers J. The assessment of L- and M-cone specific electroretinographical signals in the normal and abnormal retina. Prog Retin Eye Res. 2003;22:579–605.
22. Shapiro AG, Pokorny J, Smith VC. Cone-rod receptor spaces with illustrations that use the CRT phosphor and light-emitting-diode spectra. J Opt Soc Am A. 1996;13:2319–28.
23. Cao D, Nicandro N, Barrionuevo PA. A five-primary photostimulator suitable for studying intrinsically photosensitive retinal ganglion cell functions in humans. J Vis. 2015;15(1):15.1.27. doi:10.1167/15.1.27.
24. Huchzermeyer C, Schlomberg J, Welge-Lussen U, Berendschot TT, Pokorny J, Kremers J. Macular pigment optical density measured by heterochromatic modulation photometry. PLoS One. 2014;9(10):e110521. doi:10.1371/journal.pone.0110521.
25. Bone RA, Landrum JT, Cains A. Optical density spectra of the macular pigment in vivo and in vitro. Vision Res. 1992;32(1):105–10.
26. Kremers J, Usui T, Scholl HPN, Sharpe LT. Cone signal contributions to electroretinograms in dichromats and trichromats. Invest Ophthalmol Vis Sci. 1999;40:920–30.
27. Usui T, Kremers J, Sharpe LT, Zrenner E. Flicker cone electroretinogram in dichromats and trichromats. Vision Res. 1998;38(21):3391–6.
28. Kremers J, Stepien MW, Scholl HPN, Saito CA. Cone selective adaptation influences L- and M-cone driven signals in electroretinography and psychophysics. J Vis. 2003;3:146–60.
29. Kremers J, Parry NR, Panorgias A, Murray IJ. The influence of retinal illuminance on L- and M-cone driven electroretinograms. Vis Neurosci. 2011;28:129–35. doi:10.1017/S0952523810000556. S0952523810000556 [pii].
30. Park JC, Cao D, Collison FT, Fishman GA, McAnany JJ. Rod and cone contributions to the dark-adapted 15-Hz flicker electroretinogram. Doc Ophthalmol. 2015;130(2):111–9. doi:10.1007/s10633-015-9480-3.
31. Hecht S, Haig C, Chase AM. The influence of light adaptation on subsequent dark adaptation of the eye. J Gen Physiol. 1937;20(6):831–50.
32. Silveira LCL, Grünert U, Kremers J, Lee BB, Martin PR. Comparative anatomy and physiology of the primate retina. In: Kremers J, editor. The primate visual system; a comparative approach. Chichester: Wiley; 2005. p. 127–60.
33. Polyak SL. The retina. Chicago: University of Chicago Press; 1941.

34. Boycott BB, Dowling JE. Organization of the primate retina: light microscopy. Philos Trans R Soc Lond B. 1969;255:109–84.
35. Boycott BB, Wässle H. Morphological classification of bipolar cells of the primate retina. Eur J Neurosci. 1991;3:1069–88.
36. Silveira LC, Perry VH. The topography of magnocellular projecting ganglion cells (M-ganglion cells) in the primate retina. Neuroscience. 1991;40(1):217–37.
37. Röhrenbeck J, Wässle H, Boycott BB. Horizontal cells in the monkey retina: immunocytochemical staining with antibodies against calcium binding proteins. Eur J Neurosci. 1989;1:407–20.
38. Martin PR, Grünert U. Spatial density and immunoreactivity of bipolar cells in the macaque monkey retina. J Comp Neurol. 1992;323(2):269–87. doi:10.1002/cne.903230210.
39. Kouyama N, Marshak DW. Bipolar cells specific for blue cones in the macaque retina. J Neurosci. 1992;12:1233–52.
40. Grünert U, Greferath U, Boycott BB, Wässle H. Parasol (Pα) ganglion-cells of the primate fovea: immunocytochemical staining with antibodies against GABA A-receptors. Vision Res. 1993;33(1):1–14.
41. Wässle H, Grünert U, Martin PR, Boycott BB. Immunocytochemical characterization and spatial distribution of midget bipolar cells in the macaque monkey retina. Vision Res. 1994;34:561–79.
42. Watanabe M, Rodieck RW. Parasol and midget ganglion cells of the primate retina. J Comp Neurol. 1989;289:434–54.
43. Dacey DM, Petersen MR. Dendritic field size and morphology of midget and parasol ganglion cells of the human retina. Proc Natl Acad Sci U S A. 1992;89:9666–70.
44. Dacey DM, Lee BB. The 'blue-on' opponent pathway in primate retina originates from a distinct bistratified ganglion cell type. Nature. 1994;367:731–5.
45. Dacey DM, Lee BB, Stafford DM, Smith VC, Pokorny J. Horizontal cells of the primate retina: cone specificity without cone opponency. Science. 1996;271:656–8.
46. Dacey DM, Peterson BB, Robinson FR, Gamlin PD. Fireworks in the primate retina: in vitro photodynamics reveals diverse LGN-projecting ganglion cell types. Neuron. 2003;37(1):15–27.
47. Kolb H, Mariani A, Gallego A. A second type of horizontal cell in the monkey retina. J Comp Neurol. 1980;189:31–44.
48. Kolb H, Linberg K, Fisher SK. Neurons of the human retina: a Golgi study. J Comp Neurol. 1992;318:147–87.
49. Kolb H, Fernandez E, Schouten J, Ahnelt P, Linberg KA, Fisher SK. Are there three types of horizontal cell in the human retina? J Comp Neurol. 1994;343:370–86.
50. Ahnelt PK, Kolb H. Horizontal cells and cone photoreceptors in human retina: a Golgi-electron microscopic study of spectral connectivity. J Comp Neurol. 1994;343:406–27
51. Boycott BB, Kolb H. The horizontal cells of the rhesus monkey retina. J Comp Neurol. 1973;148:91–114.
52. Boycott BB, Hopkins JM, Sperling HG. Cone connections of the horizontal cells of the rhesus monkey's retina. Proc R Soc Lond B. 1987;229:345–79.
53. Ahnelt P, Kolb H. Horizontal cells and cone photoreceptors in primate retina: a Golgi-light microscopic study of spectral connectivity. J Comp Neurol. 1994;343(3):387–405. doi:10.1002/cne.903430305.
54. Goodchild AK, Chan TL, Grünert U. Horizontal cell connections with short-wavelength-sensitive cones in macaque monkey retina. Vis Neurosci. 1996;13:833–45.
55. dos Reis JW, de Carvalho WA, Saito CA, Silveira LC. Morphology of horizontal cells in the retina of the capuchin monkey, Cebus apella: how many horizontal cell classes are found in dichromatic primates? J Comp Neurol. 2002;443(2):105–23.
56. Dos Santos SN, Dos Reis JWL, da Silva-Filho M, Kremers J, Silveira LCL. Horizontal cell morphology in nocturnal and diurnal primates: a comparison between owl-monkey (Aotus) and capuchin monkey (Cebus). Vis Neurosci. 2005;22:405–15.

57. Peichl L. Morphology of interneurons: horizontal cells. In: Dartt DA, editor. Encyclopedia of the eye. Oxford: Academic; 2010. p. 74–82.
58. Dacheux RF, Raviola E. Horizontal cells in the retina of the rabbit. J Neurosci. 1982;2:1486–93.
59. Bloomfield SA, Miller RF. A physiological and morphological study of the horizontal cell types of the rabbit retina. J Comp Neurol. 1982;208:288–303.
60. Dorgau B, Herrling R, Schultz K, Greb H, Segelken J, Stroh S, et al. Connexin50 couples axon terminals of mouse horizontal cells by homotypic gap junctions. J Comp Neurol. 2015. doi:10.1002/cne.23779.
61. Thoreson WB, Mangel SC. Lateral interactions in the outer retina. Prog Retin Eye Res. 2012;31(5):407–41. doi:10.1016/j.preteyeres.2012.04.003.
62. Chan TL, Goodchild AK, Martin PR. The morphology and distribution of horizontal cells in the retina of a New World monkey, the marmoset Callithrix jacchus: a comparison with macaque monkey. Vis Neurosci. 1997;14:125–40.
63. Chan TL, Grunert U. Horizontal cell connections with short wavelength-sensitive cones in the retina: a comparison between New World and Old World primates. J Comp Neurol. 1998;393(2):196–209.
64. Svaetichin G, MacNichol EF. Retinal mechanisms for chromatic and achromatic vision. Ann N Y Acad Sci. 1958;74:385–404.
65. Ammermüller J, Kolb H. Functional architecture of the turtle retina. Prog Retin Eye Res. 1996;15:393–433.
66. Kolb H, Boycott BB, Dowling JE. A second type of midget bipolar cell in the primate retina. Philos Trans R Soc Lond B. 1969;255:177–80.
67. Kolb H. Organization of the outer plexiform layer of the primate retina: electron microscopy of Golgi-impregnated cells. Philos Trans R Soc Lond B. 1970;258:261–83.
68. Mariani AP. Giant bistratified bipolar cells in monkey retina. Anat Rec. 1983;206:215–20.
69. Mariani AP. Bipolar cells in monkey retina selective for the cone likely to be blue-sensitive. Nature. 1984;308:184–6.
70. Chan TL, Martin PR, Clunas N, Grunert U. Bipolar cell diversity in the primate retina: morphologic and immunocytochemical analysis of a new world monkey, the marmoset Callithrix jacchus. J Comp Neurol. 2001;437(2):219–39.
71. Grünert U, Martin PR, Wassle H. Immunocytochemical analysis of bipolar cells in the macaque monkey retina. J Comp Neurol. 1994;348(4):607–27. doi:10.1002/cne.903480410.
72. Weltzien F, Percival KA, Martin PR, Grünert U. Analysis of bipolar and amacrine populations in marmoset retina. J Comp Neurol. 2015;523(2):313–34. doi:10.1002/cne.23683.
73. Chan TL, Martin PR, Grunert U. Immunocytochemical identification and analysis of the diffuse bipolar cell type DB6 in macaque monkey retina. Eur J Neurosci. 2001;13(4):829–32.
74. Silveira LCL, Lee BB, Yamada ES, Kremers J, Hunt DM. Post-receptoral mechanisms of colour vision in new world primates. Vision Res. 1998;38(21):3329–37.
75. Lameirao SV, Hamassaki DE, Rodrigues AR, DE Lima SM, Finlay BL, Silveira LC. Rod bipolar cells in the retina of the capuchin monkey (Cebus apella): characterization and distribution. Vis Neurosci. 2009;26(4):389–96. doi:10.1017/S0952523809990186.
76. Martin PR, Lee BB, White AJ, Solomon SG, Ruttiger L. Chromatic sensitivity of ganglion cells in the peripheral primate retina. Nature. 2001;410(6831):933–6. doi:10.1038/35073587.
77. Rodieck RW. The primate retina. Comparative primate biology. Neuroscience. 1988;4:203–78.
78. Joo HR, Peterson BB, Haun TJ, Dacey DM. Characterization of a novel large-field cone bipolar cell type in the primate retina: evidence for selective cone connections. Vis Neurosci. 2011;28(1):29–37. doi:10.1017/S0952523810000374.
79. Haverkamp S, Grünert U, Wassle H. The cone pedicle, a complex synapse in the retina. Neuron. 2000;27(1):85–95.

80. Puthussery T, Venkataramani S, Gayet-Primo J, Smith RG, Taylor WR. $Na_V1.1$ channels in axon initial segments of bipolar cells augment input to magnocellular visual pathways in the primate retina. J Neurosci. 2013;33(41):16045–59. doi:10.1523/JNEUROSCI.1249-13.2013.
81. Puthussery T, Percival KA, Venkataramani S, Gayet-Primo J, Grünert U, Taylor WR. Kainate receptors mediate synaptic input to transient and sustained OFF visual pathways in primate retina. J Neurosci. 2014;34(22):7611–21. doi:10.1523/JNEUROSCI.4855-13.2014.
82. Puthussery T, Gayet-Primo J, Taylor WR, Haverkamp S. Immunohistochemical identification and synaptic inputs to the diffuse bipolar cell type DB1 in macaque retina. J Comp Neurol. 2011;519(18):3640–56. doi:10.1002/cne.22756.
83. Lee SC, Jusuf PR, Grunert U. S-cone connections of the diffuse bipolar cell type DB6 in macaque monkey retina. J Comp Neurol. 2004;474(3):353–63. doi:10.1002/cne.20139.
84. Lee SC, Grünert U. Connections of diffuse bipolar cells in primate retina are biased against S-cones. J Comp Neurol. 2007;502(1):126–40. doi:10.1002/cne.21284.
85. Hartline HK. The response of single optic nerve fibers of the vertebrate eye to illumination of the retina. Am J Physiol. 1938;121:400–15.
86. Kuffler SW. Discharge patterns and functional organization of mammalian retina. J Neurophysiol. 1953,16:37–68.
87. Nakanishi S. Second-order neurones and receptor mechanisms in visual- and olfactory-information processing. Trends Neurosci. 1995;18(8):359–64.
88. Slaughter MM, Miller RF. An excitatory amino acid antagonist blocks cone input to sign-conserving second-order retinal neurons. Science. 1983;219(4589):1230–2.
89. Slaughter MM, Miller RF. 2-Amino-4-phosphonobutyric acid: a new pharmacological tool for retina research. Science. 1981;211(4478):182–5.
90. Lee BB, Martin PR, Valberg A. The physiological basis of heterochromatic flicker photometry demonstrated in the ganglion cells of the macaque retina. J Physiol. 1988;404:323–47.
91. Kaiser PK, Lee BB, Martin PR, Valberg A. The physiological basis of the minimally distinct border demonstrated in the ganglion cells of the macaque retina. J Physiol. 1990;422:153–83.
92. Kremers J, Lee BB, Kaiser PK. Sensitivity of macaque retinal ganglion cells and human observers to combined luminance and chromatic modulation. J Opt Soc Am A. 1992;9:1477–85.
93. Maunsell JHR, Nealey TA, DePriest DD. Magnocellular and parvocellular contributions to responses in the middle temporal visual area (MT) of the macaque monkey. J Neurosci. 1990;10(10):3323–34.
94. Lee BB, Wehrhahn C, Westheimer G, Kremers J. Macaque ganglion cell responses to stimuli that elicit hyperacuity in man: detection of small displacements. J Neurosci. 1993;13(3):1001–9.
95. Lee BB, Wehrhahn C, Westheimer G, Kremers J. The spatial precision of macaque ganglion cell responses in relation to Vernier acuity of human observers. Vision Res. 1995;35(19):2743–58.
96. Ruttiger L, Lee B, Sun H. Transient cells can be neurometrically sustained: the positional accuracy or retinal signals to moving targets. J Vis. 2002;2(3):232–42. doi:10.1167/2.3.3.
97. Lee BB, Martin PR, Valberg A, Kremers J. Physiological mechanisms underlying psychophysical sensitivity to combined luminance and chromatic modulation. J Opt Soc Am A. 1993;10:1403–12.
98. Lee BB, Kremers J, Yeh T. Receptive fields of primate retinal cells studied with a novel technique. Vis Neurosci. 1998;15:161–75.
99. Martin PR, Blessing EM, Buzas P, Szmajda BA, Forte JD. Transmission of colour and acuity signals by parvocellular cells in marmoset monkeys. J Physiol. 2011;589(Pt 11):2795–812. doi:10.1113/jphysiol.2010.194076.
100. Livingstone MS, Hubel DH. Psychophysical evidence for separate channels for the perception of form, color, movement, and depth. J Neurosci. 1987;7(11):3416–68.

101. Livingstone MS, Hubel DH. Segregation of form, color, movement, and depth: anatomy, physiology, and perception. Science. 1988;240:740–9.
102. Martin PR. Colour processing in the primate retina: recent progress. J Phsyiol (London). 1998;513(3):631–8.
103. Wiesel TN, Hubel DH. Spatial and chromatic interactions in the lateral geniculate body of the rhesus monkey. J Neurophysiol. 1966;29:1115–56.
104. McCulloch DL, Marmor MF, Brigell MG, Hamilton R, Holder GE, Tzekov R, et al. ISCEV Standard for full-field clinical electroretinography (2015 update). Doc Ophthalmol. 2015;130(1):1–12. doi:10.1007/s10633-014-9473-7.
105. Frishman LJ. Origins of the electroretinogram. In: Heckenlively JR, Arden GB, editors. Principles and practice of clinical electrophysiology of vision. Cambridge, London: The MIT Press; 2006. p. 139–83.
106. Sustar M, Hawlina M, Brecelj J. ON- and OFF-response of the photopic electroretinogram in relation to stimulus characteristics. Doc Ophthalmol. 2006;113:43–52.
107. Pangeni G, Lammer R, Tornow RP, Horn FK, Kremers J. On- and off-response ERGs elicited by sawtooth stimuli in normal subjects and glaucoma patients. Doc Ophthalmol. 2012. doi:10.1007/s10633-012-9323-4.
108. Viswanathan S, Frishman LJ, Robson JG. The uniform field and pattern ERG in macaques with experimental glaucoma: removal of spiking activity. Invest Ophthalmol Vis Sci. 2000;41(9):2797–810.
109. Viswanathan S, Frishman LJ, Robson JG, Harwerth RS, Smith Iii EL. The photopic negative response of the macaque electroretinogram: reduction by experimental glaucoma. Invest Ophthalmol Vis Sci. 1999;40(6):1124–36.
110. Viswanathan S, Frishman LJ, Robson JG, Walters JW. The photopic negative response of the flash electroretinogram in primary open angle glaucoma. Invest Ophthalmol Vis Sci. 2001;42(2):514–22.
111. Armington JC. The electroretinogram. New York: Academic; 1974.
112. Jacobs GH. The discovery of spectral opponency in visual systems and its impact on understanding the neurobiology of color vision. J Hist Neurosci. 2014;23(3):287–314. doi:10.1080/0964704X.2014.896662.
113. Riggs LA, Johnson EP, Schick AM. Electrical responses of the human eye to changes in wavelength of the stimulating light. J Opt Soc Am. 1966;56:1621–7.
114. Riggs LA, Sternheim CE. Human retinal and occipital potentials evoked by changes of the wavelength of the stimulating light. J Opt Soc Am. 1969;59(5):635–40.
115. Sperling HG, Harwerth RS. Red-green cone interaction in the increment-threshold spectral sensitivity of primates. Science. 1971;172:180–4.
116. Harwerth RS, Sperling HG. Effects of intense visible radiation on the increment-threshold spectral sensitivity of the rhesus monkey eye. Vision Res. 1975;15:1193–204.
117. Mills SL, Sperling HG. Red/green opponency in the rhesus macaque ERG spectral sensitivity is reduced by bicuculline. Vis Neurosci. 1990;5:217–21.
118. Sperling HG, Mills SL. Red-green interactions in the spectral sensitivity of primates as derived from ERG and behavioral data. Vis Neurosci. 1991;7:75–86.
119. van Norren D, Baron WS. Increment spectral sensitivities of the primate late receptor potential and b-wave. Vision Res. 1977;17(7):807–10.
120. King-Smith PE, Carden D. Luminance and opponent-color contributions to visual detection and adaptation and to temporal and spatial integration. J Opt Soc Am. 1976;66:709–17.
121. Bush RA, Sieving PA. Inner retinal contributions to the primate photopic fast flicker electroretinogram. J Opt Soc Am A. 1996;13(3):557–65.
122. Armington JC. Chromatic and short term dark adaptation of the human electroretinogram. J Opt Soc Am. 1959;49:1169–75.
123. DeValois RL, Abramov I, Jacobs GH. Analysis of response patterns of LGN cells. J Opt Soc Am. 1966;56:966–77.
124. Baron WS. Cone difference signal in foveal local electroretinogram of primate. Invest Ophthalmol Vis Sci. 1980;19(12):1442–8.

125. Donovan WJ, Baron WS. Identification of the R-G-cone difference signal in the corneal electroretinogram of the primate. J Opt Soc Am A. 1982;72(8):1014–20.
126. Jacobs GH. Primate photopigments and primate color vision. Proc Natl Acad Sci U S A. 1996;93:577–81.
127. Jacobs GH, Deegan Ii JF. Spectral sensitivity of macaque monkeys measured with ERG flicker photometry. Vis Neurosci. 1997;14:921–8.
128. Jacobs GH, Deegan IJS, Moran JL. ERG measurements of the spectral sensitivity of common chimpanzee (Pan troglodytes). Vision Res. 1996;36(16):2587–94.
129. Jacobs GH, Neitz J. Electrophysiological estimates of individual variation in the L/M cone ratio. In: Drum B, editor. Colour vision deficiencies XI. Dordretch: Kluwer; 1993. p. 107–12.
130. Jacobs GH, Neitz J, Krogh K. Electroretinogram flicker photometry and its applications. J Opt Soc Am A. 1996;13(3):641–8.
131. Neitz J, Jacobs GH. Electroretinogram measurements of cone spectral sensitivity in dichromatic monkeys. J Opt Soc Am A. 1984;1:1175–80.
132. Kremers J, Scholl HPN, Knau H, Berendschot TTJM, Usui T, Sharpe LT. L/M cone ratios in human trichromats assesed by psychophysics, electroretinograpy, and retinal densitometry. J Opt Soc Am. 2000;17:517–26.
133. Brainard DH, Roorda A, Yamauchi Y, Calderone JB, Metha AB, Neitz M, et al. Functional consequences of the relative numbers of L and M cones. J Opt Soc Am A. 2000;17(3):607–14.
134. Hofer H, Carroll J, Neitz J, Neitz M, Williams DR. Organization of the human trichromatic cone mosaic. J Neurosci. 2005;25(42):9669–79.
135. Hagstrom SA, Neitz J, Neitz M. Ratio of M/L pigment gene expression decreases with retinal eccentricity. In: Cavonius CR, editor. Colour vision deficiencies XIII. Dordrecht: Kluwer; 1997. p. 59–65.
136. Hagstrom SA, Neitz J, Neitz M. Variation in cone populations for red-green color vision examined by analysis of mRNA. Neuroreport. 1998;9:1963–7.
137. Hagstrom SA, Neitz M, Neitz J. Cone pigment gene expression in individual photoreceptors and the chromatic topography of the retina. J Opt Soc Am A Opt Image Sci Vis. 2000;17(3):527–37.
138. Kuchenbecker JA, Sahay M, Tait DM, Neitz M, Neitz J. Topography of the long- to middle-wavelength sensitive cone ratio in the human retina assessed with a wide-field color multifocal electroretinogram. Vis Neurosci. 2008;25(3):301–6.
139. Jacob MM, Pangeni G, Gomes BD, Souza GS, Da Silva Filho M, Silveira LCL, et al. The spatial properties of L- and M-cone inputs to electroretinograms that reflect different types of post-receptoral processing. PLoS One. 2015;10(3):e0121218. doi:10.1371/journal.pone.0121218.
140. de Lange H. Research into the dynamic nature of the human fovea-cortex systems with intermittent and modulated light. I. Attenuation characteristics with white and colored light. J Opt Soc Am. 1958;48:777–84.
141. Kelly DH, Norren DV. Two-band model of heterochromatic flicker. J Opt Soc Am. 1977;67:1081–91.
142. Kremers J, Link B. Electroretinographic responses that may reflect activity of parvo- and magnocellular post-receptoral visual pathways. J Vis. 2008;8(15/11):1–14.
143. Kremers J, Pangeni G. Electroretinographic responses to photoreceptor specific sine wave modulation. J Opt Soc Am A. 2012;29(2):A309–16.
144. Kommanapalli D, Murray IJ, Kremers J, Parry NR, McKeefry DJ. Temporal characteristics of L- and M-cone isolating steady-state electroretinograms. J Opt Soc Am A Opt Image Sci Vis. 2014;31(4):A113–20. doi:10.1364/JOSAA.31.00A113.
145. Kremers J, Rodrigues AR, Silveira LCL, da Silva-Filho M. Flicker ERGs representing chromaticity and luminance signals. Invest Ophthalmol Vis Sci. 2010;51:577–87.
146. Lee BB, Sun H, Valberg A. Segregation of chromatic and luminance signals using a novel grating stimulus. J Physiol. 2011;589(Pt 1):59–73. doi:10.1113/jphysiol.2010.188862. jphysiol.2010.188862 [pii].

147. Parry NR, Murray IJ, Panorgias A, McKeefry DJ, Lee BB, Kremers J. Simultaneous chromatic and luminance human electroretinogram responses. J Physiol. 2012;590:3141–54. doi:10.1113/jphysiol.2011.226951.
148. Kremers J, Pangeni G, Tsaousis KT, McKeefry D, Murray IJ, Parry NR. Incremental and decremental L- and M-cone driven ERG responses: II. Sawtooth stimulation. J Opt Soc Am A Opt Image Sci Vis. 2014;31(4):A170–8. doi:10.1364/JOSAA.31.00A170.
149. McKeefry D, Kremers J, Kommanapalli D, Challa NK, Murray IJ, Maguire J, et al. Incremental and decremental L- and M-cone-driven ERG responses: I. Square-wave pulse stimulation. J Opt Soc Am A Opt Image Sci Vis. 2014;31(4):A159–69. doi:10.1364/JOSAA.31.00A159.
150. Murray IJ, Kremers J, Parry NRA. L- and M-Cone isolating ergs: LED versus CRT stimulation. Vis Neurosci. 2008;25:327–31.
151. Yamaguchi S, Motulsky AG, Deeb SS. Visual pigment gene structure and expression in human retinae. Hum Mol Genet. 1997;6:981–90.

Chapter 3
Functional Imaging of Cone Photoreceptors

Lawrence C. Sincich, Ramkumar Sabesan, William S. Tuten, Austin Roorda, and Wolf M. Harmening

Abstract Color pervades our visual sensory world, yet our understanding of the neural basis of color perception, starting with the retina and on through the multiple cortical areas that subserve vision, is still incomplete. The L, M, and S cone photoreceptors, being the cellular entry point for trichromatic vision in humans and primates, have been studied in a variety of ways to reveal their relative numbers, their spatial arrangement, and their anatomical connectivity. We review work in these species that has linked mapped cone mosaics directly to functional properties such as single neuron responses in the retina and color percepts arising from cone-targeted microstimulation. Technical issues that constrain access to single cone photoreceptors for functional studies are also considered.

Keywords Cone photoreceptors • Waveguides • Adaptive optics • Absorptance imaging • Microstimulation • Chromatic dispersion • Fixational eye motion • Retinal vessels • Increment threshold • Color psychophysics

3.1 Introduction

That the world is a colorful experience for most of us is a consequence of two biological feats. One is that our retina has evolved photoreceptors and circuitry that preserve wavelength specific sensitivities in their output signals. The other is that our visual cortex is able to elaborate upon those retinal inputs in a variety of enriching and advantageous ways. The confluence of these two feats is the foundation of color perception. For instance, the three primary categories of photoreceptors that

L.C. Sincich (✉)
Department of Optometry and Vision Science, University of Alabama at Birmingham, Birmingham, AL 35233, USA
e-mail: sincich@uab.edu

R. Sabesan • W.S. Tuten • A. Roorda
School of Optometry, University of California, Berkeley, CA, USA

W.M. Harmening
Department of Ophthalmology, University of Bonn, Bonn, Germany

J. Kremers et al. (eds.), *Human Color Vision*, Springer Series in Vision Research, DOI 10.1007/978-3-319-44978-4_3

populate the eye in humans arose through an evolutionary process that was neurally independent of the brain—as there are no known neural projections from the brain to the retina—yet it is clear that rather than ignoring the emergence of trichromatic capability, cortical processing took advantage of what the retina had to offer. Perhaps an extreme counterexample of this point is found in mantis shrimp, which have 12 photopigments yet have poor color discrimination that is roughly trichromatic [1]. The aim of this chapter is to examine the handshake between the eye's photoreceptors and the brain, based on a variety of techniques, some new, some time-proven, that are giving us fresh access to the cellular basis of color vision.

Our particular focus is on cone photoreceptors, how their spectral sensitivities can be identified in humans and primates, and how their individual response properties can be studied at the perceptual level. This last effort is a special challenge because cones are undeniably small. Depending on the individual, cone inner segment diameters can be 1.5–2 μm at the fovea, though they increase in size rapidly with eccentricity, reaching a typical diameter of 6–8 μm that nearly plateaus at about 5° eccentricity from the fovea [2]. Cone inner segment diameter is commonly used as the anatomical determinant of photoreceptor size because it sets a primary optical constraint on light capture. To study such small cells individually, microscopic access is required. As we review here, an extensive body of work has used retinal tissue removed from the eye and kept physiologically maintained in vitro for functional studies at the cellular scale. Recent developments in ocular imaging have now also provided direct microscopic access to cones in vivo [3, 4], paving the way for combined biophysical and psychophysical studies of single cones in the living eye.

3.2 Optical Constraints on Imaging Cones

Before beginning our survey of functional cone imaging, it is important to appreciate a few of the technical constraints that shape the experimental strategies used for studying cones. This section is necessarily brief given the long history of the issues, but comprehensive references are provided.

3.2.1 *Photoreceptor Waveguiding*

In order to see the photoreceptors at the microscopic scale in an intact eye, a beam of light is usually projected onto the retina and the reflected light is detected by some means. It is useful to realize that only about 1 % or less of the light entering the eye is reflected back out, with the exact percentage being dependent on wavelength [5]. The remainder is either scattered or absorbed by tissues, and of the latter only a fraction is actually captured by the photopigments that allow us to see [6]. For example, of just the light that enters a cone's inner segment, it has been estimated that about 30 % is absorbed by photopigment [7]. For the 1 % of the light that is reflected back from the eye, this occurs in part because the somewhat planar tissue of the retina acts like a weak mirror, having a very noticeable directional component. Much of this direction-dependency is due to the fact that photoreceptors act like optical waveguides, able to funnel light efficiently along their length [8–11] (but see [12] for an alternative theoretical interpretation).

Once light enters the cone inner segment, internal reflections channel the light down the long axis of the cell. This has several consequences. One is that light is captured more efficiently than if there was no waveguiding, because it can be collected over a broader range of beam positions and angles. Light entering straight into the face of the inner segment has the best chance of being captured by the photoreceptor, unsurprisingly, while light hitting the face at increasing larger angles can still be captured, though with decreasing effectiveness. The light coupling efficiency is a two-dimensional function of the angle of light entry and can be pictured as having an approximately Gaussian shape. This angular sensitivity of light capture can be measured perceptually in a variety of ways (referred to as the Stiles-Crawford effect [13–15]), one of which leads to an estimate of the dimensions of the light capture profile for single cones [16, 17]. For light heading into the eye, waveguiding helps it to be absorbed more readily by cone outer segments where the photopigments reside.

Waveguiding is equally effective for light reflected back through the photoreceptor, since optical systems are reversible. The cones themselves are generally thought to have two major reflective structures: the anatomical junction between the inner and outer segment, and the junction between the outer segment and its contact with retinal pigment epithelium [18–20]. Thus, a second consequence of waveguiding is that when light hitting these junctions is scattered, internal reflections can direct some of this light back along the path of the entering light; this portion would otherwise be lost if simply coming off a scattering surface. Such a waveguide effect is good news for cone imaging based on reflectance, as it leads to peaks of light intensity that approximately correspond to the center of each photoreceptor (Fig. 3.1). As we shall see in the next sections, the pointillism of retinal reflection allows each photoreceptor to be identified and returned to day after day for functional studies, when suitable methods are used.

3.2.2 Use of Adaptive Optics Technology for In Vivo Imaging

While waveguiding is a helpful optical property of cones—both for improving a cone's light capture and for making retinal imaging easier—it still remains a technical challenge to image cones at the microscopic scale in vivo. Under a limited set of circumstances in humans, individual cones can be seen using the natural optics of the eye [21–24]. But to attain diffraction-limited imaging of cones with wide fields of view, to achieve this in most subjects, and to do so in real-time as is needed for psychophysical testing, a method for rapidly measuring and correcting the normal optical aberrations of the eye is necessary. Beginning in the 1990s, the first practical

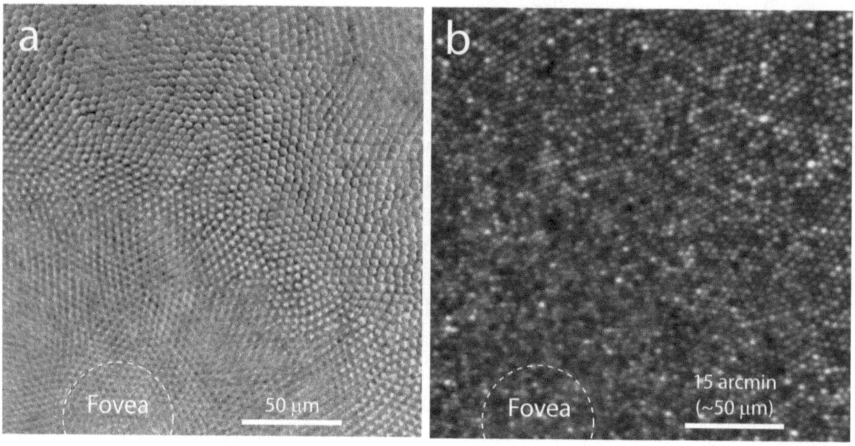

Fig. 3.1 Perifoveal mosaic of cone inner segments in macaque retina. (**a**) *En face* view of an unstained macaque retina, imaged with phase-contrast microscopy. Because the length of the inner segments increases rapidly just outside the fovea (*dashed line*), only an arc of them are in focus at the level of the inner limiting membrane. Also evident is the increase in inner segment diameter with distance from the fovea. (**b**) Adaptive optics image taken in vivo of the same region of retina from a different macaque eye. Most cones in the mosaic are well resolved, except within ~30 arcmin of the fovea, where the reflectance image is compromised by patterns arising from light interference. Imaging wavelength = 710 nm

means of making optical aberration correction for human eyes brought such imaging within reach (reviewed in Ref. [3]). The method entails two main features: a wavefront sensing device to measure the ocular aberrations, and a deformable mirror that can compensate for those aberrations. Such devices have been termed adaptive optics (AO) ophthalmoscopes because the wavefront measurements and compensations are updated regularly to accommodate aberration changes that are constantly occurring in the living eye [4, 25].

Ophthalmoscopes that have used AO for imaging the retina at high resolution have generally been configured in one of two ways. The first was a flood illuminated system, using a flash of light and a CCD camera to create single full-frame images [26]. The second used a raster-based scanning laser system to create a continuous stream of images that could be easily recorded as a movie [27]. Each of these systems has advantages and disadvantages with respect to studying human color vision. Flood AO systems are optimal for capturing an instantaneous retinal image that is free of distortions due to eye movements. Because of this, they have been used extensively for absorptance imaging of the cone mosaic (described below). Wavefront correction, however, has been time consuming with flood systems, and the retinal locus of any stimulus delivery is uncertain at the cellular scale. AO-based scanning laser ophthalmoscopes (AOSLOs) enable rapid wavefront correction and tracking of stimulus delivery locations, but the unavoidable eye motion distortions require extensive effort to overcome. To the degree that they have been overcome—as we detail later in this chapter—AOSLO imaging has shown promise for probing color vision at the individual cellular level.

3.3 Biophysical Cone Imaging

A very substantial literature exists on the biophysical characterization of cone photoreceptors, beginning with spectrometric studies of the isolated photopigments [28, 29] to the electrophysiological characterization of spectral sensitivity functions measured via photocurrents [30]. In this section, we limit ourselves to experiments that have drawn on imaging approaches to answer questions specific to intact retinas. What proportion of the retina is occupied by each cone class? Do these proportions vary between individuals? Are the cones randomly arranged?

Because cone photoreceptors have broad sensitivities to wavelengths of light, they are usually referred to by their long, medium or short wavelength peak sensitivities: L cones (traditionally associated with "red" signaling), M cones ("green"), and S cones ("blue"). Although it is discussed in detail elsewhere in this book (see Chaps. 2), we note here that the peak sensitivities of three cone classes can differ depending on how they are measured. For macaque cones measured electrophysiologically, the peak wavelength sensitivities are: L cones = 561 nm, M cones = 531 nm, and S cones = 430 nm [30, 31]. Because these values agree reasonably well with physiological and microspectrophotometric values obtained from human L and M cones [32–36], as well as psychophysically measured sensitivity peaks when derived from a nomogram fit (Eq. 8 of [37]) that take into account preretinal absorption, we use these peak sensitivity values in this chapter without adding further qualifications.

3.3.1 Absorptance Classification of Cones In Vitro

One of the earliest in situ maps of L, M, and S cones was made by applying microspectrophotometric methods developed for use in isolated photoreceptors [38] to classify cones in a piece of flattened retinal tissue (Fig. 3.2a). This revealed a mosaic that appeared to have a random distribution of L and M cones, although the mapped set of cones was perhaps too small to learn definitively if cones of like type were nonrandomly clustered together. The method required the cones to be measured one at a time, limiting the number that could be studied. Shortly after, a video-based method was devised to measure differential photopigment bleaching over an entire microscopic field of view in one pass (Fig. 3.2b). This yielded a nonrandom arrangement of L and M cones [7]. At the time of these studies, the distribution of S cones had already been shown to be fairly regularly distributed across the retina, except in the foveola where they are absent, using histochemistry or antibodies to the S cone opsin [39–42]. With no antibody that can currently distinguish between L and M cones, imaging based on differential absorption of spectral light remains the primary biophysical means for mapping the cones by class. By the mid-1990s, with only a few small cone maps available, any order in the spatial arrangement of L and M cones still remained an open question.

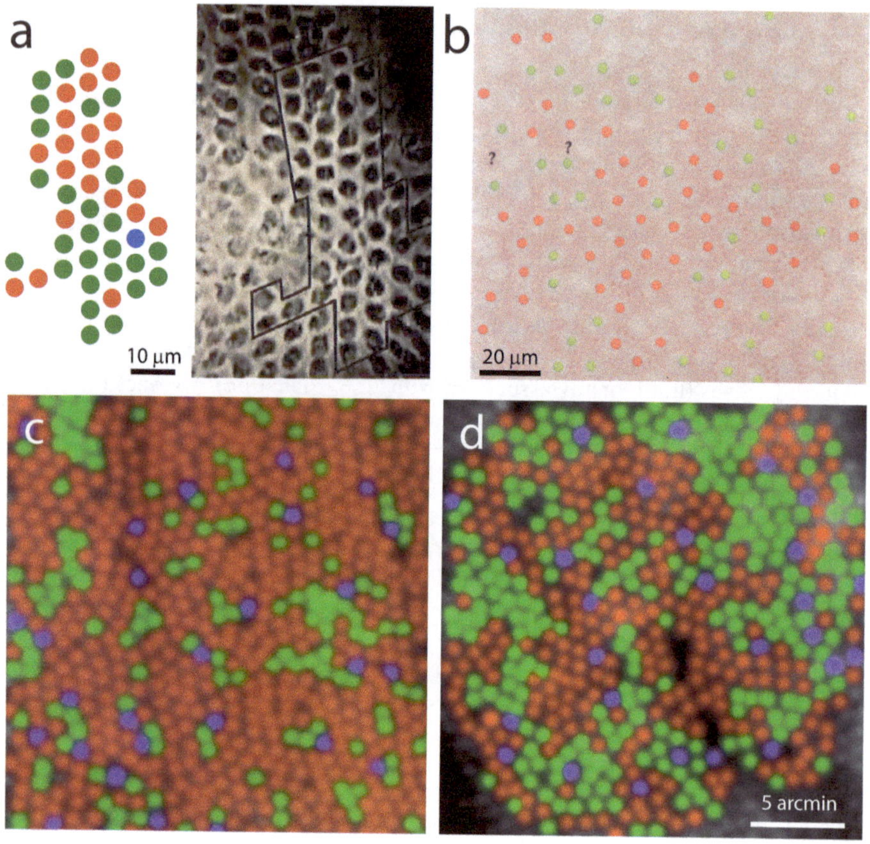

Fig. 3.2 Photopigment absorptance imaging of the trichromatic cone mosaic. L, M, and S cones are represented by red, green, and blue, respectively, in all panels. (**a**) Microspectrophotometric mapping of cones from a freshly dissected talapoin foveal retina. Adapted from [138]. (**b**) Photopigment transmittance mapping in an excised macaque peripheral retina. Candidate S cones are indicated by a question mark to the left. Adapted from [7]. (**c**, **d**) Pseudocolor images of the cone mosaic in two human subjects, mapped in the living eye via differential photopigment bleaching in conjunction with AO imaging. Retinal areas are 1° from the fovea. Statistical analysis showed a random distribution of L and M cones in (**c**), and a nonrandom distribution in (**d**), though it may have arisen from optical blur in this subject. Adapted from [61]

Why is it important to know the distribution of L and M cones? The main reason has been to learn how color signals are initially established. For a color signal to be perceived within the red-green portion of the spectrum, there must be a comparison at the neural level between activity arising from L cones and M cones. Such L/M opponency is generally considered to be established via receptive field center/surround antagonism, and transmitted by midget retinal ganglion cells to the rest of the brain [43] (also see Chap. 4). If ganglion cells receive inputs from all the cones lying within their dendritic field [44–50], a random arrangement of L and M cones would often lead to nearly equal L and M drive between receptive field centers and surrounds, producing an achromatic signal. If, however, L and M cones had nonrandom distributions, the centers and surrounds would more likely manifest differences in L and M composition, thereby producing a chromatic signal.

The foregoing assumes that many cones subserve the receptive field center and surround, which is the case in the retinal periphery. As one moves closer to the fovea, the field sizes get smaller, and consequently, the cone composition of the receptive field centers, in particular, are more likely to be of a single class if the cones are arranged nonrandomly. To confer a color signal near the fovea, therefore, the field surrounds must receive input from cones of the opposite type from those in the field center. There has been a controversy over the purity of the cone composition in the surround. Some physiological studies have concluded that field surrounds of most cells are cone pure [51–54]. Others have found that surrounds can have varying degrees of mixed cone inputs [55–58]. Without knowing the cone composition of the individual receptive fields recorded, these mixed results could be attributed to either random or nonrandom cone arrangements, or to varying ratios of L and M cones. Unraveling this dilemma is essential for understanding color vision because these midget ganglion cells also carry the highest spatial information and represent more than 80 % of all ganglion cells in the retina [59, 60]. A move toward more comprehensive mapping of the cone mosaic would help to clear up the functional role of this cell class.

3.3.2 *Absorptance Classification of Cones In Vivo*

The chance to classify larger fields of cones was created with AO-based imaging of the retina, where it became possible to examine hundreds of cones simultaneously in an intact eye. Differential absorptance imaging was first used in humans [61] (and later in a macaque [62]) where it was determined that the L and M cone distributions appeared random and had varying ratios between individuals (Fig. 3.2c, d). The extent of the variation in L:M cone ratio was confirmed and expanded in a later study of a larger population of human retinas [63]. The L:M cone ratio varied from 1.1:1 to 16.5:1 in this group of male subjects with normal color vision, with a median ratio of about 1.9:1. Interestingly, statistical analysis of the spatial arrangement of the cones found that 5 of the subjects had randomly dispersed L and M cones, whereas the remaining three subjects had nonrandom arrangements, with two of these having significant local clustering of cones of like type. In one retina, the mosaics examined from opposite sides of the fovea had different L:M cone ratios (1.24:1 and 1.77:1). There is also a notable increase in the relative proportion of L cones with distance from the fovea [64]. With many individual cone mosaics now classified, the picture that has emerged from these data—as it so often happens in biology—is that the phenotypes of cone arrangements are truly mixed. The L and M cone distributions can be random or nonrandom, and will very much depend on both subject and retinal locus.

What does this mean for color vision studies? Functional tests that use small chromatic stimuli will be most affected. If stimuli are to be targeted to small regions of cones, it will be harder to rely on assumptions about L and M cone ratios or local spatial distributions. The variation in these factors will impact certain types of experiments: the relative efficiency of detecting spectral differences in small spots [65–67], the appearance of briefly flashed spots [68–71], hyperacuity derived from stimuli with chromatic differences [72], the appearance of high-spatial frequency gratings (both chromatic and achromatic) [73], and, in physiology, the spectral responses of neurons in the early visual pathways [74, 75]. The use of AO-imaged cone mosaics in conjunction with functional testing has already begun to firm up our insights about how percepts are shaped by the specific cones being stimulated [76], and as we will see later in the chapter, this work is continuing.

That the cone mosaics classified in vivo actually do align with a physiological measure has been shown empirically. Heterochromatic flicker photometry has been used extensively to estimate L:M cone ratios [77]. The idea behind the method is to flicker light combinations that drive either L cone or M cones in isolation, with the magnitude of the response under each condition being proportional to the number of L or M cones. Comparison of the AO imaging data with flicker-photometric electroretinograms (ERGs) in the same subjects found that variation in ERGs between subjects is well correlated with the L:M cone ratios [63, 78]. This result helps to explain the wide range of ERGs that are found even when photopigment spectra are known [79, 80]. However, there remains a discrepancy between the AO-derived L:M ratios and the ERGs, as the relationship is not unitary. The authors suggest that this may be due to a ~1.5-fold larger contribution to the ERG signal from each M cone versus each L cone [63], but the mechanism is also likely to involve differential cone adaptation [81]. Even with a proportionality constant greater than 1, the good correlation between imaging data and ERG is a strong indicator that functional differences are keyed to specific L:M cone ratios.

3.3.3 *Physiological Classification of Cones In Vitro*

Biophysical imaging of cones is not limited to the photopigment absorptance approaches reviewed so far. There have also been many physiological studies that tap into the electrical responses following light stimulation to characterize retinal tissue. These methods provide opportunities to answer additional questions about retinal function that cannot be addressed by absorptance imaging. What is the cone composition of the receptive fields of neurons downstream from the cones? Is the functional weighting of each cone the same? How many different ganglion cells does each cone feed a signal to?

Much has been learned about the relative proportion of each cone type's input to the main cell classes found in the primate retina, using single electrode recordings combined with cell fills (e.g., bipolar cells [82], horizontal cells [83], ganglion cells [84]). In this section, however, we focus on studies that have used imaging techniques in their experimental approach to cone function.

The first cones mapped physiologically were S cones [85]. Using a flat multielectrode array, it was possible to record from many retinal ganglion cells simultaneously in a small piece of explant macaque retina while projecting a randomly flickering colored stimulus pattern onto the tissue. Given the spatial resolution of the stimulus and the relatively wide spacing of the S cones, the spike-triggered averaged responses of the blue-ON/yellow-OFF ganglion cells revealed an activity map of individual S cones. This functional map of a single cone class unveiled a few important themes that have since been demonstrated for all cone classes [86, 87]. First, for a given ganglion cell, the functional strength of each cone providing input can differ markedly. For example, among six S cones that fed onto one blue/yellow ganglion cell, there was a nearly threefold difference in excitatory input strength between cones (see Fig. 4a in Ref. [85]). Second, cones can provide input to more than one ganglion cell of the same type, suggesting that, at least in peripheral retina, receptive field centers can have some spatial overlap. Notably, when the same cone does connect to separate ganglion cells, the input strengths will differ to each ganglion cell. This indicates that there are genuine differences in synaptic strength from one cone to separate ganglion cells, rather than there being especially sensitive cones passing large signals to all downstream partners. Third, it was found that S cone signals combined linearly. Regardless of the relative activation of two S cones, the spiking output of the blue/yellow cell was a function of the summed input.

In all of these ganglion cells, the yellow-OFF response, arising from combined L and M cone signals (see Chap. 4), was quantified, but the spatial resolution of the stimulus did not permit the mapping of individual cones. That resolution limit was overcome in a later study where mosaics of all three cone types were revealed in unprecedented functional detail for all of the major ganglion cell types [86]. Examples of complete cones maps for one ON and one OFF midget ganglion cell are shown in Fig. 3.3a. Each cone could be classified by the relative spike-triggered activity produced by the three color primaries of the stimulus display. Cones with ON responses appear as brighter values in these maps, while cones with OFF responses yield relatively darker values. As is evident in these maps of the receptive field centers, each cone can be discretely identified (this was also confirmed anatomically), and each has a different functional weight. In the set of eyes studied, the L:M cone ratio was 2:1, as had been estimated from previous macaque studies, and the S cones were 8 % of the total population.

Because an entire array of cones was mapped simultaneously by recording from almost every ganglion cell in the field, it was possible to create not only a nearly complete mosaic of cones (Fig. 3.3b), but also to define the underlying functional connectivity between cones and ganglion cells. Cones supplying input to a receptive field center had the largest influence on ganglion cell spiking—as expected—while cones serving the field surround had much smaller weights of opposing magnitude (Fig. 3.3c). Complete sets of such connectivity diagrams for the ON and OFF varieties of both midget and parasol ganglion cells characterized several features of wiring specificity. One is that every cone provides input to each major ganglion cell class. This finding was later confirmed with single cone stimulation in vitro [87]. This means that sampling of the cone mosaic by each cell type is without gaps, and

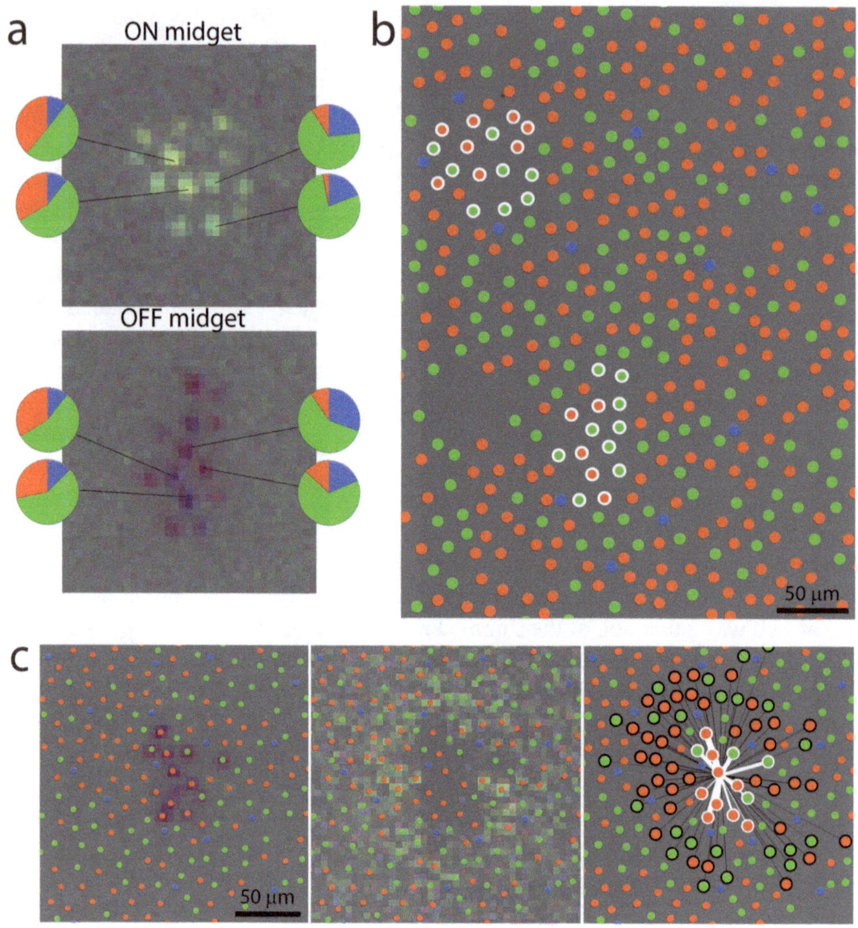

Fig. 3.3 Physiological cone classification and receptive field mapping of macaque retina in vitro. (**a**) The spectral sensitivity of cones providing input to the receptive field centers of two retinal ganglion cells is represented by the relative magnitude of the *red, green*, and *blue* pixel values in the image, each corresponding to spike-triggered average values (pie-diagrams). (**b**) For every cone in one recording, these values are converted into an index that discriminates L (*red*), M (*green*), and S (*blue*) cones. The cones identified from the retinal ganglion cells mapped in (**a**) are circled in *white*. (**c**) A cone mosaic from an OFF midget retinal ganglion, overlaid on normalized spike-triggered average maps, showing the strength of the cones defining the receptive field center (*left*) and surround (*middle*). A connectivity diagram (*right*), with line thickness proportional to the strength of each cone input, distinguishes the center response (*white*) from the surround (*black*, line thickness × 5 for visibility). Adapted from Ref. [86]

as a consequence, the visual field is represented contiguously by each ganglion cell type. In some instances, along the borders between the cone fields of ganglion cells, cone inputs are shared, indicating a slight overlap in the tiling of receptive fields. Another feature is that S cones provide input to more than just small bistratified ganglion cells [88]. About 10 % of midget and parasol ganglion cells also sample from S cones, with the exception that about 60 % of OFF midget cells receive input from at least one S cone. Therefore, S cone signals appear to be transmitted, in varying degrees, by all of the major ganglion cell types.

A third feature revealed in these data is how the cone composition of midget ganglion cell receptive fields undergirds L/M cone opponency, an issue that has been controversial [84]. Quantifying the relative strength of the L and M cone inputs, it was found that the midget ganglion cells exhibited red-green color opponency more often than predicted by random sampling. A statistical analysis of the number of L and M cones composing the field center and surround further showed that cones within the field center significantly favored connections of like type, whereas connectivity appeared random with cones in the surround. The bias in cone types for field centers was not due to cones of the same type being clumped together (as is occasionally seen in human retinas; see above). Instead, the cone opponency arose from more frequent connections of ganglion cells to cones of similar type, in addition to a stronger weighting of the same cones.

In summary, in vitro studies have offered exquisitely detailed information about cone function in the retina. Cones provide divergent and differently weighted inputs to all the major classes of ganglion cells, and there are connectivity biases that may boost red-green color perception, at least in the peripheral visual field. It is likely that the features revealed so far are also present in central retina, where the cone weight variation and wiring specificity is likely to amplify color signaling in many cells. Functional cone mapping closer to the fovea is desirable in this regard, especially for in vivo work, and is the topic we move to next.

3.4 Practical Constraints on Functional Cone Imaging In Vivo

Testing individual cones perceptually can lead to a number of insights about how cone signals are combined and ultimately generate color vision. With the advent of AO-based retinal imaging, investigators have begun to probe cones one by one for functional assessment. However, there are a number of technical challenges that any experimentalist needs to be aware of when testing vision at the cellular scale, especially with respect to color stimuli. In this section, we delve into the more prominent hurdles that arise when trying to measure response properties that originate from one cone versus another.

3.4.1 Fixational Eye Motion

Vision testing with macroscopic stimuli can usually ignore the relatively small eye movements that occur during steady fixation, but when working at the cellular scale, the movements cannot be dismissed. Such eye motion has been classified into several

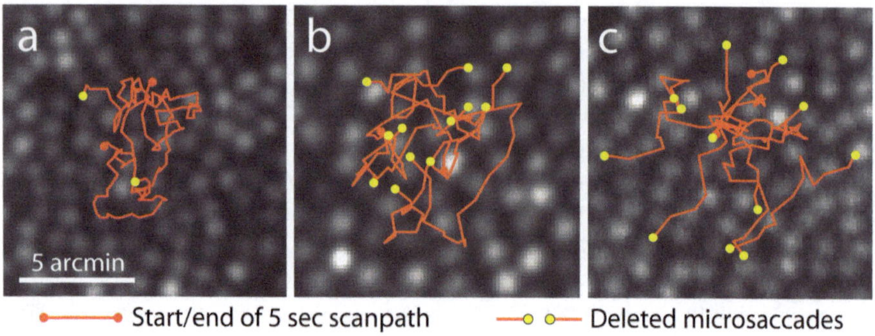

Fig. 3.4 Retinal motion in fixating subjects ranges over many cones. Each panel shows the cone mosaic and the reconstructed motion path from a 5 s stabilized AOSLO movie recorded during active fixation at 30 Hz. Scanpaths (*red*) have had microsaccades deleted (*yellow* endpoints). Some subjects, as in (**a**), exhibit small frame-by-frame eye motion and rare microsaccades, while other subjects can have larger drifts and more saccades (**b**). Subjects can also manifest persistent drift along one axis, such as diagonal (**c**), with each drift canceled by a compensating microsaccade. As long as eye motion remains less than about one-third of the frame width (5× larger than these cropped images), video stabilization can track the position of cones targeted for stimulation. Eccentricities: **a** = 1.9°; **b** = 3.7°; **c** = 3.3°

varieties: typically microsaccades, tremor, and drift (reviewed in Ref. [89]). It suffices for the present purposes to simply emphasize that, even while actively fixating, a subject's eye is constantly moving. Most importantly, this motion is substantial when considering the cone mosaic, because a stimulus will be translated over many cones during even a brief bout of concentrated fixation (Fig. 3.4). To get a sense of the magnitude of this eye motion, a subject with good fixation (Fig. 3.4a) had a mean shift of 0.36 arcmin every 30 ms. If such shifts all went in one direction, as they sometimes do, this subject's drift would have moved a stimulus from one cone to its neighbor within 100 ms (this may be an overestimate, as the sampling rate for this data was low, 30 Hz, compared to other eye-tracking methods). Without continuous monitoring of eye position, and in the absence of microsaccades (rapid movements >2 arcmin), the only practical means of delivering stimuli to the same cone repeatedly is if stimuli are presented in rapid succession, less than 10 ms apart. Given the relatively slow photocurrent responses of cones, this will result in temporal summation and appear as a single stimulus [90, 91], limiting the utility of such a method. Because eye drift is akin to a random walk during any episode of fixation, stimuli presented over periods of time longer than 10 ms will frequently land on different cones. Video-based methods for tracking fixational eye motion in AO systems have been developed and are described below.

Eye motion is also present in anesthetized animals undergoing neuromuscular blockade, when the ordinarily suppressed cardiac and respiratory movement of the eye is released [92, 93]. Retinal motion under these conditions can occasionally be low, with excursions less than 10 μm, but this is still greater than the cone spacing in the fovea and perifovea (see Supplemental Fig. 2 in Ref. [94]). Thus, there is very little chance that a small stimulus presented at one location in visual space can land on the same cone under any fixation condition, whether in humans or in animals with paralyzed eye muscles.

3.4.2 *Chromatic Dispersion*

From a purely optical standpoint, the performance of a normal eye is sometimes considered to be relatively poor, as the lens and cornea introduce high-order aberrations that interfere with high-resolution imaging, both spatially and temporally [26, 95]. Unlike defocus and astigmatism, the high-order aberrations cannot be eliminated with standard corrective lenses. As mentioned earlier, AO ophthalmoscopy has been fairly successful in ameliorating the distortions introduced by such aberrations, and presently enables imaging of both rod and cone photoreceptors in vivo [96]. Once these aberrations have been corrected, there still remains the problem of chromatic dispersion. Simply put, a white point source of light will undergo dispersion by the ocular media and be projected onto the retina as a miniscule yet perceptible "rainbow," one with variably blurred colors. A means of accounting for chromatic dispersion of the eye is therefore a critical step. Without it, delivery of different wavelengths of light to a targeted location on the retina—at the micron scale of a single cone—is unfeasible. Such stimuli require chromatic correction because an infrared image is typically used as the reference image for AO correction, eye tracking, and target selection, whereas more visible wavelengths are used for stimulation. The shorter visible wavelengths have advantages especially for color vision studies: very high contrasts can be achieved, and wavelengths that optimally differentiate between cone types can be employed.

To correct for the chromatic dispersion of the eye, two optical parameters must be considered: longitudinal chromatic aberration (LCA) and transverse chromatic aberration (TCA). Multiwavelength light originating from a single point will land on the retina with poor focus due to LCA, and in different locations because of TCA. It has been shown that LCA in the human eye is relatively consistent between individuals, yet substantial [97, 98]. For instance, the focal difference between a commonly used wavelength for imaging (840 nm) and a wavelength near the peak of visual sensitivity (540 nm) is 1 diopter. Given this relationship (described by Eq. 5a in Ref. [98]), LCA can generally be corrected by appropriate static positioning of an instrument's optics for the wavelengths of interest, to bring them all into equal focus on the retina (Fig. 3.5a). It is worth noting that the LCA of some individuals may not fall on the population curve defined by Atchison and Smith [98]; one may wish to check for equal focus of imaging and stimulation channels by examining a retinal image containing a bipartite field of infrared and visible light.

TCA correction is more challenging because it depends on the position of the imaging beam relative to the pupil and to the achromatic axis of the eye [97, 99], and varies with retinal eccentricity. Thus, for each experiment, TCA must be corrected for a given pupil position and gaze direction. TCA is primarily caused by misalignment of the imaging beams relative to the eye's achromatic axis, along which TCA is zero by definition. However, finding the achromatic axis is difficult. It is known that the position of the achromatic axis relative to the pupil center is highly idiosyncratic, and so must be found empirically for every eye [100, 101]. Instead of trying to pinpoint the achromatic axis and aligning all the beams to it, the

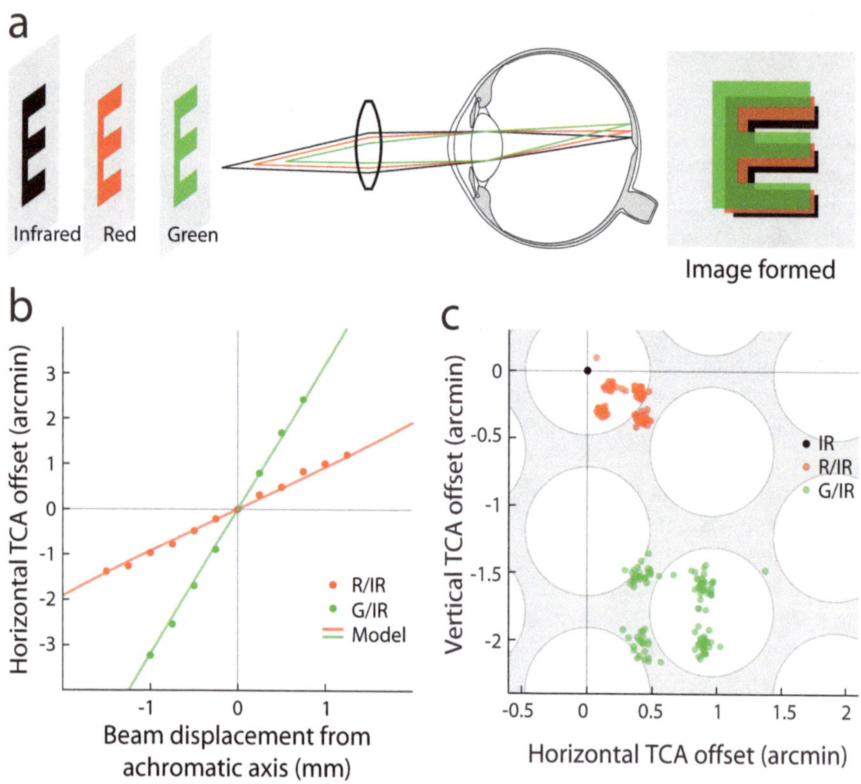

Fig. 3.5 Longitudinal and transverse chromatic dispersion in multiwavelength retinal imaging. (**a**) Schematic of how longitudinal dispersion is corrected in an AOSLO, by setting different focal distances for each wavelength channel (computed according to [98]); this leaves transverse image offsets on the retina that must be measured and compensated. (**b**) Transverse chromatic offset measurements made from retinal imaging during horizontal pupil displacements (dots, means of 20 measurements) are compared to offsets computed from a standard chromatic eye model (lines). Pupil displacements of 0.25 mm produce offsets in the green channel that are more than twice the size of typical foveal cones (~0.4 arcmin). (**c**) Frame-by-frame measurements of transverse offsets (relative to an infrared channel, IR, at zero) during sequential fixation on four corners of a 1° square. Background circles represent 5 μm diameter cones. Panels (**b**) and (**c**) adapted from Ref. [102]

problem can be solved more simply by measuring the offsets caused by TCA and beam misalignments in the resulting images directly [102]. Offsets measured this way are independent of the actual beam paths and the placement of the imaging detectors. Offsets are actually displayed on the retina, and are thus preserved as spatial information in the acquired images. Measurement of TCA offsets from retinal images fits well with offsets calculated from a standard chromatic eye model (Fig. 3.5b). TCA can also be demonstrated for small gaze shifts on a frame-by-frame basis (Fig. 3.5c). In this illustrative example, if TCA was not corrected and a cone was selected in the infrared channel for stimulation with a green spot, the stimulus would have landed on a different cone, about 2 cone diameters away.

Image-based TCA measurements have been validated psychophysically by comparing the offsets to a person's ability to assess small positional shifts of colored squares at the fovea. Because psychophysical thresholds in such tasks are lower than the sampling capacities of the cone mosaic (a perceptual feat usually called hyperacuity [103]), chromatic offsets measured this way ought to match those calculated from an image-based approach. Such a match was found; the average difference between psychophysical and image-based TCA offsets were ~8 arcsec, equivalent to 1 pixel in the cone images [102]. This result shows that image-based TCA measurements are functionally identical to that of conventional subjective TCA measurements [97, 99, 104]. One notable advantage of an image-based method is that TCA can be measured in peripheral retina where subjective methods fail, because visual acuity is not precise enough outside the fovea. From all of the foregoing, it should be clear that multiwavelength light delivery onto single cones in the living eye hinges on measuring and correcting TCA rapidly and with good spatial fidelity.

3.4.3 *Vascular Interference*

When the eye is thought of as an electronic camera, the fact that blood vessels and capillaries cast a sinewy net over the photoreceptors is overlooked. Except in the small avascular zone centered on the fovea, the vessel beds can cover more than half of the inner retinal surface area [105]. These vessels can interfere with light capture and retinal imaging in different ways. One is simply by casting shadows. Ordinarily this passes unnoticed, much in the way that the 5° hole in the retina created by the optic nerve is not perceived. Such a visual scotoma can be probed easily to reveal the perceptual gap in the lateral visual field of each eye. Similarly, scotomas associated with large vessels coursing out from the optic nerve head can also be mapped, when probed with fine enough test spots [106, 107]. These angioscotomas have even been shown to modify the local circuitry in primary visual cortex [108, 109]. A direct test of how shadowing by small blood vessels can raise thresholds by at least a factor of 2 has been demonstrated with vessel-targeted microperimetry [110]. From all of this evidence, there is little doubt that light being blocked by vessels perturbs the visual system at many levels, sometimes enigmatically so.

Shadowing is not the only form of light interference however. There is also light path distortion, arising from the clear cylindrical vessel walls and the pulsatile passage of blood cells through the narrow vessel lumen. These effects are more likely to operate on the scale of single cones. We and others, for instance, have noticed the transient alteration of reflected light from individual cones when a leukocyte passes over them [111–113]. Light reflecting back from a cone varies from bright to dark over time, and is sometimes displaced. This is reminiscent of the blue-field entoptic effect that allows one's own retinal blood cells to be seen [114]. The irregular shapes of leukocytes, especially when squeezing through capillaries, make the effect of light path distortion unpredictable. Nonetheless, these light changes have been used to map out the anatomical position of the vessels themselves [115–117].

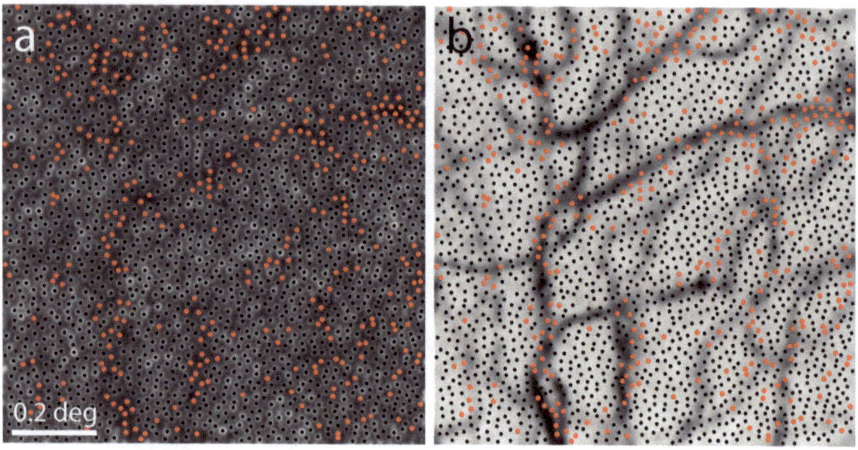

Fig. 3.6 Cones maps and retinal vasculature. (**a**) AOSLO cone image of a human retina, with gray levels scaled logarithmically to facilitate identification of poorly reflective cones. Cones brighter than the mean image reflectivity are marked with black dots ($n=1,833$), and those with reflectivity below the mean are indicated with red dots ($n=346$). (**b**) Vasculature map derived from motion contrast imaging [117] of same retinal area with cone centers from (**a**) superimposed, showing that most—but not all—dark cones are associated with blood vessels. Many other cones are situated under vessels but their reflectance is not appreciably affected. Adapted from Ref. [139]

A comparison of cone reflectance against an overlying vascular map reveals not only the effect of shadowing, as many of the darkest cones are underneath vessel lumens, but also that some of the darker cones lie where the wall of the vessel—which is not seen in the vessel maps—may be steering light away from the imaging detection path (Fig. 3.6). Comparison of the two panels in this figure make it evident that many of the capillaries are not visible with the AO-corrected infrared light. As one might imagine, vascular interference can influence the functional imaging of cones, particularly if threshold measurements are of interest. The psychophysical testing described in the next section was conducted after first making vessel maps, thereby allowing cones to be selected for microstimulation where the potential hazard of light interference was minimized.

3.4.4 *Characterization of Delivered Microstimuli*

As with any optical device, diffraction will limit the spatial resolution that can be achieved for any delivered light in an AOSLO. Because retinal images as well as stimuli are built up as pixels defined by the scanned laser, the images do not actually represent the light intensity profile of the beam landing on the retina. In confocal systems, out of focus light is discarded in image reconstruction, yet such light remains present in the stimulus itself. To get a better idea of the true geometry for micron-scale stimuli, the point-spread function (PSF) of the optical system needs to be taken into account. For a typical field size used in our psychophysical experiments (~1.2° square), the sampling resolution is high enough for each photoreceptor to be imaged within about 10 pixels (Fig. 3.7a, b). A stimulus smaller than this, defined in image pixels, can theoretically be placed within the visible margins of a single cone.

One can estimate the actual stimulus shape by convolving the stimulus defined in pixels with the PSF. With AO correction, the incident beam aperture (5.6 mm in the example figure) yields a PSF with full width at half maximum of 24 arcsec, calculated with a 543 nm stimulus wavelength. Expressed in image space, this corresponds to a diameter of 2.6 pixels, or about 1.9 μm on the retina. Thus, with optimal wavefront correction, a nominal 3×3 pixel stimulus convolved with the PSF will produce a light intensity profile where the 5 % intensity contour corresponds to an approximately circular area 7.3 pixels across (~5.3 μm on the retina), roughly matching the diameter of imaged cone apertures at 3.1° eccentricity (Fig. 3.7c). If we integrate the light falling within this 5 % intensity contour, it represents ~80 % of all light in the stimulus. Because cones vary in diameter with distance from the fovea, stimuli can be scaled accordingly to match the cone diameters, if the appropriate PSF is used. This is the first of two steps used in characterizing the delivered microstimulus.

With any psychophysical testing, repeated stimulation under controllable conditions is necessary. Consequently, the second step in characterizing light delivery must take into account the spatial delivery errors that occur over each set of stimulus trials, after fixational eye movements had been compensated for with real-time eye tracking. Tracking the motion of the retina while a subject fixates involves reading the incoming video raster, comparing select portions of the raster to a reference frame, and then, as the raster approaches the site on the retina where stimuli are to be delivered, predicting the movement of the eye just prior to stimulus delivery. Thus delivery accuracy depends, in part, on how far ahead in time the predicted location can be computed. Software operating at video frame rates in an AOSLO was developed to perform this eye tracking and stimulus delivery task [118]. In the current incarnation of this software, the prediction time can be as short as 3 ms, yielding a standard deviation of 0.15 arcmin in the stimulus delivery positional error [119].

To illustrate how the eye tracking enables psychophysics based on microstimulation to be realized, Fig. 3.7d shows the delivery accuracy of repeated trials of a 3×3 pixel stimulus onto a targeted cone. The subject fixated a small luminous target superimposed into the visual field through a pellicle beam splitter. TCA offsets were measured using the method described previously. The center of one cone was selected in the infrared channel as the stimulus location, and TCA offsets were used to compensate for the lateral displacement of the green light. In the movie recorded during each trial, a fiducial cross is written into the frame at the location of the delivered stimulus. From this, we can recover each stimulus delivery location. In the given example, 7 out of 22 deliveries landed exactly at the targeted image pixel, with the remaining 15 deliveries distributed over the immediately neighboring pixels. From a series of similar experiments, the average standard deviation of delivery jitter was 2.2 pixels (in both x and y image coordinates), representing about 1.6 μm

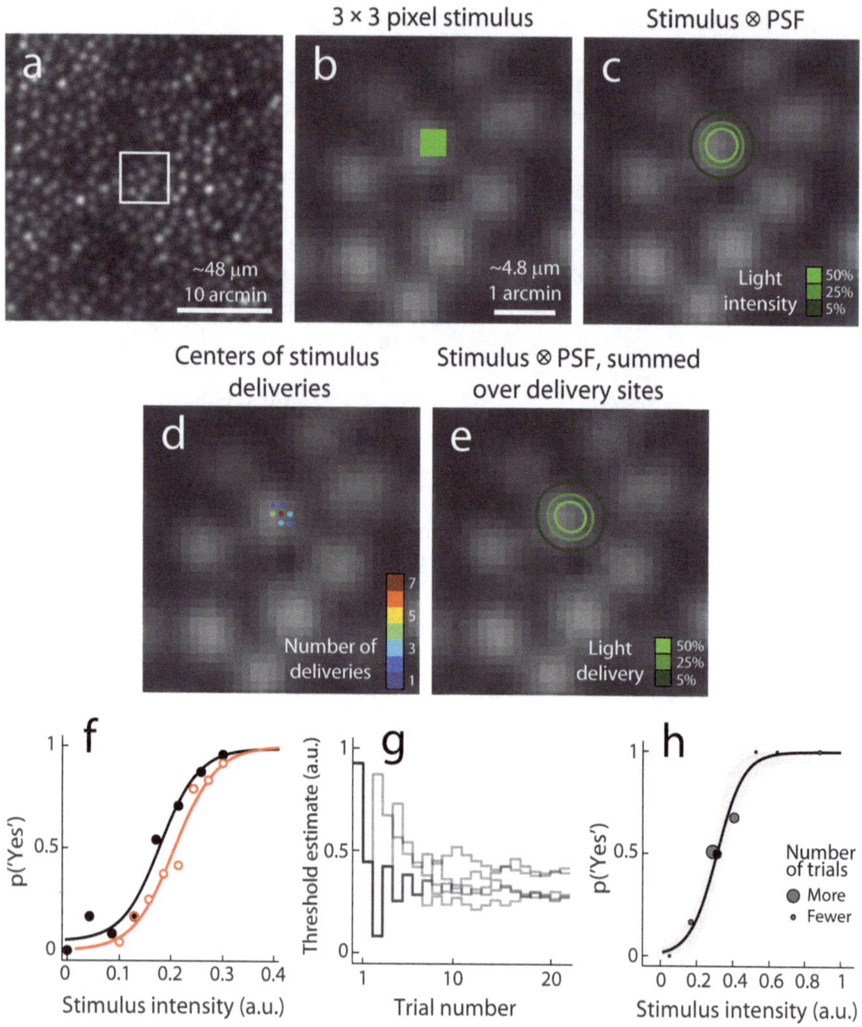

Fig. 3.7 Stimulus geometry, delivered light distribution, and psychometric results from cone-scale microstimulation. (**a**) AOSLO image of a human cone mosaic at 3.1° eccentricity, with outlined area scaled up in (**b–e**). (**b**) Cone reflectance profiles at this eccentricity span ~7 pixels, nearly 5 μm in diameter. Microstimuli are specified in image pixels; here, a 3×3 pixel square. (**c**) Light intensity delivered to the retina is estimated by convolving the stimulus geometry with the diffraction-limited point-spread function of the eye. Intensity contours show that the light spreads over a broader area than the 3×3 specification. (**d**) Plot of actual delivery locations of the stimulus center relative to the targeted cone for a 22-trial psychophysical run. Positional delivery errors in eye motion correction causes stimulus deliveries to be jittered from trial to trial. (**e**) Cumulative distribution of light delivery on the retina during the run in (**d**), derived from the diffraction-limited stimulus integrated over the actual delivery locations. (**f**) Psychometric frequency-of-seeing data and logarithmic fits from 2 subjects for the 3×3 pixel stimulus, obtained using a method-of-constant-stimuli approach ($n=20$ trials per stimulus intensity). Because of daily instrument fluctuations in light levels, stimulus intensity is given in arbitrary units (a.u.). Eccentricity was 2.1° (*black* data) and 2.5° (*red* data). (**g**) Five runs of a staircase approach to measure luminance increment thresholds from one subject, illustrating some variability in the final threshold estimate (at last trial). (**h**) Staircase data in (**g**) converted into a frequency-of-seeing psychometric function (*black line*) based on boot-strapped fits (*gray lines*). Panels (**a–e**) adapted from Ref. [120]

on the retina [120]. To make the final calculation of the light delivery profile, we sum the PSF-convolved nominal stimulus across the actual delivery locations (Fig. 3.7e), yielding the best estimate possible of the light distribution during one experiment. Integrating the light distribution within the 5 % intensity contour shows that 82 % of all the delivered light fell within this contour. Although this suggests that, even after repeated presentations, most of the delivered light was confined to a retinal area the size of a single cone, we note that this does not take into account uncontrollable light scatter (see below).

Under these conditions, we have found that robust psychophysical threshold functions can be measured for such cone-sized stimuli, whether using a classic method-of-constant-stimuli (Fig. 3.7f) or a Bayesian staircase approach (Fig. 3.7g, h). Out to about 5° eccentricity, subjects can be tested while targeting single cones. Beyond that eccentricity, we have found that more than one cone needs to be stimulated, at least for a luminance increment threshold task operating within the range of light intensity we can deliver [120]. This increase in threshold with eccentricity is consistent with prior studies using a constantly sized stimulus [121, 122]. Summation of input over multiple cones is likely to be required beyond 5° for stimuli to be effective at the perceptual level [123–125].

Further evidence that microstimuli can be utilized to probe single cones comes from the observation that thresholds are about 50 % higher when stimuli land between cones, a finding that is largely explained by a linear model of the geometry of cone light capture [120]. The Gaussian profile of a cone's light capturing ability predicts that coupling efficiency will be maximal at the cone's center and decline with increasing distance from the center. Such a prediction has been validated empirically at the level of the retina [86], at the main neural target of retinal projections, the lateral geniculate nucleus [94], and now at the perceptual level. The visual system is thus exquisitely sensitive to the exact position of delivered stimuli, down to the micron scale. Characterizing microstimuli as we have, and with robust psychometric measurements now feasible, we are nearly poised to address how color percepts originate from the activity of single cones.

3.4.5 Psychophysical Testing and Variability

The evidence from the previous section suggests that the delivery accuracy for cone-sized stimuli can be good enough to resolve the spatial grain of the photoreceptor mosaic. Given the difficulty of directing stimuli to exactly the same location on a cone repeatably in the living eye, it is clear that a major source of variability in any perceptual task will be caused by stimulus delivery errors. Is any other source of psychophysical variability even detectable given the large effect of positional variability?

One way to address this issue is to test the same group of cones over multiple days. By measuring luminance increment thresholds for each cone using interleaved trials, one can control for factors such as daily variation in subject performance and in instrument light levels. Threshold values can be normalized against the mean threshold of the group to see how much variation occurs simply with repeated measurements, whether thresholds are consistent from day to day, and, perhaps most interestingly, whether cones exhibit different intrinsic thresholds. Figure 3.8 illustrates a few cone triplets that were studied this way, using the staircase method. In Fig. 3.8a we measured thresholds multiple times on three separate days for one triplet and found no difference in threshold among any of the cones. It is evident that repeated measurements of thresholds are inherently noisy, as they can vary by as much as 60 % from one measurement to the next. Some of this variation must be noise associated with doing psychophysical threshold tasks, and some is undoubtedly due to positional delivery error (an error that includes transient TCA shifts that cannot be measured during the course of the experiment). An examination of threshold versus delivery location revealed that about 50 % of the variance in the measured thresholds come from experimental errors, if it is assumed that cone thresholds are identical [120].

Physiological data, however, have shown that cone thresholds may not all be the same. As we noted earlier, when the functional weighting of cones were measured in macaque retina, it was found that each retinal ganglion cell was receiving input from a handful of cones expressing a range of synaptic weights [86, 87]. Such a result raises the question of whether differential cone weighting in the retina could be propagated to the perceptual level. Figure 3.8b provides evidence that such functional weighting can be measured psychophysically. Here, a triplet of cones was measured repeatedly over several days, with one cone having a significantly higher threshold than the other two. The high-threshold cone needed about 40 % more light in the stimulus in order to be detected. Although there is no way to determine if all the cones in this triplet are connected to the same retinal ganglion cell, this result suggests that differential functional weighting—either from cones to ganglion cells or between two ganglion cells—can be detected perceptually. The important point is to realize that no two L or M cones can be assumed to have the same luminance increment threshold.

Additional evidence that microstimulation can generate cone-specific responses comes from the occasional encounter with a candidate S cone. The stimulus wavelength band used during the two previous examples was 543 ± 11 nm, a range that is absorbed equally well by L and M cone opsins. For an S cone to produce an equivalent response with this wavelength, it would need ~400 times more light than an L or M cone [37]. This amount of light exceeds the range our current AOSLO can deliver. So, under these experimental conditions, if an S cone was probed, the subject would respond "not seen" during most trials and drive the staircase above the deliverable range. One of the cones in the triplet illustrated in Fig. 3.8c had just such an outcome. Given that the other two cones had reliable thresholds measured across two days, this result suggests that the cone with unmeasurable threshold was an S type. Additional data given in the next section confirms that such psychophysical outcomes can be associated with S cones.

Although prone to a high degree of variability, most likely caused by positional errors in stimulus delivery, reproducible perceptual data can be acquired with cone-targeted microstimulation. The fact that relative increment thresholds between neighboring cones are consistent from day to day, and that they can be consistently distinct, implies that the perceptual discriminations are essentially driven by signals arising from individual cones.

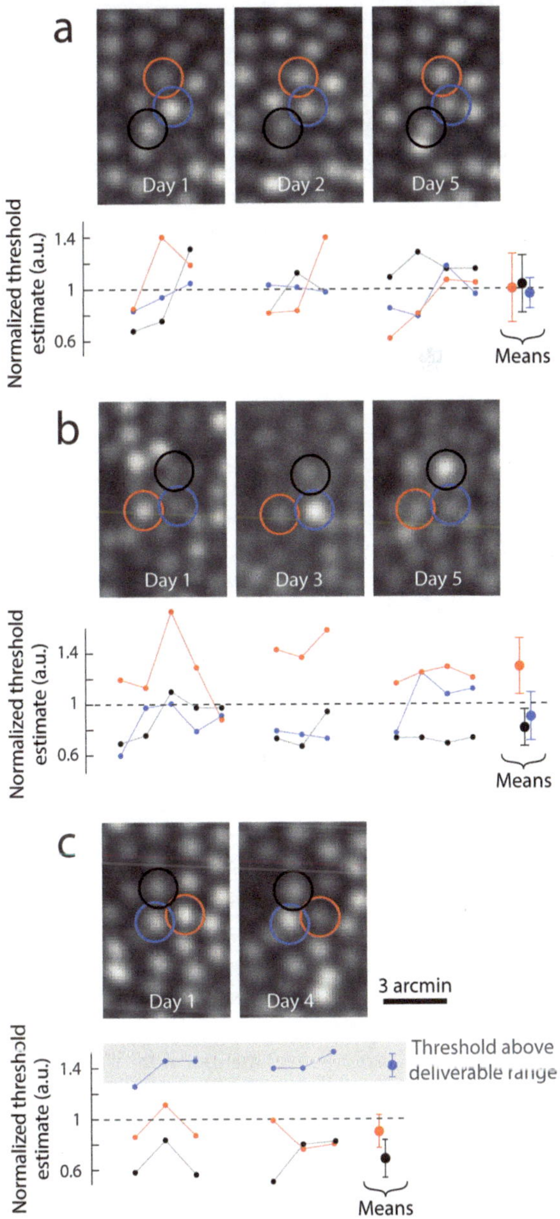

Fig. 3.8 Luminance increment thresholds measured with microstimulation are associated with specific cones. The AOSLO images show the human cone triplets studied over multiple days in different subjects. Cones targeted for stimulation and single threshold measurements from each are coded by color. Data are grouped by test day and were normalized to the mean threshold of the triplet (to control for small day-to-day changes in stimulus light levels). Each cone was tested 3–5 times per experiment (small dots), using 543 ± 11 nm light, a wavelength equally absorbed by L and M cones. Mean single-cone thresholds (±1 SD) across all days are shown on the right within each graph. In (**a**) all cones had similar increment thresholds, while in (**b**) the cone circled in red had consistently higher thresholds than the other two cones in its triplet. In (**c**) the cone circled in blue had thresholds beyond the range of deliverable light (indicated by shading), which suggests it is an S cone, given the stimulus conditions. Adapted from Ref. [140]

3.5 Psychophysical Cone Imaging

Having discussed the main experimental constraints that need to be faced when trying to extract perceptual data from microstimulation, we now review a number of previous studies, as well as some work in progress, that have revealed cone-specific responses in the human retina. Although considerable insights have been made in relating inferred cone mosaics to color phenomenon without the use of AO-based imaging (e.g., [66, 126, 127]), we focus here on work that has relied on imaged cone mosaics to strategically test for cone-driven percepts.

3.5.1 *S Cone Testing*

Once it was appreciated that S cones are relatively uncommon yet regularly spaced members of the cone mosaic, it was realized that it may be possible to map them out psychophysically by probing with small S-cone-isolating stimuli. Williams et al. [128] set out to do just that. Presenting a 1.1 arcmin spot of 420 nm light for 50 ms over a grid of locations at and around the fovea, they measured thresholds in the presence of a background light that adapted out L and M cone responses [128]. Sensitivity for these stimuli were relatively low at the center of fixation, and increased by about 1 log unit within 15 arcmin of the fovea. Control experiments showed that this sensitivity profile was not due to light absorption by the macular pigment lying in front of the photoreceptors. Such perceptual data is consistent with the known absence of S cones in the foveola [41, 42]. More tellingly, the sensitivity terrain around the fovea contained peaks and valleys with about the same spacing as that subsequently observed histologically for S cones. Repeat measurements over a 2 year timespan suggested that the retinal location of these S cone sensitivities were stable. A model of light capture under their experimental conditions, assuming single S cones were the detectors, was also consistent with the psychometric data produced by varying stimulus intensities. Taken together, the most parsimonious explanation of their results was that S cones could indeed be mapped, though in small numbers. Such experiments are taxing, as the authors noted, and having to probe randomly across a coarse grid, the chances of stimuli landing directly on a cone are low. Nonetheless, this early study held promise that single cone activity could be detected perceptually, and helped to encourage the development of techniques for imaging cones and enable cone-targeted stimulation.

3.5.2 *Dysfunctional Cone Testing*

After AO-corrected retinal imaging became available, finer spatial testing of perception could be achieved because the stimuli impinging on the retina were no longer blurred by optical aberrations. In a situation somewhat converse to the S cone mapping, sensitivity losses might be expected in cone mosaics with lacunae—if stimuli were truly small enough. To learn if such microscotomas could be detected, a retina was studied in a deuteranopic subject that had dysfunctional cones due to a mutant M photopigment [129]. In AO images from this subject, about 30 % of the cones appeared relatively dark compared to normals, suggesting that the mutation either damaged the cones or lead to outright cone loss. Cone-sized stimuli of 550 nm were flashed for 46 ms at various locations all 0.5° from a fixation spot, in an effort to see if a suitable proportion of these AO-corrected stimuli would not be perceived. Comparison of the deuteranope's frequency-of-seeing curves to those of control subjects showed that mutant cones did cause a lower sensitivity and slope in the curve, as predicted by a model of cone loss. Larger stimuli, such as those used in clinical exams, showed no difference between the M-cone-compromised subject and controls. Microstimuli, therefore, appeared to unmask microscotomas. This was another piece of evidence telling us that perceptual effects can rest on the activity—or inactivity in this case—of single cones.

3.5.3 Cone Spectral Identification and Match with Absorptance Imaging

In a previous section, we elaborated on the considerations that need to be made in developing techniques for single cone microstimulation. One of the obvious interests in these techniques is to learn if individual cones can be classified by spectral type psychophysically. Given the many stages of post-receptoral processing that intervene between cones and perception, it is not necessarily the case that activation of a single cone will lead to a veridical representation of that cone's spectral class. Retinal circuitry imposes an opponency signal between L and M cones that may variably dilute the signal from either of these cone types (see Chaps. 2 and 4). Cortical circuitry—about which much less is known (see Chap. 7)—may also be altering the sensory input. To put such psychophysical experiments on firmer footing, therefore, a comparison with an objective method such as absorptance imaging seems sensible. Here we describe our initial efforts in comparing the biophysical and psychophysical mapping of the same set of cones.

The experiment was designed to optimize distinguishing L from M cones. With a sensitivity difference between these two cones of ~0.05 log units at their peak wavelengths [37], it was unlikely that comparing responses between stimuli of 561 and 531 nm would yield reliable classification given the variability of measuring thresholds with microstimulation (Fig. 3.8). Instead, the strategy was to isolate responses from L cones alone, with M cones identified, in part, by default. First, we used a 710 nm spot stimulus for measuring increment thresholds, as this wavelength represents the maximum sensitivity difference between L and M cones. Second, we presented the microstimuli against a constant L-cone-isolating background of 470 nm light [130]. Taking the remaining light in the imaging field into

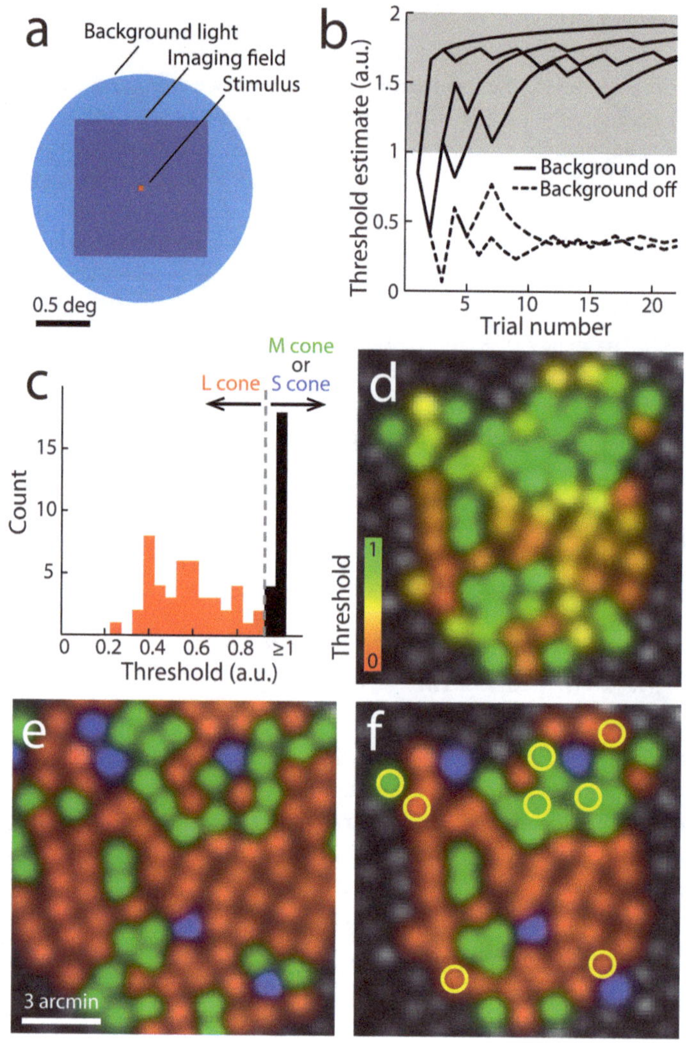

Fig. 3.9 Comparison of functional cone classification methods in the living eye. (**a**) Schematic of L cone isolating conditions from the subject's perspective. A 2° background light (peak wavelength = 470 nm) presented in Maxwellian view is used to counterbalance light adaptation caused by instrument light leak in the imaging field. This *blue* light adapts M and S cones preferentially, leaving L cones >20 times more sensitive to a 710 nm test flash (small *red square*). (**b**) Example data used to classify an M cone (1.5° eccentricity). With the adapting field turned on, intensity staircases (*black lines*) consistently climbed above the upper limit of deliverable light intensity (*shaded area*). With the background light switched off, staircases yielded measurable thresholds (*green lines*), indicating that the cone was M type. L cones, in contrast, produced such thresholds when the background light was on. (**c**) Histogram of mean increment thresholds for a contiguous array (see **d**) of cones tested as shown in (**a**). *Dashed line* represents the threshold cutoff that produced the best agreement between perceptual and biophysical classification methods; for this subject, cones with a mean threshold less than 0.95 arbitrary units (a.u.) were classified as L type. (**d**) Increment threshold data for 67 cones, color-coded by the mean threshold values in (**c**). (**e**) Pseudocolor images of the same cone mosaic as shown in (**d**), but derived from AOSLO imaging and retinal densitometry. *Red, green*, and *blue* labels correspond to L, M, and S cones, respectively. (**f**) A comparison of the cones classified biophysically with those measured psychophysically revealed a mismatch of 8 (*yellow circles*) out of 68 cones between the two methods. Adapted from Ref. [134]

account (Fig. 3.9a), these stimulus conditions are predicted to have an L:M sensitivity ratio of about 24:1, and S cones would be unresponsive [131]. Thus, over the range of deliverable light intensity (scaled 0–1 in arbitrary units), if the average L cone had a threshold above 0.05, then any M or S cones encountered would register thresholds greater than 1. To avoid issues concerning subject fatigue that might alter threshold, 2 or 3 cones were tested simultaneously by randomly interleaving trials, with independent staircases running on each selected cone. When cones with off-scale thresholds were encountered, the adapting background light was then switched off and the threshold remeasured. If the sensitivity recovered, such cones were classified as M type (Fig. 3.9b).

Increment thresholds measured under L-cone-isolating conditions for a group 68 cones mapped over several days showed a bimodal distribution, with a wide peak corresponding to candidate L cones and sharp peak near and above 1 representing likely M or S cones (Fig. 3.9c). The broad range of L cone thresholds arises from several factors. Two we have already highlighted in this chapter: intrinsic variability of threshold among cones, and positional noise in stimulus delivery. A third factor is a contribution from cones surrounding the one targeted for stimulation. L and M cones are electrically coupled via gap junctions [132, 133]. A Monte Carlo simulation of the effect of varying the composition of the surrounding cones suggests that ~33 % of the threshold range could be due to adaptation state differences in those surrounding cones [134]. One can appreciate the possible impact of surrounding cones in a map where each cone is color-coded by its mean threshold value, as there are a number of instances where a gentle gradation of threshold differences exists between cones (Fig. 3.9d). On the whole, however, such a map is more remarkable for the many cases where neighboring cones have sizable threshold differences—often shifting from a value 0.5 to "unseen" from one cone to the next.

To confirm that microstimulation can classify cones psychophysically, the same set of cones in this subject was also classified using absorptance imaging. The method used here was modified because a scanning AO system was used rather than a flood AO system, but the underlying principles were the same. Dynamic differential cone bleaching was performed under two different conditions to distinguish S from L or M cones, and separately, L from M cones [135]. The resulting cone map revealed L and M cones in a 2.1:1 ratio (Fig. 3.9e). To make a comparison between the threshold maps and absorptance images, the former's thresholds need to be converted into a true 3-cone classification. This was done using Cohen's kappa coefficient, a statistic of agreement for categorized data [136]. For this subject, a threshold criterion of 0.95 (dashed line in Fig. 3.9c) yielded the best agreement between the two maps for L and M cones (Fig. 3.9f). The four S cones in the field, which are more reliably identified by differential bleaching [61, 63, 135], all had thresholds exceeding 1, as expected for these stimuli. Of the remaining L and M cones, there was 88 % agreement between the two methods—a reasonably good match between subjective and objective methods of cone classification. Under these stimulus conditions, at least near the fovea, signals from individual cones do indeed seem to propagate from the retina to the perceptual level, carrying their spectral identity with them. Such data effectively redefine the groundwork for experiments asking elementary questions about the cellular basis of color percepts, a topic we touch on next.

3.5.4 Color Appearance of Microstimuli

Early efforts at using small stimuli and standard optical correction to probe color appearance revealed that subjects often required a range of hue options to categorize monochromatic lights (reviewed in Ref. [76]). A genuine impediment for an adequate interpretation of these results is that the underlying cone mosaics were not known, for both the relative numbers and the spatial arrangement of the cones are likely to produce different color signals depending on where the stimuli land from trial to trial. In addition, uncorrected optical blur will broaden any stimuli, and particularly at the fovea this would activate many more than one cone. Both of these hurdles were removed in a study that used AO correction along with classified cone mosaics to see how color sensations were generated [70]. Briefly flashed microstimuli of ~0.3 arcmin were presented in a retinal locus near 1° where the cones had been mapped by absorptance imaging. The stimuli varied in two important ways: by wavelength (500, 550, or 600 nm) and by position (5 sites within a 14 arcmin square, with fixational eye movements uncompensated). These parameters allowed the authors to distinguish between the influence of wavelength versus cone composition on color appearance. All subjects required white plus up to seven hues to categorize their percepts. The range of hues was generally independent of stimulus wavelength, but did depend on L:M cone ratio. When the ratio was weighted toward M cones, more greenish hues and fewer reddish hues were needed to describe the colors, and vice versa in subjects with more L cones. Blue and purple categories were also required, which the authors suggest may arise through strongly activated M cones mimicking the L:M excitation ratio of bluish light.

Because this result implies that different color sensations originate from the stimulation of one cone, the role of the surrounding cones naturally comes into question. If one cone was stimulated within a small field of identical cones, the L/M opponent mechanism may not be driven and a "white" response would be predicted (and was indeed frequently reported). A statistical approach was pursued to see if this and the other reported hues could be explained by a stimulated cone surrounded by different proportions of other cone types [73]. The resulting model broadly captured the color-naming data from Hofer et al. [70], and is consistent with the idea that the contributions of individual cones to color percepts is dependent to some degree on the local arrangement of cones. However, the psychophysical data could never be predicted exactly because the precise location of the stimulus on a trial-by-trial basis was not known.

To correlate each trial's response with the cone being tested requires cone-targeted microstimulation, like what we have already described. Preliminary data from such an experiment also revealed a family of color appearances, even when the same cone

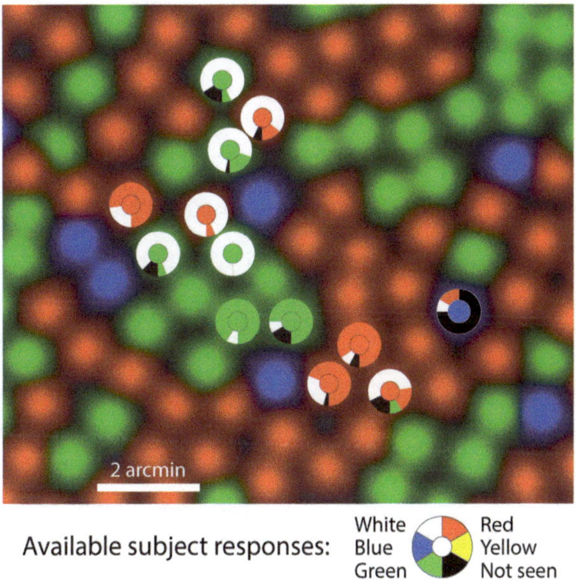

Fig. 3.10 Comparison of cone type and color naming responses. A field of human cones classified by AOSLO absorptance imaging (as described in Fig. 3.9) was tested for color categorization with 0.5 s retinally stabilized flashes of 543 nm cone-sized microstimuli, in the presence of a neutral white background. The subject was allowed to respond in one of six ways (shown in legend), and the proportion of the responses are illustrated in the circular histograms surrounding each tested cone (20 trials apiece). The data shown here are compiled from multiple sessions where several cones were stimulated with interleaved trials. In most instances, stimulation of L cones lead to *red* or *white* responses, while *green* or *white* was reported when M cones were stimulated. *Blue* or *yellow* percepts were never reported, even for the one S-cone tested (where most stimuli were not seen). In this field, the L:M cone ratio was 1.8:1, eccentricity = 1.5°. Adapted from Ref. [141]

was stimulated repeatedly with 543 nm light on a neutral white background (Fig. 3.10). Veridical percepts would appear green in this paradigm only if the stimulus wavelength was mediating the color sensation. In the results shown here, the subject used only red, green, white, and "not seen" among the available response categories, which included blue and yellow. Notably, in trials when a color other than white was reported, L cones most frequently lead to red responses, and M cones to reports of green, with a high consistency in these reports. White responses were more common than in the subject with a similar L:M cone ratio in Hofer et al. [70], but this may have been due to differences in methods. More cones will need to be tested to learn if surrounding cones have any influence on perceived color from a targeted cone. Several of the cones were tested again on separate days, and often yielded the same responses. The main indications from this work are that color appearance is testable at the cellular scale and depends on the cone being stimulated. Because the underlying cell types that are processing these signals are unknown for this task (e.g., midget or parasol retinal ganglion cells?), a more comprehensive picture of how color percepts emerge from the retina still awaits further work.

3.6 Conclusions and Caveats

The central points of this chapter bear on how the cone mosaic in humans and primates can be fruitfully explored, and what factors might guide the interpretation of experimental results. Here we summarize those points:

- Light absorption by photopigments can be used to objectively map cone arrays in a variety of imaging modalities.
- Cone composition varies widely from subject to subject, and regionally within a single retina; random arrangement cannot be assumed.
- The functional weighting of each cone is not a constant; at both the ganglion cell and perceptual levels, the strength of each cone's input varies.
- Cellular-scale testing in vivo requires compensation of the eye's optics, motion, and vasculature for the most uncompromised delivery of microstimuli.
- Color percepts that ensue from cone-sized stimuli appear to be most influenced by signals from individual cones.

Given the psychophysical results described in this chapter, it may be tempting to think that "single-cone psychophysics" has been realized. We have not used the phrase, for it carries the unfortunate implication that a percept is being determined by light absorption in only one cone, and nothing more. There is no direct evidence that only one cone is being activated by microstimuli, and it seems unlikely this can ever be garnered from in vivo studies. In fact, by optical diffraction alone, the delivered light profile cannot be restricted to one cone. The point-spread function, calculated from a model eye, includes a first Airy ring with an amplitude of ~1 % of the central peak. From the light intensity profile shown in Fig. 3.7c, this would mean that the immediately neighboring cones situated around a cone targeted for stimulation would have a chance to absorb a few percent of the total light. Intraocular scatter occurring along the light path in front of the retina would divert another small fraction of light onto these surrounding cones (reviewed in Ref. [137]). Together with the point-spread function, a more realistic light distribution profile might have ~20 % of all the light falling outside the diameter of the inner segment of a targeted cone. It is important to realize, however, that this skirt of light is distributed over a very broad area and would activate most cones weakly. Some of this activity may be modulating percepts to varying degrees, depending on stimulus conditions, as we have seen.

The light profile itself is not the only factor that affects percepts resulting from a micron-scaled stimulus. Inner retinal wiring, ongoing levels of activity in the cones, varying functional weighting from cones to ganglion cells, and which ganglion cell type is actually propagating the signal to cortex for a particular stimulus condition—these all come into play and are the objects of future research. Functional maps of the cone mosaic will ultimately come in different flavors, given each cone's divergent input to many cell types as well as the stimulus conditions being faced. For psychophysicists, AO-corrected microstimulation may become a useful tool for making such maps, because it simply allows researchers to pour more light into any one cone of interest. Like drops hitting an ocean, they make ripples we can occasionally see—were it the visual system, the drops would sometimes appear filled with color.

Acknowledgements We thank J. K. Bowmaker, K. S. Bruce, E. J. Chichilnisky, G. D. Field, J. D. Mollon, and B. Schmidt for generously providing materials for figures. For improving the text, we are grateful to K. S. Bruce, T. W. Kraft, M. S. Loop, and A. S. McKeown. Our work has been supported by the National Eye Institute (L.C.S., W.S.T., A.R.), the Eyesight Foundation of Alabama (L.C.S.), Fight for Sight (R.S.), the American Optometric Foundation (W.S.T.), and German Research Council grant Ha 5323/5-1 (W.M.H.). Ramkumar Sabesan holds a Career Award at the Scientific Interfaces from the Burroughs Wellcome Fund.

References

1. Thoen HH, How MJ, Chiou TH, Marshall J. A different form of color vision in mantis shrimp. Science. 2014;343(6169):411–3.
2. Curcio CA, Sloan KR, Kalina RE, Hendrickson AE. Human photoreceptor topography. J Comp Neurol. 1990;292(4):497–523.
3. Williams DR. Imaging single cells in the living retina. Vision Res. 2011;51(13):1379–96.
4. Roorda A. Adaptive optics for studying visual function: a comprehensive review. J Vis. 2011;11(7).
5. Berendschot TT, DeLint PJ, van Norren D. Fundus reflectance—historical and present ideas. Prog Retin Eye Res. 2003;22(2):171–200.
6. Rodieck RW. The first steps in seeing. Sunderland, MA: Sinauer; 1998.
7. Packer OS, Williams DR, Bensinger DG. Photopigment transmittance imaging of the primate photoreceptor mosaic. J Neurosci. 1996;16(7):2251–60.
8. Enoch JM, Tobey FL. Vertebrate photoreceptor optics. Berlin: Springer; 1981.
9. Marcos S, Burns SA. Cone spacing and waveguide properties from cone directionality measurements. J Opt Soc Am A Opt Image Sci Vis. 1999;16(5):995–1004.
10. Vohnsen B, Iglesias I, Artal P. Guided light and diffraction model of human-eye photoreceptors. J Opt Soc Am A Opt Image Sci Vis. 2005;22(11):2318–28.
11. Lakshminarayanan V, Enoch JM. Biological waveguides. In: Bass M, Enoch JM, Lakshminarayanan V, editors. Handbook of optics. New York: McGraw-Hill; 2010.
12. Vohnsen B. Directional sensitivity of the retina: a layered scattering model of outer-segment photoreceptor pigments. Biomed Opt Express. 2014;5(5):1569–87.
13. Stiles CW, Crawford BH. The luminous efficiency of rays entering the eye pupil at different points. Proc R Soc Lond B Biol Sci. 1933;112:428–50.
14. Applegate RA, Lakshminarayanan V. Parametric representation of Stiles-Crawford functions: normal variation of peak location and directionality. J Opt Soc Am A. 1993;10(7):1611–23.
15. Chen B, Makous W. Light capture by human cones. J Physiol. 1989;414:89–109.
16. MacLeod DI, Williams DR, Makous W. A visual nonlinearity fed by single cones. Vision Res. 1992;32(2):347–63.
17. Chen B, Makous W, Williams DR. Serial spatial filters in vision. Vision Res. 1993;33(3):413–27.
18. Gao W, Cense B, Zhang Y, Jonnal RS, Miller DT. Measuring retinal contributions to the optical Stiles-Crawford effect with optical coherence tomography. Opt Express. 2008;16(9):6486–501.
19. Jonnal RS, Besecker JR, Derby JC, Kocaoglu OP, Cense B, Gao W, Wang Q, Miller DT. Imaging outer segment renewal in living human cone photoreceptors. Opt Express. 2010;18(5):5257–70.
20. Jonnal RS, Kocaoglu OP, Zawadzki RJ, Lee SH, Werner JS, Miller DT. The cellular origins of the outer retinal bands in optical coherence tomography images. Invest Ophthalmol Vis Sci. 2014;55(12):7904–18.
21. Miller DT, Williams DR, Morris GM, Liang J. Images of cone photoreceptors in the living human eye. Vision Res. 1996;36(8):1067–79.

22. Wade A, Fitzke F. In vivo imaging of the human cone-photoreceptor mosaic using a confocal laser scanning ophthalmoscope. Lasers Light Ophthal. 1998;8(3):129–36.
23. Vohnsen B, Iglesias I, Artal P. Directional imaging of the retinal cone mosaic. Opt Lett. 2004;29(9):968–70.
24. Pircher M, Baumann B, Gotzinger E, Hitzenberger CK. Retinal cone mosaic imaged with transverse scanning optical coherence tomography. Opt Lett. 2006;31(12):1821–3.
25. Porter J, Queener H, Lin J, Thorn K, Awwal A. Adaptive optics for vision science. Hoboken, NH: Wiley-Interscience; 2006.
26. Liang J, Williams DR, Miller DT. Supernormal vision and high-resolution retinal imaging through adaptive optics. J Opt Soc Am A Opt Image Sci Vis. 1997;14(11):2884–92.
27. Roorda A, Romero-Borja F, Donnelly 3rd WJ, Queener H, Hebert T, Campbell M. Adaptive optics scanning laser ophthalmoscopy. Opt Express. 2002;10(9):405–12.
28. Rushton WA. Pigments and signals in colour vision. J Physiol. 1972;220(3):1P-P
29. Marks WB, Dobelle WH, Macnichol Jr EF. Visual pigments of single primate cones. Science. 1964;143(3611):1181–3.
30. Baylor DA, Nunn BJ, Schnapf JL. Spectral sensitivity of cones of the monkey Macaca fascicularis. J Physiol. 1987;390:145–60.
31. Schnapf JL, Kraft TW, Nunn BJ, Baylor DA. Spectral sensitivity of primate photoreceptors. Vis Neurosci. 1988;1(3):255–61.
32. Schnapf JL, Kraft TW, Baylor DA. Spectral sensitivity of human cone photoreceptors. Nature. 1987;325(6103):439–41.
33. Dartnall HJ, Bowmaker JK, Mollon JD. Human visual pigments: microspectrophotometric results from the eyes of seven persons. Proc R Soc Lond B Biol Sci. 1983;220(1218):115–30.
34. Stockman A, Sharpe LT, Merbs S, Nathans J. Spectral sensitivities of human cone visual pigments determined in vivo and in vitro. Methods Enzymol. 2000;316:626–50.
35. Bowmaker JK, Dartnall HJ. Visual pigments of rods and cones in a human retina. J Physiol. 1980;298:501–11.
36. Nork TM, McCormick SA, Chao GM, Odom JV. Distribution of carbonic anhydrase among human photoreceptors. Invest Ophthalmol Vis Sci. 1990;31(8):1451–8.
37. Stockman A, Sharpe LT. The spectral sensitivities of the middle- and long-wavelength-sensitive cones derived from measurements in observers of known genotype. Vision Res. 2000;40(13):1711–37.
38. Bowmaker JK, Dartnall HJ, Mollon JD. Microspectrophotometric demonstration of four classes of photoreceptor in an old world primate, Macaca fascicularis. J Physiol. 1980;298:131–43.
39. Marc RE, Sperling HG. Chromatic organization of primate cones. Science. 1977;196(4288):454–6.
40. de Monasterio FM, McCrane EP, Newlander JK, Schein SJ. Density profile of blue-sensitive cones along the horizontal meridian of macaque retina. Invest Ophthalmol Vis Sci. 1985;26(3):289–302.
41. Wikler KC, Rakic P. Distribution of photoreceptor subtypes in the retina of diurnal and nocturnal primates. J Neurosci. 1990;10(10):3390–401.
42. Curcio CA, Allen KA, Sloan KR, Lerea CL, Hurley JB, Klock IB, Milam AH. Distribution and morphology of human cone photoreceptors stained with anti-blue opsin. J Comp Neurol. 1991;312(4):610–24.
43. Lennie P, Movshon JA. Coding of color and form in the geniculostriate visual pathway. J Opt Soc Am A Opt Image Sci Vis. 2005;22(10):2013–33.
44. Kolb H, Dekorver L. Midget ganglion cells of the parafovea of the human retina: a study by electron microscopy and serial section reconstructions. J Comp Neurol. 1991;303(4):617–36.
45. Dacey DM. The mosaic of midget ganglion cells in the human retina. J Neurosci. 1993;13(12):5334–55.

46. Wassle H, Grunert U, Martin PR, Boycott BB. Immunocytochemical characterization and spatial distribution of midget bipolar cells in the macaque monkey retina. Vision Res. 1994;34(5):561–79.
47. Wassle H, Grunert U, Rohrenbeck J, Boycott BB. Cortical magnification factor and the ganglion cell density of the primate retina. Nature. 1989;341(6243):643–6.
48. Calkins DJ, Sterling P. Absence of spectrally specific lateral inputs to midget ganglion cells in primate retina. Nature. 1996;381(6583):613–5.
49. Dacey DM, Lee BB, Stafford DK, Pokorny J, Smith VC. Horizontal cells of the primate retina: cone specificity without spectral opponency. Science. 1996;271(5249):656–9.
50. Jusuf PR, Martin PR, Grunert U. Synaptic connectivity in the midget-parvocellular pathway of primate central retina. J Comp Neurol. 2006;494(2):260–74.
51. Smith VC, Lee BB, Pokorny J, Martin PR, Valberg A. Responses of macaque ganglion cells to the relative phase of heterochromatically modulated lights. J Physiol. 1992;458:191–221.
52. Lee BB, Kremers J, Yeh T. Receptive fields of primate retinal ganglion cells studied with a novel technique. Vis Neurosci. 1998;15(1):161–75.
53. Martin PR, Lee BB, White AJ, Solomon SG, Ruttiger L. Chromatic sensitivity of ganglion cells in the peripheral primate retina. Nature. 2001;410(6831):933–6.
54. Reid RC, Shapley RM. Space and time maps of cone photoreceptor signals in macaque lateral geniculate nucleus. J Neurosci. 2002;22(14):6158–75.
55. De Monasterio FM, Gouras P. Functional properties of ganglion cells of the rhesus monkey retina. J Physiol. 1975;251(1):167–95.
56. Lankheet MJ, Lennie P, Krauskopf J. Distinctive characteristics of subclasses of red-green P-cells in LGN of macaque. Vis Neurosci. 1998;15(1):37–46.
57. Diller L, Packer OS, Verweij J, McMahon MJ, Williams DR, Dacey DM. L and M cone contributions to the midget and parasol ganglion cell receptive fields of macaque monkey retina. J Neurosci. 2004;24(5):1079–88.
58. Buzas P, Blessing EM, Szmajda BA, Martin PR. Specificity of M and L cone inputs to receptive fields in the parvocellular pathway: random wiring with functional bias. J Neurosci. 2006;26(43):11148–61.
59. Silveira LC, Perry VH. The topography of magnocellular projecting ganglion cells (M-ganglion cells) in the primate retina. Neuroscience. 1991;40(1):217–37.
60. Grunert U, Greferath U, Boycott BB, Wassle H. Parasol (P alpha) ganglion-cells of the primate fovea: immunocytochemical staining with antibodies against GABAA-receptors. Vision Res. 1993;33(1):1–14.
61. Roorda A, Williams DR. The arrangement of the three cone classes in the living human eye. Nature. 1999;397(6719):520–2.
62. Roorda A, Metha AB, Lennie P, Williams DR. Packing arrangement of the three cone classes in primate retina. Vision Res. 2001;41(10-11):1291–306.
63. Hofer H, Carroll J, Neitz J, Neitz M, Williams DR. Organization of the human trichromatic cone mosaic. J Neurosci. 2005;25(42):9669–79.
64. Neitz M, Balding SD, McMahon C, Sjoberg SA, Neitz J. Topography of long- and middle-wavelength sensitive cone opsin gene expression in human and Old World monkey retina. Vis Neurosci. 2006;23(3-4):379–85.
65. Cicerone CM, Nerger JL. The relative numbers of long-wavelength-sensitive to middle-wavelength-sensitive cones in the human fovea centralis. Vision Res. 1989;29(1):115–28.
66. Vimal RL, Pokorny J, Smith VC, Shevell SK. Foveal cone thresholds. Vision Res. 1989;29(1):61–78.
67. Wesner MF, Pokorny J, Shevell SK, Smith VC. Foveal cone detection statistics in color-normals and dichromats. Vision Res. 1991;31(6):1021–37.
68. Krauskopf J. Color appearance of small stimuli and the spatial distribution of color receptors. J Opt Soc Am. 1964;54:1171.
69. Krauskopf J, Srebro R. Spectral sensitivity of color mechanisms: derivation from fluctuations of color appearance near threshold. Science. 1965;150(3702):1477–9.

70. Hofer H, Singer B, Williams DR. Different sensations from cones with the same photopigment. J Vis. 2005;5(5):444–54.
71. Koenig DE, Hofer HJ. Do color appearance judgments interfere with detection of small threshold stimuli? J Opt Soc Am A Opt Image Sci Vis. 2012;29(2):A258–67.
72. Gowdy PD, Cicerone CM. The spatial arrangement of the L and M cones in the central fovea of the living human eye. Vision Res. 1998;38(17):2575–89.
73. Brainard DH, Williams DR, Hofer H. Trichromatic reconstruction from the interleaved cone mosaic: Bayesian model and the color appearance of small spots. J Vis. 2008;8(5):15, 1–23.
74. Lennie P, Haake PW, Williams DR. The design of chromatically opponent receptive fields. In: Landy MS, Movshon JA, editors. Computational modeling of visual processing. Cambridge, MA: MIT Press; 1991. p. 71–82.
75. Forte JD, Blessing EM, Buzas P, Martin PR. Contribution of chromatic aberrations to color signals in the primate visual system. J Vis. 2006;6(2):97–105.
76. Hofer H, Williams DR. Color vision and the retinal mosaic. In: Werner JS, Chalupa LM, editors. The new visual neurosciences. Cambridge, MA: MIT Press; 2014. p. 469–83.
77. Jacobs GH, Neitz J, Krogh K. Electroretinogram flicker photometry and its applications. J Opt Soc Am A Opt Image Sci Vis. 1996;13(3):641–8.
78. Brainard DH, Roorda A, Yamauchi Y, Calderone JB, Metha A, Neitz M, Neitz J, Williams DR, Jacobs GH. Functional consequences of the relative numbers of L and M cones. J Opt Soc Am A Opt Image Sci Vis. 2000;17(3):607–14.
79. Carroll J, McMahon C, Neitz M, Neitz J. Flicker-photometric electroretinogram estimates of L:M cone photoreceptor ratio in men with photopigment spectra derived from genetics. J Opt Soc Am A Opt Image Sci Vis. 2000;17(3):499–509.
80. Carroll J, Neitz J, Neitz M. Estimates of L:M cone ratio from ERG flicker photometry and genetics. J Vis. 2002;2(8):531–42.
81. Kremers J, Stepien MW, Scholl HP, Saito C. Cone selective adaptation influences L- and M-cone driven signals in electroretinography and psychophysics. J Vis. 2003;3(2):146–60.
82. Euler T, Haverkamp S, Schubert T, Baden T. Retinal bipolar cells: elementary building blocks of vision. Nat Rev Neurosci. 2014;15(8):507–19.
83. Thoreson WB, Mangel SC. Lateral interactions in the outer retina. Prog Retin Eye Res. 2012;31(5):407–41.
84. Lee BB, Martin PR, Grunert U. Retinal connectivity and primate vision. Prog Retin Eye Res. 2010;29(6):622–39.
85. Chichilnisky EJ, Baylor DA. Receptive-field microstructure of blue-yellow ganglion cells in primate retina. Nat Neurosci. 1999;2(10):889–93.
86. Field GD, Gauthier JL, Sher A, Greschner M, Machado TA, Jepson LH, Shlens J, Gunning DE, Mathieson K, Dabrowski W, Paninski L, Litke AM, Chichilnisky EJ. Functional connectivity in the retina at the resolution of photoreceptors. Nature. 2010;467(7316):673–7.
87. Li PH, Field GD, Greschner M, Ahn D, Gunning DE, Mathieson K, Sher A, Litke AM, Chichilnisky EJ. Retinal representation of the elementary visual signal. Neuron. 2014;81(1):130–9.
88. Dacey DM, Crook JD, Packer OS. Distinct synaptic mechanisms create parallel S-ON and S-OFF color opponent pathways in the primate retina. Vis Neurosci. 2014;31(2):139–51.
89. Rolfs M. Microsaccades: small steps on a long way. Vision Res. 2009;49(20):2415–41.
90. Schnapf JL, Nunn BJ, Meister M, Baylor DA. Visual transduction in cones of the monkey Macaca fascicularis. J Physiol. 1990;427:681–713.
91. Cao LH, Luo DG, Yau KW. Light responses of primate and other mammalian cones. Proc Natl Acad Sci U S A. 2014;111(7):2752–7.
92. Linsenmeier RA, Hertz BG. Eye movements in paralyzed cats induced by drugs and sympathetic stimulation. Vision Res. 1979;19(11):1249–52.
93. Forte J, Peirce JW, Kraft JM, Krauskopf J, Lennie P. Residual eye-movements in macaque and their effects on visual responses of neurons. Vis Neurosci. 2002;19(1):31–8.
94. Sincich LC, Zhang Y, Tiruveedhula P, Horton JC, Roorda A. Resolving single cone inputs to visual receptive fields. Nat Neurosci. 2009;12(8):967–9.

95. Hofer H, Artal P, Singer B, Aragon JL, Williams DR. Dynamics of the eye's wave aberration. J Opt Soc Am A Opt Image Sci Vis. 2001;18(3):497–506.
96. Dubra A, Sulai Y, Norris JL, Cooper RF, Dubis AM, Williams DR, Carroll J. Noninvasive imaging of the human rod photoreceptor mosaic using a confocal adaptive optics scanning ophthalmoscope. Biomed Opt Express. 2011;2(7):1864–76.
97. Thibos LN, Bradley A, Still DL, Zhang X, Howarth PA. Theory and measurement of ocular chromatic aberration. Vision Res. 1990;30(1):33–49.
98. Atchison DA, Smith G. Chromatic dispersions of the ocular media of human eyes. J Opt Soc Am A Opt Image Sci Vis. 2005;22(1):29–37.
99. Simonet P, Campbell MC. The optical transverse chromatic aberration on the fovea of the human eye. Vision Res. 1990;30(2):187–206.
100. Rynders M, Lidkea B, Chisholm W, Thibos LN. Statistical distribution of foveal transverse chromatic aberration, pupil centration, and angle psi in a population of young adult eyes. J Opt Soc Am A Opt Image Sci Vis. 1995;12(10):2348–57.
101. Marcos S, Burns SA, Prieto PM, Navarro R, Baraibar B. Investigating sources of variability of monochromatic and transverse chromatic aberrations across eyes. Vision Res. 2001;41(28):3861–71.
102. Harmening WM, Tiruveedhula P, Roorda A, Sincich LC. Measurement and correction of transverse chromatic offsets for multi-wavelength retinal microscopy in the living eye. Biomed Opt Express. 2012;3(9):2066–77.
103. Westheimer G, McKee SP. Spatial configurations for visual hyperacuity. Vision Res. 1977;17(8):941–7.
104. Marcos S, Burns SA, Moreno-Barriusop E, Navarro R. A new approach to the study of ocular chromatic aberrations. Vision Res. 1999;39(26):4309–23.
105. Snodderly DM, Weinhaus RS, Choi JC. Neural-vascular relationships in central retina of macaque monkeys (Macaca fascicularis). J Neurosci. 1992;12(4):1169–93.
106. Schiefer U, Benda N, Dietrich TJ, Selig B, Hofmann C, Schiller J. Angioscotoma detection with fundus-oriented perimetry. A study with dark and bright stimuli of different sizes. Vision Res. 1999;39(10):1897–909.
107. Remky A, Beausencourt E, Elsner AE. Angioscotometry with the scanning laser ophthalmoscope. Comparison of the effect of different wavelengths. Invest Ophthalmol Vis Sci. 1996;37(11):2350–5.
108. Adams DL, Horton JC. Shadows cast by retinal blood vessels mapped in primary visual cortex. Science. 2002;298(5593):572–6.
109. Adams DL, Horton JC. A precise retinotopic map of primate striate cortex generated from the representation of angioscotomas. J Neurosci. 2003;23(9):3771–89.
110. Tuten WS, Tiruveedhula P, Roorda A. Adaptive optics scanning laser ophthalmoscope-based microperimetry. Optom Vis Sci. 2012;89(5):563–74.
111. Nishiwaki H, Ogura Y, Kimura H, Kiryu J, Miyamoto K, Matsuda N. Visualization and quantitative analysis of leukocyte dynamics in retinal microcirculation of rats. Invest Ophthalmol Vis Sci. 1996;37(7):1341–7.
112. Tam J, Tiruveedhula P, Roorda A. Characterization of single-file flow through human retinal parafoveal capillaries using an adaptive optics scanning laser ophthalmoscope. Biomed Opt Express. 2011;2(4):781–93.
113. Uji A, Hangai M, Ooto S, Takayama K, Arakawa N, Imamura H, Nozato K, Yoshimura N. The source of moving particles in parafoveal capillaries detected by adaptive optics scanning laser ophthalmoscopy. Invest Ophthalmol Vis Sci. 2012;53(1):171–8.
114. Sinclair SH, Azar-Cavanagh M, Soper KA, Tuma RF, Mayrovitz HN. Investigation of the source of the blue field entoptic phenomenon. Invest Ophthalmol Vis Sci. 1989;30(4):668–73.
115. Martin JA, Roorda A. Direct and noninvasive assessment of parafoveal capillary leukocyte velocity. Ophthalmology. 2005;112(12):2219–24.
116. Martin JA, Roorda A. Pulsatility of parafoveal capillary leukocytes. Exp Eye Res. 2009;88(3):356–60.

117. Tam J, Martin JA, Roorda A. Noninvasive visualization and analysis of parafoveal capillaries in humans. Invest Ophthalmol Vis Sci. 2010;51(3):1691–8.
118. Arathorn DW, Yang Q, Vogel CR, Zhang Y, Tiruveedhula P, Roorda A. Retinally stabilized cone-targeted stimulus delivery. Opt Express. 2007;15(21):13731–44.
119. Yang Q, Arathorn DW, Tiruveedhula P, Vogel CR, Roorda A. Design of an integrated hardware interface for AOSLO image capture and cone-targeted stimulus delivery. Opt Express. 2010;18(17):17841–58.
120. Harmening WM, Tuten WS, Roorda A, Sincich LC. Mapping the perceptual grain of the human retina. J Neurosci. 2014;34(16):5667–77.
121. Wilson ME. Invariant features of spatial summation with changing locus in the visual field. J Physiol. 1970;207(3):611–22.
122. Lie I. Visual detection and resolution as a function of retinal locus. Vision Res. 1980;20(11):967–74.
123. Anderson SJ, Mullen KT, Hess RF. Human peripheral spatial resolution for achromatic and chromatic stimuli: limits imposed by optical and retinal factors. J Physiol. 1991;442:47–64.
124. Volbrecht VJ, Shrago EE, Schefrin BE, Werner JS. Spatial summation in human cone mechanisms from 0 degrees to 20 degrees in the superior retina. J Opt Soc Am A Opt Image Sci Vis. 2000;17(3):641–50.
125. Drasdo N, Millican CL, Katholi CR, Curcio CA. The length of Henle fibers in the human retina and a model of ganglion receptive field density in the visual field. Vision Res. 2007;47(22):2901–11.
126. Nerger JL, Cicerone CM. The ratio of L cones to M cones in the human parafoveal retina. Vision Res. 1992;32(5):879–88.
127. Otake S, Cicerone CM. L and M cone relative numerosity and red-green opponency from fovea to midperiphery in the human retina. J Opt Soc Am A Opt Image Sci Vis. 2000;17(3):615–27.
128. Williams DR, MacLeod DI, Hayhoe MM. Punctate sensitivity of the blue-sensitive mechanism. Vision Res. 1981;21(9):1357–75.
129. Makous W, Carroll J, Wolfing JI, Lin J, Christie N, Williams DR. Retinal microscotomas revealed with adaptive-optics microflashes. Invest Ophthalmol Vis Sci. 2006;47(9):4160–7.
130. Yeh T, Smith VC, Pokorny J. The effect of background luminance on cone sensitivity functions. Invest Ophthalmol Vis Sci. 1989;30(10):2077–86.
131. Cole GR, Hine T. Computation of cone contrasts for color vision research. Behav Res Meth Instrum Comput. 1992;24:22–7.
132. DeVries SH, Qi X, Smith R, Makous W, Sterling P. Electrical coupling between mammalian cones. Curr Biol. 2002;12(22):1900–7.
133. Hornstein EP, Verweij J, Schnapf JL. Electrical coupling between red and green cones in primate retina. Nat Neurosci. 2004;7(7):745–50.
134. Tuten WS, Harmening WM, Sabesan R, Sincich LC, Roorda A. Functional mapping of the trichromatic cone mosaic in vivo. In: Fall vision meeting abstracts. Philadelphia: Optical Society of America; 2014.
135. Sabesan R, Tuten WS, Roorda A. Mapping the human trichromatic cone mosaic with an AOSLO. In: ARVO. 2014.
136. Cohen J. Weighted kappa: nominal scale agreement with provision for scaled disagreement or partial credit. Psychol Bull. 1968;70(4):213–20.
137. van den Berg TJ, Franssen L, Kruijt B, Coppens JE. History of ocular straylight measurement: a review. Z Med Phys. 2013;23(1):6–20.
138. Mollon JD, Bowmaker JK. The spatial arrangement of cones in the primate fovea. Nature. 1992;360(6405):677–9.
139. Bruce KS, Harmening WM, Langston BR, Tuten WS, Roorda A, Sincich LC. Normal perceptual sensitivity arising from weakly reflective cone photoreceptors. Invest Ophthalmol Vis Sci. 2015;56(8):4431–8.
140. Bruce KS, Harmening WM, Roorda A, Sincich LS. Cone-by-cone threshold variability in the human retina. In: Society for neuroscience conference. Washington, DC; 2014.
141. Sabesan R. Studying cone-by-cone contributions to color vision. In: Fall vision meeting. J Vis. 2014:17.

Chapter 4
Cone Opponency: An Efficient Way of Transmitting Chromatic Information

Barry B. Lee and Luiz Carlos L. Silveira

Abstract Cone or color opponency provides visual neurons with a spectral selectivity more sharply tuned than the absorption spectra of the cone photoreceptors. It is achieved by subtractive interaction of inputs from different cones. In different species, there are different varieties of cone and color opponency, but in human and primates, there are two cone-opponent pathways, one with subtractive input from the middle- and long-wavelength cones and the other with subtractive inputs from the short-wavelength (S) cone and some combination of the other two. These pathways are distinctive anatomically as well as physiologically, and the S-cone pathway is phylogenetically the older. We review these pathways. This way of coding for color is thought to be efficient from an informational standpoint; we argue that these pathways are specialized for transmitting the surface characteristics of objects.

Keywords Cone • Ganglion cell • Parvocellular • Magnocellular • Koniocellular • Lateral geniculate nucleus • Receptive fields

B.B. Lee (✉)
School of Optometry, State University of New York, New York, NY, USA

Max Planck Institute for Biophysical Chemistry, Göttingen, Germany
e-mail: blee@sunyopt.edu

L.C.L. Silveira (deceased)

J. Kremers et al. (eds.), *Human Color Vision*, Springer Series in Vision Research, DOI 10.1007/978-3-319-44978-4_4

4.1 Introduction

4.1.1 Cone and Color Opponency in Vertebrates

Color opponency relates to selectivity of a neural response to the wavelength of a light stimulus, beyond what might be expected from the absorption spectra of single cone photoreceptors. Usually, a color-opponent cell gives an excitatory response to some wavelengths and is inhibited by others. This is due to antagonistic input from cone types with different spectral sensitivities. Although well-developed color vision is common in vertebrates, in mammals it is restricted to primates, and, in this review, we are mainly concerned with primate retinal ganglion cells or cells from the lateral geniculate nucleus (LGN). Although cone pigment sensitivities have been measured in many vertebrate species, much more is known of post-receptoral color vision physiology of primates than in nonmammalian vertebrates. One exception is the horizontal cells of, for example, fish and turtle. Some of these cells show cone and color opponency [1–3]. But in the primate, horizontal cells show cone selectivity, but no cone opponency; horizontal cells may receive stronger input from one cone type than from another but there is no antagonism between cone inputs [4]. After the horizontal cells, there are intriguing indications that the responses of ganglion cells (and probably beyond) also differ in fishes and turtles compared to primates; for example, ganglion cells of the goldfish show "double opponency" [5, 6] but such cells are not found in the primate until the cortex. This difference indicates that different vertebrate groups may have adopted different physiological substrates for cone and color opponency, and perhaps different strategies for color vision. In this review, we focus on cone and color opponency in the primate retina and the LGN. However, we emphasize that it is a good idea to consider color opponent systems in both functional and evolutionary contexts (see Chap. 11).

4.1.2 Color Opponency in Human and Primates

Humans and primates demonstrate considerable functional similarities in their processing of simple colors and patterns [7, 8], or indeed in more complex abilities such as face recognition [9]. There is a good deal of psychophysical evidence for luminance and two chromatic (red-green and blue yellow) channels in human vision [10]. Much early work stressed the likelihood that visual sensitivity of the different channels is established early in the visual system, in the retina [11, 12], and there is now much electrophysiological evidence from primates linking physiological channels in the retina to psychophysical achromatic and chromatic channels. An alternative concept is that such psychophysical mechanisms are somehow synthesized in the cortex [13]. One aspect of color opponency must be a cortical construct. Retinal/LGN mechanisms [14] and psychophysical detection thresholds for color [10, 15] map closely onto one another, but suprathreshold hue perception (the unique hues) do not fit well with either [16] (see Chap. 5). Nevertheless, we shall cite supporting evidence for the idea that luminance and chromatic psychophysical detection channels have a firm retinal basis in the course of this review. Also, to paraphrase Alexander Pope ("The proper study of mankind is man" [17]), the proper study of human vision is that of the primate, rather than the mouse [18]. The primate retina, alone among mammals, is highly specialized to transmit a red-green dimension of color vision; other mammalian models may be useful for basic synaptology or clinical models, but their relevance to primate trichromatic capabilities remains to be demonstrated.

There are three cone types in the primate retina, the long (L), middle (M), and short (S) wavelength sensitive cones. Spectral sensitivities are well established [19–21], and their genetic coding known [22] (see Chap. 1); the L and M cone opsin genes form a tandem array on the X chromosome and the S cone opsin gene is located on chromosome 7 [23, 24]. The tandem array of opsin genes on the X chromosome is characteristic of catarrhines (Old-World monkeys, apes, and humans); opsin coding in platyrrhines (New-World monkeys) is different [23, 24] and the neurophysiological implications we discuss later.

The first recordings from the primate visual system to demonstrate cone and color opponency were those of Russel DeValois and his collaborators [25–27], closely followed by Thorsten Wiesel and David Hubel [28] and Peter Gouras and his coworkers [29, 30]. A historical perspective of the development of ideas about the physiological basis of human color vision can be found elsewhere [31]. A major advance in physiological studies of cone and color opponency was provided by more sophisticated stimulus techniques, in which color and luminance is modulated around a mean background level in such a way as to carefully isolate the influence of either the M-, L-, or S-cones, and to study post-receptoral combinations of their signals [32, 33] (see also Chap. 2). In parallel to physiology, major advances in knowledge of anatomical connectivity in the primate retina have permitted comparison of physiological responses with potential anatomical substrates [34–36]. The ideal approach is to record from a retinal cell and then recover its morphology by intracellular staining; this route has been impressively exploited by Dennis Dacey and his coworkers [37–43]. The anatomical and physiological basis of cone opponency thus now stands on firmer ground; but there are still contentious issues, some of which we will stress.

4.1.3 Cone Opponency and Color Opponency

In this review we refer to M/L-cone opponency and S/ML-cone opponency; these dimensions may be directly related to chromatic detection mechanisms. After the receptors, cone signals are added and subtracted to produce three major pathways providing input to the LGN and cortex. The magnocellular (MC) pathway begins in the parasol ganglion cell (or in the diffuse bipolar cell) and passes through the magnocellular layers of the LGN; these cells receive summed input from M and L cones. The parvocellular (PC) pathway begins in the midget ganglion cell (or in the midget bipolar cell) and passes through the parvocellular layers of the LGN; its cells receive antagonistic, subtractive input from the M and L cones. S/ML cone opponent cells form part of the koniocellular (KC) pathway, and signals of these cells are relayed through the koniocellular layers of the LGN. Signals of S cones are antagonistic to some combination of those of the M and L cones. Most of these pathways have sub-pathways deriving inputs from inner (on) and outer (off) sub-laminae of the inner plexiform layer (IPL). Details of this morphology can be found elsewhere [44], but it is important to stress that the fine structure of the lamination and connectivity with amacrine cells is likely to be critical for the elaboration of cone opponency. Lastly, the M/L-cone and S/ML-cone opponent pathways are distinguished by very sustained responses; this is probably desirable if cells transmit information about surfaces, such as color. On the other hand, MC cells are very transient and respond best to luminance edges, acting as contrast detectors.

This review focuses on cone opponency, but functional interpretation is bound up with the fact that primate visual system is not only good at color discrimination but at processing luminance contrast patterns; we shall frequently refer to the properties of the MC pathway for comparative purposes. More general reviews of primate retinal anatomy [44] and function [45] may be found elsewhere.

4.2 The S/ML Cone Opponent System

The information provided by the eye's optical system is sampled by the L-, M-, and S-cone mosaics and distributed to a series of second order neurons, comprising two classes of horizontal cells and several classes of bipolar cells. Retinal circuits formed by photoreceptors and associated interneurons are the origin of parallel pathways comprising ganglion cells, neurons from the LGN, and neurons located further downstream, in a series of visual areas of the cerebral cortex (see for example Chap. 7 for those involved in color vision). Retinal ganglion cells and their LGN relay cells exhibit many similarities in their physiological properties and information obtained from these two loci can be combined to describe the division of labor between retino-geniculo-cortical visual streams.

Cells with properties other than those of MC or PC cells have been recorded in studies of ganglion cells [30] and of LGN [28, 32, 33, 46, 47]. However, while neurons of the MC and PC pathways are relatively uniform regarding their main physiological properties, cells of the KC pathway comprise a diversified group of achromatic (non-cone-opponent) and S/ML cone opponent neurons, endowed with a variety of spatial and temporal properties. An important tool to identify S/ML cone opponent neurons is to measure their responses to stimuli that excite differentially S cones while leaving the excitations of L cones and M cones constant [32, 33] (i.e., S-cone isolating stimuli using at least double silent substitution; see Chap. 2 for details). These stimuli are located on tritanopic confusion lines, i.e., they appear identical to a tritanope lacking S cones, and differ for normal trichromats due to differential S cone excitation.

From studies of receptive field properties of LGN KC relay neurons, neurotracer injections in the KC layers, and subsequent intracellular recordings of labeled ganglion cells, out of the numerous retinal ganglion cell classes described five exhibit S/ML cone opponency [48, 49]. As well as the S-ON/ML-OFF (+S−ML) small bistratified cell, there are a +S−ML large bistratified, a S-OFF/ML-ON (−S+ML) large sparse, and two −S+ML giant sparse ganglion cells. If and how this variety of cells may contribute to visual experience and color vision in particular is a matter of debate. A more detailed review of these cells can be found elsewhere [50].

4.2.1 S-ON/LM-OFF; Small Bistratified Cells

The small bistratified cells are the shrub cells of Polyak [51]; their morphology was further defined by retrograde staining after LGN injections [52] and direct retinal staining [53]. In vitro physiological recordings [37] convincingly demonstrated +S−ML response properties. Small bistratified cells have inner dendritic tiers in the most vitreal layer of the inner plexiform layer and outer dendritic tiers in the outer half of the inner plexiform layer (Fig. 4.1a, b). They give a vigorous S-ON tonic response to S cone isolating stimuli [37] (Fig. 4.1c); they give ON responses to blue and OFF responses to yellow light.

The two dendritic tiers of the small bistratified cells have similar sizes but different dendritic branching densities (the inner tier is more ramified than the outer tier). There have been various studies of receptive field structure of these cells in retina [54, 55] and LGN [56–58]. Generally, the LM-OFF mechanism has been found to be slightly larger than the S-ON mechanism, typically by about 30 % in area. This is, strictly speaking, not a Type II cell, defined as cells with centers and surrounds with equal sizes (see below), [28], but simple modeling can show that such a minimal difference in center-surround size does not have much effect on spatial tuning curves (even for the study with the largest difference in size between yellow-off and blue-on fields (yellow-off 50 % larger than blue on; Field et al. [54]). Additionally, the two fields are sometimes displaced relative to one another and anisotropic, giving some orientation specificity and other more complex properties [57].

Despite these anisotropies, recordings from small bistratified cells in in vitro retinal preparations to drifting sinusoidal S cone and LM cone isolating gratings, varying in spatial frequency, indicated that both the S and LM spatial frequency responses were low-pass with comparable peak responses and similar high spatial frequency cutoffs [55]. This indicates that the receptive field of both mechanisms had similar sizes (i.e., they were spatially co-extensive; Fig. 4.1d, e). In addition, other properties of the S-ON and LM-OFF mechanisms of small bistratified cells have been characterized: contrast response functions, temporal sensitivities, latencies, and weights of the two mechanisms were found to be very similar, as in vivo experiments [59]. Small bistratified cells thus have +S and −ML receptive field mechanisms bearing very similar physiological properties, spatially co-extensive or close to it, an arrangement considered to be ideal for cells that are specialized for color information processing.

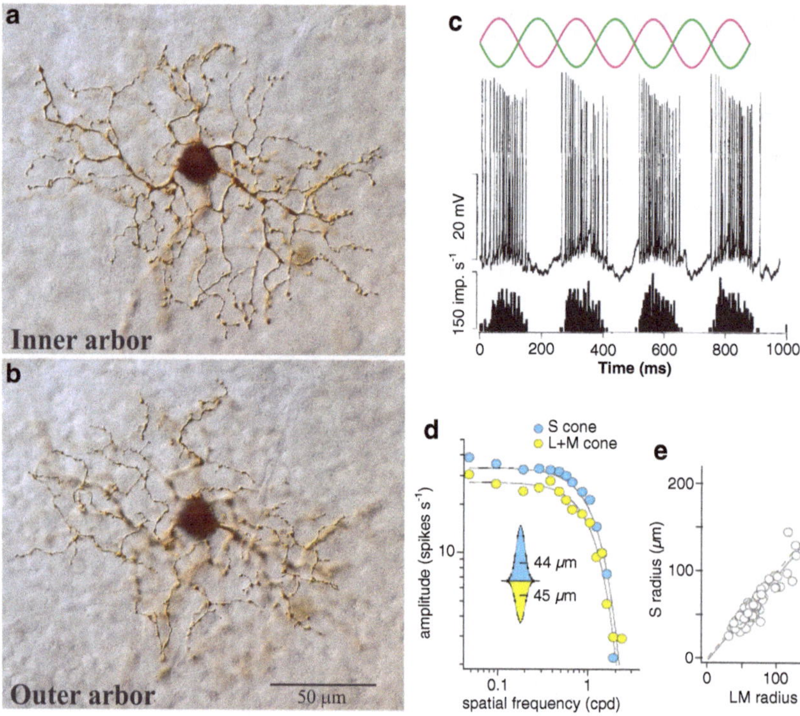

Fig. 4.1 Morphology and physiology of S-ON/ML-OFF small bistratified cells of macaque retina. The in vitro isolated retina-choroid preparation was used. The membrane potential of retinal ganglion cells was recorded with intracellular microelectrodes while light stimuli were delivered through the camera port of the microscope. Recorded cells were labeled by intracellular injection of neurobiotin. (**a**, **b**) Inner (**a**) and outer (**b**) tiers of an S-ON/ML-OFF small bistratified cell (ca. 30° eccentricity). The inner tier is located in the inner plexiform layer near the ganglion cell layer and receives the S-ON input provided by BB bipolar cells, while the outer tier is located near the inner nuclear layer and receives the ML-OFF input provided by DB2 and DB3 bipolar cells. (**c**) An S-ON/ML-OFF small bistratified cell responds strongly to stimuli modulated along a tritanopic confusion line where only S cones are modulated while the stimulation of M and L cones is kept constant. From top to bottom: stimulus composition (100 % contrast), intracellular recording, peristimulus time histogram averaged over 20 stimulus presentation. (**d**) Spatial frequency tuning of a small bistratified cell in response to S cone and ML cone isolating gratings drifting in the cell receptive field. Data were fit with a Gaussian model (inset). S-ON and ML-OFF receptive fields were spatially coextensive, i.e., Gaussians of comparable radii. (**e**) Estimated Gaussian radii for the receptive fields of S-ON and ML-OFF mechanisms of 55 small bistratified cells (*open circles*). S-ON/M-OFF ratio = 0.92 ± 0.13 µm (mean ± s.d.). The dashed line shows the unity slope and the solid line the actual fit to the data (slope = 0.97; $R^2 = 0.90$). (**a**, **b**, **d**, **e**) Reproduced with permission from [67] (**c**) Reproduced with permission from [37]

4.2.2 S-ON/LM-OFF Synaptic Circuit

The small bistratified cells are the output neurons of a retinal circuit comprising photoreceptors, horizontal cells, bipolar cells, and amacrine cells. S cones contact bipolar cells and horizontal cells. Of at least nine different classes of cone bipolar cells (two midget bipolars (flat (FMB) and invaginating (IMB)) and six classes of diffuse bipolars (DB1-6 cells)), there is a single class of S cone bipolars (BB cells) [36]. M- and L-cones make excitatory or inhibitory glutamatergic synapses with bipolar cells: OFF bipolars are hyperpolarized by light and thus conserve the sign of photoreceptor responses, while ON bipolars invert the sign of photoreceptor responses to light through metabotropic glutamate receptors. The presence of specific glutamate receptors in the bipolar dendrites allows for the pharmacological dissection of the mechanisms of ganglion cell responses to light, such as the +S−ML responses of small bistratified cells. All bipolar cell axon terminals make excitatory glutamatergic synapses with their targets in the inner plexiform layer, either ganglion cells or amacrine cells.

S cones preferentially contact the BB cells [36, 60–63]. BB cells are easily identified by their long, smoothly curved dendrites, which are horizontally oriented in the outer plexiform layer and contact between one and three cone pedicles. In macaques, there is a maximum of 2800 BB cells/mm^2 at 1–2 mm eccentricity, where S-cone density reaches a maximum, and this density falls to 280–600 BB cells/mm^2 at the far retinal periphery [61]. BB cells form a dedicated pathway conveying information from S cones to the inner plexiform layer where their relatively large axon terminals stratify in the deepest stratum, close to the ganglion cell layer, where they make excitatory synapses with the small bistratified cell dendrites [62].

Pharmacological dissection of the S-ON and ML-OFF mechanisms of small bistratified cells has been attempted by using receptor antagonists [54, 55]. The S-ON response was blocked by L-AP-4, an antagonist for the metabotropic glutamate receptor, suggesting that the S cone response is transmitted by these bipolars to the inner plexiform layer, and then relayed to small bistratified cells. The ML-OFF response, however, remained unblocked with the combined application of L-AP-4 and antagonists to inhibitory transmitters, suggesting that it was transferred by OFF bipolar cells (possibly DB2 and DB3) via ionotropic glutamate synapses.

4.2.3 S-ON/ML-OFF Large Bistratified Cells

Little is known about the +S−ML large bistratified cells. Large bistratified cells have a distinct morphology (large, sparsely branched dendritic trees) but the dendrites branch in similar levels of the inner plexiform layer compared to the small bistratified cell, and they have roughly similar responses to light [48, 64]. Due to their similarity in the level of branching, it is possible that large and small bistratified cells share at least partially the +S−ML synaptic circuit described in the previous section. Due to the inverse relation between dendritic tree size and cell density observed in most retinal neurons, large bistratified cells probably occur at lower density than small bistratified cells at all retinal locations. Thus, it seems that S-ON information is sent from the retina to higher visual centers using two systems with different spatial resolutions.

4.2.4 S-OFF/ML-ON Cells

A group of retinal ganglion cells with –S+ML opponency, complementing the +S–ML cells, has been sought, to fulfill the "standard model" of color opponency. Such cells are readily found in the LGN [32, 46], but their retinal origin has not been firmly identified.

–S+ML cells with large, sparse dendritic trees have been recorded in macaque retina but detailed information is lacking; however, their –S+ML responses seem similar to those of LGN cells recorded in vivo [48, 65]. Their low encounter rate in in vivo retinal recordings ([30]; BB Lee, unpublished observations) may be due to a low retinal density or other factors (e.g., their locus in the ganglion cell layer); in LGN perhaps 3S-ON cells are recorded for each S-OFF [32, 46, 58].

Another intriguing possibility is an S-OFF pathway associated with the midget system. Synapses between S cones and FMB cells near the fovea have been suggested based on electron microscopy [66]. This raises the possibility of the existence of OFF midget ganglion cells with receptive field centers driven by single S cones, [67]. However, FMB cells can be selectively stained, and no such cells could be found contacting S-cones in the marmoset [68], while the other cones were contacted. In addition, with the differences in receptive field size, structure, and spacing between putative S-OFF midget cells and small bistratified cells, these two cell classes would form a spatially very asymmetrical channel to provide S cone information. Such asymmetry is not usual in major ganglion cell classes with paired complementary ON and OFF cells.

–S+ML giant sparse cells containing melanopsin were recorded and labeled in vitro in the macaque retina [39]. They are the largest ganglion cells in the primate retina at any eccentricity and occur in two subclasses, but both subclasses respond similarly to light: –S+ML responses to cone isolating stimuli, ON responses to scotopic stimuli that excite only rods, and melanopsin-dependent, intrinsic ON responses. However, the receptive field sizes and spatial density distribution of –S+ML giant sparse cells make it unlikely that they play a role in blue-yellow vision; their density relative to PC, MC, and small bistratified cells is very low in central retina, where color vision is at its most acute.

Lastly, an S-OFF pathway has recently been discovered in the ground squirrel [69, 70]. These cells receive signals from the S-cone ON-bipolar, inverted by a specific amacrine cell type. Whether this circuit exists in the primate, and whether it corresponds to the S-OFF class mentioned above, remains an open question.

4.2.5 S/ML Cone Opponency in the LGN

+S–ML cells were found in early recordings from primate LGN [28, 32, 33]. These studies generally agree on the properties of S-ON cells: receptive field sizes larger than M/L cone opponent cells, co-extensive S-ON and ML-OFF mechanisms, and tonic responses to S cone isolating stimuli, resembling retinal findings. More recent studies have determined their location in specific LGN layers. The favorable LGN anatomy of marmosets, with broad koniocellular layers, made this New-World primate particularly suitable for selective recordings from LGN laminae and this permitted the first clear assignment of +S–ML cells to the KC layers [47]; more recent work confirmed this finding in macaques [71].

In recordings from the macaque LGN, a small proportion of sustained color opponent cells were strongly inhibited by S cones [32, 46, 71, 72]. An excellent example was shown by Wiesel and Hubel ([28] their Fig. 4.6). Although earlier work suggested fewer –S+ML than +S–ML cells, more equal proportions of have been reported in a recent study [57]. The –S+ML LGN cells were found in the KC layers, together with the +S–ML cells and a variety of achromatic ON and OFF cells.

In the majority of LGN studies, the receptive field centers and surrounds of both +S–ML cells and –S+ML cells were found to be co-extensive. Other receptive field properties of S-OFF cells substantially differ from S-ON cells: the S-cone input is weaker and more variable in S-OFF cells with lower S cone contrast sensitivity. Spatial and temporal frequency sensitivities of S-OFF cells may be lower for S cone modulation and higher for achromatic modulation than in S-ON cells [57].

A one-to-one correspondence between S-cone LGN neurons and their retinal afferents has yet to be established. The number and properties of S-ON and S-OFF cells so far distinguished in the LGN do not easily match S-ON and S-OFF ganglion cells so far specified. However, since classification in both retina and LGN is still somewhat vague, the mismatch between retina and LGN could be due to insufficient information.

4.3 The M/L Cone Opponent System

Only primates among mammals possess both L and M cones and so the M/L cone opponent system is novel in primates. The two opsins diverged close to the beginning of primate evolution (Chap. 1), and so the necessary retinal machinery to extract an M/L opponent signal must have evolved in response to the opportunity provided by dual opsins. How and why this exactly came about has been the object of much speculation; we comment from a physiological perspective in a later section (see also Chaps. 1 and 11). In any event, the midget bipolar and ganglion cell systems are unique among primates and closely involved in the generation of an M/L opponent signal. Early work [25–27] found ganglion and LGN cells whose firing was excited by light of some wavelengths and inhibited by others; in terms of receptive field structure they were often described as having centers of one cone type and surrounds of the other (e.g., red on-centers and green off-surrounds) in various permutations. But before discussing physiological properties, we first describe midget anatomy.

4.3.1 Morphological Properties of PC Cells: The Midget Anatomy

In central retina, a single midget bipolar has synapses with a single cone and a single midget bipolar contacts a single midget ganglion cell [51, 73]. Examples of such cells are shown in Fig. 4.2a, taken from a Golgi preparation of a cross section of the retina [73]. The receptors are located toward the upper edge of the panel. The upper-left cell is a midget bipolar, with a dendritic tuft reaching upward toward a single cone pedicle; bipolar dendritic tuft and cone axonal dimensions match closely. Below and to the right is a midget ganglion cell that gets its input from another midget bipolar. The bipolar axonal arbor matches the ganglion cell dendritic tuft. An electron microscopic study confirmed that this connection is highly specific [74].

The midget morphology breaks down toward the retinal periphery. Firstly, at very high eccentricities (12 mm, ca. 60°) some midget bipolars are "double headers" receiving input from two cones. Photomicrographs of this arrangement are shown in Fig. 4.2b, which shows two examples of the apical dendrites of midget bipolar cells from peripheral retina viewed in horizontal section; one (on the left) has only one tuft of apical dendrites, and the other has two, with a separation matching cone separation at that eccentricity; some midget bipolars even may get input from more than two cones [75]. Such bipolars always contact neighboring cones, so the unproven inference is that such cells may receive nonselective cone input [75]. However, even at the retinal rim single cone contacts can still be found. It is difficult to avoid the conclusion that transmitting an M,L cone specific signal into the inner retina is an important aspect of this morphology [36]. Convergence of midget bipolar onto midget ganglion cells begins closer to the fovea. Beyond ca.10° eccentricity, some midget ganglion cells have dendritic trees that contact two or more axonal arborizations of midget bipolar cells. Dendritic trees can show multiple tufts and have an irregular, patchy morphology. Figure 4.2c–h replicates these results; midget ganglion cells from nasal retina are shown at several eccentricities, viewed from above, stained after horseradish peroxidase injection in the optic tract (P. R. Martin & B. B. Lee, unpublished data). Figure 4.2c, d shows cells 600 μm (3°) from the foveola; each cell has a single tuft of dendrites corresponding to the axonal bouquet of a single midget bipolar. Figure 4.2e–g shows cells at a 3.6 mm (18°) eccentricity, at which distinct tufts (usually 2–4) can be seen, each putatively contacting a single midget bipolar. In the further periphery, dendritic

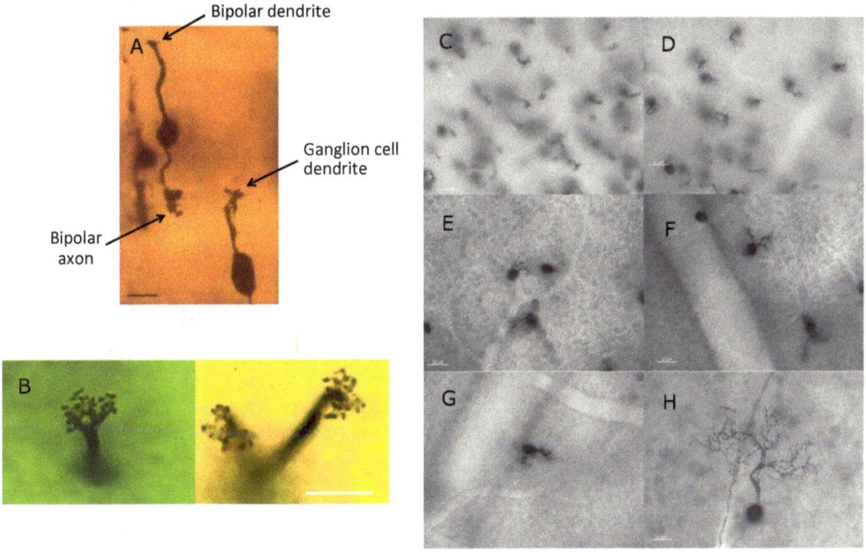

Fig. 4.2 Principles of midget morphology. (**a**) Golgi-stained midget bipolar (*upper left*) and ganglion (*lower right*) cells. The midget bipolar cell dendritic tree (*top*) contacts a single cone pedicle and its axonal tuft matches in size the dendritic tree of the neighboring ganglion cell. This implies one-to-one connectivity. Eccentricity ca. 8°, scale bar 20 μm. (**b**) Left-hand panel shows the typical single dendritic tuft of a midget bipolar in central retina. At high eccentricities (>50°; right-hand panel) a small proportion of bipolar cells have two tufts indicating contacts to two cones. (**c–h**) Ganglion cell dendritic morphology at different eccentricities. Cells were stained by retrograde transport of horseradish peroxidase following injection into the optic tract. Micrographs show different eccentricities (3°, 18°, >40°). The central cells have small dendritic trees consistent with midget morphology. At the intermediate eccentricity some cells' dendritic trees show a multiple tuft morphology consistent with input from few bipolars. At higher eccentricities (H) dendritic trees become larger and irregular in shape and receive input from multiple bipolars. (**a**, **b**) modified from [44], original micrographs provided by Heinz Wässle. (**c–h**) unpublished data, (P.R. Martin & B.B. Lee)

trees become larger, covering the axonal arbors of many midget bipolar cells, but as noted by Dacey [76], and seen in Fig. 4.2h, they are frequently irregular and anisotropic, in comparison, for example, with parasol ganglion cells, for which the dendritic trees are much more circular.

Although there had been a suggestion that there are other L/M-cone opponent ganglion cells [52], this has not been confirmed. Thus, L/M-cone opponency is thought to be based on the midget morphology just described. This means that a single cone could provide an input to the receptive field center in central retina, which automatically confers a degree of cone opponency [77] (see also Chap. 3). For instance, with an L cone and an ON midget bipolar, this could provide a "red ON-center." Where does the surround come from? This has been a contentious issue; is it cone selective (just the M cone for a green off-surrond) or is there some mixture of M and L cones providing input? Before addressing this further, we first describe basic physiological properties.

4.3.2 *Physiological Properties of PC Receptive Fields*

Modulation of the chromaticity (and luminance) of a stimulus about a white point in some kind of cone space is a powerful means of specifying the cone inputs to a cell, and their relative weights. Earlier work where different wavelength targets were presented on a dark or black background left the adaptation state of the retina indeterminate, and isolating cone inputs by differential chromatic adaptation leaves cone weighting under neutral adaptation conditions uncertain (see also Chap. 2).

We first describe the features of opponency with large-field stimuli. Figure 4.3 shows responses of a +M-L PC retinal ganglion cell to modulation about a mean luminance and chromaticity corresponding to equal energy white (2.4 Hz, 4°, 2000 td), with the L and M cones modulated in different proportions. In Fig. 4.3a response amplitudes are plotted in a L,M coordinate space. The axes represent L or M

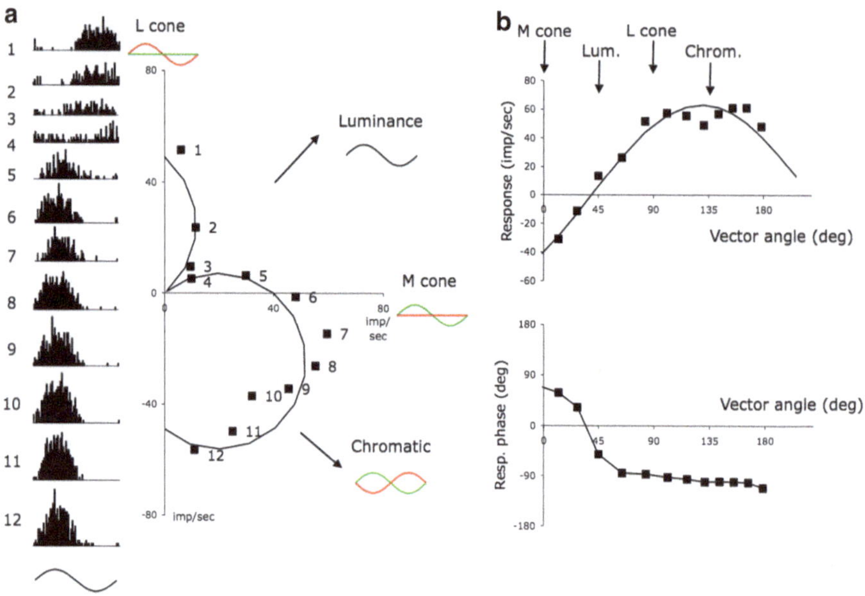

Fig. 4.3 Basic physiology of M,L cone opponency. Modulation of a stimulus in different vector directions in an L,M cone space reveals properties of cone inputs. Different vector directions are indicated by the numbers associated with each point/histogram (**a**). These directions can be achieved by changing the relative modulations of red and green light sources [80]. Luminance and chromatic modulation directions are indicated, together with examples of stimulus waveforms. +M–L cell shown. Each histogram represents response to one cycle of 2 Hz modulation, mean cone contrast has been adjusted to 28 % for all conditions. Response histograms show a minimum close to the luminance vector, associated with a phase reversal, and a maximum to the chromatic vector. Response amplitudes in the vector diagram confirm this pattern. (**b**) An alternative method of data presentation [32]. Response amplitude and phase are plotted against vector angle. Response amplitude follows a sinusoidal function, and the rapid change in response phase can be seen; the phase change is not a step function because of center-surround latency difference. Solid curves represent fits of a model [80]

vectors, and luminance and chromatic vectors are also indicated, with stimulus waveforms. At left, response histograms are shown. Response amplitudes were derived from these waveforms by Fourier analysis. There is a vigorous response to most modulation conditions, but a response minimum close to the luminance axis. This is associated with a response phase reversal. The weak response (or low contrast gain, defined as the increase of response amplitude as a function of stimulus contrast) of PC cells to luminance modulation was remarked upon in early studies of these cells [78, 79]. The curve shown in Fig. 4.3b is derived from another way of analyzing the data; this analysis closely resembles that of Derrington et al. [32]. Response amplitude (and phase; lower graph) is plotted as a function of vector angle. Response amplitudes are made negative at the phase reversal. If the cone responses are sinusoidal and add linearly, the data should be described by a sinusoid, and this is the case. The sinusoid crosses the x axis close to 45°, the luminance vector. Response amplitude through the chromatic quadrant (90–180°) is vigorous. When expressed in terms of cone contrast sensitivity, it is found that PC cells are just as sensitive to M/L cone modulation as are MC cells to luminance modulation, except that PC cells are maximally sensitive at low temporal frequency whereas MC cells are maximally sensitive at ca. 10 Hz.

We have plotted the responses of this +M-L cell in these two ways to emphasize that responses go through a minimum close to the luminance vector. Very similar results are obtained from +L-M cells, but with reversed phase. A more extensive description of results obtained with this approach is found elsewhere [80]. A response minimum close to the luminance direction implies the L and M cone inputs are in close balance, so that with a luminance target the opponent L and M signals cancel. Estimating L/M cone weighting over a large cell sample, Derrington et al. [32] confirmed this to be a general feature of the PC cell population. Using stimuli with much greater modulations about the white point [33], it is necessary to incorporate response saturation into modeling, but cone signals still add linearly and remain well balanced. This means that center and surround in PC cells (the opponent L and M cone inputs) are closely balanced. This is an unusual ganglion cell feature; for example in X and Y retinal ganglion cells in the cat or MC cells in the primate, the center is dominant over the surround.

We now discuss center and surround structure in relation to cone opponency in more detail. In DeValois' pioneering work, large field stimuli were used. Wiesel and Hubel [28] were the first to point out that M/L cone opponent inputs to receptive fields may have spatial structure, i.e., center-surround organization. They described a spectrum of spatial organization, from Type I to Type II (Fig. 4.4a); in the former, there is center-surround structure, with one cone type providing input to the center and the other to the surround, as sketched in the upper example. For Type II cells, the cone-opponent inputs are co-extensive. Cone-opponent cells can be divided into these categories; as described in the preceding section, S-cone opponent cells have Type II opponent structure. For M/L cone opponent cells, Derrington et al. [32] and others [81] found that there was a continuum of spatial structure between the two extremes, i.e., Type I and Type II form extremes of a continuum. We stress this point here since it has important functional implications. It is convenient to assess spatial receptive field structure using spatial

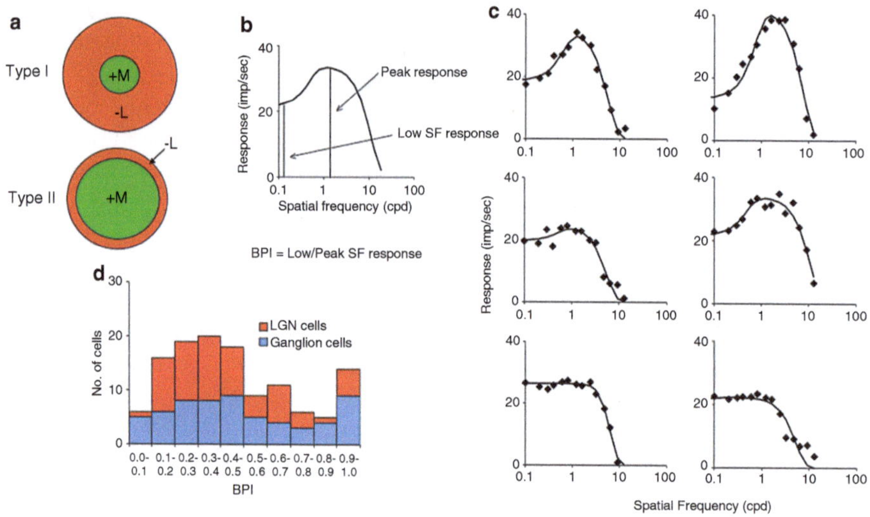

Fig. 4.4 Receptive field structure of M/L cone opponent cells. (**a**) Wiesel and Hubel [28] described Type I and Type II receptive field structure. S-cone cells show Type II structure. M, L cone-opponent cells show more variability. This can be quantified from spatial frequency tuning curves to achromatic gratings. (**b**, **c**). Tuning curves are shown (**c**) for PC retinal ganglion cells (2 Hz, 50 % contrast). A variety of tuning curve shapes can be seen; some show low spatial frequency attenuation, indicating center-surround receptive field structure, and others do not, corresponding to Type II structure. To quantify this difference, a band-pass index (BPI) was calculated, the ratio of the low spatial frequency to the peak spatial frequency response (**b**). (**d**) Distribution of BPI from samples of retinal ganglion cells and LGN neurons [82]. Some cells show clear Type I structure (e.g., BPI 0.4–0.6) but there is a continuum toward Type II (BPI 0.9–1.0). This is unusual for ganglion cells and calls into question the role of these cells in contrast enhancement

frequency tuning curves; a bandpass tuning curve indicates a center-surround structure, whereas a lowpass tuning curve indicate little or no center-surround structure, i.e., Type II. In Derrington and Lennie [14] Fig. 4.3, the variability in center-surround structure can be seen, and a set of similar data for macaque retinal ganglion cells are shown in Fig. 4.4c. Each panel shows responses of a PC ganglion cell to achromatic gratings as a function of spatial frequency, each curve fitted with a difference-of-Gaussians model. The range of spatial frequency tuning is apparent. To quantify this, we calculate a band-pass index (BPI), the ratio of the response at the lowest spatial frequency response to the response at the peak the tuning curve (Fig. 4.4b). A ratio of one indicates a Type II structure and a ratio of zero a very bandpass tuning. Figure 4.4d shows distributions of the BPI for samples of retinal ganglion and LGN PC cells [82]; distributions are similar. About 25 % cells have a BPI above 0.7. This indicates the variation in spatial structure of the M/L-cone opponent inputs to PC cells, which has functional implications. It is usually thought that center-surround structures in ganglion cell receptive fields help to provide a contrast border signal, but if center-surround structure is minimal or absent in a substantial fraction of cells, this cannot be the case. It severely restricts the ability of PC cells to transmit an achromatic contrast border signal.

Although the basic properties of M/L cone opponency in these cells are well established, there are several unresolved issues concerning the generation of opponency in this pathway. We now briefly point up these issues; their full discussion is beyond the scope of this review.

4.4 Peculiarities of the Parvocellular Pathway

4.4.1 Center Size of PC Receptive Fields

One puzzling issue about the PC pathway is center size. It is often supposed that the center sizes of midget ganglion cells near the fovea must be very small. The midget anatomy, with a single cone connection to a single ganglion cell via a midget bipolar seemed an obvious substrate for this (see above). However, measured center sizes are much larger (reviewed in Lee [45]), scarcely smaller than the centers of MC cells. The primary reason for this is likely to be optical blur. Foveal cone diameter is ca. 0.5 min of arc [83] but the point spread function for white light is several minutes of arc across [84]. Functionally, centers cannot be smaller than this. In this context, the midget system itself is unusual. Usually, the major site of spatial convergence from receptors via bipolars to ganglion cells is the ganglion cell dendritic tree, and this largely defines center size. This is the case in cat X and Y cells and MC cells [44, 85–87]. But midget, PC cell deviate from this pattern. Lastly, measurements of PC cell centers with interference fringes [88] or adaptive optics [89] indicate a more complex center structure than suggested by midget anatomy; if this is born out, the apparent simplicity of midget anatomy might be misleading.

4.4.2 Origin of the PC Receptive Field Surrounds

A second issue is the origin of the PC cells' surround. In early work, this was usually supposed to be cone specific. However, it was pointed out that just having a cone-specific center is enough to give cone opponency; with random, mixed M/L-cone input to the surround, cone opponency will still be present [77, 90]. This concept is attractive for two reasons. When dual M/L cone opsins evolved in primates, random connectivity had to be enough to generate an opponent signal. Secondly, a physiological substrate for cone-specific surrounds has not been found. The H1 horizontal cell receives mixed, additive input just from M and L cones [4], and there is no obvious physiological substrate for a cone-specific surround in the IPL [91]. Inhibitory pathways in the IPL do not seem to be involved [43]. Yet physiological measurements clearly indicate a considerable degree of cone specificity to the surround, though not always complete [72, 81, 82, 92].

It can be argued that at least some degree of specificity improves the signal-to-noise ratio of the opponent signal in the presence of receptor noise [93]. Perhaps cone-specific surround wiring evolved from a random pattern for this purpose. In any event, it is of note that S-cone input to the surround is not present [94]. MC cells also lack S-cone input. The retina can thus make cone specific wiring if it wants. The functional significance of the avoidance of S cones by the PC and MC pathways is unknown (see Chap. 3 for results that imply S-cone input to all ganglion cells).

4.4.3 M/L-Cone Opponency of PC Receptive Fields Toward Retinal Periphery

A further issue is the degree of M/L-cone opponency toward the retinal periphery. Although the red-green dimension of color vision is still obvious in the periphery [95, 96], there is some evidence that psychophysical detection along the M/L-cone axis has higher thresholds than in the fovea, even after appropriate scaling of stimulus size [97] (see also Chaps. 3 and 5). As noted above, the convergence of midget bipolars onto midget ganglion cells becomes substantial in the periphery, although the single cone to midget bipolar selectivity is maintained (see above). It was suggested that this convergence might cause a loss of M/L opponency, accounting for the psychophysical results [98–100]. However, those psychophysical changes in sensitivity seem to be clearly apparent at 10–15° eccentricity, where the midget morphology is well maintained. Also, a physiological study showed that, with large targets, many midget ganglion cells at high eccentricities are as responsive to chromatic, M/L-cone modulation as central cells [101, 102]. This would mean that a loss of psychophysical sensitivity occurs at a central site. The authors of this earlier study suggested that the anisotropic midget ganglion morphology at high eccentricity might underlie selective connectivity [101, 102]. However, this has been shown not be the case [43]. In the in vivo preparation, spatial frequency tuning curves were measured for gratings modulating either the L or M cones. At high spatial frequencies, responses to both cones were in phase, confirming summed input, consistent with summed M/L-cone input to the dendritic tree of the peripheral midget cell. Yet at low spatial frequencies, strongly opponent responses were found. It is difficult to avoid the conclusion that some mechanism in the IPL is devoted to generating a balanced M/L opponent signal. In any event, the changes of M/L opponency with eccentricity and associated perceptual changes involve a complex of anatomical, physiological and central factors.

In this context, it is constructive to compare the balanced M/L cone inputs to PC cells to the variable weighting of M/L cone inputs to centers of MC cells. These inputs sum, but the spectral sensitivity of the center is variable, consistent with the center drawing input at random from the underlying cone matrix. Spectral sensitivity corresponds to ca. a 2:1L:M cone ratio. This contrasts with the consistent 1:1 opponent M/L balance in PC cells [103]. How this is achieved in the presence of the frequent patches of M or L cones even in random cone arrays [104, 105] has received little attention. One possibility might be that the variability in center surround structure of PC cells has to do with patchiness in the underlying cone mosaic; techniques where responses can be recorded from PC cells while stimulating individual cones [89] could tell more about the details of PC cell receptive field structure.

Although it is often assumed that the functional architecture of the PC pathway is well understood, we stress here that it has several unusual and unexplained features. Many of these may result from its specialization for extracting an M/L opponent signal.

4.5 Color Opponency in New World Monkeys

4.5.1 Allelic or Incomplete Trichromacy in New World Monkeys

In most platyrrhines (New World monkeys), other than in catarrhines (Old World monkeys, apes, and humans), only a proportion of females are trichromats. The remaining females and all males are dichromats. In each individual there are only two genes for coding cone opsins, one on the X-chromosome that codes for an M, L, or ML-opsin (with intermediate spectral sensitivity between M and L opsins), and another on chromosome 7 that codes for an S opsin (see Chap. 1). In males, with only one X-chromosome, and homozygous females, with two identical X-chromosomes, the resulting phenotype is dichromacy, while in heterozygous females, gene polymorphism permits trichromacy; there is a variable number and frequency of alleles in different species [23, 106]. This "incomplete" or "allelic" form of trichromacy found in New World monkeys was well documented in a series of genetic, MSP, ERG, and behavioral experiments [21, 23, 107]. One possible scenario for primate evolution was that the common ancestor of catarrhines and platyrrhines was a dichromat and independent evolution of "incomplete trichromacy" in platyrrhines and "full trichromacy" in catarrhines occurred after the split of these two groups about 40 mya (million years ago) [23]. Another possibility is that the common ancestor of catarrhines and platyrrhines had already enjoyed "incomplete trichromacy"; opsin polymorphism is also found in some prosimians [108, 109]. In this case, catarrhines evolved later to "full trichromacy" while platyrrhines conserved "incomplete trichromacy." Evolution of color vision is treated elsewhere in this volume (Chap. 11).

Several studies on the visual system of New World monkeys have been performed in squirrel monkeys (*Saimiri*), capuchins (*Cebus*), and marmosets (*Callithrix*), all platyrrhines with "incomplete trichromacy." Major retinal anatomical and physiological characteristics are generally conserved in comparison with Old World primates, although there may be differences in detail [110, 111]. Single unit recordings of retinal ganglion cells and LGN neurons in these primates have confirmed the existence of M/L cone opponent PC cells in heterozygous females while homozygous females and males only have achromatic, non cone-opponent PC cells [112–114]. PC cells of dichromats are very similar to M/L cone opponent PC cells, except that they are "red-green color blind." As expected, +S−ML and −S+ML cells in dichromats and trichromats are also similar with the exception that the ML mechanism of dichromats is driven by a single cone type [47, 58, 115, 116].

4.5.2 *Other New World Monkeys*

There are two important exceptions for this pattern of opsin expression among New World monkeys. One is the howler monkey (*Alouatta*), the only New World monkey endowed with full trichromacy [117]. ERG and genetic investigations showed that the retinas of howlers contain two spectrally distinct M and L opsins and MSP identified the corresponding M and L cone classes with absorption peak values similar to those of catarrhines [117, 118]. Genetic studies suggested that howlers evolved trichromacy independently from catarrhines much more recently [23, 119]. Other aspects similar to catarrhines are the opsin absorption peaks and absence of gene polymorphism. Single unit recordings found M/L cone-opponent PC ganglion cells that respond similarly to PC cells of other trichromats both catarrhines and platyrrhines [118]. Appropriate behavioral tests with howlers are limited but they confirmed that howlers display behavioral trichromacy ([120]; LD Henriques, PRK Goulart, DMO Bonci, LCL Silveira, OF Galvão, and DF Ventura, unpublished observations).

The other exception is the owl monkey (*Aotus*), the only living nocturnal monkey, a monochromat with a single cone class, as revealed by genetic, anatomical, and ERG studies [117, 121, 122]. In the owl monkey retina, the S opsin gene is not expressed and all cones express the same ML opsin. In addition, no ML polymorphism was described in this primate and it seems that the opsin found in the retina of all individuals is the same. It is not clear what evolutionary advantages the owl monkey and other nocturnal primates obtained by adopting monochromacy, but this form of vision is also found in other nocturnal and marine mammals and vertebrates and thus seems to be imposed by the necessity to catch light at low levels of retinal illuminance [123] (see also Chap. 11). PC cells lacking cone opponency have been found in the owl monkey retina [110, 124] and LGN [125].

We have stressed studies on New World primates since it is hoped that work on these species (and prosimians) might help us understand how cone opponency evolved. So far these comparative studies have raised more questions than have been answered. Nevertheless, it is remarkable that when dual M, L opsins are present in an individual, cone opponent properties are apparent physiologically, as is behavioral trichromacy. The necessary retinal machinery must be present in the developing retina to exploit this eventuality.

4.6 Cone Opponency Through the Visual System and Functional Implications

Beyond the LGN, the strict separation of cone and color opponent pathways into L,M opponent and S vs. LM systems becomes much less clear. This is puzzling in view of the care with which S cone input is excluded from the PC and MC pathways below the cortex, as discussed in a previous section [94]. Early studies [126] suggested quite substantial mixing of S-cone opponent signals with those of other pathways in V1 but more recent results suggest a more complex situation (reviewed in Shapley and Hawken [127]). In particular, cells with what appear to be mixed PC and MC inputs are found. However, there are strongly opponent cells that resemble L,M opponent and S vs. LM cells of the LGN. These may even be segregated into different blobs [128]; these are small regions of cortex, arranged in regular arrays across the cortical surface and thought to receive predominantly PC input. Whether these different varieties of color opponency in the cortex, including double opponent cells, represent distinct processing streams remains an open question.

Another issue in cortical processing concerns how far it is possible to identify neurophysiological correlates of psychophysical color opponency; as mentioned in a previous section, physiological cone opponency does not precisely match the perceptual dimensions of color opponency, such as the unique hues [16] (for an explanation for why this may be see also Chap. 7 and Sect. 7.3.2). Although it has been suggested that responses of cells in inferotemporal cortex may fit with perceptual opponency [129], this is controversial [130]; it is not clear whether a substrate for unique hues at a single cell level can be expected, even if apparent with fMRI [131].

From a theoretical viewpoint, transmission of chromatic information in an achromatic and two cone-opponent channels has been shown to be the most efficient way of coding spectral composition. This was first shown by Buchsbaum and Gottschalk [132] based on pigment spectra; a similar conclusion was reached by MacLeod and von der Twer [133] in a more neurophysiological context. In natural scenes, variation in luminance is very much larger than variation in the |L–M| or |S–ML| signal, and these latter authors suggest that the MC, PC, and S-cone pathways are well adapted to information transmission requirements given natural scene statistics.

Although unique hues may not match cone-opponent pathways, detection of color changes does map well onto detection by these pathways. This is best studied in a cone opponent space such as that in Fig. 4.2, which can be made three-dimensional by including an S-cone axis. If, instead of response amplitude as in Fig. 4.2, psychophysical thresholds are plotted, the resulting surface consists of planar segments [10, 134]. This is consistent with modulation along different vectors in the space being detected by either a luminance or one of two chromatically opponent channels. Equivalent physiological analysis strongly suggests that these correspond to the MC pathway, PC pathway and S-cone system [80].

A psychophysical luminance channel is thought to underlie human performance on psychophysical tasks defining the luminosity function, such as heterochromatic flicker photometry (HFP) or the minimally distinct border (MDB). In these psychophysical tasks, two lights of different spectral composition alternate in time or space, and the human observer has to adjust the relative intensity of the two lights until either the sensation of flicker or border distinctness is minimized. There is good evidence that the physiological substrate for performance on these tasks rests on parasol ganglion cells of the MC pathway [135, 136]. It has been suggested that luminance is a cortical construct [13], built up somehow from PC pathway activity. This seems unlikely for two reasons. This suggestion is based on earlier analyses, in which an achromatic signal might be built up by adding activities of different PC cell classes, so that the chromatic signals cancel to leave an achromatic component [137, 138]. This is superficially plausible, but although PC cells are linear in their properties, they display one major linearity which makes untenable the construction of an achromatic signal from their responses [103]; this is the rectification associated with the impossibility of negative firing rates. In addition, many details of HFP and MDB, such as odd effects of changing the relative phase of the modulated lights in HFP, are also found in MC cells [139, 140].

The discussion often focuses around the role of cone-opponent cells, especially PC cells, in achromatic pattern vision at high spatial frequencies. This discussion goes beyond the scope of the current review, but it should be pointed out the information-based arguments mentioned above should hold whatever the spatial frequency. Also, recent studies using patterns with a complex spatial frequency spectrum with both luminance and chromatic components, strongly suggest a highly specific segregation of luminance and chromatic information in MC and PC pathways respectively, independent of spatial frequency [141, 142].

Signals from cone-opponent pathways are not only concerned with color. Chromatic composition is an important component in spatial vision [143]; texture and depth perception can utilize color cues (although binocular, disparity-based depth perception seems to be luminance driven [144]; see Chap. 8).

Another, related role of cone-/color-opponent channels may be to distinguish surface characteristics of objects (brightness and spectral composition) from spatial structure. In the example shown below [145], responses of cells to Mondrian patterns drifted across the receptive field in different directions are shown (see Fig. 4.5). The examples shown are from a +S−ML cell, +M−L, and +L−M PC cells from a MC on-center cell. The MC cells respond primarily to luminance contrast edges within the Mondrian whereas the cone opponent cells appear to represent spectral composition of surfaces within the Mondrian. For example, the +S−ML cell responds primarily to the blue area. One reason for the vigorous response of MC cells to edges is that when patterns are drifted across the retina, high spatial frequency components in the image (i.e., edges) are preferentially coded in the MC pathway, since their bandpass temporal response selects high spatial frequency components in the image [146]. Computer algorithms for analysis of visual patterns have attempted to distinguish surface characteristics of objects from shape and shading using similar algorithms. It seems apparent that spatial information derived from color extends the range of the visual system.

In a similar vein, it has been hypothesized that MC and PC cells perform simultaneous and overlapping representations of the visual field using different

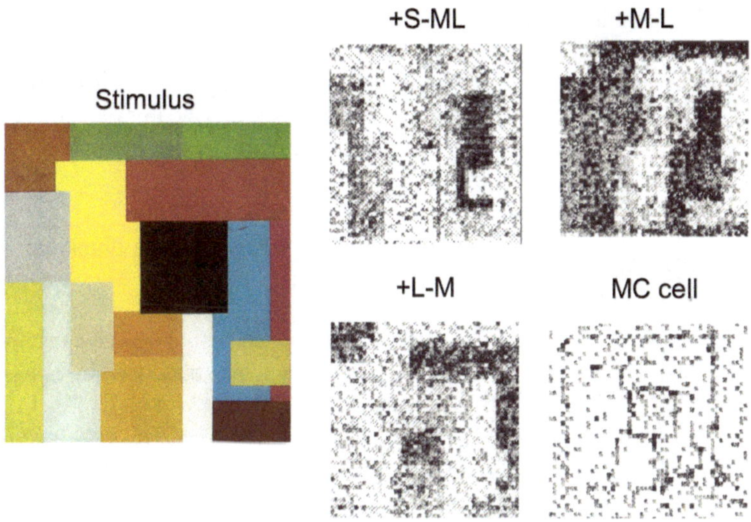

Fig. 4.5 Roles of KC, PC, and MC cells in transmission of chromatic and pattern information. A Mondrian pattern was drifted in different directions across the receptive field of different cell types [145, 153]. The raster plots show responses of different cell types to the pattern. Cone-opponent cells show sustained responses to different parts of the Mondrian, whereas the MC cell responds predominantly to contrast edges. All cell types deliver spatial information, but these data graphically demonstrate that cone-opponent cells are able to deliver information about chromatic and brightness composition of surfaces

strategies to minimize "entropy" [147, 148], defined as the uncertainty involved in simultaneously transmitting spatial, temporal and chromatic information. This would enable higher order visual neurons [149] to selectively combine MC and PC inputs in different ways, and could explain why MC and PC inputs need to converge at the high levels of the visual system after being kept separate at the subcortical levels. For example, psychometric functions were used to estimate the joint entropy for spatial extent and spatial frequency discrimination [150]. At low and intermediate spatial frequencies and high contrasts, joint entropy reached levels below the theoretical minimum, suggesting nonlinear interactions between two or more visual mechanisms such as the MC and PC pathways.

4.7 Summary

Cone opponency and its associated mechanisms play a critical role in primate vision. Trichromacy re-evolved in primates and superimposed itself on an existing set of mammalian visual pathways. One way this might have occurred is that a high-resolution ganglion cell class, with midget morphology, was taken over to provide M,L color opponency [151]. We argue above that optical and anatomical factors make this unlikely. Shapley and Perry [152] reached similar conclusions. In any event, cone opponency and its utility in visual processing is the theme of other chapters in this volume; if a clear hypothesis as to the evolution of trichromacy in primates, and the associated pathways, were to emerge, a clearer division of labor between MC, PC, and KC systems might become more apparent.

4.8 Perspectives

There are still many lacunae in a comprehensive description of cells and mechanisms responsible for cone opponency and how they contribute to color vision [31]. These include (1) How is PC M/L opponency generated? (2) Which retinal circuits and cells are involved in the −S+ML mechanisms? (3) Are LGN cells just retinal relay cells or do they modify their retinal inputs? (4) How is retino-geniculate input translated into visual cortical activity and color perception? These are achievable goals using the new techniques available for neuronal stimulation, recording, and labeling both in vitro and in vivo.

Acknowledgements Luiz Carlos da Silva Silveira passed away on 10th July 2016. This work was partially supported by NEI 13112 and CNPq 442191/2014-6. LCLS is a CNPq research fellow.

References

1. Svaetichin G, MacNichol EF. Retinal mechanisms for chromatic and achromatic vision. Ann N Y Acad Sci. 1958;74:385–404.
2. Tomita T. Electrophysiological study of the mechanisms subserving color coding in the fish retina. Cold Spring Harb Symp Quant Biol. 1965;30:559–66.
3. Twig G, Perlman I. Homogeneity and diversity of color-opponent horizontal cells in the turtle retina: consequences for potential wavelength discrimination. J Vis. 2004;4:403–14.
4. Dacey DM, Lee BB, Stafford DK, Pokorny J, Smith VC. Horizontal cells of the primate retina: cone specificity without spectral opponency. Science. 1996;271:656–9.
5. Daw NW. Goldfish retina: organization for simultaneous color contrast. Science. 1967;158:942–4.
6. Daw NW. Colour-coded ganglion cells in the goldfish retina: extension of their receptive fields by means of new stimuli. J Physiol. 1968;197:567–92.
7. DeValois RL, Morgan C, Polson MC, Mead WR, Hull EM. Psychophysical studies of monkey vision-I. Macaque luminosity and color vision tests. Vision Res. 1974;14:53–67.
8. DeValois RL, Morgan H, Snodderly DM. Psychophysical studies of monkey vision—III. Spatial luminance contrast sensitivity tests of macaque and human observers. Vision Res. 1974;14:75–81.
9. Yovel G, Freiwald WA. Face recognition systems in monkey and human: are they the same thing? F1000Prime Rep. 2013;5:10.
10. Cole GR, Hine T, McIlhagga W. Detection mechanisms in L-, M- and S-cone contrast space. J Opt Soc Am A. 1993;10:38–51.
11. Barlow HB, Levick WR. Three factors limiting the reliable detection of light by retinal ganglion cells of the cat. J Physiol. 1969;200:1–24.

12. Barlow HB. Single units and sensation: a neuron doctrine for perceptual psychology? Perception. 1972;1:371–94.
13. Lennie P, Movshon JA. Coding of color and form in the geniculostriate visual pathway (invited review). J Opt Soc Am A. 2005;22:2013–33.
14. Derrington AM, Lennie P. Spatial and temporal contrast sensitivities of neurones in lateral geniculate nucleus of macaque. J Physiol. 1984;357:219–40.
15. Krauskopf J, Williams DR, Heeley DW. Cardinal directions of color space. Vision Res. 1982;22:1123–31.
16. Mollon JD, Jordan G. On the nature of unique hues. In: Dickinson C, Murray I, Carden D, editors. John Dalton's colour vision legacy. London: Taylor & Francis; 1997. p. 381–92.
17. Pope A. An essay on man. Moral essays and satires. London, UK: Cassell & Company, Limited; 1734. Edition of 1891.
18. Baker M. Through the eyes of a mouse. Nature. 2013;502:156–8.
19. Bowmaker JK, Dartnall HJA. Visual pigments of rods and cones in a human retina. J Physiol. 1980;298:501–11.
20. Bowmaker JK, Dartnall HJA, Mollon JD. Microspectrophotometric demonstration of four classes of photoreceptor in an old world primate, macaca fascicularis. J Physiol. 1980;298:131–43.
21. Mollon JD, Bowmaker JK, Jacobs GH. Variations in colour vision in a new-world monkey can be explained by a polymorphism of retinal photopigments. Proc R Soc B. 1984;222:373–89.
22. Nathans J, Piantanida TP, Eddy RL, Shows TB, Hogness DS. Molecular genetics of inherited variation in human color vision. Nature. 1986;232:203–10.
23. Hunt DM, Dulai KS, Cowing JA, Julliot C, Mollon JD, Bowmaker JK, Li W-H, Hewett-Emmett D. Molecular evolution of trichromacy in primates. Vision Res. 1998;38:3299–306.
24. Dulai KS, von Dornum M, Mollon JD, Hunt DM. The evolution of trichromatic color vision by opsin gene duplication in New World and Old World primates. Genome Res. 1999;9:629–38.
25. DeValois RL, Smith CJ, Kitai ST, Karoly AJ. Response of single cells in monkey lateral geniculate nucleus to monochromatic light. Science. 1958;127:238–9.
26. Devalois RL, Jacobs GH, Abramov I. Responses of single cells in visual system to shifts in the wavelength of light. Science. 1964;146:1184–6.
27. DeValois RL, Abramov I, Jacobs GH. Analysis of response patterns of LGN cells. J Opt Soc Am. 1966;56:966–77.
28. Wiesel T, Hubel DH. Spatial and chromatic interactions in the lateral geniculate body of the rhesus monkey. J Neurophysiol. 1966;29:1115–56.
29. Gouras P. Identification of cone mechanisms in monkey ganglion cells. J Physiol. 1968;199:533–47.
30. de Monasterio FM, Gouras P. Functional properties of ganglion cells of the rhesus monkey retina. J Physiol. 1975;251:167–95.
31. Lee BB. Color coding in the primate visual pathway: a historical view. J Opt Soc Am A. 2014;31:A103–112.
32. Derrington AM, Krauskopf J, Lennie P. Chromatic mechanisms in lateral geniculate nucleus of macaque. J Physiol. 1984;357:241–65.
33. Lee BB, Valberg A, Tigwell DA, Tryti J. An account of responses of spectrally opponent neurons in macaque lateral geniculate nucleus to successive contrast. Proc R Soc B. 1987;230:293–314.
34. Boycott BB, Kolb H. The horizontal cells of the rhesus monkey retina. J Comp Neurol. 1973;148:91–114.
35. Boycott BB. Horizontal cells of mammalian retinae. Neurosci Res. 1988;8:97–111.
36. Boycott BB, Wässle H. Morphological classification of bipolar cells of the primate retina. Eur J Neurosci. 1991;3:1069–88.
37. Dacey DM, Lee BB. The blue-ON opponent pathway in primate retina originates from a distinct bistratified ganglion cell type. Nature. 1994;367:731–5.

38. Dacey DM. Circuitry for color coding in the primate retina. Proc Natl Acad Sci. 1996;93:582–8.
39. Dacey DM, Liao HW, Peterson BB, Robinson FR, Smith VC, Pokorny J, Yau KW, Gamlin PD. Melanopsin-expressing ganglion cells in primate retina signal colour and irradiance and project to the LGN. Nature. 2005;433:749–54.
40. Diller L, Packer OS, Verweij J, McMahon MJ, Williams DR, Dacey DM. L and M cone contributions to the midget and parasol ganglion cell receptive fields of macaque monkey retina. J Neurosci. 2004;24:1079–88.
41. Crook JD, Peterson BB, Packer OS, Robinson FR, Gamlin PD, Troy JB, Dacey DM. The smooth monostratified ganglion cell: evidence for spatial diversity in the Y-cell pathway to the lateral geniculate nucleus and superior colliculus in the macaque monkey. J Neurosci. 2008;28:12654–71.
42. Crook JD, Peterson BB, Packer OS, Robinson FR, Troy JB, Dacey DM. Y-cell receptive field and collicular projection of parasol ganglion cells in macaque monkey retina. J Neurosci. 2008;28:11277–91.
43. Crook JD, Manookin MB, Packer OS, Dacey DM. Horizontal cell feedback without cone type-selective inhibition mediates "red-green" color opponency in midget ganglion cells of the primate retina. J Neurosci. 2011;31:1762–72.
44. Lee BB, Martin PR, Grünert U. Retinal connectivity and primate vision. Prog Retin Res. 2010;29:622–39.
45. Lee BB. Visual pathways and psychophysical channels in the primate. J Physiol. 2011;589:41–7.
46. Valberg A, Lee BB, Tigwell DA. Neurons with strong inhibitory S-cone inputs in the macaque lateral geniculate nucleus. Vision Res. 1986;26:1061–4.
47. Martin PR, White AJR, Goodchild AK, Wilder HD, Sefton AE. Evidence that blue-on cells are part of the third geniculocortical pathway in primates. Eur J Neurosci. 1997;9:1536–41.
48. Dacey DM, Packer OS. Colour coding in the primate retina: diverse cell types and cone-specific circuitry. Curr Opin Neurobiol. 2003;13:421–7.
49. Dacey DM, Peterson BB, Robinson FR. Identification of an S-cone opponent OFF pathway in the macaque retina: morphology, physiology and possible circuitry. Invest Ophthalmol Vis Sci. 2002;43:E-Abstract 2983.
50. Martin PR, Lee BB. Distribution and specificity of S-cone ("blue cone") signals in subcortical visual pathways. Vis Neurosci. 2014;31:177–87.
51. Polyak SL. The Retina. Chicago: University of Chicago Press; 1941.
52. Rodieck RW. Which cells code for color? In: Valberg A, Lee BB, editors. From pigments to perception. New York: Plenum; 1991.
53. Dacey DM. Morphology of a small-field bistratified ganglion cell type in the macaque and human retina. Vis Neurosci. 1993;10:1081–98.
54. Field GD, Sher A, Gauthier JL, Greschner M, Shlens J, Litke AM, Chichilnisky EJ. Spatial properties and functional organization of small bistratified ganglion cells in primate retina. J Neurosci. 2007;27:13261–72.
55. Crook JD, Davenport CM, Peterson BB, Packer OS, Detwiler PB, Dacey DM. Parallel ON and OFF cone bipolar inputs establish spatially coextensive receptive field structure of blue-yellow ganglion cells in primate retina. J Neurosci. 2009;29:8372–87.
56. Forte JD, Hashemi-Nezhad M, Dobbie WJ, Dreher B, Martin PR. Spatial coding and response redundancy in parallel visual pathways of the marmoset Callithrix jacchus. Vis Neurosci. 2005;22:479–91.
57. Tailby C, Solomon SG, Lennie P. Functional asymmetries in visual pathways carrying S-cone signals in macaque. J Neurosci. 2008;28:4078–87.
58. Tailby C, Szmajda BA, Buzas P, Lee BB, Martin PR. Transmission of blue (S) cone signals through the primate lateral geniculate nucleus. J Physiol. 2008;586:5947–67.
59. Yeh T, Lee BB, Kremers J. Temporal contrast sensitivities measured in retinal parvocellular cells of the macaque. Perception. 1992;21 Suppl 2:9.

60. Mariani AP. Bipolar cells in monkey retina selective for the cone likely to be blue-sensitive. Nature. 1984;308:184–6.
61. Kouyama N, Marshak DW. Bipolar cells specific for blue cones in the macaque retina. J Neurosci. 1992;12:1233–52.
62. Calkins DJ, Tsukamoto Y, Sterling P. Microcircuitry and mosaic of a blue-yellow ganglion cell in the primate retina. J Neurosci. 1998;18:3373–85.
63. Herr S, Klug K, Sterling P, Schein S. Inner S-cone bipolar cells provide all of the central elements for S cones in macaque retina. J Comp Neurol. 2003;457:185–201.
64. Percival KA, Martin PR, Grünert U. Organisation of koniocellular-projecting ganglion cells and diffuse bipolar cells in the primate fovea. Eur J Neurosci. 2013;37:1072–89.
65. Dacey DM, Peterson BB, Robinson FR, Gamlin PD. Fireworks in the primate retina: in vitro photodynamics reveals diverse LGN-projecting ganglion cell types. Neuron. 2003;37:15–27.
66. Klug K, Herr S, Ngo IT, Sterling P, Schein S. Macaque retina contains an S-cone OFF midget pathway. J Neurosci. 2003;23:9881–7.
67. Dacey DM, Crook JD, Packer OS. Distinct synaptic mechanisms create parallel S-ON and S-OFF color opponent pathways in the primate retina. Vis Neurosci. 2014;31:139–51.
68. Lee SC, Telkes I, Grunert U. S-cones do not contribute to the OFF-midget pathway in the retina of the marmoset, Callithrix jacchus. Eur J Neurosci. 2005;22:437–47.
69. Sher A, DeVries SH. A non-canonical pathway for mammalian blue-green color vision. Nat Neurosci. 2012;15:952–3.
70. Chen S, Li W. A color-coding amacrine cell may provide a blue-off signal in a mammalian retina. Nat Neurosci. 2012;15:954–6.
71. Roy S, Jakakumar J, Martin PR, Reher BD, Saalmann YB, Hu D, Vidyasaga TR. Segregation of short-wavelength-sensitive (S) cone signals in the macaque dorsal lateral geniculate nucleus. Eur J Neurosci. 2009;30:1517–26.
72. Reid RC, Shapley RM. Space and time maps of cone photoreceptor signals in macaque lateral geniculate nucleus. J Neurosci. 2002;22:6158–75.
73. Boycott BB, Dowling JE. Organization of the primate retina: light microscopy. Phil Trans R Soc B. 1969;255:109–84.
74. Kolb H, Dekorver L. Midget ganglion cells of the parafovea of the human retina: a study by electron microscopy and serial section reconstructions. J Comp Neurol. 1991;303:617–36.
75. Wässle H, Grünert U, Martin PR, Boycott BB. Color coding in the primate retina: predictions and constrants from anatomy. In: Albowitz B, Albus K, Kuhnt U, Nothdurft HC, Wahle P, editors. Structural and functional organization of the neocortex. A symposium in the memory of Otto D. Creutzfeldt. Berlin: Springer; 1994. p. 94–104.
76. Dacey DM. The mosaic of midget ganglion cells in the human retina. J Neurosci. 1993;13:5334–55.
77. Lennie P, Haake PW, Williams DR. The design of chromatically opponent receptive fields. In: Landy MS, Movshon JA, editors. Computational models of visual processing. Cambridge, MA, USA: MIT Press; 1991. p. 71–82.
78. Kaplan E, Shapley RM. X and Y cells in the lateral geniculate nucleus of the macaque monkeys. J Physiol. 1982;330:125–43.
79. Hicks TP, Lee BB, Vidyasagar TR. The responses of cells in macaque lateral geniculate nucleus to sinusoiodal gratings. J Physiol. 1983;337:183–200.
80. Lee BB, Martin PR, Valberg A, Kremers J. Physiological mechanisms underlying psychophysical sensitivity to combined luminance and chromatic modulation. J Opt Soc Am A. 1993;10:1403–12.
81. Lee BB, Kremers J, Yeh T. Receptive fields of primate ganglion cells studied with a novel technique. Vis Neurosci. 1998;15:161–75.
82. Lee BB, Shapley RM, Hawken MJ, Sun H. The spatial distributions of cone inputs to cells of the parvocellular pathway investigated with cone-isolating gratings. J Opt Soc Am A. 2012;29(2):A223–32.

83. Packer O, Hendrickson AE, Curcio CA. Photoreceptor topography of the retina in the adult pigtail macaque (macaca nemestrina). J Comp Neurol. 1989;288:165–83.
84. Campbell FW, Gubisch RW. Optical quality of the human eye. J Physiol. 1966;186:558–78.
85. Peichl L, Wässle H. Morphological identification of on- and off-centre brisk transient (Y) cells in the cat retina. Proc R Soc B. 1981;212:139–56.
86. Wässle H, Peichl L, Boycott BB. Dendritic territories of cat retinal ganglion cells. Nature. 1981;292:344–5.
87. Wässle H, Peichl L, Boycott BB. Morphology and topography of on- and off-alpha cells in the cat retina. Proc R Soc B. 1981;212:157–75.
88. McMahon MJ, Lankheet MJ, Lennie P, Williams DR. Fine structure of parvocellular receptive fields in the primate fovea revealed by laser interferometry. J Neurosci. 2000;20:2043–53.
89. Sincich LC, Zhang Y, Tiruveedhula P, Horton JC, Roorda A. Resolving single cone inputs to visual receptive fields. Nat Neurosci. 2009;12:967–9.
90. Paulus W, Kröger-Paulus A. A new concept of retinal colour coding. Vision Res. 1983;23:529–40.
91. Calkins DJ, Sterling P. Absence of spectrally specific lateral inputs to midget ganglion cells in primate retina. Nature. 1996;381:613–5.
92. Reid RC, Shapley RM. Spatial structure of cone inputs to receptive fields in primate lateral geniculate nucleus. Nature. 1992;356:716–8.
93. Lee BB. Neural models and physiological reality. Vis Neurosci. 2008;251:231–41.
94. Sun H, Smithson H, Zaidi Q, Lee BB. Specificity of cone inputs to macaque ganglion cells. J Neurophysiol. 2006;95:837–49.
95. Abramov I, Gordon J. Color vision in the peripheral retina. I. Spectral sensitivity. J Opt Soc Am. 1977;67:195–202.
96. Abramov I, Gordon J, Chan H. Color appearance in the peripheral retina: effects of stimulus size. J Opt Soc Am A. 1991;8:404–14.
97. Anderson SJ, Mullen KT, Hess RF. Human peripheral spatial resolution for achromatic and chromatic stimuli: limits imposed by optical and retinal factors. J Physiol. 1991;442:47–64.
98. Mullen KT, Kingdom FA. Losses in peripheral colour sensitivity predicted from "hit or miss" post-receptoral cone connections. Vision Res. 1996;36:1995–2000.
99. Mullen KT, Kingdom FA. Differential distributions of red-green and blue-yellow cone opponency across the visual field. Vis Neurosci. 2002;19:109–18.
100. Mullen KT, Sakurai M, Chu W. Does L/M cone opponency disappear in human periphery? Perception. 2005;34:951–9.
101. Martin PR, Lee BB, White AJ, Solomon SG, Rüttiger L. Chromatic sensitivity of ganglion cells in peripheral primate retina. Nature. 2001;410:933–6.
102. Solomon SG, Lee BB, White AJ, Ruttiger L, Martin PR. Chromatic organization of ganglion cell receptive fields in the peripheral retina. J Neurosci. 2005;25:4527–39.
103. Valberg A, Lee BB, Kaiser PK, Kremers J. Responses of macaque ganglion cells to movement of chromatic borders. J Physiol. 1992;458:579–602.
104. Roorda A, Metha AB, Lennie P, Williams DR. Packing arrangement of the three cone classes in primate retina. Vision Res. 2001;41:1291–306.
105. Roorda A, Williams DR. The arrangement of the three cone classes in the living human eye. Nature. 1999;397:520–2.
106. Jacobs GH. New World monkeys and color. Int J Primatol. 2007;28:729–59.
107. Jacobs GH. Within-species variations in visual capacity among squirrel monkeys (Saimiri Sciureus): colour vision. Vision Res. 1984;24:1267–77.
108. Tan Y, Li W-H. Trichromatic vision in prosimians. Nature. 1999;402:36.
109. Jacobs GH, Deegan JF, Tan YC, Li WH. Opsin gene and photopigment polymorphism in a prosimian primate. Vision Res. 2003;42:11–8.
110. Silveira LC, Saito CA, Lee BB, Kremers J, da Silva Filho M, Kilavik BE, Yamada ES, Perry VH. Morphology and physiology of primate M- and P-cells. Prog Brain Res. 2004;144:21–46.

111. Silveira LCL, Grünert U, Kremers J, Lee BB, Martin PR. Comparative anatomy and physiology of the primate retina. In: Kremers J, editor. Structure, function and evolution of the primate visual system. Chichester, England: Wiley; 2005. p. 127–60.
112. Yeh T, Lee BB, Kremers J, Cowing JA, Hunt DM, Martin PR, Troy JB. Visual responses in the lateral geniculate of dichromatic and trichromatic marmosets (*Callithrix jacchus*). J Neurosci. 1995;15:7892–904.
113. Lee BB, Silveira LCL, Yamada E, Kremers J. Parallel pathways in the retina of old and new-world primates. Braz J Biol. 1996;1996:323–38.
114. Lee BB, Silveira LCL, Yamada ES, Hunt DM, Kremers J, Martin PR, Troy JB, Silva-Filho MD. Visual responses of ganglion cells of a New-World primate, the capuchin monkey, *Cebus apella*. J Physiol. 2000;528:573–90.
115. Silveira LC, Lee BB, Yamada ES, Kremers J, Hunt DM, Martin PR, Gomes FL. Ganglion cells of a short-wavelength-sensitive cone pathway in New World monkeys: morphology and physiology. Vis Neurosci. 1999;16:333–43.
116. Szmajda BA, Buzas P, Fitzgibbon T, Martin PR. Geniculocortical relay of blue-off signals in the primate visual system. Proc Natl Acad Sci. 2006;103:19512–7.
117. Jacobs GH. Primate photopigments and primate color vision. Proc Natl Acad Sci. 1996;93:577–81.
118. Silveira LCL, Saito CA, Silva MHD, Kremers J, Bowmaker JK, Lee BB. Alouatta trichromatic color vision: cone spectra and physiological responses studied with microspectrophotometry and single unit retinal electrophysiology. PLOS One. 2014;9.
119. Kainz PM, Neitz J, Neitz M. Recent evolution of uniform trichromacy in a new-world monkey. Vision Res. 1998;38:3315–20.
120. Araujo AC, Didonet JJ, Araujo JS, Saletti PG, Borges TR, Pessoa FV. Color vision in the black howler monkey (Alouatta caraya). Vis Neurosci. 2008;25:243–8.
121. Wikler KC, Rakic P. Distribution of photoreceptor subtypes in the retina of diurnal and nocturnal primates. J Neurosci. 1990;10:3390–401.
122. Jacobs GH, Deegan JF, Neitz J, Crognale MA, Neitz M. Photopigments and color vision in the nocturnal monkey, *Aotus*. Vision Res. 1993;33:1773–83.
123. Bowmaker JK. Evolution of vertebrate visual pigments. Vision Res. 2008;48:2022–41.
124. Saito CA, Lee BB, Kremers J, Silveira LCL, Silva MFD, Kilavek BE. Rod-cone interactions in the ganglion cell response: studies using the diurnal capuchin-monkey and the nocturnal owl-monkey. Invest Ophthalmol Vis Sci. 2001;42:S676.
125. Kilavek BE, Silveira LCL, Kremers J. Spatial receptive field properties of lateral geniculate cells in the owl monkey (Aotus azarae) at different contrasts: a comparative study. Eur J Neurosci. 2007;26:992–1006.
126. Lennie P, Krauskopf J, Sclar G. Chromatic mechanisms in striate cortex of macaque. J Neurosci. 1990;10:649–69.
127. Shapley R, Hawken MJ. Color in the cortex: single- and double-opponent cells. Vision Res. 2011;51:701–17.
128. Ts'o D, Gilbert CD. The organization of chromatic and spatial interactions in the primate striate cortex. J Neurosci. 1988;8:1712–27.
129. Stoughton CM, Conway BR. Neural basis for unique hues. Curr Biol. 2008;18:R698–9.
130. Mollon JD. A neural basis for unique hues? Curr Biol. 2009;19:R441–2. author reply R442–443.
131. Brouwer GJ, Heeger DJ. Categorical clustering of the neural representation of color. J Neurosci. 2013;25:1454–15465.
132. Buchsbaum G, Gottschalk A. Trichromacy, opponent colours coding and optimum colour information transmission in the retina. Proc R Soc B. 1983;220:89–113.
133. MacLeod DIA, Twer TVD. The pleistochrome: optimal opponent codes for natural colours. In: Mausfeld R, Heyer D, editors. Color perception: mind and the physical world. Oxford: Oxford University Press; 2003.
134. Cole GR, III CFS, Kronauer RE. Visual interactions with luminance and chromatic stimuli. J Opt Soc Am A. 1990;7:128–40.

135. Lee BB, Martin PR, Valberg A. The physiological basis of heterochromatic flicker photometry demonstrated in the ganglion cells of the macaque retina. J Physiol. 1988;404:323–47.
136. Kaiser PK, Lee BB, Martin PR, Valberg A. The physiological basis of the minimally distinct border demonstrated in the ganglion cells of the macaque retina. J Physiol. 1990;422:153–83.
137. Ingling Jr CR, Martinez-Uriegas E. The spatiotemporal properties of the r-g X-cell channel. Vision Res. 1985;25:33–8.
138. Lennie P, D'Zmura MD. Mechanisms of color vision. CRC Crit Rev Neurobiol. 1988;3:333–400.
139. Smith VC. Origin in the visual system of perceptually measured phase shifts. In: Valberg A, Lee BB, editors. From pigments to perception. London: Plenum Press; 1991. p. 401–12.
140. Smith VC, Lee BB, Pokorny J, Martin PR, Valberg A. Responses of macaque ganglion cells to the relative phase of heterochromatically modulated lights. J Physiol. 1992;458:191–221.
141. Lee BB, Sun H, Valberg A. Segregation of chromatic and luminance signals using a novel grating stimulus. J Physiol. 2011;589:59–73.
142. Cooper B, Sun H, Lee BB. Psychophysical and physiological responses to gratings with luminance and chromatic components of different spatial frequencies. J Opt Soc Am A. 2012;29:314–23.
143. Shevell SK, Kingdom FA. Color in complex scenes. Annu Rev Psychol. 2008;59:143–66.
144. Livingstone M, Hubel DH. Segregation of form, color, movement and depth: anatomy, physiology and perception. Science. 1988;240:740–9.
145. Nothdurft HC, Lee BB. Responses to coloured patterns in the macaque lateral geniculate nucleus; pattern processing in single neurones. Exp Brain Res. 1982;48:43–54.
146. Cooper B, Lee BB. Spatiotemporal properties of macaque retinal ganglion cells: an harmonic analysis and relationships to psychophysical data. J Vis. 2014;14:1411.
147. Silveira LCL. Joint entropy loci of M and P cells: a hypothesis for parallel processing in the primate visual system. Rev Bras Biol. 1996;56 Suppl 1:345–67.
148. Silveira LCL, de Mello Jr HD. Parallel pathways of the primate vision: sampling of the information in the Fourier space by M and P cells. In: Chalupa LM, Finlay BL, editors. Development and organization of the retina: from molecules to function. New York: Plenum Press; 1998. p. 173–99.
149. Rosa MG, Palmer SM, Gamberini M, Burman KJ, Yu HH, Reser DH, Bourne JA, Tweedale R, Galletti C. Connections of the dorsomedial visual area: pathways for early integration of dorsal and ventral streams in extrastriate cortex. J Neurosci. 2009;29:4548–63.
150. Silveira VA, Souza GS, Rodrigues AR, Silveira LCL. Joint entropy for space and spatial frequency domains estimated from psychometric functions of achromatic discrimination. PLoS One. 2014;9:e86579.
151. Wässle H, Boycott BB. Functional architecture of the mammalian retina. Physiol Rev. 1991;71:447–80.
152. Shapley R, Perry VH. Cat and monkey retinal ganglion cells and their visual functional roles. Trends Neurosci. 1986;9:229–35.
153. Nothdurft HC, Lee BB. Responses to coloured patterns in the macaque lateral geniculate nucleus; analysis of receptive field properties. Exp Brain Res. 1982;48:55–65.

Chapter 5
Psychophysical Correlates of Retinal Processing

Rigmor C. Baraas and Andrew J. Zele

Abstract Color is a psychological construct of our visual experience that represents an interaction between the physical properties of objects in the environment, the illuminant, and our nervous system. This chapter describes how the psychophysically measured subjective experiences that arise from signals originating in the retina lead to chromatic and achromatic visual perception. We explore how the processing of cone signals under photopic illumination, the interaction between cones and rods under mesopic illumination, and rod signaling under scotopic illumination give rise to human color vision by examining links to retinal physiology and the effect that individual variability has on visual perception.

Keywords Color vision • Trichromacy • Color opponency • Hue perception • L:M cone ratio • Psychophysics • Cone fundamentals • Spectral sensitivity

5.1 Introduction

> It is remarkable that psychophysical performance at threshold is constrained at a retinal level. There are a lot of synapses between the retina and a behavioral response, yet links between retinal physiology and psychophysics can be robust; the "nothing mucks it up" principle. [1]

A central question in visual neuroscience is to understand *how we see color*. Color is a psychological construct of our visual experience that represents an interaction

R.C. Baraas, Ph.D. (✉)
National Centre for Optics, Vision and Eye Care, Department of Optometry and Visual Science, Faculty of Health Sciences, University College of Southeast Norway (USN), Hasbergsvei 36, 3616 Kongsberg, Norway
e-mail: Rigmor.Baraas@hbv.no

A.J. Zele, Ph.D.
School of Optometry and Vision Science and Institute of Health and Biomedical Innovation, Queensland University of Technology (QUT),
60 Musk Avenue, Brisbane, QLD 4059, Australia
e-mail: andrew.zele@qut.edu.au

J. Kremers et al. (eds.), *Human Color Vision*, Springer Series in Vision Research, DOI 10.1007/978-3-319-44978-4_5

between the physical properties of the objects in the environment, the illuminant (the sun and/or artificial illuminants), and our nervous system. This subjective color experience shows between-individual variability, yet the quality of any hue of the reported range of discriminable colors (10,000–10 million [2]) can be sufficiently described by 11 basic color terms, including the 4 unique hues (red, green, blue, yellow) in addition to orange, purple, brown, and pink plus three achromatic color terms, black, white, and grey [3, 4]. Some of this individual variability is related to the fact that people do not all share the same number and combination of cone photoreceptor pigments [5, 6] (see Chap. 1 for details); here, we focus on normal color vision. A male with normal trichromatic color vision has three types of cone photoreceptors where each photoreceptor expresses a photopigment with a different spectral sensitivity. A female has two X chromosomes and may therefore have the genetic code to express more than one photopigment per cone type [7]. Both males and females also express the rhodopsin photopigment in rod photoreceptors and the melanopsin photopigment on intrinsically photosensitive retinal ganglion cells (for review see Ref. [8]).

Psychophysical methods quantify visual function and performance by measuring an observer's response to carefully controlled changes in the properties (e.g., spatial, temporal, or spectral) of known physical light stimuli. By doing so, quantitative relations can be established between subjective visual experience (perception) and the physical stimuli that produce the sensation [9]. For a specific set of experimental conditions the psychophysical data may then be inferred to represent the activities of specific neurons in the visual pathways. This may have applications in the development of clinical vision tests and be used to interpret how functional losses detected with these tests reflect changes in the properties of specific visual processes and pathways. Some of the major discoveries in the history of vision science, including the determination of the rod and cone spectral sensitivities [10, 11], rod and cone dark adaptation functions [12] and the directional sensitivity of the photoreceptors [13] were made utilizing psychophysics. Importantly, these discoveries reflect neural processes in the retina: If there is perceptual change then the underlying physiological response has changed, as embodied in the quote by Teller at the head of this chapter.

5.2 Psychophysical Correlates of Retinal Processing: Linking Hypotheses

In the broadest framework, the standard psychophysical model explains a particular aspect of perception, such as color, in terms of the activity of a hypothesized set of underlying (psychophysical) mechanisms using psychophysical paradigms to measure detection, discrimination, matching, identification, categorization, and appearance at threshold and suprathreshold levels. These model constructs are measured along multiple stimulus dimensions including wavelength, temporal and spatial frequency, retinal locus and the adaptation level. The physiology experiments record neuronal activity of in vitro or in vivo preparations in response to specific input signals along stimulus dimensions that parallel the psychophysical experiments. The models define how and where information is transmitted, filtered and processed within the visual system. This physiological characterization of neurons is an important issue when developing psychophysiological models because the cell's response properties depend on an experimenter's criterion for classification [14–16], the specificity (or randomness) of connections between retinal cells [17] and the stimulus conditions. A single cell can respond across many dimensions (e.g., Ref. [18]) to multiplex luminance and color signals for example [17, 19, 20] and a null response may also reveal important information (e.g., color-opponent cells; see Sect. 5.5). It is not only the response of a single cell, but also the pattern of activity among neuronal populations that contributes to perception and in some instances, measureable physiological responses may not be observable with (existing) psychophysical paradigms.

The psychophysiological model linking visual function and physiologically measured neural activity proposes a direct correspondence between the perceptual response of the observer on a specific psychophysical task to the activity of a class (or classes) of neurons and the pattern of activity of interacting and non-interacting neurons [21, 22]. Because noise exists in both the stimulus and in the neural response, linking statements must recognize the statistical, rather than absolute, nature of the system [17, 21]; repeated presentation of the same stimulus generates variable neuronal responses from trial to trial, with higher variability evident in central neurons under natural viewing conditions [23–25]. Disease can further modify the statistical properties of the physiological responses. The linking hypotheses are therefore complex [17, 26, 27] and require the parallel exploration of multiple stimulus dimensions in both psychophysics and physiology [17]. Implicit in these hypotheses are assumptions as to the signal transformations performed within and between multiple stages of the visual system and the range of applicability across other stimulus and viewing conditions [28]. Both the neuronal computations and their representations must be defined because neural convergence at successive stages of the visual system reduces large sets of stimuli to smaller sets of neural outputs.

Linking assumptions and psychophysiological models will continue to be subject to revision and development as new relationships between visual function and physiological responses are established. In the following sections, we will emphasize psychophysical measures of visual function interpreted in terms of current physiological knowledge, and highlight cases where the physiological interpretation of commonly used measures of visual function is inconsistent with existing physiological evidence.

5.3 Color Vision

Color vision is the capacity to discriminate between wavelengths irrespective of their irradiance. Color discrimination and color matching data provide fundamental insights into the trichromatic nature of vision. Chromatic discrimination is the ability to distinguish between lights measured in terms of wavelength, purity or chromaticity. Color matching is the quantification of visually metameric matches (two different spectral reflectance curves with the same appearance) made between a test light and a combination of three suitable primary lights that can be described by Grassman's laws of additivity at photopic luminance levels (from several cd m^{-2} to ~10^8 cd m^{-2}). The amount of energy in each primary required to match a test light (the tristimulus value) is represented by the color matching functions, with negative values indicating the amount of the primary light added to the test light. These color-matching functions are then displayed in chromaticity diagrams as a function of two primaries (e.g., the r,g chromaticity diagram); if three primaries are normalized relative to their sum, then two primaries are sufficient to display the color matching data; however, regions of the chromaticity diagram will be specified using negative values (colors). To overcome the problem of negative values, the 1931 CIE chromaticity diagram adopted imaginary primaries (X, Y, Z) with tristimulus values that do not correspond to actual primary lights so that all chromaticities were positive, and the Y (imaginary) primary was linked to $V(\lambda)$ in the photometric system (see Sect. 5.8). Standardized CIE functions representing the average (younger) trichromatic male are used for all measurements related to color science, including characterizing visual displays for presenting stimuli for psychophysics and physiological experiments. Importantly, neither chromatic discrimination nor color matching consider what color is perceived, that is, the color appearance of the lights, and so they have direct application in the development of clinical color vision tests for the detection and grading of color vision deficiencies.

5.4 Cone Spectral Sensitivities and Trichromacy

The simplest model of retinal processing that leads to color perception includes two neural stages, the first being the trichromatic stage followed by a second opponent stage. The physiological substrates of the first stage are the three types of cone photoreceptors (see Chap. 2 for details) with pigments of overlapping broadband sensitivities spanning the visible spectrum [11, 29, 30]; the short-, medium- and long-wavelength sensitive cones (L, M, and S cones; Fig. 5.1) [For details about the genetics of different photoreceptor pigments see Chap. 1]. Because the probability of photon absorption varies with both wavelength and irradiance as defined by the photopigment spectral sensitivity, information about wavelength is lost; this is the principle of univariance [31, 32]. In other words, individual cone types are color blind. Knowing the spectral sensitivity functions of the L, M, and S cones is essential for understanding and measuring color vision.

The favored and most successful method for deriving human photopigment spectral sensitivities is to measure them psychophysically. Young's theory of trichromacy (that color vision was based on the three different photoreceptor cells) [33] was in line with the first psychophysical measures of cone spectral sensitivities [34]. König and Dieterici derived the cone spectral sensitivities from measurements with normal trichromats or dichromats, the latter lacking one of the three cone types. In this method the wavelength and radiance of the background field is

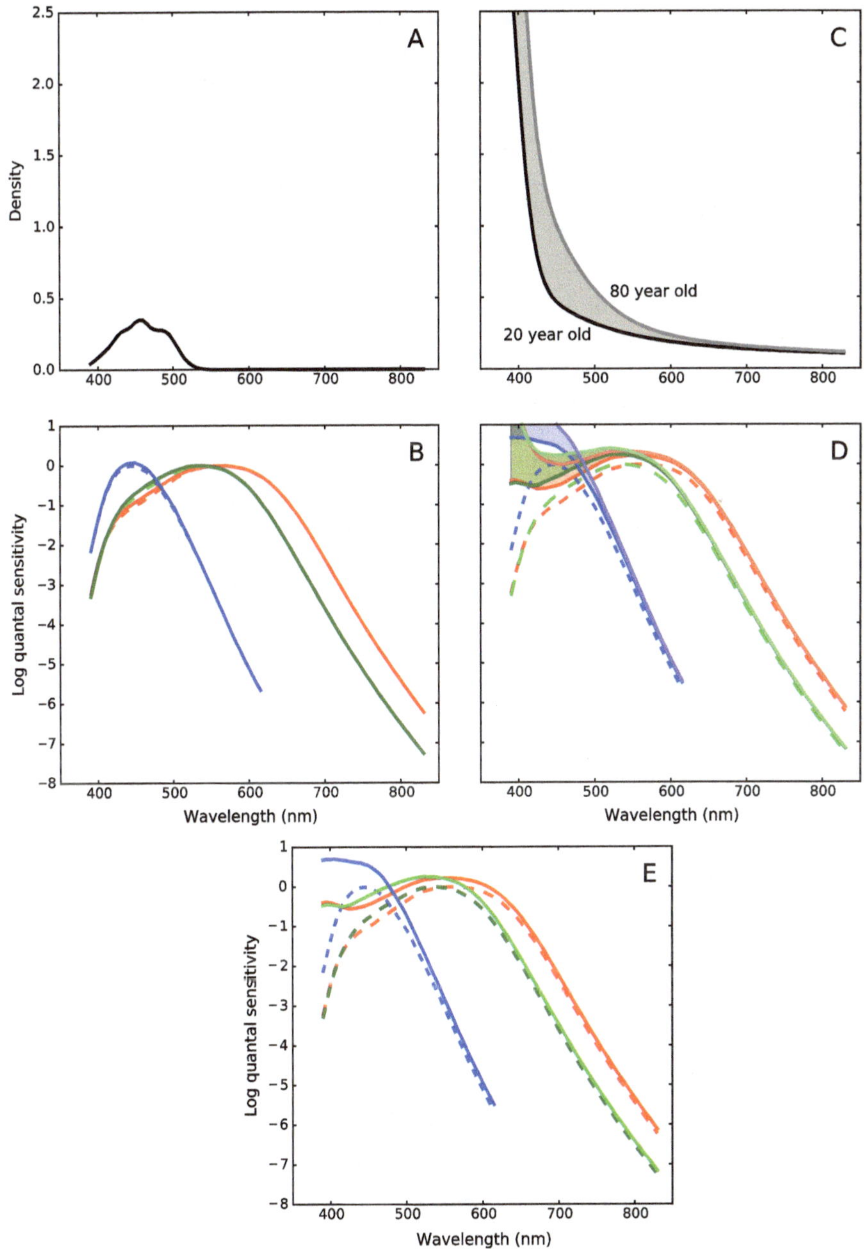

Fig. 5.1 Photoreceptor spectral sensitivities and the effects of lens and macular pigment. Photoreceptor sensitivities can be derived from corneal spectral sensitivities (*dashed lines* throughout) by removal of the changes introduced by the macular pigment and the density of the lens (lens yellowing) from Ref. [11, 29, 30]. The *left column* shows the macular pigment density spectrum (**a**) and its effect (*solid line*) upon corneal sensitivities (**b**) as a function of wavelength. The *right column* shows an age-based lens density model for ages 20 and 80 (**c**) as a function of wavelength, and the commensurate effect (*solid lines* and *shaded areas*) on corneal sensitivities (**d**). *Bottom figure* (**e**) shows both macular pigment and lens density effect removed, revealing the derived retinal photoreceptor sensitivities (*solid line*)

held constant and is chosen to desensitize the two unwanted cone types in normal trichromats or the remaining cone type in dichromats [35]. The test field radiance is adjusted until the test is detected for a range of wavelengths spanning the visible spectrum. Isolation of the targeted cone is obtained when adaptation of the non-targeted cone(s) to the background is stronger than its selective stimulation by the test field [36, 37]. Isolation of L-, M-, and S-cones is only obtained by heterochromatic flicker photometry (HFP) with the test field flickering in opposite phase to the background field, typically at high frequencies (>15 Hz), as done by Smith and Pokorny [11] and Stockman and Sharpe [29, 30]. The spectral sensitivities measured by Smith and Pokorny have been the favored ones for years. A difference between the two estimates is that the Smith–Pokorny fundamentals are a transformation of the Judd-Vos corrected CIE 1931 functions with specific assumptions that S-cones do not contribute to luminance and that L- and M-cones sum to the luminous efficiency function $V(\lambda)$, while the Stockman–Sharpe fundamentals are a transformation of the Stiles and Burch 10° data adjusted to 2° (for further details see Ref. [38]). These measured sensitivities are prime examples of accurate psychophysical measurement of a physiological process and the Smith–Pokorny data were confirmed through in vitro microspectrophotometric measurements of human cone photoreceptors taking into account pre-retinal absorption [39]. Importantly, Stockman and Sharpe measured spectral sensitivities in both single-gene dichromats and normal male trichromats with known OPN1LW and OPN1MW genotypes and the most recent version of the standard CIEXYZ color matching functions are linear transformations of the Stockman–Sharpe sensitivities [40].

These standardized CIE functions must be considered in relation to the fact that there are about seven amino acids that affect the peak wavelengths of the L- and M-cone pigment absorption spectra [41]. That is, we do not all share the same cone spectral sensitivity functions (see Chap. 1 for details). Color matching is also influenced by the observational conditions, including field size, illuminance, choice of the primary lights and methodology. There are additional physiological bases for these variations. For example, the photopigment optical density influences color matching with different field sizes. Both the light path through a photoreceptor and its photopigment concentration determine the effective photopigment optical density (Beers Law), and so the higher the photopigment optical density the broader the absorptance spectrum and thus the spectral sensitivity curve. Broader sensitivity curves will in general terms, capture more photons, whilst narrowing of the spectral sensitivity curves (spectral sharpening) may give rise to improved color discrimination [42–44]. Photoreceptor length decreases with increasing retinal eccentricity, thus, optical density decreases with increasing field size [45]. The pre-receptoral filtering of light by the ocular lens and macular pigments give rise to variations in color vision due to preferential absorption of short wavelength light that significantly differs between subjects (Fig. 5.1). Lens density increases (and light transmission decreases) with age [46] and the macular pigments lutein and zeaxanthin, with peak absorption at about 460 nm [47], contribute to between-individual variation in pigment density across the macula to about 3° eccentricity.

5.5 Cone Opponency and Unique Hues

Color appearance can be quantified along three dimensions of hue, saturation and brightness. In natural viewing environments, observers have no way of judging the spectral composition of lights, either those lights emitted from a source or those reflected from objects, based solely on their color appearance. This is because the color appearance is subject to factors including induction, where nearby lights in space or time shift the appearance of the light away from the appearance of the inducing light (chromatic contrast) or towards it (chromatic assimilation) [48] (for details see Chap. 7). Here, the focus is on aperture colors presented against a uniform adapting background (or in the dark) which limits the effects of the color shifts experienced in complex environments.

Photopic color vision requires comparison of output of a minimum of two cones having different spectral sensitivities (Fig. 5.2). It arises because the post-receptoral neural circuitry compares the outputs of two or more cone types when these have overlapping spectral sensitivities. The circuitry therefore distinguishes between the two factors (wavelength and irradiance) affecting the probability of photon absorption. This is often referred to as the second neural stage—the opponent stage. The first evidence of opponency came from psychophysical measures of human observers [49]; some pairs of hues can coexist in a single color sensation (e.g., red and yellow or blue and green), but pairs of opponent hues such as red and green or blue and yellow cannot [50]. These four colors are termed the unique hues. The zero-crossings of the two opponency curves, that is, the null points of the two opponent dichromatic subsystems, have been argued to represent the spectral position of unique hues [49, 51]. Hurvich and Jameson proposed a linear model for cone signal inputs to the opponent stage where all three cone types contributed to both the red-green and the blue-yellow opponent mechanisms, implying that the hue mechanism was subserved at the same neural stage.

Cone-opponent processing is demonstrated psychophysically in different ways. For example, when measuring spectral sensitivity functions with chromatic spots (1°, 1 Hz) presented on a neutral white adapting background [52, 53] the basic color appearance is reliably identified at detection threshold [54, 55], as first noted by King-Smith and Carden [52]. Werner and Wooten [56] performed hue naming, unique hue determination and opponent hue cancelation experiments and concluded that hue naming is also predicted from the opponent hue cancelation experiments, except that their data for the blue-yellow mechanism was better fitted with a nonlinear version of Jameson and Hurvich's model providing independent weighting of the L- and M-cone photopigments (see below).

Red-green opponency has also been demonstrated when measuring detection thresholds of reddish or greenish changes from white before and after adaptation to a red-green background. Similar measurements of bluish and yellowish changes from white did, however, not reveal the same blue-yellow opponency as Hurvich and Jameson, but the representation of the tritanopic confusion line [57]. These results indicated that cone signals were organized along two cardinal cone-opponent axes

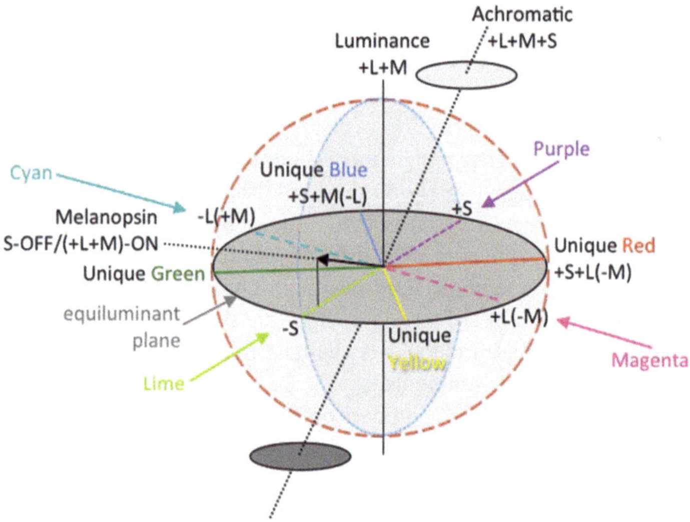

Fig. 5.2 Three-dimensional color space showing the cardinal directions and unique hues in an equiluminant plane. The center of the equiluminant plane represents the background adaptation level (e.g., a *white point*). The *cyan* and *magenta lines* represent an axis along which changes in stimulus chromaticity vary the relative excitation of L- and M-cones to modulate color opponent L- and M-cone inputs to the midget cells of the parvocellular (PC) pathway without changing S-cone excitation. The *purple* and *lime lines* represent an orthogonal axis to the L- and M-cones and along which changes in stimulus chromaticity varies the relative S-cone excitation to modulate color opponent inputs to the bistratified ganglion cells of the koniocellular (KC) pathway. Rotations parallel to the equiluminant plane that are away from these cardinal directions alters the sign and relative L-, M- and S-cone excitations to produce hypothetical cone combinations giving rise to the four unique hues (red, green, blue, and yellow). A rotation angle perpendicular to the equilumiant plane that produces an increase (or decrease) in L- and M-cone excitation (without changing S-cone excitation) modulates additive cone inputs to the magnocellular (MC) pathway by changing stimulus luminance and chromaticity; the simultaneous increase (or decrease) in the three cone photoreceptor excitations produce an achromatic stimulus that consists of a change in luminance but not in chromaticity. The melanopsin-expressing intrinsically photoreceptive retinal ganglion cells have a color opponent receptive field that may provide signals for image forming visual perception in addition to their non-image forming roles

where the red-green mechanism utilizes the difference of weighted responses from L and M cones with constant S-cone excitation, whereas the yellow-blue mechanism utilizes the difference of weighted responses from S cone versus the sum of responses from L and M cones. Given that color-matching functions are linear transformations of the cone spectral sensitivities, this opponency can be represented in cone chromaticity spaces with cone spectral sensitivities as their physiological primaries to link colorimetry to physiological mechanisms [57, 58]. These spaces express lights in terms of cone excitations ascribing the neural correlate of cone-opponency to the retina. The cardinal axes, however, are rotated compared with the hue axes (Fig. 5.2), leaving the neural correlate of the hue mechanism as an unknown [59, 60]. The discrepancy may be down to the stimuli used to derive the cardinal opponent-color space (see Sect. 7.3.2 for a discussion on transformation of color signals). Opponent cone signaling is observed in several neural loci, not only the retina and LGN. To match hue perception the three cone signals must be reconstructed differently to that represented at the opponent stage, in the cortex [61, 62] and/or within the retina [63]. A new model of color perception based on retinal signaling, proposes that feed-forward processing from S-cones to a small subset of midget bipolars via the HII horizontal cell forms four distinct pathways: The ON midget ganglion cells with S-cone OFF inputs can signal yellow when receptive field centers receive L-cone inputs, and signal green when the centers receive M-cone inputs; the OFF midget ganglion cells with S-cone ON inputs can signal blue when receptive field centers receive L-cone inputs, and signal red when the centers receive M-cone inputs [63].

It is interesting to consider the individual differences in performance between the few observers in all the abovementioned experiments that demonstrate cone-opponency [49, 52, 55–57]. Such differences are also common in psychophysical experiments for determining unique hues. Hues that appears as unique blue to one observer may appear as unique green to another [60]. Concurrently, colors appear both blue and red in the short wavelength region of the spectrum. When neutral white or chromatic backgrounds are used, the performance of some observers indicates that S-cone signals contribute to red-green color sensations [56, 64, 65]. The variability is greatest for the perception of unique green, requiring varying degrees of S-cone excitation [51, 66, 67]. A small contribution of S-cone signals to both cardinal mechanisms irrespective of eccentricity has also been reported for some, but not all observers [65, 68–70]. Others have reported M-cone excitation contributing to perception of blue [71–75]. This nonlinearity of the yellow–blue mechanism and the violation of colinearity of unique red and unique green [51, 64–67, 74], was also observed by Werner and Wooten [56]; they remarked that it was evident in one of the two observers in the Jameson and Hurvich study too. Some observed differences between studies are methodology dependent. For example, detection or discrimination threshold experiments can be determined by lower order mechanisms (retina) as compared with supra-threshold categorization experiments that require higher order mechanisms (visual cortex). In a study using the same stimuli to examine detection thresholds and categorization of color, Cropper et al. [76] observed no difference in the actual peak of the hue functions, but individuals had a lower criterion for categorizing. As a whole, based on psychophysical evidence there is reason to argue that both the red-green and blue-yellow opponent stage mechanisms receive input from S-, M-, and L-cone types (see also Ref. [77, 78]). There is physiological evidence consistent with this interpretation [79, 80] (see Fig. 5.3, as well as Chap. 3).

Standard neural models of color perception [51, 57, 62] do little to account for the observed nonlinearities or between-individual differences. In the early studies it was argued that hue naming could be predicted from the opponent hue cancelation experiments, implying retinal mechanisms [49]. Some have argued that perception of unique hues has no specific neural basis, but that it is a soft-wired individualized experience [81, 82], or just phenomenological [83]. Others have proposed an alternative neural model of color perception which is consistent with physiological evidence, where the main bulk of color processing is present at a retinal level, including the first labeling of unique hues [63, 84].

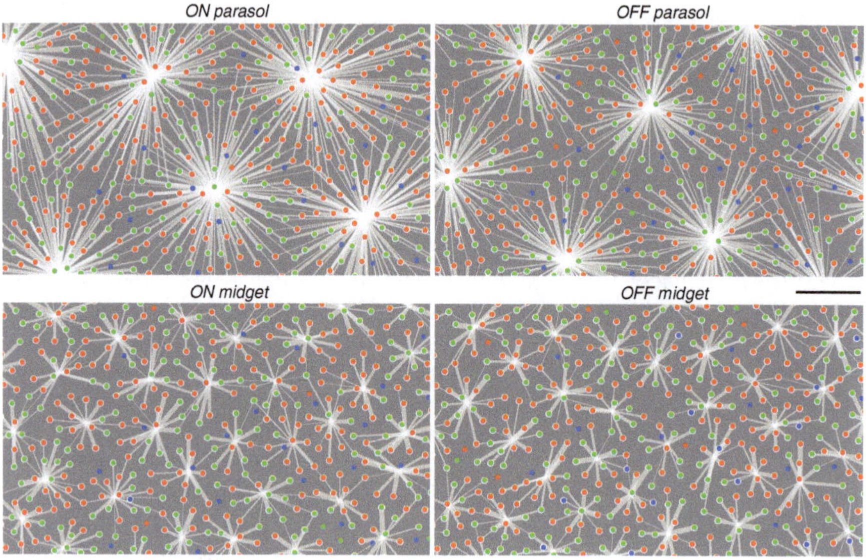

Fig. 5.3 Sampling of L, M, and S cone types by the ON and OFF midget and parasol ganglion cells in the macaque retina. Cones are color coded according to type and those highlighted with a *white annulus* provide input to at least one retinal ganglion cell. The L–M–S cone ratio is about 8:4:1. The *white lines* in each of the four panels show how the different receptive field centers of each retinal ganglion cell class sample the same cones. At this retinal eccentricity (7 mm temporal), the parasol and midget cells sample from 40 to 80 and 5 to 15 cones, respectively, with variation in the purity of cone inputs to the receptive field center. The ON and OFF midget and parasol cells sample from S cones, but S-cone connectivity is most frequent with OFF midget cells. The *black scale bar* is 50 μm. Figure is adapted by permission from Macmillan Publishers Ltd: Nature [80], copyright 2010

5.6 Peripheral Aspects and Cone-Selective Versus Random Wiring

Many have reported diminished color perception at more peripheral eccentricities (e.g., Ref. [85, 86]) with better preservation of blue-yellow as compared with red-green opponent signaling if stimulus size does not change with eccentricity [87–90]. The argument for this difference is typically related to the anatomical distribution of photoreceptor types and their densities across the retina. All three cone-types have been shown to have a random distribution, with S-cones being more evenly distributed throughout the retina than L- and M-cones, as the density of these cones decrease with increasing eccentricity away from the fovea [91, 92] and there is a tendency for the same cone types to clump together [93, 94]. In vivo high-resolution imaging has identified a random arrangement of L- and M-cones near the fovea in humans and the functional effect of this arrangement is observed in psychophysical experiments, both in the fovea and peripheral retina (17° eccentricity) [95, 96]. This arrangement plays a vital role for understanding post-receptoral signaling and whether the wiring of cones to midget ganglion cells (with center-surround receptive fields) is selective or random. If wiring is random then the prediction is that the strength of the red-green opponent response will be reduced in the periphery because retinal patches with a single cone type would give rise to surround signals whose spectral weighting would vary between receptive fields with all M-cones to all L-cones. If wiring is cone selective then the prediction is that red-green opponency will be similar in the fovea and periphery. There is evidence from physiological recordings in primate retina for both random (e.g., Ref. [97]), and selective wiring (e.g., Ref. [98]), but the measurements were at different eccentricities, thus the findings are difficult to compare (see Chap. 4 for details). The reported losses of color opponency in peripheral vision (e.g., Ref. [88, 89, 99]), however, does not persist if the peripheral stimuli are scaled in size to compensate for anatomical changes in receptive field size with increasing eccentricity [100]. Vakrou et al. [100] argued that these observations support cone-selective wiring, but consider the following: If midget cells draw their inputs randomly from both L and M cones, patches of cones of a single type would increase the probability for the center of a midget cell's receptive field to be driven by cones of just one type when a larger area of the retina is stimulated. If so, this would lend further support to random wiring.

Perception of unique hues has also been reported to be invariant irrespective of retinal eccentricity if the size of the stimulus is scaled appropriately [101]. Interestingly, blue, yellow, and red hues are reported to be more robust than green, irrespective of retinal eccentricity [102–104]. The relative number of S-cones has been shown to affect the perception of green with increasing eccentricity [104]. With small stimuli (2°) green is also the color that appears to be the most desaturated when test conditions allow for rod intrusion [105]. These results are compatible with electrophysiological measurements of L and M cone sensitivity across the retina, revealing an increase in the relative numbers of L versus M cones more peripherally [106]. Additionally, the perception of green has been reported to be associated with the relative number of L versus M cones in the retina (L:M cone ratio). Lower L:M ratios were observed in those that had unique green closer to yellow, while those with higher ratios had unique green closer to blue [63]. The L:M ratio, however, does not seem to have an impact on an individual's choice of unique yellow [107–109].

5.7 L:M Cone Ratios

DeVries [110] first psychophysically determined the L:M cone ratios and reported between-individual variation ranging between 0.59 and 4.0. The psychophysical measurement of L:M cone ratios has been replicated by many and verified utilizing molecular [111], electrophysiological methods [112], microspectrophotometry [113], retinal densiometry [114, 115] and direct optical imaging [109, 116]. The standard psychophysical method for revealing the L:M cone ratio is with heterochromatic flicker photometry with a temporal frequency greater than about 15 Hz

[110, 112, 117] to bias detection to the magnocellular (MC) pathway. The L:M ratio obtained with this method is based on the assumption that the luminance mechanism mainly receives input from the sum of activity in the L and M cones [118] corresponding with signaling through the MC pathway without additional neural weighting between the cone photoreceptors and the ganglion cells [119].

The L:M cone ratio is important for chromatic processing, but has been difficult to reveal. An explanation of this may be related to the neural wiring of cones to midget ganglion cells, as discussed earlier. Evidence from non-human primates show that wiring is random in the fovea; the cone type that feeds into the center dominates the receptive field response which is not much affected by the cone input to the surround [120]. In the parafovea and near periphery, however, receptive field responses appear balanced between the center and surround [120, 121]. If this is the same in humans, then it would yield an L:M cone ratio close to unity for a foveal stimulus and an L:M cone ratio that deviates from unity in the parafovea and near periphery. This implies that there are two important parameters that affect the L:M cone ratio in the chromatic pathway: size and location. And it should be possible to retrieve the underlying L:M cone ratio through the chromatic pathway if the stimulus is large enough or presented in the parafovea or near periphery. Danilova et al. [122] reported a difference in both the high frequency cut off and overall contrast sensitivity for L- and M-cone isolating stimuli for a spatial acuity task (finding the threshold of resolving a Landolt C) at 5° eccentricity, revealing L:M cone ratios that ranged from 1.1:1 to 2.6:1 for normal trichromats. The range of L:M cone ratios, however, is narrower than that observed with direct optical imaging [109, 116]. Even if a correlation between L-cone isolating Sloan E threshold and cone density has been observed, it implies that there is a third parameter that affects the L:M cone ratio for local spatial tasks; the organization of the underlying sub-mosaics of L and M cones [123]. The hue detection threshold of a cone isolating letter is better than the spatial discrimination threshold (you can still perceive the hue of the stimuli even if it is too small for you to resolve the letter). Hence, signaling is mainly through the chromatic pathway even if it involves processing of high spatial frequency information. Signals from individual cone photoreceptors input to both chromatic and luminance pathways through parallel neural circuits in the retina [80], and both spatial and temporal properties of the stimuli will dictate which pathway that will be responsible for the main part of the processing (Fig. 5.3).

5.8 Luminance

The sum of L- and M-cone inputs is generally thought to subserve the luminance mechanism as is evident from physiological recordings of macaque retina [61, 124, 125]. The sensitivity of the luminance mechanism is represented by the photopic spectral luminous efficiency function $V(\lambda)$ and is typically measured using heterochromatic flicker photometry, but psychophysical techniques such as minimally distinct border, heterochromatic brightness matching and the critical flicker frequency (CFF) may also be used (for review see Ref. [118]). The assumption has been that

S-cones do not contribute to the luminance mechanism [118], thus psychophysical derivation of $V(\lambda)$ uses spatial and temporal frequencies chosen to limit signaling from S-cones and signals in other chromatic pathways. Careful psychophysical measurements, however, have revealed that S-cones may play a small, but robust role in perceiving luminance [126] (see Fig. 5.3). A constraint of the standard $V(\lambda)$ function is that its derivation from the results of a large number of observers means it represents an average, normal observer, and therefore not precisely represent a given individual observer. Thus, in experimental psychophysics it is common to measure individual's flicker nulls with for example HFP and to make adjustments for this if a more precise measure is needed. Additionally, luminance noise can be used to mask the effect of any unwanted luminance information as exemplified by the pseudo-isochromatic plate tests of color vision (see Chap. 10).

The visual effectiveness of a radiant spectral power distribution of any broad-band light source is estimated using the International Commission on Illumination (CIE) photopic spectral luminous efficiency function $V(\lambda)$. The $V(\lambda)$ function peaks at 555 nm and applies only to cone vision with centrally fixated fields when rods are in saturation (>3 cd m^{-2}, but this transition value depends on the viewing conditions). The photopic luminous efficiency function is defined for two field sizes (2° and 10°); for fields larger than about 4° in diameter, or small fields viewed extra-foveally, the 10° field is more applicable due to the increased sensitivity to short wavelength light, partly resultant from decreased macular pigmentation (CIE, 1994). Importantly the measurement of (near) monochromatic light sources with $V(\lambda)$ does not accurately predict their brightness, but for many practical purposes this is minor (CIE 1994). For radiances below cone threshold ($\lesssim 10^{-3}$ cd m^{-2}) the 20° scotopic spectral luminous efficiency function $V'(\lambda)$ is based on the output of a single photoreceptor class, namely rods. The peak sensitivity of the $V'(\lambda)$ is at 507 nm and so it is more sensitive than $V(\lambda)$ at short wavelengths and less sensitive at longer wavelengths (the Purkinje shift). Currently there are no accepted luminous efficiency functions for off-axis measurements or for mesopic illuminations when rods and cones both are operational (for review see Ref. [127]).

5.9 Color Vision Under Dim Illumination

With decreasing illumination, the reductions in cone photoreceptor sensitivities are paralleled by increased rod photoreceptor sensitivity. Grassman's laws of trichromatic color matching are violated under mesopic illumination and in dichromats, the rods act as a third photoreceptor class with color matching performance more akin to that of normal trichromats at photopic illumination [128]. The progressive increase in rod sensitivity markedly affects perceived hue, saturation and brightness. Rod activity shifts the wavelength of unique red, green and blue hues to longer wavelengths [129], decreases the saturation of spectral lights [130], enhances brightness [131] and causes brightness induction [132]. Rod activity also degrades cone mediated chromatic discrimination [133–135].

Under twilight illumination, von Kries [136] and Nagel [137] reported that vision contained a bluish hue. Later reports included green-blue and cyan-blue, but with little agreement between observers, in part due to the subtlety of these desaturated color percepts, the achromatic or colorless ("farblos") prevailed [137]. More recently, advances in colorimetric instrumentation using four-primary lights to independently control rod, L-cone, M-cone and S-cone excitations and their post-receptoral signaling [138] has allowed the quantification of mesopic rod hues in terms of cone excitations: blue-green rod color percepts are equivalent to a decrease in L/(L+M) (i.e., parvocellular [PC] pathway) excitation and an increase in both S/(L+M) (i.e., blue yellow koniocellular [KC] pathway) and L+M (i.e., magnocellular [MC] pathway) excitation [139]. On the other hand, reddish rod color percepts produced by a decrease in rod excitation (a rod decrement) relative to the adaptation level are equivalent to an increase in L/(L+M) excitation, a decrease in L+M excitation and little or no change in S/(L+M) excitation [135]. There is a linear relationship between rod contrast and the level of inferred MC, PC, and KC pathway activation [140] with the relative rod inputs to the three pathways dependent on the temporal properties of the rod signal [141].

Central to the explanation of twilight color percepts is the state of chromatic adaptation of rods and cones [142–144]. Scotopic contrast hues produced by rod activation are predominantly opponent to the pre-adaptation hue, with a preferential displacement towards blue. An implication is that rod signaling is modified by the pre-adaptation state of cone opponent processes to give rise to multiple scotopic hues. With rods and cones sharing the same retinal pathways, rod-cone interactions then determine these perceptual variations in color vision under dim illumination. For example, multicolored percepts can be generated by the interaction of signals originating in the rods and L-cones [145].

The principle of univariance predicts that a single perceptual quality, namely brightness, should be experienced under scotopic illumination when rods are the only operational photoreceptor class. That this is not the case points to a cortical locus for rod color. Normal trichromats experience variegated hue percepts that are related to the scotopic lightness of test samples and depend on the viewing context; this is known as relational rod color [146] that is also perceptible in complex natural scenes [147]. Color naming by dichromats shows good agreement with trichromats at photopic illumination [148] yet rod hue perception of dichromats is inconsistent with the sample scotopic lightness, potentially due to the reduced dichromatic color gamut at photopic illumination [146]. With only one photoreceptor class operating at scotopic levels, cortical mechanisms are required to integrate previous experience in daylight conditions to estimate probable color appearance, with a bottom-up rod signaling required to generate rod hue percepts [146, 147].

5.10 Temporal and Spatial Processing

Psychophysical measurement of amplitude or modulation sensitivity as a function of temporal frequency (Hz) or spatial frequency (c/°) of a stimulus light changing in luminance or chromaticity over time (i.e., a flickering light) or space (i.e., a grating) provides a measure of the temporal [149, 150] and spatial [151] contrast sensitivity functions of the visual system, respectively. By adjusting the relative amplitude and phase of stimulus lights to find the threshold for perceiving flicker, temporal sensitivity can be measured for both chromatic and achromatic stimulation revealing broad sensitivity functions with low- and band-pass characteristics, respectively. An increase in temporal frequency of the stimulus above 1 Hz gradually reduces the contribution of the chromatic mechanism and favors the luminance detection mechanism, whose overall peak sensitivity is around 10–15 Hz, depending on mean luminance (higher frequencies at higher luminances) [152]. This peak, in its residual form, is still perceptible even if a foveally presented spot is made isoluminant with respect to the background, a condition that guarantees a pure chromatic modulation [153]. The persistence of this peak at about 10 Hz can be ascribed to the sensitivity of magnocellular neurons to isoluminant red-green borders [154]. It is evident that the temporal sensitivity function is an envelope of three filters when stimuli are chosen to favor certain components, or when participants perform discrimination tasks with achromatic [155–157], red-green [158, 159] and blue-yellow stimuli [160]. For achromatic stimuli the three filters are all band-pass; the first with peak around 0.6–1 Hz, the second with peak around 4–8 Hz, and the third with peak 9–16 Hz) [155–157]. The peak of the intermediate temporal filter is most readily revealed in luminance-modulated elongated patterns such as gratings [161], and does not contribute to the detection of chromatic spots. Detection of spectral spots is predominantly processed by the cone-opponent mechanism at low temporal frequencies, and by the luminance mechanism at very high temporal frequencies. The involvement of an intermediate filter for detecting elongated chromatic stimuli [158–160] implies that this filter is important for perception of chromatic spatial patterns [162]. Ganglion cells are reported to have temporal sensitivity functions that are similar to those obtained with psychophysics [79, 163, 164].

With increasing temporal frequency a perceptual transition criterion (from perceiving flickering lights to perceiving a uniform even field) is reached—this is the critical flicker frequency (CFF). Beyond this CFF the apparent brightness of the flickering stimulus is equal to its time average luminance (as per the Talbot-Plateau Law). The CFF approaches 60 Hz under photopic illuminations [149, 150] whereas the rod CFF is between 20 and 30 Hz under mesopic conditions [132, 165]. In normal trichromats the CFF is independent of wavelength when normalized to the photopic luminosity function [166]. The CFF is proportional to the logarithm of the stimulus area (Granit Harper Law), higher in the fovea than in the periphery at photopic illuminances and higher in the periphery at scotopic illuminations [167].

Similar to temporal thresholds, the detection thresholds for gratings can be measured by adjusting the contrast of a sinusoidal grating at different spatial frequencies. The spatial contrast sensitivity functions are broad with low- and band-pass characteristics for chromatic and achromatic gratings, respectively [168]. Signaling of photopic spatial information, regardless of whether it is achromatic or chromatic, is conveyed by the cone photoreceptors, but midget ganglion cell receptive fields limit spatial resolution [169, 170]. Thus, the distinct fall-off in sensitivity with increasing spatial frequency for both chromatic and achromatic functions [168] is associated with both blurring by the optics of the eye and with center-surround ganglion cell receptive fields having poor sensitivity to high spatial frequencies. The optics of the eye does not affect sensitivity at low spatial frequencies. Any change in this region of the contrast sensitivity function is related to neural changes only. If contrast sensitivity is measured for isochromatic gratings, where an individual's luminosity function and its variation across the visual field is accounted for, then large between-individual variations are observed [171]. This variation was found to be associated with polymorphisms at position 180 of the L-opsin gene, but cone and midget cell density may also play a role [172]. Thus, the perception of chromatic spatial patterns is very much dependent on the properties of retinal photoreceptors and post-receptoral processing.

5.11 Contrast Processing in the Post-receptoral MC, PC, and KC Pathways

There have been numerous different methods developed to psychophysically separate and define visual functions inferred to represent the properties of the MC, PC, and KC retinogeniculate pathways. Many of these methods were developed based on physiological differences in the temporal, chromatic, achromatic and contrast response properties of non-human primate ganglion cells. For example, chromatic discrimination is linked to the PC and KC pathways [173] and visual function measured at equiluminance [174] has been inferred to differentiate between PC and MC pathway processing based on the threshold response. This psychophysical separation of visual function has implications for measuring the properties of retinogeniculate pathways in humans, and for defining the locus of visual deficits in clinical patients (see Chap. 10). The "pedestal" paradigms [175] were introduced to differentiate the MC and PC pathway achromatic contrast responses by comparing the slopes of their psychophysical contrast discrimination functions with physiological recordings in primate ganglion cells. The stimulus is a simple 4-square spatial array (the pedestals) that is set within a large uniform adapting surround; only the pre- and post-adaptation differs between the three pedestal paradigms. Plotting the contrast discrimination functions for each pedestal paradigm as a function of the incremental and decremental pedestal retinal illuminance produces characteristic functions (Fig. 5.4); thresholds for increments and decrements allow psychophysical comparison with the ON-and OFF-pathways. In the steady-pedestal paradigm, the 4-square array is continuously present and the observer's task is to discriminate the one test square with a higher (or lower) retinal illuminance; the contrast discrimination function is monotonic (Fig. 5.4; circular symbols) and the data described by a line with a slope of unity and a single scaling constant that is inferred to represent steady-state MC pathway sensitivity.

For the pulsed-pedestal paradigm, the 4-square array appears only during the trial period during which time the observer's task is to discriminate the test square with a higher (or lower) retinal illuminance: the data for this paradigm return a shallow V-shape (Fig. 5.4; triangular symbols) and the slopes of the best-fitting contrast satu-

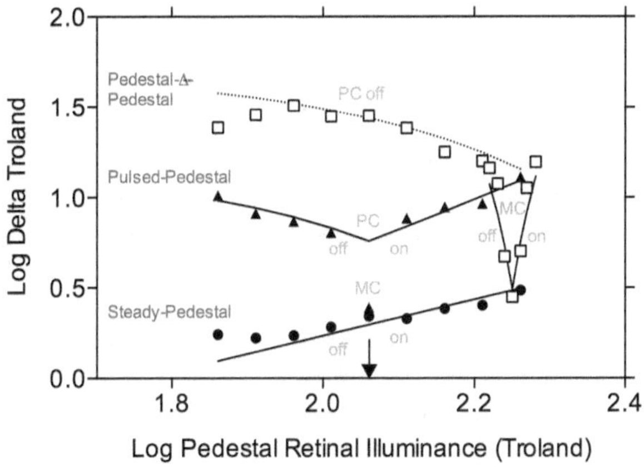

Fig. 5.4 Separation of MC and PC pathway achromatic contrast discrimination using the pedestal paradigms [175]. *Lines* show the best-fitting contrast saturation functions (pulsed pedestal and pedestal-Δ-pedestal paradigms) and the unit slope (steady-pedestal paradigm). The *arrow* defines the retinal illuminance at the mean adaptation level. See text for details. Modified from Pokorny [176] with permission from the Association for Research in Vision and Ophthalmology (ARVO)

ration functions are in the range of values for individual PC cells and inferred to represent PC pathway contrast gain. The pedestal-Δ-pedestal paradigm is the same as the steady-pedestal paradigm except during the trial period when all 4-squares are incremented (or decremented) and the test square is incremented (or decremented) by a different amount; the data for this paradigm show a steep V-shape (Fig. 5.4; square symbols) with values for the contrast saturation function in the range of individual MC cells and inferred to represent MC pathway contrast gain. At higher pedestal-Δ-pedestal contrasts (>0.02 log units) the MC-pathway saturates and the slope of the contrast saturation function is consistent with values for the PC pathway. An extensive series of psychophysical investigations using modifications of the pedestal paradigms have defined the spatial and temporal summation properties of the two pathways and the technique has been successfully translated to the clinic for the assessment of retino-geniculate function in eye disease (for review see Ref. [176]).

5.12 Future Perspectives

Understanding how we perceive colors through the mosaic of L, M, and S cones is central to explaining the micro-cosmos of processes within the visual system that are required to generate a perceptual representation of the environment. Associations between the organization, pigments and relative numbers of cones may also contribute to the overall understating of how we perceive form, depth and motion, and how this changes as we age [5, 6, 171]. Questions related to whether this also affects susceptibility to outer retina (photoreceptor-based) diseases will become important for developing new diagnostic tests. The net contributions of the three cone types to perceiving colors are not fixed, and will depend on the abundance and combination of cone photoreceptor pigments. Advances in understanding of the initial stages of cone based color vision are left with unresolved problems of how and where this information is transmitted, filtered and processed within the visual system to lead to visual perception. Future investigations will also need to look at how the rod photoreceptors and melanopsin containing intrinsically photosensitive retinal ganglion cells contribute to color vision. The determination of between-individual variation in color perception experienced among normal trichromats [5, 6, 171] will facilitate the expansion of experimental strategies to quantify how differences in anatomical structures and the range of physiological response variability's that exist within and between neuronal populations leads to these individual differences in visual function.

Acknowledgements Supported by the Australian Research Council Discovery Projects ARC-DP140100333 (A.J.Z.).

References

1. Teller DY. Locus questions in visual science. In: Harris CS, editor. Visual coding and adaptability. Hillsdale, NJ: Lawrence Erlbaum Associates; 1980. p. 151–76.
2. Masaoka K, Berns RS, Fairchild MD, Abed FM. Number of discernible object colors is a conundrum. J Opt Soc Am A. 2013;30(2):264–77.
3. Berlin B, Kay P. Basic color terms: their universality and evolution. Berkeley: University of California Press; 1969.
4. Boynton RM, Olson CX. Salience of chromatic basic color terms confirmed by three measures. Vision Res. 1990;30:1311–7.
5. Smith VC, Pokorny J. Chromatic-discrimination axes, CRT phosphor spectra, and individual variation in color vision. J Opt Soc Am A. 1995;12(1):27–35.
6. Neitz J, Neitz M. The genetics of normal and defective color vision. Vision Res. 2011;51(7):633–51. doi:10.1016/j.visres.2010.12.002. S0042-6989(10)00569-9 [pii].
7. Neitz M, Kraft TW, Neitz J. Expression of L cone pigment gene subtypes in females. Vision Res. 1998;38(21):3221–5. S0042-6989(98)00076-5 [pii].
8. Feigl B, Zele AJ. Melanopsin-expressing intrinsically photosensitive retinal ganglion cells in retinal disease. Optom Vis Sci. 2014;91(8):894–903. doi:10.1097/OPX.0000000000000284.
9. Fechner GT. Elemente der Psychophysik. The classical psychologists. Boston: Houghton Mifflin; 1860.
10. Abney WD, Watson W. The threshold of vision for different coloured lights. Philos Trans R Soc Lond A. 1916;216:91.
11. Smith VC, Pokorny J. Spectral sensitivity of the foveal cone photopigments between 400 and 500 nm. Vision Res. 1975;15(2):161–71. 0042-6989(75)90203-5 [pii].
12. Hecht S, Haig C, Wald G. The dark adaptation of retinal fields of different size and location. J Gen Physiol. 1935;19(2):321–37.
13. Stiles WS, Crawford BH. The luminous efficiency of rays entering the eye pupil at different points. Proc R Soc Lond B. 1933;112:428–50.
14. deMonasterio FM, Gouras P, Tolhurst DJ. Concealed colour-opponency in ganglion cells of the rhesus monkey retina. J Physiol (London). 1975;251:251.
15. Finkelstein MA, Hood DC. Opponent-color cells can influence detection of small brief lights. Vision Res. 1982;22:89–95.

16. Wiesel T, Hubel DH. Spatial and chromatic interactions in the lateral geniculate body of the rhesus monkey. J Neurophysiol. 1966;29:1115–56.
17. Lee BB. Neural models and physiological reality. Vis Neurosci. 2008;25(3):231–41. doi:10.1017/S0952523808080140.
18. Ingling Jr CR, Grigsby SS. Perceptual correlates of magnocellular and parvocellular channels: seeing form and depth in afterimages. Vision Res. 1990;30:823–8.
19. Ingling Jr CR, Martinez-Uriegas E. Simple-opponent receptive fields are asymmetrical: G-cone centers predominate. J Opt Soc Am. 1983;73(11):1527–32.
20. Lennie P, D'Zmura MD. Mechanisms of color vision. CRC Crit Rev Neurobiol. 1988;3:333–400.
21. Barlow HB. Dark and light adaptation: psychophysics. In: Jameson D, Hurvich LM, editors. Handbook of sensory physiology, visual psychophysics. Berlin: Springer; 1972. p. 1–28.
22. Hood DC, Finkelstein MA. A case for the revision of textbook models of color vision: the detection and appearance of small brief lights. In: Mollon JD, Sharpe LT, editors. Colour vision: physiology and psychophysics. London: Academic; 1983. p. 385–98.
23. Croner LJ, Purpura K, Kaplan E. Response variability in retinal ganglion cells of primates. Proc Natl Acad Sci U S A. 1993;90:8128–30.
24. Gur M, Beylin A, Snodderly DM. Response variability of neurons in primary visual cortex (V1) of alert monkeys. J Neurosci. 1997;17(8):2914–20.
25. Uzzell VJ, Chichilnisky EJ. Precision of spike trains in primate retinal ganglion cells. J Neurophysiol. 2004;92(2):780–9. doi:10.1152/jn.01171.2003.
26. Brindley GS. Physiology of the retina and the visual pathway. London: Arnold; 1960.
27. Teller DY. Linking propositions. Vision Res. 1984;24:1233–46.
28. Lamb TD. Spontaneous quantal events induced in toad rods by pigment bleaching. Nature. 1980;287(5780):349–51.
29. Stockman A, Sharpe LT. The spectral sensitivities of the middle- and long-wavelength-sensitive cones derived from measurements in observers of known genotype. Vision Res. 2000;40(13):1711–37. S0042-6989(00)00021-3 [pii].
30. Stockman A, Sharpe LT, Fach C. The spectral sensitivity of the human short-wavelength sensitive cones derived from thresholds and color matches. Vision Res. 1999;39(17):2901–27. S0042-6989(98)00225-9 [pii].
31. Rushton WA. Pigments and signals in colour vision. J Physiol. 1972;220(3):1P-P.
32. Graham CH, Hartline HK. The response of single visual sense cells to lights of different wave lengths. J Gen Physiol. 1935;18(6):917–31.
33. Young T. On the theory of light and colours. Philos Trans Lond. 1802;92:12–48.
34. König A, Dieterici C. Die Grundempfindungen und ihre Intensitäts-Vertheilung im Spectrum. Berlin: Sitzungsberichte Akademie der Wissenschaften; 1886.
35. Stiles WS. Incremental thresholds and the mechanisms of colour vision. Doc Ophthalmol. 1949;3:138–63.
36. King-Smith PE, Webb JR. The use of photopic saturation in determining the fundamental spectral sensitivity curves. Vision Res. 1974;14(6):421–9. doi:0042-6989(74)90240-5 [pii].
37. Stockman A, Mollon J. The spectral sensitivities of the middle- and long-wavelength cones: an extension of the two-colour threshold technique of W S Stiles. Perception. 1986;15(6):729–54.
38. Brainard DH, Stockman A. Colorimetry. In: Bass M, Enoch JM, Lakshminarayanan V, editors. Handbook of optics, Vision and vision optics, vol. 3. New York: McGraw Hill; 2010. p. 1–49.
39. Dartnall HJ, Bowmaker JK, Mollon JD. Human visual pigments: microspectrophotometric results from the eyes of seven persons. Proc R Soc Lond B Biol Sci. 1983;220(1218):115–30.
40. CIE. Fundamental chromaticity diagram with physiological axes—Part 1. Vienna: Central Bureau of the Commission Internationale de l' Éclairage; 2006.
41. Asenjo AB, Rim J, Oprian DD. Molecular determinants of human red/green color discrimination. Neuron. 1994;12(5):1131–8. 0896-6273(94)90320-4 [pii].
42. Neitz J, Neitz M, He JC, Shevell SK. Trichromatic color vision with only two spectrally distinct photopigments. Nat Neurosci. 1999;2(10):884–8. doi:10.1038/13185.

43. He JC, Shevell SK. Variation in color matching and discrimination among deuteranomalous trichromats: theoretical implications of small differences in photopigments. Vision Res. 1995;35(18):2579–88. 0042-6989(95)00007-M [pii].
44. Thomas PB, Formankiewicz MA, Mollon JD. The effect of photopigment optical density on the color vision of the anomalous trichromat. Vision Res. 2011;51(20):2224–33. doi:10.1016/j.visres.2011.08.016. S0042-6989(11)00305-1 [pii].
45. Pokorny J, Smith VC. Effect of field size on red-green color mixture equations. J Opt Soc Am. 1976;66:705–8.
46. van de Kraats J, van Norren D. Optical density of the aging human ocular media in the visible and the UV. J Opt Soc Am A. 2007;24(7):1842–57.
47. Snodderly DM, Brown PK, Delori FC, Auran JD. The macular pigment. I. Absorbance spectra, localization, and discrimination from other yellow pigments in primate retinas. Invest Ophthalmol Vis Sci. 1984;25:660–73.
48. Shevell SK, Kingdom FA. Color in complex scenes. Annu Rev Psychol. 2008;59:143–66. doi:10.1146/annurev.psych.59.103006.093619.
49. Hurvich LM, Jameson D. An opponent-process theory of color vision. Psychol Rev. 1957;64(6 Pt 1):384–404.
50. Hering E. Outlines of a theory of the light sense (L. M. Hurvich & D. Jameson, Transl.). Cambridge, MA: Harvard University Press; 1964.
51. De Valois RL, De Valois KK, Switkes E, Mahon L. Hue scaling of isoluminant and cone-specific lights. Vision Res. 1997;37(7):885–97. S0042698996002349 [pii].
52. King-Smith PE, Carden D. Luminance and opponent-color contributions to visual detection and adaptation and to temporal and spatial integration. J Opt Soc Am. 1976;66(7):709–17.
53. Kranda K, King-Smith PE. Detection of coloured stimuli by independent linear systems. Vision Res. 1979;19(7):733–45. 0042-6989(79)90149-4 [pii].
54. Calkins DJ, Thornton JE, Pugh Jr EN. Monochromatism determined at a long-wavelength/middle-wavelength cone-antagonistic locus. Vision Res. 1992;32(12):2349–67. 0042-6989(92)90098-4 [pii].
55. Mullen KT, Kulikowski JJ. Wavelength discrimination at detection threshold. J Opt Soc Am A. 1990;7(4):733–42.
56. Werner JS, Wooten BR. Opponent chromatic mechanisms: relation to photopigments and hue naming. J Opt Soc Am. 1979;69(3):422–34.
57. Krauskopf J, Williams DR, Heeley DW. Cardinal directions of color space. Vision Res. 1982;22(9):1123–31.
58. MacLeod DI, Boynton RM. Chromaticity diagram showing cone excitation by stimuli of equal luminance. J Opt Soc Am. 1979;69(8):1183–6.
59. Jordan G, Mollon JD. On the nature of unique hues. In: Dickinson C, Murray I, Carden D, editors. John Dalton's colour vision legacy. London: Taylor and Francis; 1997. p. 381–92.
60. Kuehni RG. Variability in unique hue selection: a surprising phenomenon. Color Res Appl. 2004;29:158–62.
61. De Valois RL, Abramov I, Jacobs GH. Analysis of response patterns of LGN cells. J Opt Soc Am A. 1966;56(7):966–77.
62. De Valois RL, De Valois KK. A multi-stage color model. Vision Res. 1993;33(8):1053–65. 0042-6989(93)90240-W [pii].
63. Schmidt BP, Touch P, Neitz M, Neitz J. Circuitry to explain how the relative number of L and M cones shapes color experience. J Vis. 2016;16(8):18. doi:10.1167/16.8.18.
64. Shevell SK, Humanski RA. Color perception under chromatic adaptation: red/green equilibria with adapted short-wavelength-sensitive cones. Vision Res. 1988;28(12):1345–56.
65. Stromeyer III CF, Chaparro A, Rodriguez C, Chen D, Hu E, Kronauer RE. Short-wave cone signal in the red-green detection mechanism. Vision Res. 1998;38(6):813–26. S0042-6989(97)00231-9 [pii].
66. Burns SA, Elsner AE, Pokorny J, Smith VC. The Abney effect: chromaticity coordinates of unique and other constant hues. Vision Res. 1984;24(5):479–89.

67. Webster MA, Mollon JD. The influence of contrast adaptation on color appearance. Vision Res. 1994;34(15):1993–2020. 0042-6989(94)90028-0 [pii].
68. Cole GR, Hine T, McIlhagga W. Detection mechanisms in L-, M-, and S-cone contrast space. J Opt Soc Am A. 1993;10(1):38–51.
69. Sankeralli MJ, Mullen KT. Postreceptoral chromatic detection mechanisms revealed by noise masking in three-dimensional cone contrast space. J Opt Soc Am A. 1997;14(10):2633–46.
70. Sakurai M, Mullen KT. Cone weights for the two cone-opponent systems in peripheral vision and asymmetries of cone contrast sensitivity. Vision Res. 2006;46(26):4346–54. doi:10.1016/j.visres.2006.08.016. S0042-6989(06)00388-9 [pii].
71. Drum B. Hue signals from short- and middle-wavelength-sensitive cones. J Opt Soc Am A. 1989;6(1):153–7.
72. Guth SL. Model for color vision and light adaptation. J Opt Soc Am A. 1991;8(6):976–93.
73. Valberg A. Unique hues: an old problem for a new generation. Vision Res. 2001;41(13):1645–57. S0042-6989(01)00041-4 [pii].
74. Webster MA, Miyahara E, Malkoc G, Raker VE. Variations in normal color vision. I. Cone-opponent axes. J Opt Soc Am A. 2000;17(9):1535–44.
75. Boynton RM, Nagy AL, Olson CX. A flaw in equations for predicting chromatic differences. Color Res Appl. 1983;8:69–74.
76. Cropper SJ, Kvansakul JG, Little DR. The categorisation of non-categorical colours: a novel paradigm in colour perception. PLoS One. 2013;8(3):e59945. doi:10.1371/journal.pone.0059945. PONE-D-12-39574 [pii].
77. Schefrin BE, Werner JS. Loci of spectral unique hues throughout the life span. J Opt Soc Am A. 1990;7(2):305–11.
78. Wachtler T, Dohrmann U, Hertel R. Modeling color percepts of dichromats. Vision Res. 2004;44(24):2843–55. doi:10.1016/j.visres.2004.06.016. S0042-6989(04)00329-3 [pii].
79. Tailby C, Solomon SG, Lennie P. Functional asymmetries in visual pathways carrying S-cone signals in macaque. J Neurosci. 2008;28(15):4078–87. doi:10.1523/JNEUROSCI.5338-07.2008. 28/15/4078 [pii].
80. Field GD, Gauthier JL, Sher A, Greschner M, Machado TA, Jepson LH, et al. Functional connectivity in the retina at the resolution of photoreceptors. Nature. 2010;467(7316):673–7. doi:10.1038/nature09424. nature09424 [pii].
81. Delahunt PB, Webster MA, Ma L, Werner JS. Long-term renormalization of chromatic mechanisms following cataract surgery. Vis Neurosci. 2004;21(3):301–7. S0952523804213025 [pii].
82. Mollon J. Monge: the Verriest lecture, Lyon, July 2005. Vis Neurosci. 2006;23(3–4):297–309. doi:10.1017/S0952523806233479. S0952523806233479 [pii].
83. Panorgias A, Kulikowski JJ, Parry NR, McKeefry DJ, Murray IJ. Phases of daylight and the stability of color perception in the near peripheral human retina. J Vision. 2012;12(3). doi:10.1167/12.3.1. 12.3.1.
84. Neitz J, Neitz M. Colour vision: the wonder of hue. Curr Biol. 2008;18(16):R700–2. doi:10.1016/j.cub.2008.06.062. S0960-9822(08)00819-1 [pii].
85. Moreland J, Cruz A. Colour perception with the peripheral retina. Opt Acta. 1959;6:117–51.
86. Mullen KT, Kingdom FA. Differential distributions of red-green and blue-yellow cone opponency across the visual field. Vis Neurosci. 2002;19(1):109–18. S0952523802191103 [pii].
87. Anderson SJ, Mullen KT, Hess RF. Human peripheral spatial resolution for achromatic and chromatic stimuli: limits imposed by optical and retinal factors. J Physiol. 1991;442:47–64.
88. Mullen KT, Sankeralli MJ, Hess RF. Color and luminance vision in human amblyopia: shifts in isoluminance, contrast sensitivity losses, and positional deficits. Vision Res. 1996;36(5):645–53. 0042-6989(95)00159-X [pii].
89. Mullen KT, Sakurai M, Chu W. Does L/M cone opponency disappear in human periphery? Perception. 2005;34(8):951–9.
90. Stromeyer III CF, Lee J, Eskew Jr RT. Peripheral chromatic sensitivity for flashes: a post-receptoral red-green asymmetry. Vision Res. 1992;32(10):1865–73.

91. Williams DR, MacLeod DI, Hayhoe MM. Punctate sensitivity of the blue-sensitive mechanism. Vision Res. 1981;21(9):1357–75. 0042-6989(81)90242-X [pii].
92. Curcio CA, Allen KA, Sloan KR, Lerea CL, Hurley JB, Klock IB, et al. Distribution and morphology of human cone photoreceptors stained with anti-blue opsin. J Comp Neurol. 1991;312(4):610–24. doi:10.1002/cne.903120411.
93. Roorda A, Williams DR. The arrangement of the three cone classes in the living human eye. Nature. 1999;397(6719):520–2. doi:10.1038/17383.
94. Roorda A, Metha AB, Lennie P, Williams DR. Packing arrangement of the three cone classes in primate retina. Vision Res. 2001;41(10–11):1291–306. S0042-6989(01)00043-8 [pii].
95. Gowdy PD, Cicerone CM. The spatial arrangement of the L and M cones in the central fovea of the living human eye. Vision Res. 1998;38(17):2575–89. S0042-6989(97)00416-1 [pii].
96. Otake S, Gowdy PD, Cicerone CM. The spatial arrangement of L and M cones in the peripheral human retina. Vision Res. 2000;40(6):677–93. S0042-6989(99)00202-3 [pii].
97. Dacey DM. Parallel pathways for spectral coding in primate retina. Annu Rev Neurosci. 2000;23:743–75. doi:10.1146/annurev.neuro.23.1.743.
98. Reid RC, Shapley RM. Space and time maps of cone photoreceptor signals in macaque lateral geniculate nucleus. J Neurosci. 2002;22(14):6158–75. 20026444 22/14/6158 [pii].
99. Newton JR, Eskew Jr RT. Chromatic detection and discrimination in the periphery: a postreceptoral loss of color sensitivity. Vis Neurosci. 2003;20(5):511–21. S0952523803205058 [pii].
100. Vakrou C, Whitaker D, McGraw PV, McKeefry D. Functional evidence for cone-specific connectivity in the human retina. J Physiol. 2005;566(Pt 1):93–102. doi:10.1113/jphysiol.2005.084855. jphysiol.2005.084855 [pii].
101. Abramov I, Gordon J, Chan H. Color appearance in the peripheral retina: effects of stimulus size. J Opt Soc Am A. 1991;8(2):404–14.
102. McKeefry DJ, Murray IJ, Parry NR. Perceived shifts in saturation and hue of chromatic stimuli in the near peripheral retina. J Opt Soc Am A. 2007;24(10):3168–79. 141432 [pii].
103. Parry NR, McKeefry DJ, Murray IJ. Variant and invariant color perception in the near peripheral retina. J Opt Soc Am A. 2006;23(7):1586–97. 90375 [pii].
104. Volbrecht VJ, Nerger JL, Imhoff SM, Ayde CJ. Effect of the short-wavelength-sensitive-cone mosaic and rods on the locus of unique green. J Opt Soc Am A. 2000;17(3):628–34.
105. Volbrecht VJ, Nerger JL. Color appearance at +/−10 degrees along the vertical and horizontal meridians. J Opt Soc Am A. 2012;29(2):A44–51. 226498 [pii].
106. Kuchenbecker JA, Sahay M, Tait DM, Neitz M, Neitz J. Topography of the long- to middle-wavelength sensitive cone ratio in the human retina assessed with a wide-field color multifocal electroretinogram. Vis Neurosci. 2008;25(3):301–6. doi:10.1017/S0952523808080474. S0952523808080474 [pii].
107. Pokorny J, Smith VC. L/M cone ratios and the null point of the perceptual red/green opponent system. Farbe. 1987;34:53–7.
108. Miyahara E, Pokorny J, Smith VC, Baron R, Baron E. Color vision in two observers with highly biased LWS/MWS cone ratios. Vision Res. 1998;38(4):601–12. S0042-6989(97)88334-4 [pii].
109. Brainard DH, Roorda A, Yamauchi Y, Calderone JB, Metha A, Neitz M, et al. Functional consequences of the relative numbers of L and M cones. J Opt Soc Am A. 2000;17(3): 607–14.
110. De Vries HL. The luminosity curve of the eye as determined by measurements with the flickerphotometer. Physica. 1949;14:319–33.
111. Yamaguchi T, Motulsky AG, Deeb SS. Visual pigment gene structure and expression in human retinae. Hum Mol Genet. 1997;6(7):981–90. dda133 [pii].
112. Kremers J, Scholl HP, Knau H, Berendschot TT, Usui T, Sharpe LT. L/M cone ratios in human trichromats assessed by psychophysics, electroretinography, and retinal densitometry. J Opt Soc Am A. 2000;17(3):517–26.
113. Bowmaker JK, Parry JW, Mollon J. The arrangement of L and M cones in human retina and a primate retina. In: Mollon J, Pokorny J, Knoblauch K, editors. Normal and defective colour vision. New York: Oxford University Press; 2003.

114. Rushton WA, Baker HD. Red–green sensitivity in normal vision. Vision Res. 1964;4(1):75–85.
115. Berendschot TT, van de Kraats J, van Norren D. Foveal cone mosaic and visual pigment density in dichromats. J Physiol (London). 1996;492(Pt 1):307–14.
116. Hofer H, Carroll J, Neitz J, Neitz M, Williams DR. Organization of the human trichromatic cone mosaic. J Neurosci. 2005;25(42):9669–79. doi:10.1523/JNEUROSCI.2414-05.2005. 25/42/9669 [pii].
117. Gunther KL, Dobkins KR. Individual differences in chromatic (red/green) contrast sensitivity are constrained by the relative number of L- versus M-cones in the eye. Vision Res. 2002;42(11):1367–78. S0042698902000433 [pii].
118. Lennie P, Pokorny J, Smith VC. Luminance. J Opt Soc Am A. 1993;10(6):1283–93.
119. Diller L, Packer OS, Verweij J, McMahon MJ, Williams DR, Dacey DM. L and M cone contributions to the midget and parasol ganglion cell receptive fields of macaque monkey retina. J Neurosci. 2004;24(5):1079–88. doi:10.1523/JNEUROSCI.3828-03.2004. 24/5/1079 [pii].
120. Martin PR, Blessing EM, Buzas P, Szmajda BA, Forte JD. Transmission of colour and acuity signals by parvocellular cells in marmoset monkeys. J Physiol. 2011;589(Pt 11):2795–812. doi:10.1113/jphysiol.2010.194076. jphysiol.2010.194076 [pii].
121. Crook JD, Manookin MB, Packer OS, Dacey DM. Horizontal cell feedback without cone type-selective inhibition mediates "red-green" color opponency in midget ganglion cells of the primate retina. J Neurosci. 2011;31(5):1762–72. doi:10.1523/JNEUROSCI.4385-10.2011. 31/5/1762 [pii].
122. Danilova MV, Chan CH, Mollon JD. Can spatial resolution reveal individual differences in the L:M cone ratio? Vision Res. 2013;78:26–38. doi:10.1016/j.visres.2012.12.006. S0042-6989(12)00396-3 [pii].
123. Baraas RC, Gjelle JV, Finstad EB, Jacobsen SB, Gilson SJ. The relationship between perifoveal achromatic, L- and M-cone acuity and retinal structure as assessed with multimodal high resolution imaging. Vision Res. 2016. doi:10.1016/j.visres.2016.06.005.
124. Kaiser PK, Lee BB, Martin PR, Valberg A. The physiological basis of the minimally distinct border demonstrated in the ganglion cells of the macaque retina. J Physiol (London). 1990;422:153–83.
125. Lee BB, Martin PR, Valberg A. The physiological basis of heterochromatic flicker photometry demonstrated in the ganglion cells of the macaque retina. J Physiol (London). 1988;404:323–47.
126. Stockman A, MacLeod DI, Lebrun SJ. Faster than the eye can see: blue cones respond to rapid flicker. J Opt Soc Am A. 1993;10(6):1396–402.
127. Zele AJ, Cao D. Vision under mesopic and scotopic illumination. Front Psychol. 2015;5:1594. doi:10.3389/fpsyg.2014.01594.
128. Smith VC, Pokorny J. Large-field trichromacy in protanopes and deuteranopes. J Opt Soc Am. 1977;67:213–20.
129. Buck SL, Knight R, Fowler G, Hunt B. Rod influence on hue-scaling functions. Vision Res. 1998;38:3259–63.
130. Lythgoe R. Dark-adaptation and the peripheral colour sensations of normal subjects. Br J Ophthalmol. 1931;15:193–210.
131. Benimoff N, Schneider S, Hood DC. Interactions between rod and cone channels above threshold: a test of various models. Vision Res. 1982;22:1133–40.
132. Sun H, Pokorny J, Smith VC. Brightness Induction from rods. J Vis. 2001;1:32–41. doi:10.1167/1.1.4.
133. Knight R, Buck SL, Pereverzeva M. Stimulus size affects rod influence on tritan chromatic discrimination. Color Res Appl. 2001;26:S65–8.
134. Walkey HC, Barbur JL, Harlow JA, Makous W. Measurements of chromatic sensitivity in the mesopic range. Color Res Appl. 2001;26:S36–42.
135. Cao D, Zele AJ, Pokorny J. Chromatic discrimination in the presence of incremental and decremental rod pedestals. Vis Neurosci. 2008;25(3):399–404. doi:10.1017/S0952523808080425
136. von Kries J. Uber die Funktion der Netzhautstabchen. Z Psychol Physiol Sinnesorg. 1896;9:81–123.

137. Nagel W. Appendix: Adaptation, twilight vision and the duplicity theory. Helmholtz's treatise on physiological optics. Translated from the Third German Edition by J.P.C. Southall. Third German Edition ed. Rochester, New York: Optical Society of America; 1924. p. 313-43.
138. Pokorny J, Smithson H, Quinlan J. Photostimulator allowing independent control of rods and the three cone types. Vis Neurosci. 2004;21(3):263–7.
139. Cao D, Pokorny J, Smith VC. Matching rod percepts with cone stimuli. Vision Res. 2005;45(16):2119–28. doi:10.1016/j.visres.2005.01.034.
140. Cao D, Pokorny J, Smith VC, Zele AJ. Rod contributions to color perception: linear with rod contrast. Vision Res. 2008;48(26):2586–92. doi:10.1016/j.visres.2008.05.001
141. Zele AJ, Maynard ML, Feigl B. Rod and cone pathway signaling and interaction under mesopic illumination. J Vis. 2013;13(1):1–19. doi:10.1167/13.1.21.
142. Willmer EN. Interaction between lights of different wave-length in the central fovea. J Physiol. 1950;111(1–2):69–80.
143. Lie I. Dark adaptation and the photochromatic interval. Doc Ophthalmol. 1963;17:411–510.
144. Stabell U, Stabell B. Scotopic contrast hues triggered by rod activity. Vision Res. 1975;15:1119–23.
145. McCann JJ. Rod-cone interactions: different color sensations from identical stimuli. Science. 1972;176(4040):1255–7.
146. Pokorny J, Lutze M, Cao D, Zele AJ. The color of night: surface color perception under dim illuminations. Vis Neurosci. 2006;23:525–30. doi:10.1017/S0952523806233492.
147. Elliott SL, Cao D. Scotopic hue percepts in natural scenes. J Vis. 2013;13(13):15. doi:10.1167/13.13.15.
148. Nagy AL, Boynton RM. Large-field color naming of dichromats with rods bleached. J Opt Soc Am. 1979;69(9):1259–65.
149. de Lange H. Research into the dynamic nature of the human fovea-cortex systems with intermittent and modulated light. I Attenuation characteristics with white and coloured light. J Opt Soc Am. 1958;48:777–84.
150. Kelly DH. Visual responses to time-dependent stimuli: I. Amplitude sensitivity measurements. J Opt Soc Am. 1961;51:422–9.
151. Schade Sr OH. Optical and photoelectric analog of the eye. J Opt Soc Am. 1956;46(9):721–39.
152. Snowden RJ, Hess RF, Waugh SJ. The processing of temporal modulation at different levels of retinal illuminance. Vision Res. 1995;35(6):775–89. 0042-6989(94)00158-I [pii].
153. Swanson WH, Ueno T, Smith VC, Pokorny J. Temporal modulation sensitivity and pulse-detection thresholds for chromatic and luminance perturbations. J Opt Soc Am A. 1987;4(10):1992–2005.
154. Lee BB, Martin PR, Valberg A. Sensitivity of macaque retinal ganglion cells to chromatic and luminance flicker. J Physiol (London). 1989;414:223–43.
155. King-Smith PE, Kulikowski JJ. Pattern and flicker detection analysed by subthreshold summation. J Physiol (London). 1975;249:519–48.
156. Mandler MB, Makous W. A three channel model of temporal frequency perception. Vision Res. 1984;24:1881–7.
157. Hess RF, Plant GT. Temporal frequency discrimination in human vision: evidence for an additional mechanism in the low spatial and high temporal frequency region. Vision Res. 1985;25:1493–500.
158. Metha AB, Mullen KT. Temporal mechanisms underlying flicker detection and identification for red-green and achromatic stimuli. J Opt Soc Am A. 1996;13:1969–80.
159. Metha AB, Mullen KT. Red-green and achromatic temporal filters: a ratio model predicts contrast-dependent speed perception. J Opt Soc Am A. 1997;14:984–96.
160. Baraas RC, Kulikowski JJ, Muldoon MR. Bar-like S-cone stimuli reveal the importance of an intermediate temporal filter. J Opt Soc Am A Opt Image Sci Vis. 2010;27(4):766–80. 196741 [pii].
161. Kulikowski JJ, Tolhurst DJ. Psychophysical evidence for sustained and transient detectors in human vision. J Physiol. 1973;232(1):149–62.
162. Clifford CWG, Spehar B, Solomon SG, Martin PR, Zaidi Q. Interactions between color and luminance in the perception of orientation. J Vis. 2003;3:106–15.

163. Smith VC, Pokorny J, Lee BB, Dacey DM. Sequential processing in vision: the interaction of sensitivity regulation and temporal dynamics. Vision Res. 2008;48(26):2649–56.
164. Yeh T, Lee BB, Kremers J. Temporal response of ganglion cells of the macaque retina to cone-specific modulation. J Opt Soc Am A Opt Image Sci Vis. 1995;12(3):456–64.
165. Conner JD, MacLeod DI. Rod photoreceptors detect rapid flicker. Science. 1977; 195(4279):698–9.
166. Hecht S, Smith EL. Intermittent stimulation by light. IV. Area and the relation between critical frequency and intensity. J Gen Physiol. 1936;19:979–91.
167. Hecht S, Verrijp CD. The influence of intensity, color and retinal location on the fusion frequency of intermittent illumination. Proc Natl Acad Sci U S A. 1933;19:522–35.
168. Mullen KT. The contrast sensitivity of human colour vision to red-green and blue-yellow chromatic gratings. J Physiol. 1985;359:381–400.
169. Thibos LN, Cheney FE, Walsh DJ. Retinal limits to the detection and resolution of gratings. J Opt Soc Am A. 1987;4(8):1524–9.
170. Rossi EA, Roorda A. The relationship between visual resolution and cone spacing in the human fovea. Nat Neurosci. 2010;13(2):156–7. doi:10.1038/nn.2465. nn.2465 [pii].
171. Dees EW, Gilson SJ, Neitz M, Baraas RC. The influence of L-opsin gene polymorphisms and neural ageing on spatio-chromatic contrast sensitivity in 20–71 year olds. Vision Res. 2015. doi:10.1016/j.visres.2015.08.015. S0042-6989(15)00284-9 [pii].
172. Williams D, Sekiguchi N, Brainard D. Color, contrast sensitivity, and the cone mosaic. Proc Natl Acad Sci U S A. 1993;90(21):9770–7.
173. Zaidi Q, Shapiro A, Hood D. The effect of adaptation on the differential sensitivity of the S-cone color system. Vision Res. 1992;32:1297–318.
174. Livingstone MS, Hubel DH. Psychophysical evidence for separate channels for the perception of form, color, motion and depth. J Neurosci. 1987;7:3416–68.
175. Pokorny J, Smith VC. Psychophysical signatures associated with magnocellular and parvocellular pathway contrast gain. J Opt Soc Am A. 1997;14:2477–86.
176. Pokorny J. Review: steady and pulsed pedestals, the how and why of post-receptoral pathway separation. J Vis. 2011;11(5):1–23. doi:10.1167/11.5.7.

Chapter 6
Color Constancy and Contextual Effects on Color Appearance

Maria Olkkonen and Vebjørn Ekroll

Abstract Color is a useful cue to object properties such as object identity and state (e.g., edibility), and color information supports important communicative functions. Although the perceived color of objects is related to their physical surface properties, this relationship is not straightforward. The ambiguity in perceived color arises because the light entering the eyes contains information about both surface reflectance and prevailing illumination. The challenge of color constancy is to estimate surface reflectance from this mixed signal. In addition to illumination, the spatial context of an object may also affect its color appearance. In this chapter, we discuss how viewing context affects color percepts. We highlight some important results from previous research, and move on to discuss what could help us make further progress in the field. Some promising avenues for future research include using individual differences to help in theory development, and integrating more naturalistic scenes and tasks along with model comparison into color constancy and color appearance research.

Keywords Color perception • Color constancy • Color appearance • Context • Psychophysics • Individual differences

6.1 Introduction

Color is a useful cue to object properties such as object identity and state (e.g., edibility), and color information supports important communicative functions [1]. Although the perceived color of objects is related to their physical surface properties,

M. Olkkonen, M.A. (Psych), Dr. rer. nat. (✉)
Department of Psychology, Science Laboratories, Durham University,
South Road, Durham DH1 3LE, UK

Institute of Behavioural Sciences, University of Helsinki,
Siltavuorenpenger 1A, 00014 Helsinki, Finland
e-mail: maria.olkkonen@durham.ac.uk; maria.olkkonen@helsinki.fi

V. Ekroll, Dipl.-Psych., Dr. habil.
Laboratory of Experimental Psychology, University of Leuven (KU Leuven),
Tiensestraat 102, box 3711, Leuven 3000, Belgium
e-mail: vebjorn.ekroll@ppw.kuleuven.be

J. Kremers et al. (eds.), *Human Color Vision*, Springer Series in Vision Research, DOI 10.1007/978-3-319-44978-4_6

this relationship is not straightforward. In this chapter, we focus on how viewing context and illumination affect color percepts.

The development of the Young-Helmholtz trichromatic theory, according to which color sensations rely on three mechanisms with sensitivities in different parts of the electromagnetic spectrum, was essentially complete more than a century ago [2, 3]. Most research efforts in color vision thereafter have either been directed towards obtaining a better understanding of (a) the physiology and genetics underlying the trichromatic theory (also see Chaps. 1, 3, and 4) or (b) the many aspects of color perception which are beyond the explanatory scope of trichromatic theory. Many significant advances have been made in the first line of research: after the trichromatic theory was corroborated with direct measurement of retinal photoreceptors [4], our understanding of the early processing of color signals has evolved to the stage where treatment of color blindness seems realistic in the near future [5, 6]. This chapter is about central aspects of the second line of research, namely color constancy and the context-dependence of perceived color. Here, progress has been more modest, despite intense and often quite ingenious research efforts. For that reason, we organize this chapter around the question of why progress has been so slow and what might be done to remedy it rather than providing a comprehensive state-of-the art review of the literature (for those, see Refs. [7–10]).

We start by reviewing the problem of color constancy and context effects on color appearance, and then move on to review commonly used methods. We further discuss a few more novel methods that hold much promise in advancing the field. We conclude the chapter by laying out some outstanding questions and suggestions for future research.

6.2 Contextual Effects on Color Appearance

Contextual effects on color perception have traditionally been studied under two relatively separate rubrics: color constancy and color induction/color appearance. In general, studies in color constancy aim to understand how constant descriptors of surface color are extracted from variable sensory signals, whereas studies in color induction aim to understand how changes in background properties affect the color appearance of a target stimulus. In fact, many studies could be classified under either rubric, making the distinction rather arbitrary. But because studies in the two traditions often differ in how fundamental problems are formulated, we introduce them separately. However, we do not make a hard distinction between the two rubrics when discussing methodology and theoretical approaches later in the chapter.

6.2.1 Color Constancy

Imagine stepping out on the patio with your favorite cup in hand; the surface color of the cup does not appear to change even though the illumination impinging on the cup changes quite dramatically. This ability to perceive stable object colors in varying illumination is called color constancy. Although seemingly effortless, constancy involves complex visual processing: as the illumination on the cup changes, the light reflected to the eye from the cup changes as well; there is no unique signal for any given surface color (Fig. 6.1 illustrates this for a clover). In order to use object color as a cue to object identity, the visual system faces the challenge of parsing the retinal light signal into object and illumination components ("inverse optics" [11]). This mathematically underdetermined estimation problem cannot be solved without constraints, for instance from prior knowledge about the nature of surfaces and illuminants [12].

Although vision scientists agree to some extent about the nature of the computational problem, there is little agreement about whether the visual system actually inverts the calculation (and estimates reflectance and illumination) (e.g., [13–16]), or rather bypasses the inverse problem by using heuristics or image cues to make decisions about color [17–19]. The first computational models solved the inverse problem by assuming certain regularities in the visual scene, for instance

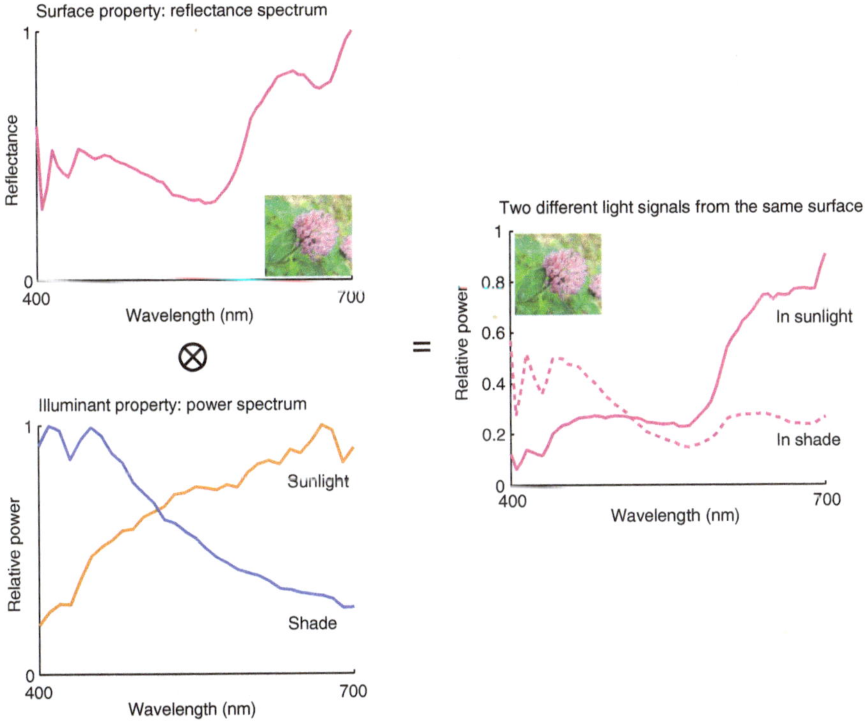

Fig. 6.1 Color constancy is hard because the light reflected off a surface depends both on the reflectance properties of the surface and the light illuminating the surface. The *upper left panel* shows the reflectance of a clover as a function of wavelength. The *lower left panel* shows the power spectrum of two different daylight illuminants: direct sunlight and skylight (shade). The light reflected to the eyes off a clover, shown in the *right panel*, depends on whether the clover is illuminated by sunlight or skylight. In order to perceive the surface color accurately, the visual system has to estimate the surface reflectance from the combined light signal

that the mean chromaticity of a scene is neutral or at least known [20], or that illumination changes more slowly across a visual scene than surface reflectance [21–23]. Another class of models relied on the fact that natural surface reflectances and daylight illuminants can be represented with a limited number of basis functions, greatly simplifying the inverse calculation (e.g., [20, 24, 25]). These models showed that the illuminant in relatively simple scenes can be estimated and thus discounted based on cues from mean chromaticity across the scene [20], specular highlights on surfaces [24, 26], and mutual illumination [27].

Of course, that these models work for some visual scenes does not mean that they accurately model human color constancy, but they have been useful in developing experimental hypotheses. In a seminal study, Kraft and Brainard [28] tested three common theoretical assumptions: that color constancy is determined by adaptation to mean luminance; adaptation to local contrast; or adaptation to the brightest surface in the scene. Their results were not consistent with any single mechanism, but rather with a combination of several mechanisms. In a series of studies taking advantage of computer rendering techniques, Maloney and colleagues tested the role of different image cues for color constancy. They found that observers used information from several cues, such as highlights, cast shadows, and depth cues when estimating surface color [18, 29–31].

Color constancy is usually quantified by measuring changes in color appearance caused by a change in viewing context. Reduced to a two-dimensional world, this can be something like the simultaneous contrast illusion in Fig. 6.2a. In the top panel, the surrounds affect perceived lightness of the target: the two physically identical middle-gray patches look either dark or light depending on the surround. In the chromatic case depicted in the middle panel of Fig. 6.2a, the two physically identical targets appear either bluish or yellowish depending on the surround. One can measure the magnitude of the context effect on target color appearance and use mathematical methods to quantify the "compensation" for the difference in surrounds in terms of color constancy (*color constancy index*, see Ref. [8]). As the stimuli in Fig. 6.2a (top and middle) are very simple compared to natural scenes, contemporary studies commonly use more complex backgrounds (Fig. 6.2a, bottom) or computer renderings of three-dimensional scenes (Fig. 6.2b). The idea behind all of these visual displays is the same, however: the overall color difference between the two sides of the display simulates an illumination change, and the extent to which the observer compensates for this difference when judging target color appearance is a measure of color constancy.

Contextual effects on color perception have been extensively quantified during the past century with scenes similar to the ones depicted in Fig. 6.2a, and we now have some understanding about the regularities of color perception in such scenes. To a first approximation, observers are able to partially compensate for illumination changes when judging object color appearance across illumination variation, especially with more "natural" tasks and realistic displays (for a comprehensive review, see Ref. [8]). However, the degree of color constancy depends on the instructions and task [34, 35], the display, and the realism of the stimuli in ways that are not well-understood (for a review, see Ref. [7]). In order to uncover the mechanisms of color constancy, it is important to develop general models that predict both successes and failures of color constancy in a broad range of experimental situations.

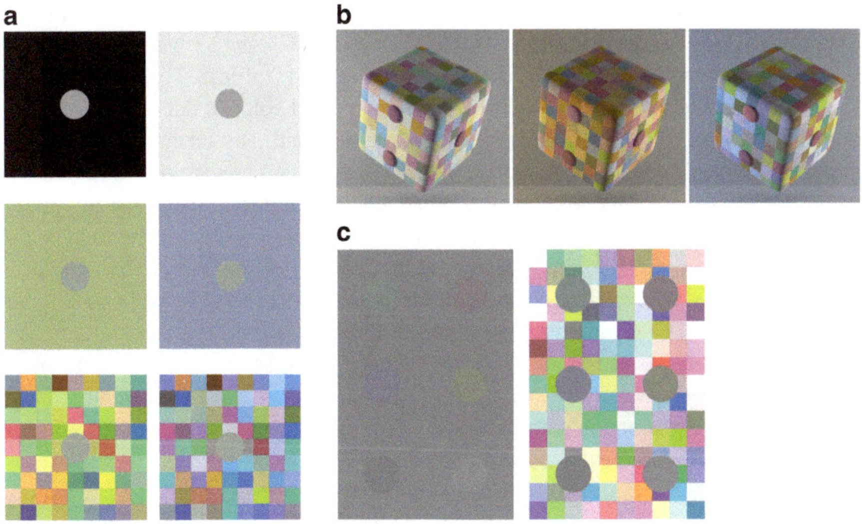

Fig. 6.2 Typical displays used in the study of color constancy and color induction. (**a**) A demonstration of the classical simultaneous brightness/color contrast illusion. The two central disks are identical in terms of reflected light, but viewing them embedded in different surrounds make their colors appear different. *Top*: achromatic; *middle*: chromatic; *bottom*: chromatic with different spatial properties in the targets and surrounds. (**b**) Here, the observer's task is to choose the button on the left side of a cube that matches the button on the right side. The cube on the *left* is illuminated by a standard light on both sides; the two other cubes are illuminated by a standard light on the right and a yellowish (*center*) or bluish (*right*) light on the left. Figure reproduced with permission of Association for Research in Vision and Ophthalmology from [32] via Copyright Clearance Center. (**c**) A demonstration of Brown and MacLeod's [33] gamut expansion effect. The six colored disks embedded in the uniform grey surround (*left*) are printed in the same ink as the six disks embedded in the variegated surround (*right*), yet they appear more saturated (or colorful). This demonstrates that target color appearance is influenced by the variance of surround colors even with the same average surround color

6.2.2 Color Induction

Stimulus context plays a crucial role in enabling the organism to solve the problem of color constancy. It has been proposed that observers derive information about the illuminant from the surround [36], and some models achieve color constant representations from edge contrasts between a target and its surround (e.g., [37, 38]). Accordingly, several authors have suggested that color constancy and various effects of context on color appearance are essentially two sides of the same coin. Both von Helmholtz [12] and Hering [39] agreed on this point, although their theories were rather different [40, 41].

Helmholtz famously proposed that the simultaneous contrast effect is produced by an error of judgement (or "unconscious inference"), where the target is mistakenly assumed to be illuminated with a light having the color of the surround (also see Fig. 4 in Ref. [42]). Due to the discounting of the illumination, the perceived color of the target shifts in a direction opposite in color space to the color of the surround. Thus, a grey target appears greenish if embedded in a red surround, bluish if embedded in a yellow surround, and so on (for a computational implementation of this idea, see Ref. [16]). Hering, on the other hand, explained both color constancy and simultaneous contrast in terms of lateral inhibition between neighboring receptors at the retina. Helmholtz and Hering's ideas have been hugely influential and most current theories of color constancy and simultaneous contrast can be regarded as modern incarnations of their theories. Consequently, their shared basic assumption that simultaneous contrast is due to mechanisms subserving color constancy is broadly adopted in the field [43–47]. Indeed, the stimuli and tasks in experiments on simultaneous color constancy and simultaneous contrast are often quite similar or even identical.

A priori, though, any given observable induction effect may be due to a host of different mechanisms, subserving different functional goals (or even be spandrels serving no particular functional purpose at all [48]). The widespread idea that simultaneous contrast is a side effect of mechanisms correcting for the influence of the prevailing illumination is therefore by no means necessary. For instance, it has been proposed that mechanisms are involved that serve to counteract intraocular glare [49] or to infer the color of transparent media [50]. Importantly, if such alternative mechanisms contribute, empirical measurements of the effect would not directly reflect the pure effect of constancy mechanisms correcting for the prevailing illumination. More specifically, the transparency mechanism discussed by Ekroll and Faul [50] seems to be triggered only when the surround is uniform in the vicinity of the target. Thus, experiments using variegated surrounds such as in Fig. 6.2b, c may provide a better, uncontaminated estimate of the effect due to mechanisms correcting for the illuminant [51].

6.2.3 Type I and Type II Constancy

An interesting challenge to the popular idea that both simultaneous contrast and color constancy are simply due to mechanisms encoding difference (contrast) information at the borders between surfaces [45, 52–54] is the observation that when an object moves across a multicolored background, we hardly ever experience the strong changes in its perceived color that this idea would predict [53, 55]. This observation of a "natural background independence" has led researchers to propose that in addition to the mechanisms that provide color constancy across illumination changes (Type I constancy), the visual system also possesses mechanisms providing color constancy across background changes (Type II constancy) [53, 56]. According to Whittle, the initial coding of color based on differences at edges (presumed to take place in the retina or early visual cortex) is counteracted by a subsequent stage of processing in which the differences at edges are integrated across space, much as in the Retinex model by Land and McCann [23]. While retinal processing can be understood as mathematical differentiation of the retinal image, later processing is essentially conceived of as mathematical integration. The concatenation of these two operations yields the original image up to an unknown additive constant (which could be determined based on some kind of anchoring rule, see Gilchrist et al. [17]). Thus, perfect integration would lead to Type II constancy (background independence). From this perspective, simultaneous contrast effects (a failure of background independence) are to be understood as failures of integration [53]. Indeed, the Retinex model, which integrates luminance across edges, does not "see" the classical simultaneous contrast illusion.

It has been proposed that the integration process depends on perceptual mechanisms that classify edges in the visual input as reflectance edges or illumination edges: It would make sense for the processes responsible for the computation of surface color to integrate only across reflectance edges and to disregard illumination edges [57]. In a certain sense, this differentiation-integration perspective redefines the problem of understanding color perception. From this perspective, color constancy and strong simultaneous contrast effects are not very mysterious. Rather, the central problem becomes to understand exactly how the postulated integration and edge classification processes work [53, 57]. As pointed out by Koffka ([58], cited in Ref. [57]), "a complete answer to this question would probably supply the key to the complete theory of color perception in the broadest sense." It should be pointed out, however, that the problematic observation of "a natural background independence" is also amenable to an alternative explanation. There is evidence that simultaneous color contrast is sometimes considerably stronger for the uniform backgrounds typically used in many studies than in more naturalistic variegated backgrounds [43, 51, 59], and as pointed out by Evans ([60], p. 210) uniform backgrounds come "close to being contrary to the laws of nature." Thus, strong simultaneous contrast effects may be the result of the activation of special-purpose mechanisms [50, 61] rather than general principles of retinal coding. An informal observation consistent with this proposal is that when an object moves in front of different uniform surfaces (such as a bird crossing the clouds of the twilight sky), quite strong simultaneous contrast effects can be observed (see also [43]). From this perspective, the often-made implicit or explicit assumption that "the centre-surround configuration is a particularly important one" [53] may be misleading.

6.3 How Are Contextual Effects on Color Perception Quantified?

6.3.1 Current Situation in the Field

Although the amount of literature on color constancy and color induction is nothing short of daunting, our understanding of the fundamental mechanisms and processes in contextual color perception is still far from complete. As one delves deeper into the literature, the lack of convergence into general quantitative models becomes tangible [53]. There also remains disagreement about fundamental theoretical issues; for instance, whether the perceptual representation of surface color is equivalent to the physical property "surface reflectance," and whether illumination is explicitly represented [48, 62, 63].

Another important issue is to what extent color perception can be understood independently from other aspects of perception such as material properties, scene layout, depth relationships, and perceptual organization in general [61, 64–76]. It is also unclear what the relevant dimensions of color experience are [50, 60, 77–82]. On a more practical level, there is much debate about what is the best way to characterize color constancy, and whether it is even possible to measure it in "objective" or artifact-free ways. Many different methods have been employed in the past [83], but they may often provide conflicting results [53, 84]. An unfortunate consequence of these issues is that it is not always clear which results should be incorporated into theories, and which are artifacts of inadequate measurement methods or misleading assumptions about the phenomenal structure of color experience.

In the following, we briefly review the most popular classic methods, along with a few interesting more novel approaches.

6.3.2 *Classic Methods*

6.3.2.1 Asymmetric Color Matching

Asymmetric matching has traditionally been a popular method to measure color constancy and color induction [34, 85]. Here, the observer is presented with two target patches embedded in different surrounds and asked to adjust the color coordinates of one of the targets such that it appears identical to the other one. The difference between the color coordinates of the two targets at the perceptual match can then be taken as a measure of the combined effect of the two surrounds on the color appearance of the targets. To quantify the amount of color constancy, the observed color match is often compared to a perfectly color constant match, whereby the ratio between the two matches indicates the degree of color constancy (*color constancy index*). A further option is to infer the observer's illuminant estimate from the color match and use this to quantify color constancy [86]. This approach has the advantage of offering a direct link to computational models of constancy that estimate the illuminant.

It would often be more interesting to measure the effect of each surround on the color appearance of a target separately, rather than the combined effect of the two surrounds. Unfortunately, there seems to be no principled way for decomposing the measured combined effect into single surround-specific effects without relying on theoretical assumptions. An easy and popular, but potentially questionable way out of this dilemma is to posit that some particular surround is functionally neutral (i.e., that it has no effect on the perceived color of targets embedded in it). That way, the effect of any surround of interest can be directly measured by adjusting targets in the functionally neutral surround to match those presented in the surround of interest. It is often assumed that a completely dark surround or phenomenologically neutral surrounds (i.e., surrounds that appear achromatic) are functionally neutral, but this is a purely theoretical assumption that may well be incorrect [87]. For instance, one often speaks of a completely dark surround as "no surround," but the absence of physical stimulation does not imply that such a surround is functionally neutral. If, for instance, the difference between target and surround is the essential determinant of the target's perceived color [46, 47], a completely dark surround affects perceived color in essentially the same way as any other surround by contributing to the critical target-surround difference.

To gain a clearer idea of the inherent ambiguity of asymmetric matching data, a simple formal consideration may be helpful. Models of color appearance typically posit a three-dimensional color code [88], i.e., a function f of the tristimulus vectors[1] t, s of the target T and the surround S which yields a triplet of numbers representing the perceived color of the target. Any such model predicts that two targets T_1, T_2 embedded in surrounds S_1, S_2 should match whenever $f(t_1, s_2)=f(t_2, s_2)$. However, since any other color code $h:=g(f)$ obtained by concatenation of the function f with an arbitrary invertible function g makes exactly the same predictions, many rather different models of color appearance are compatible with the same set of matching data. This makes it evident that the shape of the function relating target coordinates to color appearance for targets in a certain surround cannot be inferred from asymmetric matching data. Once it is known (say, based on scaling or threshold data, see Sect. 6.3.2.5) for one surround, however, the matching data can be used to infer it for another surround [90].

It is also important to realize that the direction of the induction effect produced by a surround cannot be inferred from matching data without making (potentially incorrect) assumptions about what constitutes a functionally neutral surround.

6.3.2.2 Asymmetric Matching with Haploscopically Superimposed Displays

Based on an idea of Hering [91], Whittle [47, 53] pioneered the use of a special matching paradigm in which one target-surround stimulus is presented separately to each eye such that, through binocular fusion, the observer experiences the two targets to be matched as embedded in the same surround, although their monocular surrounds are different (haploscopically superimposed displays, HSD). An important advantage of this technique is that the subjective matching problems often reported in experiments using conventional side-by-side displays seem to be absent or at least significantly reduced. This technique also tends to produce much stronger induction effects than experiments with side-by-side displays [92]. Whittle [54] argued that this is because the results obtained with these techniques reflect the consequences of retinal mechanisms that essentially only register color differences at edges more directly than the results obtained with conventional side-by-side matching: The weaker effects obtained

[1] Colorimetric specification of a light stimulus is often made in terms of tristimulus values, which are the intensity values of three reference lights needed to match the test light. Different systems, such as the [R,G,B] and [X,Y,Z] of the Commission Internationale de L'éclairage (CIE) are based on different choices of the three reference lights, but they are essentially equivalent: different tristimulus values are related to each other by a linear transform. The tristimulus vector of a light is often just referred to as its "color," but this is potentially misleading, since the perceived color depends on context as well (see [89] for a good introduction to color measurement)

with conventional matching are attributed to secondary mechanisms of integration counteracting the effects of the primary difference coding. However, alternative explanations of the special results obtained with these techniques are also possible: as the observer's eyes receive different input from the two hemifields, the haploscopic matching results could be attributed to strong temporal adaptation in each hemifield rather than to simultaneous contrast per se ([87], p. 125).

6.3.2.3 Achromatic Settings

In this technique, the observer is asked to adjust the chromaticity of a target embedded in a surround such that it appears achromatic (e.g., [93, 94]). Compared to asymmetric matching, this method has the advantage of obviating the need for presenting a second comparison target (and surround) which may influence the perceived color of the target [95]. Apart from that, however, results obtained with this method suffer from the same fundamental ambiguity as the results from asymmetric color matching. This is because the interpretation and modeling of achromatic matches implicitly or explicitly involves the comparison with targets that would appear achromatic when presented in another surround (typically gray or black). But note that this ambiguity is a problem only if the goal is to understand the quality of color appearance, instead of quantifying how constantly observers judge color appearance across illuminant changes [93].

An important disadvantage of achromatic settings is that measurements can only be made for a small subset of all possible colors (those that appear achromatic in a given surround). Given the large nonlinearities sometimes observed in asymmetric color matching experiments [51, 96–101], it does not appear advisable to draw general conclusions based only on achromatic settings.

6.3.2.4 Unique Hue Settings

The method of unique hue settings [102] is based on a central notion of opponent color theory, namely that the four unique hues red, green, yellow, and blue[2] have special properties that make them particularly useful as landmarks in color space [39, 103]. Unique hue settings have the advantage that measurements can be made not only for targets that appear achromatic in a given surround, but also for targets that appear unique red, green, blue, or yellow (see also [104] for the related method of unique hue scaling, and see also Chap. 5 about between-individual variation in unique hues). A disadvantage, however, is that interpretation and modeling requires even more assumptions than achromatic settings because the set of colors that appear in a given unique hue is a two-dimensional manifold. Like achromatic settings, unique hue settings can be regarded as a limited form of implicit asymmetric matching. Hence, this technique also suffers from the aforementioned fundamental ambiguity of asymmetric matching data.

---

[2] So called because they cannot be divided further into component hues; see Chap. 5.

6.3.2.5 Threshold and Scaling Measurements

Context does not only influence color appearance, but also the discriminability of colors [105–108]. Color discrimination may be measured with threshold measurements and various supra-threshold scaling techniques [109, 110] and in some cases threshold measurements and scaling techniques seem to yield consistent results [108]. In theory, much of the aforementioned inherent ambiguity of asymmetric matching data can be resolved based on corresponding data from discrimination experiments. Specifically, threshold or scaling measurements can be used to estimate the derivative of the color code $f(t,s)$ (target color appearance in a given surround) with respect to t (the color coordinates of the target), which cannot be determined based on asymmetric matching. Therefore, employing both appearance and discrimination measurements in conjunction [90, 111, 112] can be particularly useful for constraining models of color appearance.

6.3.3 *Critical Assumptions in Measuring and Modeling Color Appearance*

The methods described above for measuring color appearance, and the modeling of the resulting data typically rely on a host of critical (implicit or explicit) assumptions that warrant consideration. We have already discussed the assumption that certain surrounds are often assumed to be "functionally neutral" and now briefly scrutinize two other potentially questionable assumptions.

6.3.3.1 The Continuity Assumption

Virtually all models of color appearance implicitly assume that perceived color (represented by a color code) is a continuous function of the color coordinates of the target and the surround. There is some evidence to suggest that this assumption, sometimes referred to as Grassmann's second law [113], is not generally valid. Rather, there is often a discontinuity in color appearance when the target color coordinates reach the coordinates of the surround, at least when the surround is uniform [47, 50, 51, 114]. This phenomenon, sometimes called "crispening," is related to the observation that color discrimination is best for target colors close to the surround color [106].

6.3.3.2 The Compensation Assumption

The core idea behind the asymmetric matching method is that the observer compensates the net effect of the surrounds on the perceived colors of the targets by adjusting the tristimulus values of one of the targets. As the observer can adjust three variables, this should be possible provided that the space of perceived colors is three-dimensional.

But if this space has more than three dimensions, it should be difficult to establish a subjectively satisfactory match. A priori, the perceived color of a target embedded in a simple uniform surround depends on six variables, namely the tristimulus values of the target plus those of the surround and hence the space of perceived colors could theoretically be up to six-dimensional even in this simple case [60]. Empirically, the evidence for or against the validity of the three-dimensionality assumption is mixed. While many studies do not mention any matching problems, other studies suggest that the problems can be quite dramatic [85, 115, 116]. Interestingly, subjective matching problems seem to be absent using the special viewing conditions of haploscopically superimposed displays [53, 92]. A general hypothesis that could account for the mixed results is that matching is possible for some combinations of target and surround colors, but not for others. For instance, subjective matching problems are often particularly evident when the contrast between target and surround is small [51] or when the two targets have a different contrast polarity [54].

6.3.4 Performance-Based Measures

Although the matching methods described above have various advantages, they are rather far-removed from everyday color tasks (compare matching two color patches in hue and saturation to selecting ripe tomatoes in the market). One potential issue with matching methods is that such an artificial task might not tap into real-world constancy mechanisms. For instance, observers might choose a variety of strategies to accomplish an asymmetric match, introducing variability unrelated to the experimental manipulation.

Due to these issues, an increasing number of laboratories use performance-based measures to characterize color constancy. Although these methods vary in many regards, they all measure observers' ability to recognize surface colors when the illuminant or viewing context is varied, rather than asking observers to match colors. Of course, the choice of task depends on the goal of the experiment; if the variable of interest is *perceptual appearance*, other methods are more suitable. But performance-based tasks are useful for quantifying how well and under what conditions observers are able to *identify surface colors* across context changes, which is useful for understanding real-world color constancy mechanisms.

6.3.4.1 Color Identification and Color Selection

In color identification and selection tasks, observers are asked to identify surfaces across illuminant changes. Because observers are not required to adjust or match colors, they often find this more intuitive. Color selection tasks are also arguably closer to everyday demands on color constancy than classic adjustment tasks; if the function of color constancy is to infer relevant object properties, such as ripeness, the outcome of the constancy process should lead to a successful selection of the target item amongst similar "distractors."

In one of the first performance-based color constancy studies, Bramwell and Hurlbert [92] used a task where observers saw a target under one simulated illuminant and were asked to pick the matching surface from a set of possible matches displayed under a different illuminant. Constancy was quantified as the distance between the true reflectance match and the match picked by the observer. The task was easy for observers and yielded measures of both color constancy and color discrimination, unlike classic matching methods.

Zaidi and colleagues [117–119] developed a method along similar lines but using real objects and illuminants. In their influential studies, observers saw four objects under two different illuminations, and were asked to pick the odd object out. By comparing observers' choices with different strategies—for instance reflectance matching and color contrast matching—Zaidi et al. showed that observers used a suboptimal strategy based on color similarity when identifying objects, instead of a more complicated reflectance estimation or "inverse-optics" approach.

Recently, Radonjić et al. [32] developed a color selection task with a similar principle to the paradigm of Zaidi and colleagues. Their use of rendered stimuli (see Fig. 6.2b) affords more flexibility in the choice of stimuli, and thus more accurate measurements of color constancy. Analyzing the selection data with a variant of maximum likelihood difference scaling, Radonjić et al. showed that observers were very color constant with complex, 3D stimuli (similar to the ones in Fig. 6.2b), but poor with simple, 2D stimuli. In a follow-up study with an even more naturalistic block-sorting task, Radonjić et al. [120] found good color constancy that was robust to manipulations of scene complexity (i.e., number of surfaces in the scene) and local contrast.

6.3.4.2 Color Naming and Categorization

One can also study the influence of context by asking subjects to categorize the color of targets in different contexts [121–125]. This has the same advantage as achromatic or unique hue settings: There is no need to display a comparison stimulus, which may influence the measurements. Color naming also has the advantage over achromatic adjustments that it can yield information about several landmarks in color space [121]. A well-known limitation is that humans are able to discriminate many more colors than they have color names for [83]. One way to overcome this issue is to use stimulus sets that span a large portion of color space and to model constancy for a group of stimuli simultaneously [121, 122, 126].

Color naming is similar to typical forced-choice paradigms in that it requires observers to categorize stimuli instead of making matches across contexts. In an early study, Jacobs and Gaylord [127] measured adaptation to spectral narrow-band lights and found color naming to be as accurate as asymmetric matching for measuring adaptation effects but more intuitive for observers. Later studies have found similar results for color constancy [124, 125, 128]. In a series of studies, Gegenfurtner and colleagues studied how the structure of color space changes under varying illumination by using a combination of color naming and mathematical modeling. They found the structure of color space to be largely stable, with small transformations in category boundaries explained by relatively simple linear models [121, 126]. By using Munsell chips with known surface reflectances, Olkkonen et al. [122] were further able to compare how consistently individual observers named surface colors across illuminants with how consistently different observers named the same surfaces under one illuminant. They found across-illuminant naming consistency to be similar to across-observer naming consistency; in other words, those surface colors that remained stable across illuminants also reached high inter-observer agreement under a baseline illuminant.

6.3.4.3 Priming

The Helmholtzian view that unconscious inference is necessary for achieving color constancy from "raw sensations" at the retina has often been taken as a given by color scientists (see [78], for further discussion). Norman et al. [129] investigated this assertion with a clever priming task, where they used metacontrast masking to display subliminal color primes to observers. Norman et al. tested whether a prime matched with the mask in terms of reflected light (proximal properties) or reflectance (distal properties) differentially facilitated the subsequent color categorization of the mask. They found that categorization was facilitated more when the prime matched the mask in terms of reflectance rather than in reflected color. This led Norman and colleagues to conclude that object color is initially represented in terms of its surface reflectance, and not the reflected color. This is notably inconsistent with the common notion that the earliest processing stages represent proximal stimulus qualities that need to be processed in order to arrive at a representation of surface color [12, 130].

6.3.4.4 Operational Color Constancy

Based on the observation that color constancy in the laboratory is often poor (e.g., [34, 131]), Foster and colleagues have advocated a more operational definition of color constancy (also see Ref. [132]). Their approach agrees with the observation that although the color appearance of objects often changes with the illuminant (a sheet of white paper often looks yellower under sunlight than in shade), we do not perceive the surface material as having changed; rather, we are able to distinguish illuminant from material changes. Consequently, the function of color constancy might be to tell illuminant changes from reflectance changes, rather than maintaining equal color appearance across illuminant changes. Foster and colleagues have shown in a series of experiments that observers are good at discriminating illuminant changes from changes in object properties (e.g., [83, 133, 134]). The advantage of this method is the naturalness and ease of the discrimination task, but as other performance-based measures, it does not inform us about *how* object color appearance changes in different illuminations. Indeed, one interesting implication of this research is that operational color constancy may be good even when *appearance* constancy fails.

In a variant of the operational color constancy task, Pearce et al. [135] quantified color constancy by measuring discrimination thresholds for illuminant changes in different color directions. They found that illumination discrimination was poorest on the daylight axis toward bluish illuminations. The fact that observers are the least sensitive to bluish illuminant changes implies that color constancy is *best* for these illuminants. This is consistent with the suggestion that color constancy should be best for natural daylight variation; a common hypothesis that has lacked empirical evidence so far (but see Ref. [15]).

6.3.5 The Role of Scene Complexity

Figure 6.2 shows an array of displays with different levels of scene complexity. It is possible to quantify the effect of color context on color appearance in a simple display such as the simultaneous contrast (Fig. 6.2a). As seen in this classical illusion, context can affect color appearance quite dramatically. It is now known, however, that the complexity of the viewing context modulates the strength of context effects. Comparing the middle and bottom panels of Fig. 6.2a reveals that a complex surround may, at least in some cases, have a smaller effect on target color appearance than a simple one, even with the same average surround chromaticity [43, 75, 136, 137]. Manipulating the number and type of surfaces in a scene also affects color constancy (e.g., [138–142]). Finally, Brown and MacLeod [33] showed that the same low-contrast targets appear more colorful on uniform surrounds than on variegated surrounds (Fig. 6.2c). Taken together, these results show that findings from simple scenes cannot be straightforwardly generalized to complex scenes and ultimately to color perception in the real world.

It is important to note that scene complexity can be increased by adding more variability in the visual scene without adding more structure. The display in the bottom panel of Fig. 6.2a has more chromatic variability than the middle panel, but not considerably more spatial structure. The scene in Fig. 6.2b, on the other hand, has both more chromatic variability and structure than the displays in Fig. 6.2a. It is conceivable that variability and structure separately modulate the strength of context effects on color appearance. In a seminal demonstration, Adelson [64] showed that lightness percepts are influenced by perceived scene structure although the pattern of luminance across the scene was held constant (see also [70]).

With the advent of physically based rendering tools, several labs have moved to using more realistic displays to study color constancy. These displays confer several advantages over more traditional 2D displays. They allow for more freedom in manipulating the objects—their shape and surface material, etc.—along with illumination properties. Such scenes allow studying the role of complex scene cues to color constancy; for instance stereo disparity [143], highlights [29, 144, 145], object pose [146, 147], and material [76]. Computer graphics also enables us to study the perception of more complex illumination and reflectance properties, such as the effect of illumination geometry on the perception of glossiness [148–151].

One might argue that it would be most ecologically valid to use real objects and illuminants, because they contain cues that are challenging to reproduce in rendered scenes. Indeed, some early studies in color constancy were conducted with real (albeit comparatively simple) stimuli under real illuminants (e.g., [36, 79, 152]). Even after the advent of digital displays, some groups still employ real surfaces and illuminants with the thought that they might tap more "natural" constancy mechanisms [85, 122, 153–158]. One notable disadvantage in using real displays is the difficulty to parametrically manipulate object or illumination properties. To overcome this limitation, some laboratories have built setups that combine real objects with projectors so that the apparent object and illuminant colors can be independently and parametrically manipulated [85, 157]. It is important to note, however, that no large differences between rendered and real scenes in terms of color constancy performance have been found [122, 159, 160], and thus it is appropriate to use the experimental setup that is most practical for a given research question.

6.4 The Role of Individual Differences in Theory Development

More than 50 years ago, Cronbach [161] pointed to a curious and unfortunate theoretical schism dividing psychology into two largely separate schools of thought he referred to as "correlational psychology" and "experimental psychology." The main difference between them is their perspective on the importance of individual differences. It is probably fair to say that research on perception has been, and to some extent still is, deeply entrenched in the one-sided perspective of the "experimental psychology" described by Cronbach and has yet to fully exploit the benefits of complementing traditional experimental studies with analyses of natural variation [162–164].

This is particularly tangible in the literature on color constancy and color appearance. Many studies have been performed with a small number of observers, quite often only the authors and perhaps a couple of naive observers. This common practice is probably more strongly rooted in tradition [165] along with practical issues[3] than based on a principled scientific deliberations. One could argue that the fair inter-observer agreement evident in many published studies suggests that individual differences are small and thus insignificant. The idea that inter-observer agreement tends to be good may, however, be a self-perpetuating prejudice caused and maintained by a publication bias. Whenever large individual differences are observed, many perceptual psychologists are probably prone to think that something went wrong in the experiment and hesitate to publish the results, particularly if data suggesting a higher precision have already been published. In the early days of perception science, large inter-individual variability was considered "prima facie evidence that the attempted isolation of critical determining factors had failed and that uncontrolled disturbing processes had supervened" ([165], p. 73) and even today, this kind of attitude may be encountered ([166], p.101).

[3] Perception experiments typically involve long sessions with a large number of stimuli presented to each observer, which poses a challenge to recruiting large samples of naive observers.

In the literature on color constancy and color induction, there is often good agreement among the few observers participating in single studies, but vastly different results across studies [54]. More recent research documents surprisingly large individual differences within studies [96, 97, 167–169], which suggests that publication bias may indeed have been a real issue. As a notable exception in color constancy research, Allen et al. [167, 168] used individual differences in working memory performance to successfully account for individual differences in color constancy, specifically in how scene complexity affects color constancy performance (see also [170]).

The tendency to neglect individual differences in studies of color constancy and color induction is not entirely irrational. First, it is well known that the phenomena under study depend crucially on a host of stimulus variables [53, 54]. Thus, even small differences in the stimuli between studies may explain why they produce rather diverging results. Second, it is also clear that different methods for measuring how color appearance depends on the stimulus can lead to dramatically different results [84]. Thus, any differences observed across different studies can plausibly be attributed to the effect of known or unknown differences in the experimental variables.

It is also well known that the results of asymmetric color matching experiments may depend on instructions ([35, 171–173]). In a similar vein, subjects sometimes find it impossible to make the targets actually appear equal in perceived color [40, 51, 85, 115, 174]. In such cases, it may be unclear to the subjects how they are supposed to proceed, and spontaneously adopt idiosyncratic criteria for complying with the impossible task. Subtle unintended demand characteristics [175] may also influence the actual settings. A further, related consideration is that the observed individual differences in color induction may be due to differences in eye movement patterns [176, 177] or the allocation of attention [178, 179] rather than genuine individual differences in color perception. Thus, individual differences observed within a single experiment may plausibly be attributed to criterion problems or subtle differences in how attention is deployed rather than to genuine individual differences in perception.

All of these considerations make it very difficult to unambiguously conclude that any observed differences in the data from different subjects reflect genuine differences in perception, but thinking in terms of biology and evolution, it appears implausible that the mechanisms underlying color constancy and color induction exhibit any less natural variability than any other parts of our biological makeup (e.g., [180]).

In 2015, interest in individual differences in color constancy and color induction was boosted by the heated discussions about "the Dress" [181–187]. When viewing this photograph of a dress (reprinted in Fig. 1D of [182]), some people say that the dress is blue-black, while others maintain that it is white-gold. It is difficult to say why precisely this phenomenon created such a stir in social media, but it appears reasonable to assume that people were particularly intrigued by the categorical differences in the colors reported by different people.

As testified by the vigorous exchanges on the Color and Vision Network (CVNet) mailing list (http://lawton.ewind.com/mailman/listinfo/cvnet), the scientific community was almost equally surprised by this phenomenon as the general public. Many interesting potentially important factors were discussed, ranging from the optics of the eye to subjective criteria, but a hypothesis that was repeatedly proposed in various guises was that the effect is due to individual differences in the strength and/or parameters of the mechanisms underlying color constancy. Importantly, though, it seems to be an entirely open question how these putative individual differences come about [181, 182]. It is probably not very far off the mark when Macknik and Martinez-Conde ([185], p.20) note that before "the discovery of The Dress, vision scientists had thought that people with normal vision experienced color illusions similarly." Some interesting experimental work directed towards elucidating the questions raised by the Dress has already been published [184, 186], and we welcome a new era in color constancy and color induction research where individual differences will no more be neglected, but rather be the main focus of interest. This would be a fortunate development, particularly because individual differences can be used as an additional tool for answering the very questions about underlying mechanisms that the field has always striven to answer with traditional experimental methods [97, 164, 188].

Ultimately, the best proof for the relevance of individual differences for perception research would be if analyzing them turns out to advance our understanding of the underlying mechanisms. While some attempts have been undertaken in this direction [97, 167, 168, 188], much remains to be done. A pressing question, therefore, is what kind of research needs to be done to turn this into a fruitful research program, what challenges it faces, and how they can be solved.

6.4.1 Individual Differences: What Needs to Be Done?

A first step would be to determine the extent and prevalence of individual differences in the susceptibility to context effects in color perception in the general population. Next, individual differences should be used to establish a general, principled, and integrative theory of the multiple visual mechanisms underlying the computation of perceived color. Here, it is important to note that meaningful individual differences are to be expected at the level of (potentially unknown) visual mechanisms, rather than at the level of directly observable effects (i.e., effects measured with psychophysical methods). There is good reason to believe that most directly observable context effects are the net result of several underlying mechanisms [28, 96, 189, 190]. In principle, it is quite possible that there does not exist any stimulus that would isolate a single mechanism, and if there are such stimuli, we can only speculate—based on preliminary and potentially misleading hypotheses—which stimuli have this property. Therefore, to be most useful and diagnostic, directly observable individual differences need to be decomposed into components attributable to specific mechanisms. Obviously, this cannot be done without explicit models, heuristic assumptions and additional empirical constraints. Thus, a one-sided focus on individual differences in themselves is unlikely to be very productive. Rather, approaches which combine the virtues of classical experimental paradigms (such as modeling of general mechanisms and how they depend on stimulus properties) with modeling of how these mechanisms may differ across individuals are called for. This not only has the advantage that a particular model can be tested against two aspects of the data, namely how the results depend on stimulus properties and how they depend on the observer, but also against how the results depend on their interaction [164]. Thus, a cleverly devised combination of the experimental approach and a correlational individual differences approach is more than the sum of its parts. If the directly observable data depend on multiple mechanisms which all may vary in their efficiency and/or influence across observers, there is no reason to expect that the effects of parametric stimulus manipulations and observer characteristics are separable. For instance, the potential pitfalls of averaging data curves across observers are well known [191, 192] and this general problem may be even more serious in asymmetric matching experiments, where reasonably simple combinations of simple mechanism may produce rather complicated-looking matching curves [96].

One of the central methodological aims of traditional psychophysical research is to isolate single mechanisms, but achieving this by purely experimental means is not likely to be simple, and may in many cases be downright impossible. Here, data from atypical individuals who lack one or more of the mechanisms in question may be particularly informative when compared to data from typical observers. This point is nicely illustrated by the important role played by data from dichromats in shaping our understanding of normal color vision [3].

Despite the intense research on color constancy and color induction over the last century, the well-developed and potentially very informative methods of behavioral genetics [193] such as twin studies [194] have hardly been employed so far, but we anticipate that they will play an important role in shaping future research in the field.

In order to use individual differences to drive theory and research on color constancy and color induction, it is important to take into account related factors that may account for the observed individual differences, such as known variations at the level of the photoreceptors, visual acuity and contrast sensitivity [195] (also see Chap. 5). It is also important to develop strategies for distinguishing between perceptual and "cognitive" or "strategic" aspects in color matching ([169], p. 761).

Using individual differences as a tool for making inferences about visual mechanisms requires using considerably larger samples of observers than has been customary. Different from the trained psychophysical observers traditionally used in studies of perception [165], these observers will, by practical necessity, have to be essentially untrained. It is therefore imperative to devise and validate methods for reducing experimental errors in psychophysical experiments associated with the use of untrained and naive observers. The methods outlined in Sect. 6.3.4 should not require much training and would thus be suitable for untrained observers.

6.5 Outstanding Questions and Suggestions for Future Research

We conclude this chapter by highlighting some outstanding questions and potential avenues for future research.

6.5.1 *Being Clear About General Research Goals*

To help place individual studies in a larger context it would be useful if researchers were more explicit about their research aims. There is a crucial difference in trying to understand the contents of color *experience* versus the *performance* of the visual system in different viewing conditions. Further, do the experiments test a specific model, and if yes, which? Do the experiments aim to contribute to developing neural or computational constancy models? If the goal is model development, does implementation matter? The answers to these questions bear upon the choice of experimental methods and analyses, and being clear about them when reporting experimental results helps the readers put the results in a larger context.

6.5.2 *Developing and Testing Theories*

There are vast amounts of data on the effects of context on color perception, especially in simple scenes, but not many models. The existing data could be used to develop theories, which could then be tested systematically with carefully designed experiments.

The key to model testing is to derive predictions for both successes and failures in constancy. As an example, Brainard and colleagues developed a Bayesian model of color constancy based on available color matching data. Crucially, the model predicted both failures and successes in color constancy depending on the scene structure (specifically, whether the background of the target objects was manipulated together with the illumination or not); these predictions were well in line with the human data [15].

Model comparison is also a useful tool for testing theories. Olkkonen et al. [16] characterized an interaction between context and memory biases in the perceived lightness of a central target. They compared two probabilistic models that were based on existing constancy theories: ratio coding and reflectance estimation. Although both models explained the separate effects of stimulus context and memory on perceived lightness, only the reflectance model accounted for their interaction.

6.5.3 Incorporating Learning and Memory into Constancy Theories

There is accumulating behavioral and neurophysiological evidence that perception is not separate from memory and learning processes (e.g., [167, 196–200]). But only a handful of studies have investigated the relationship between memory and color constancy [157, 167, 201–204], and their results have yet to be integrated into a theory of color constancy.

Adding a memory component into a color constancy task can help adjudicate between competing models. Olkkonen and Allred [203] measured the independent and joint effects of color context and short-term memory on color appearance, and found that a 2-s memory delay decreased color constancy performance compared to simultaneous matching. Based on a comparison of ideal-observer models, a reflectance-estimation strategy accounted for this result better than a contrast-coding strategy [16].

Long-term memory and expectations about object colors can also affect color appearance as shown by Gegenfurtner and colleagues (e.g., [72, 205, 206]). In these studies, observers perceived grayscale photographs of familiar objects (e.g., a banana or a mailbox) to be slightly tinted in their typical colors, showing an influence of prior knowledge on color perception. Finally, the Dress phenomenon discussed earlier may be taken to suggest that expectation about illumination geometry can affect surface color estimates [184]. These effects need to be incorporated into color constancy theories.

6.5.4 Improving Existing Measurement Methods

How can we improve existing methods for measuring color appearance or develop new ones that avoid the problems and ambiguities of more classical methods? How can we make sense of the diverging results obtained with different methods? One problem with many classical methods and experimental tasks is that they may, though seemingly "objective," fail to be properly tailored to the actual structure of phenomenal experience. As nicely illustrated by Runeson's [207] insightful discussion of how a simple measurement device may excel at performing seemingly complicated measurements while being essentially worthless in performing seemingly simpler measurements, it is evident that the experimental task should be tailored to the actual (unknown) function and format of the visual mechanisms to yield sensible results. This makes it evident that designing good methods for measuring color appearance is not a trivial task. Some useful criteria for determining whether a particular measurement method is appropriate are (a) whether the participants experience it as easy and natural and (b) whether it yields consistent and clear results. Ultimately the appropriateness of a measurement method has to be judged in terms of how helpful it is in developing our theoretical understanding.

Finally, it would be important to understand why different measurement methods sometimes yield very different results [53, 84]. Do the measurement methods typically used in the field really work the way they are intended to? How can we cleverly combine different methods to constrain possible theories and models better?

6.5.5 Understanding the Phenomenological Structure of Color

What are the natural perceptual variables of color experience? What is the dimensionality of perceptual color space? Is it possible to understand color as an attribute detached from other attributes of perceptual experience, such as material properties, shape, and space? Is the color of the illumination simply discounted by the visual system or is it sometimes part of the perceptual experience itself [119]? Although much attention was paid to these fundamental questions in the Gestalt era of color research [40, 79], they were subsequently largely neglected for decades (with a few notable exceptions, [60]). In line with a general revival of ideas and perspectives from Gestalt psychology [208, 209] and experimental phenomenology [210], these questions have recently been pursued with renewed vigor, but much remains to be done in this area.

6.5.6 Understanding the Neural Coding of Contextual Color Perception

Is color appearance fundamentally coded by edge differences at the retina or early visual cortex, such that color constancy and strong simultaneous contrast effects can be explained by this neural difference coding while background independence (Type II constancy) requires an explanation in terms of additional, complementary mechanisms of edge integration [53, 54]? Or is the retinal code in terms of absolute values instead of contrast, such that Type II constancy is easy to explain, but color constancy and simultaneous contrast effects require another explanation? Is simultaneous contrast due to a general mechanism of difference coding or to special-purpose mechanisms that only play a role in special cases [50]? The evidence for and against difference coding as a fundamental coding principle needs to be carefully evaluated, particularly now, where one of the core findings that originally inspired the development of the difference coding perspective ([54], p. 38)—namely that the eye "ceases to see" when the retinal image is artificially stabilized [211]—is being reevaluated as a potential artifact of binocular rivalry [212].

More generally, it is important to make sure that color constancy theories are consistent with the known physiology of the visual system (for instance, ganglion cells are sensitive to luminance edges rather than absolute luminance (e.g., [213], Chap. 4), and many neurons in primary visual cortex code color contrast ([214], Chap. 7), thus potentially contributing to color constancy). To start developing a complete theory of color constancy, combining psychophysical measurements with computational modeling and existing neural data is key.

Acknowledgments V.E. was supported by a grant from the Methusalem program by the Flemish Government (METH/08/02 and METH/14/02), awarded to Johan Wagemans. M.O. received support through the Academy Research Fellow program of the Academy of Finland.

References

1. Osorio D, Vorobyev M. A review of the evolution of animal colour vision and visual communication signals. Vision Res. 2008;48(20):2042–51.
2. Judd DB. Fundamental studies of color vision from 1860 to 1960. Proc Natl Acad Sci U S A. 1966;55(6):1313–30.
3. König A, Dieterici C. Die Grundempfindungen und ihre Intensitätsvertheilung im Spectrum. Juli: Sitzungsberichte der Akademie der Wissenschaften in Berlin; 1886. p. 805–29.
4. Schnapf JL, Kraft TW, Baylor DA. Spectral sensitivity of human cone photoreceptors. Nature. 1987;325(6103):439–41.
5. Jacobs GH, Williams GA, Cahill H, Nathans J. Emergence of novel color vision in mice engineered to express a human cone photopigment. Science. 2007;315(5819):1723–5.
6. Mancuso K, Hauswirth WW, Li Q, Connor TB, Kuchenbecker JA, Mauck MC, et al. Gene therapy for red-green colour blindness in adult primates. Nature. 2009;461(7265):784–7.
7. Brainard DH, Radonjić A. Color constancy. N Vis Neurosci. 2014;1:545–56.
8. Foster DH. Color constancy. Vision Res. 2011;51(7):674–700.
9. Shevell SK, Kingdom FAA. Color in complex scenes. Annu Rev Psychol. 2008;59:143–66.
10. Smithson HE. Sensory, computational and cognitive components of human colour constancy. Philos Trans R Soc Lond B Biol Sci. 2005;360(1458):1329–46.
11. Palmer SE. Vision science: photons to phenomenology. Cambridge, MA: Bradford Books, MIT Press; 1999.
12. von Helmholz H. Handbuch der Physiologischen Optik. Leipzig: Leopold Voss; 1867.
13. Adelson EH, Pentland AP. The perception of shading and reflectance. In: Knill DC, Richards W, editors. Perception as Bayesian inference, vol. 1. New York: Cambridge University Press; 1996. p. 409–23.
14. Barrow HG, Tenenbaum JM. Recovering intrinsic scene characteristics from images. Computer Vision Systems; 1978. p. 3–26.
15. Brainard DH, Longère P, Delahunt PB, Freeman WT, Kraft JM, Xiao B. Bayesian model of human color constancy. J Vis. 2006;6(11):1267–81.
16. Olkkonen M, Saarela TP, Allred SR. Perception-memory interactions reveal the computational strategy of reflectance perception. J Vis. 2016;16:38.
17. Gilchrist AL, Kossyfidis C, Bonato F, Agostini T, Cataliotti J, Xiaojun L, et al. An anchoring theory of lightness perception. Psychol Rev. 1999;106:795–834.
18. Maloney LT. Illuminant estimation as cue combination. J Vis. 2002;2(6):493–504.
19. Zaidi Q. Identification of illuminant and object colors: heuristic-based algorithms. J Opt Soc Am A. 1998;15(7):1767–76.
20. Buchsbaum G. A spatial processor model for object color perception. J Franklin Inst. 1980;310:1–26.
21. Hurlbert AC, Poggio TA. Synthesizing a color algorithm from examples. Science. 1988 ;239(4839):482–5.
22. Land EH. Recent advances in retinex theory and some implications for cortical computations: color vision and the natural image. Proc Natl Acad Sci U S A. 1983;80(16):5163–9.
23. Land EH, McCann JJ. Lightness and retinex theory. J Opt Soc Am. 1971;61(1):1–11.

24. D'Zmura M, Lennie P. Mechanisms of color constancy. J Opt Soc Am A Opt Image Sci. 1986;3(10):1662–72.
25. Maloney LT, Wandell BA. Color constancy: a method for recovering surface spectral reflectance. J Opt Soc Am A. 1986;3(1):29–33.
26. Lee HC. Method for computing the scene-illuminant chromaticity from specular highlights. J Opt Soc Am A. 1986;3(10):1694–9.
27. Funt BV, Drew M, Ho J. Color constancy from mutual reflection. Int J Comput Vis. 1991;6:5–24.
28. Kraft JM, Brainard DH. Mechanisms of color constancy under nearly natural viewing. Proc Natl Acad Sci U S A. 1999;96(1):307–12.
29. Boyaci H, Doerschner K, Maloney LT. Cues to an equivalent lighting model. J Vis. 2006;6(6):106–18.
30. Kitazaki M, Kobiki H, Maloney LT. Effect of pictorial depth cues, binocular disparity cues and motion parallax depth cues on lightness perception in three-dimensional virtual scenes. PLoS One. 2008;3(9):e3177.
31. Snyder JL, Doerschner K, Maloney LT. Illumination estimation in three-dimensional scenes with and without specular cues. J Vis. 2005;5(10):863–77.
32. Radonjić A, Cottaris NP, Brainard DH. Color constancy supports cross-illumination color selection. J Vis. 2015;15:1–19.
33. Brown RO, MacLeod DI. Color appearance depends on the variance of surround colors. Curr Biol. 1997;7(11):844–9.
34. Arend LE, Reeves A. Simultaneous color constancy. J Opt Soc Am A. 1986;3(10):1743–51.
35. Arend LE, Reeves A, Schirillo J, Goldstein R. Simultaneous color constancy: paper with diverse Munsell values. J Opt Soc Am A. 1991;8(4):661–72.
36. Helson BYH. Fundamental problems in color vision. I. The principle governing changes in hue, saturation and lightness of non-selective samples in chromatic illumination. J Exp Psychol. 1938;23:439–76.
37. Blakeslee B, McCourt ME. A multiscale spatial filtering account of the White effect, simultaneous brightness contrast and grating induction. Vision Res. 1999;39(26):4361–77.
38. Rudd ME. A cortical edge-integration model of object-based lightness computation that explains effects of spatial context and individual differences. Front Hum Neurosci. 2014;8:1–14.
39. Hering E. Grundzüge der Lehre vom Lichtsinn. Berlin: Springer; 1920.
40. Gelb A. Die "Farbenkonstanz" der Sehdinge. In: Bethe A, von Bergman G, Embden G, Ellinger A, editors. Handbuch der normalen und pathologischen Physiologie. Berlin: Springer; 1929. p. 594–687.
41. Kingdom FAA. Simultaneous contrast: the legacies of Hering and Helmholtz. Perception. 1997;26(6):673–7.
42. Lotto RB, Purves D. An empirical explanation of color contrast. Proc Natl Acad Sci U S A. 2000;97(23):12834–9.
43. Hurlbert AC, Wolf K. Color contrast: a contributory mechanism to color constancy. Prog Brain Res. 2004;144:147–60.
44. Jameson D, Hurvich LM. Essay concerning color constancy. Annu Rev Psychol. 1989;40:1–22.
45. Wallach H. Brightness constancy and the nature of achromatic colors. J Exp Psychol. 1948;38(3):310–24.
46. Walraven J. Discounting the background: the missing link in the explanation of chromatic induction. Vision Res. 1976;16(3):289–95.
47. Whittle P. Contrast colours. In: Mausfeld M, Heyer D, editors. Color perception: mind and the physical world. Oxford: Oxford University Press; 2003. p. 115–38.
48. Anderson BL. The perceptual representation of transparency, lightness, and gloss. In: Wagemans J, editor. Oxford handbook of perceptual organization. Oxford: Oxford University Press; 2015.
49. Rizzi A, McCann J. Simultaneous contrast and intraocular glare: opposing image dependent mechanisms in appearance. In Association Internationale de la Couleur (AIC). Interim Meeting in Stockholm June 15–18; 2008.
50. Ekroll V, Faul F. Transparency perception: the key to understanding simultaneous color contrast. J Opt Soc Am A Opt Image Sci Vis. 2013;30(3):342–52.

51. Ekroll V, Faul F, Niederée R. The peculiar nature of simultaneous colour contrast in uniform surrounds. Vision Res. 2004;44(15):1765–86.
52. Arend LE, Buehler JN, Lockhead GR. Difference information in brightness perception. Percept Psychophys. 1971;9(3):367–70.
53. Whittle P. Contrast brightness and ordinary seeing. In: Gilchrist AL, editor. Lightness, brightness, and transparency. Hillsdale, NJ: Erlbaum; 1994. p. 111–58.
54. Whittle P. The psychophysics of contrast brightness. In: Gilchrist AL, editor. Lightness, brightness, and transparency. Hillsdale, NJ: Erlbaum; 1994. p. 35–110.
55. Whittle P, Challands PD. The effect of background luminance on the brightness of flashes. Vision Res. 1969;9(9):1095–110.
56. Gilchrist AL. Introduction: absolute versus relative theories of lightness perception. In: Gilchrist AL, editor. Lightness, brightness, and transparency. Hillsdale, NJ: Erlbaum; 1994. p. 1–34.
57. Gilchrist AL. Lightness contrast and failures of constancy: a common explanation. Percept Psychophys. 1988;43(5):415–24.
58. Koffka K. Principles of gestalt psychology. Trench: Kegan Paul; 1936.
59. Allred SR, Olkkonen M. The effect of background and illumination on color identification of real, 3D objects. Front Psychol. 2013;4:1–14.
60. Evans RM. The perception of color. New York: Wiley; 1974.
61. Schmid AC, Anderson BL. Do surface reflectance properties and 3-D mesostructure influence the perception of lightness? J Vis. 2014;14:1–24.
62. Mausfeld R. The perception of material qualities and the internal semantics of the perceptual system. In: Albertazzi L, van Tonder GJ, Vishwanath D, editors. Perception beyond inference. The information content of visual processes. Cambridge, MA: MIT Press; 2010. p. 159–200.
63. Vishwanath D. Coplanar reflectance change and the ontology of surface perception. In: Albertazzi L, editor. Visual thought: the depictive space of perception, advances in consciousness research. Amsterdam: John Benjamins Publishing Company; 2006. p. 35–70.
64. Adelson EH. Perceptual organization and the judgment of brightness. Science. 1993;262(5142):2042–4.
65. Adelson EH. On seeing stuff: the perception of materials by humans and machines. Proc SPIE. 2001;4299:1–12.
66. Anderson BL. A theory of illusory lightness and transparency in monocular and binocular images: the role of contour junctions. Perception. 1997;26(4):419–53.
67. Anderson BL, Winawer J. Image segmentation and lightness perception. Nature. 2005;434:79–83.
68. Brainard DH, Maloney LT. Perception of color and material properties in complex scenes. J Vis. 2004;4(9):ii–v.
69. Fleming RW, Nishida S, Gegenfurtner KR. Perception of material properties. Vis Res. 2015;115:157–302.
70. Gilchrist AL. Perceived lightness depends on perceived spatial arrangement. Science. 1977 ;195(4274):185–7.
71. Mausfeld R. 'Colour' as part of the format of different perceptual primitives: the dual coding of colour. In: Mausfeld R, Heyer D, editors. Colour perception: mind and the physical world. Oxford: Oxford University Press; 2003. p. 381–429.
72. Olkkonen M, Hansen T, Gegenfurtner KR. Color appearance of familiar objects: effects of object shape, texture, and illumination changes. J Vis. 2008;8(5):13.1–16.
73. Smithson HE. Perceptual organization of colour. In: Wagemans J, editor. Oxford handbook of perceptual organization. Oxford: Oxford University Press; 2015.
74. Werner A. The influence of depth segmentation on colour constancy. Perception. 2006 ;35(9):1171–84.
75. Wollschläger D, Anderson BL. The role of layered scene representations in color appearance. Curr Biol. 2009;19(5):430–5.
76. Xiao B, Hurst B, MacIntyre L, Brainard DH. The color constancy of three-dimensional objects. J Vis. 2012;12(4):6.
77. Blakeslee B, McCourt ME. Comments and responses to "Theoretical approaches to lightness and perception". Perception. 2015;44(4):359–67.

78. Gilchrist A. Theoretical approaches to lightness and perception. Perception. 2015;44(4):339–58.
79. Katz D. Die Erscheinungsweisen der Farben und ihre Beeinflussung durch die individuelle Erfahrung. Leipzig: Barth; 1911.
80. MacLeod D. New dimensions in color perception. Trends Cogn Sci. 2003;7(3):97–9.
81. Tokunaga R, Logvinenko AD. Material and lighting hues of object colour. Ophthalmic Physiol Opt. 2010;30(5):611–7.
82. Vladusich T. Gamut relativity: a new computational approach to brightness and lightness perception. J Vis. 2013;13(1):14.
83. Foster DH. Does colour constancy exist? Trends Cogn Sci. 2003;7(10):439–43.
84. Bosten JM, Mollon JD. Kirschmann's fourth law. Vision Res. 2012;53(1):40–6.
85. Brainard DH, Brunt WA, Speigle JM. Color constancy in the nearly natural image. I. Asymmetric matches. J Opt Soc Am A. 1997;14:2091–110.
86. Brainard DH, Maloney LT. Surface color perception and equivalent illumination models. J Vis. 2011;11(5):1–18.
87. Ekroll V, Faul F. New laws of simultaneous contrast? Seeing Perceiving. 2012;25(2):107–41.
88. Mausfeld R, Niederée R. An inquiry into relational concepts of colour, based on incremental principles of colour coding for minimal relational stimuli. Perception. 1993;22(1975):427–62.
89. Koenderink JJ. Color for the sciences. Cambridge: MIT Press; 2010.
90. Hillis JM, Brainard DH. Do common mechanisms of adaptation mediate color discrimination and appearance? Uniform backgrounds. J Opt Soc Am A. 2005;22(10):2090–106.
91. Hering E. Eine Methode zur Beobachtung contrastes. Pflügers Arch. 1890;47(1):236–42.
92. Bramwell DI, Hurlbert AC. Measurements of colour constancy by using a forced-choice matching technique. Perception. 1996;25(2):229–41.
93. Brainard DH. Color constancy in the nearly natural image II. Achromatic loci. J Opt Soc Am A. 1998;17:307–25.
94. Helson BYH. Adaptation-level as a basis for a quantitative theory of frames of reference. Psychol Rev. 1948;55(6):297–313.
95. Speigle JM, Brainard DH. Predicting color from gray: the relationship between achromatic adjustment and asymmetric matching. J Opt Soc Am A. 1999;16:2370–6.
96. Ekroll V, Faul F. A simple model describes large individual differences in simultaneous colour contrast. Vision Res. 2009;49(18):2261–72.
97. Ekroll V, Faul F, Wendt G. The strengths of simultaneous colour contrast and the gamut expansion effect correlate across observers: evidence for a common mechanism. Vision Res. 2011;51(3):311–22.
98. Miyahara E, Smith VC, Pokorny J. The consequences of opponent rectification: the effect of surround size and luminance on color appearance. Vision Res. 2001;41(7):859–71.
99. Smith VC, Pokorny J. Color contrast under controlled chromatic adaptation reveals opponent rectification. Vision Res. 1996;36(19):3087–105.
100. Takasaki H. Lightness change of grays induced by change in reflectance of gray background. J Opt Soc Am. 1966;56(4):504–9.
101. Takasaki H. Chromatic changes induced by changes in chromaticity of background of constant lightness. J Opt Soc Am. 1967;57(1):93–6.
102. Arend LE. How much does illuminant color affect unattributed colors? J Opt Soc Am A Opt Image Sci Vis. 1993;10(10):2134–47.
103. Hurvich LM, Jameson D. An opponent-process theory of color vision. Psychol Rev. 1957;64(6 Pt 1):384–404.
104. Schultz S, Doerschner K, Maloney LT. Color constancy and hue scaling. J Vis. 2006 ;6(10):1102–16.
105. Giesel M, Hansen T, Gegenfurtner KR. The discrimination of chromatic textures. J Vis. 2009;9:1–28.
106. Krauskopf J, Gegenfurtner KR. Color discrimination and adaptation. Vision Res. 1992 ;32(11):2165+2175.
107. Miyahara E, Smith VC, Pokorny J. How surrounds affect chromaticity discrimination. J Opt Soc Am A Opt Image Sci. 1993;10(4):545–53.
108. Whittle P. Brightness, discriminability and the "crispening effect". Vision Res. 1992;32(8): 1493–507.

109. Kingdom FAA, Prins N. Psychophysics: a practical introduction. London: Academic; 2010.
110. Maloney LT, Yang JN. Maximum likelihood difference scaling. J Vis. 2003;3(8):573–85.
111. Abrams AB, Hillis JM, Brainard DH. The relation between color discrimination and color constancy: when is optimal adaptation task dependent? Neural Comput. 2007;19(10):2610–37.
112. Heinemann EG. The relation of apparent brightness to the threshold for differences in luminance. J Exp Psychol. 1961;61:389–99.
113. Niederée R. More than three dimensions: what continuity considerations can tell us about perceived color. In: Cohen J, Matthen M, editors. Color ontology and color science. Cambridge: MIT Press; 2010. p. 91–122.
114. Ekroll V, Faul F. Basic characteristics of simultaneous color contrast revisited. Psychol Sci. 2012;23(10):1246–55.
115. Burgh P, Grindley GC. Size of test patch and simultaneous contrast. Q J Exp Psychol. 1962;14(2):89–93.
116. Logvinenko AD, Maloney LT. The proximity structure of achromatic surface colors and the impossibility of asymmetric lightness matching. Percept Psychophys. 2006;68(1):76–83.
117. Robilotto R, Zaidi Q. Limits of lightness identification for real objects under natural viewing conditions. J Vis. 2004;4:779–97.
118. Robilotto R, Zaidi Q. Lightness identification of patterned three-dimensional, real objects. J Vis. 2006;6(1):18–36.
119. Zaidi Q, Bostic M. Color strategies for object identification. Vision Res. 2008;48(26):2673–81.
120. Radonjić A, Cottaris NP, Brainard DH. Color constancy in a naturalistic goal-directed task. J Vis. 2015;15(13):3.1–3.21.
121. Hansen T, Walter S, Gegenfurtner KR. Effects of spatial and temporal context on color categories and color constancy. J Vis. 2007;7(4):2.1–2.15.
122. Olkkonen M, Witzel C, Hansen T, Gegenfurtner KR. Categorical color constancy for real surfaces. J Vis. 2010;10(9):9.1–9.22.
123. Smithson H, Zaidi Q. Colour constancy in context: roles for local adaptation and levels of reference. J Vis. 2004;4(9):693–710.
124. Speigle JM, Brainard DH. Is color constancy task independent. In: The fourth color imaging conference: color science, systems and applications; 1996. p. 167–72.
125. Troost JM, de Weert CM. Naming versus matching in color constancy. Percept Psychophys. 1991;50(6):591–602.
126. Olkkonen M, Hansen T, Gegenfurtner KR. Categorical color constancy for simulated surfaces. J Vis. 2009;9(12):6.1–6.18.
127. Jacobs GH, Gaylord HA. Effects of chromatic adaptation on color naming. Vision Res. 1967;7(7):645–53.
128. Uchikawa K, Yokoi K, Yamauchi Y. Categorical color constancy is more tolerant than apparent color constancy. J Vis. 2004;4(8):327.
129. Norman LJ, Akins K, Heywood CA, Kentridge RW. Color constancy for an unseen surface. Curr Biol. 2014;24(23):2822–6.
130. Reeves AJ, Amano K, Foster DH. Color constancy: phenomenal or projective? Percept Psychophys. 2008;70(2):219–28.
131. Blackwell KT, Buchsbaum G. Quantitative studies of color constancy. J Opt Soc Am A Opt Image Sci. 1988;5(10):1772–80.
132. Gerhard HE, Maloney LT. Detection of light transformations and concomitant changes in surface albedo. J Vis. 2010;10(9):1–14.
133. Craven BJ, Foster DH. An operational approach to colour constancy. Vision Res. 1992;32(7):1359–66.
134. Foster DH, Nascimento SM, Amano K, Arend LE, Linnell KJ, Nieves JL, et al. Parallel detection of violations of color constancy. Proc Natl Acad Sci U S A. 2001;98(14):8151–6.
135. Pearce B, Crichton S, Mackiewicz M, Finlayson GD, Hurlbert A. Chromatic illumination discrimination ability reveals that human colour constancy is optimised for blue daylight illuminations. PLoS One. 2014;9(2):e87989.

136. Ekroll V, Faul F. Perceptual organization in colour perception: inverting the gamut expansion effect. i-Perception. 2013;4(5):328–32.
137. Shevell SK, Wei J. Chromatic induction: border contrast or adaptation to surrounding light? Vision Res. 1998;38(11):1561–6.
138. Bäuml KH. Color appearance: effects of illuminant changes under different surface collections. J Opt Soc Am A Opt Image Sci Vis. 1994;11(2):531–42.
139. Jenness JW, Shevell SK. Color appearance with sparse chromatic context. Vision Res. 1995;35(6):797–805.
140. Linnell KJ, Foster DH. Scene articulation: dependence of illuminant estimates on number of surfaces. Perception. 2002;31(2):151–9.
141. Zaidi Q, Spehar B, DeBonet J. Color constancy in variegated scenes: role of low-level mechanisms in discounting illumination changes. J Opt Soc Am A. 1997;14(10):2608–21.
142. Zemach IK, Rudd ME. Effects of surround articulation on lightness depend on the spatial arrangement of the articulated region. J Opt Soc Am A Opt Image Sci Vis. 2007;24(7):1830–41.
143. Yang JN, Shevell SK. Stereo disparity improves color constancy. Vision Res. 2002;42(16):1979–89.
144. Yang JN, Maloney LT. Illuminant cues in surface color perception: tests of three candidate cues. Vision Res. 2001;41:2581–600.
145. Yang JN, Shevell SK. Surface color perception under two illuminants: the second illuminant reduces color constancy. J Vis. 2003;3(5):369–79.
146. Boyaci H, Maloney LT, Hersh S. The effect of perceived surface orientation on perceived surface albedo in binocularly viewed scenes. J Vis. 2003;3(8):541–53.
147. Doerschner K, Boyaci H, Maloney LT. Human observers compensate for secondary illumination originating in nearby chromatic surfaces. J Vis. 2004;4(9):92–105.
148. Doerschner K, Boyaci H, Maloney LT. Estimating the glossiness transfer function induced by illumination change and testing its transitivity. J Vis. 2010;10(4):8.1–9.
149. Fleming RW, Dror RO, Adelson EH. Real-world illumination and the perception of surface reflectance properties. J Vis. 2003;3(5):347–68.
150. Olkkonen M, Brainard DH. Perceived glossiness and lightness under real-world illumination. J Vis. 2010;10(9):5.1–5.19.
151. Olkkonen M, Brainard DH. Joint effects of illumination geometry and object shape in the perception of surface reflectance. i-Perception. 2011;2(9):1014–34.
152. Judd DB. Hue saturation and lightness of surface colors with chromatic illumination. J Opt Soc Am. 1940;30(1):2–32.
153. Allred SR, Olkkonen M. The effect of memory and context changes on color matches to real objects. Atten Percept Psychophys. 2015;77(5):1608–24.
154. de Almeida VMN, Nascimento SMC. Perception of illuminant colour changes across real scenes. Perception. 2009;38(8):1109–17.
155. Granzier JJM, Vergne R, Gegenfurtner KR. The effects of surface gloss and roughness on color constancy for real 3-D objects. J Vis. 2014;14:1–20.
156. Hedrich M, Ruppertsberg AI. Color constancy improves for real 3D objects. J Vis. 2009;9:1–16.
157. Ling Y, Hurlbert A. Role of color memory in successive color constancy. J Opt Soc Am A. 2008;25(6):1215–26.
158. Radonjić A, Gilchrist AL. Depth effect on lightness revisited: the role of articulation, proximity and fields of illumination. i-Perception. 2013;4:437–55.
159. Agostini T, Bruno N. Lightness contrast in CRT and paper-and-illuminant displays. Percept Psychophys. 1996;58(2):250–8.
160. Brainard DH, Ishigami K. Factors influencing the appearance of CRT colors. In: Proceedings of the IS&T/SID 1995 Imaging Conference, Scottsdale, AZ. IS&T, Springfield, VA; 1995. p. 62–6.
161. Cronbach LJ. The two disciplines of scientific psychology. Am Psychol. 1957;12:671–84.
162. De-Wit L, Wagemans J. Individual differences in local and global perceptual organization. In: Wagemans J, editor. Oxford handbook of perceptual organization. Oxford: Oxford University Press; 2015.
163. Kanai R, Rees G. The structural basis of inter-individual differences in human behaviour and cognition. Nat Rev Neurosci. 2011;12(4):231–42.

164. Wilmer JB. How to use individual differences to isolate functional organization, biology, and utility of visual functions; with illustrative proposals for stereopsis. Spat Vis. 2008;21(6):561–79.
165. Danziger K. Constructing the subject: historical origins of psychological research. Cambridge: Cambridge University Press; 1994.
166. Navarro DJ, Griffiths TL, Steyvers M, Lee MD. Modeling individual differences using Dirichlet processes. J Math Psychol. 2006;50(2):101–22.
167. Allen EC, Beilock SL, Shevell SK. Working memory is related to perceptual processing: a case from color perception. J Exp Psychol Learn Mem Cogn. 2011;37(4):1014–21.
168. Allen EC, Beilock SL, Shevell SK. Individual differences in simultaneous color constancy are related to working memory. J Opt Soc Am A. 2012;29(2):A52–9.
169. Ripamonti C, Bloj M, Greenwald S, Maloney SI, Brainard DH. Measurements of the effect of surface slant on perceived lightness. J Vis. 2004;4(9):7.
170. Kraft JM, Maloney SI, Brainard DH. Surface-illuminant ambiguity and color constancy: effects of scene complexity and depth cues. Perception. 2002;31(2):247–63.
171. Arend LE, Goldstein R. Simultaneous constancy, lightness, and brightness. J Opt Soc Am A. 1987;4(12):2281–5.
172. Arend LE, Spehar B. Lightness, brightness, and brightness contrast: 2. Reflectance variation. Percept Psychophys. 1993;54(4):457–68.
173. Bäuml KH. Simultaneous color constancy: how surface color perception varies with the illuminant. Vision Res. 1999;39(8):1531–50.
174. Burzlaff W. Methodologische Beträge zum Problem der Farbenkonstanz. Z Psychol. 1931;119:177–235.
175. Orne MT. Demand characteristics and the concept of quasi-controls 1. In: Rosenthal R, Rosnow RL, Kazdin AE, editors. Artifacts in {behavioral} {research}. Oxford: Oxford University Press; 2009. p. 1–33.
176. Cornelissen FW, Brenner E. Simultaneous colour constancy revisited: an analysis of viewing strategies. Vision Res. 1995;35(17):2431–48.
177. Granzier JJM, Toscani M, Gegenfurtner KR. Role of eye movements in chromatic induction. J Opt Soc Am A Opt Image Sci Vis. 2012;29(2):A353–65.
178. Festinger L, Coren S, Rivers G. The effect of attention on brightness contrast and assimilation. Am J Psychol. 1970;83(2):189–207.
179. Tse PU, Reavis EA, Kohler PJ, Caplovitz GP, Wheatley T. How attention can alter appearances. In: Albertazzi L, editor. Handbook of experimental phenomenology: visual perception of shape, space and appearance. New York: Wiley; 2013. p. 291–315.
180. Hess RF, Wang G, Cooperstock JR. Stereo vision : the haves and have-nots. i-Perception. 2015;6(3):1–5.
181. Brainard DHH, Hurlbert ACC. Colour vision: understanding #TheDress. Curr Biol. 2015 ;25(13):R551–4.
182. Gegenfurtner KR, Bloj M, Toscani M. The many colours of 'the dress'. Curr Biol. 2015 ;25:R1–2.
183. Gilchrist A. Perception and the social psychology of 'The Dress'. Perception. 2015 ;44(3):229–31.
184. Lafer-Sousa R, Hermann KL, Conway BR. Striking individual differences in color perception uncovered by 'the dress' photograph. Curr Biol. 2015;25(13):R545–6.
185. Macknik SL, Martinez-Conde S. Unraveling "the dress". Sci Am Mind. 2015;26(4):19–21.
186. Winkler AD, Spillmann L, Werner JS, Webster MA. Asymmetries in blue-yellow color perception and in the color of 'the dress'. Curr Biol. 2015;25(D):2–3.
187. Why do different observers see extremely different colours in the same photo?; http://lpp.psycho.univ-paris5.fr/feel/?page_id=929. Retrieved on September 16th, 2016.
188. Bosten JM, Mollon JD. Is there a general trait of susceptibility to simultaneous contrast? Vision Res. 2010;50(17):1656–64.
189. Adelson EH. Lightness perception and lightness illusions, chapter 24. In: Gazzaniga M, editor. The new cognitive neurosciences, vol. 3. 2nd ed. Cambridge, MA: MIT Press; 2000. p. 339–51.

190. Kingdom FAA. Levels of brightness perception. In: Harris L, Jenkin M, editors. Levels of perception. New York: Springer; 2003. p. 23–46.
191. Estes WK. The problem of inference from curves based on group data. Psychol Bull. 1956;53(2):134–40.
192. Gallistel CR, Fairhurst S, Balsam P. The learning curve: implications of a quantitative analysis. Proc Natl Acad Sci U S A. 2004;101(36):13124–31.
193. Neale M, Cardon L. Methodology for genetic studies of twins and families, vol. 67. Dordrecht, The Netherlands: Kluwer; 1992.
194. Miller SM, Hansell NK, Ngo TT, Liu GB, Pettigrew JD, Martin NG, et al. Genetic contribution to individual variation in binocular rivalry rate. Proc Natl Acad Sci U S A. 2010;107(6):2664–8.
195. Peterzell DH, Teller DY. Individual differences in contrast sensitivity functions: the lowest spatial frequency channels. Vision Res. 1996;36(19):3077–85.
196. O'Herron P, von der Heydt R. Short-term memory for figure-ground organization in the visual cortex. Neuron. 2009;61(5):801–9.
197. Scocchia L, Cicchini GM, Triesch J. What's "up"? Working memory contents can bias orientation processing. Vision Res. 2013;78:46–55.
198. Serences JT, Ester EF, Vogel EK, Awh E. Stimulus-specific delay activity in human primary visual cortex. Psychol Sci. 2009;20(2):207–14.
199. Silvanto J, Soto D. Causal evidence for subliminal percept-to-memory interference in early visual cortex. Neuroimage. 2012;59(1):840–5.
200. Sreenivasan KK, Gratton C, Vytlacil J, D'Esposito M. Evidence for working memory storage operations in perceptual cortex. Cogn Affect Behav Neurosci. 2014;14(1):117–28.
201. de Fez MD, Capilla P, Luque MJ, Pérez-Carpinell J, del Pozo JC. Asymmetric colour matching: memory matching versus simultaneous matching. Color Res Appl. 2001;26(6):458–68.
202. Jin EW, Shevell SK. Color memory and color constancy. J Opt Soc Am A. 1996;13(10):1981–91.
203. Olkkonen M, Allred SR. Short-term memory affects color perception in context. PLoS One. 2014;9(1):e8648.
204. Uchikawa K, Kuriki I, Tone Y. Measurement of color constancy by color memory matching. Opt Rev. 1998;5(I):59–63.
205. Hansen T, Olkkonen M, Walter S, Gegenfurtner KR. Memory modulates color appearance. Nat Neurosci. 2006;9(11):1367–8.
206. Witzel C, Valkova H, Hansen T, Gegenfurtner KR. Object knowledge modulates colour appearance. i-Perception. 2011;2:13–49.
207. Runeson S. On the possibility of "smart" perceptual mechanisms. Scand J Psychol. 1977;18(3):172–9.
208. Wagemans J, Elder JH, Kubovy M, Palmer SE, Peterson MA, Singh M, et al. A century of Gestalt psychology in visual perception: I. Perceptual grouping and figure–ground organization. Psychol Bull. 2012;138(6):1172–217.
209. Wagemans J, Feldman J, Gepshtein S, Kimchi R, Pomerantz JR, van der Helm PA, et al. A century of Gestalt psychology in visual perception: II. Conceptual and theoretical foundations. Psychol Bull. 2012;138(6):1218–52.
210. Koenderink JJ. Experimental phenomenology. In: Wagemans J, editor. Oxford handbook of perceptual organization. Oxford: Oxford University Press; 2015.
211. Yarbus AL. Eye movements and vision. New York: Plenum Press; 1967.
212. Rozhkova GI, Nikolaev PP. Visual percepts in the cases of binocular and monocular viewing stabilized test objects, Ganzfeld stimuli, and prolonged afterimages. Perception. 2015. doi:10.1177/0301006615594957.
213. Shapley RM, Enroth-Cugell C. Visual adaptation and retinal gain controls. Prog Retin Res. 1984;3:263–345.
214. Johnson EN, Hawken MJ, Shapley R. The spatial transformation of color in the primary visual cortex of the macaque monkey. Nat Neurosci. 2001;4(4):409–16.

Chapter 7
Color in the Cortex

Elizabeth N. Johnson and Kathy T. Mullen

Abstract We begin with a discussion of the role of human color vision, asking what value the possession of color vision adds to the perception of the natural scene, both in terms of our ability to see color differences (contrast) and in color identification. We then consider the psychophysical properties of cortical color vision and what they reveal about its use in determining shape and form. We pit against each other different models accounting for how achromatic (luminance) contrast and color contrast may be linked in the determination of shape, comparing a coloring book model, in which color plays only a subordinate or minor role, an intrinsic images model in which color contrast makes an independent contribution, and an integration model in which color and luminance contrast both provide cue-invariant form information to color–luminance shape detectors. These models are also interpreted in the light of what we know about the physiological basis of color vision through primate single cell recordings, particularly in area V1. Finally, we discuss what has been revealed about human color vision in V1 and extra striate cortex from fMRI studies.

Keywords Color agnosia • Color boundaries • Color contrast • Color identification • Cone-opponent • Contrast sensitivity • Cortex • Double-opponent • Isoluminance • Orientation • Shape processing • Spatial frequency

7.1 Roles of Color Vision

A clear understanding of color vision begins with an appreciation of its goals and functions. To this end, we begin our chapter with an introduction to the roles of human color perception: what does color vision do for us? The roles of color vision

E.N. Johnson, Ph.D. (✉)
Department of Neurobiology and the Duke Institute for Brain Sciences,
Duke University School of Medicine and Duke University,
LSRC M054, 308 Research Drive, Box 91003, Durham, NC 27708, USA
e-mail: Elizabeth.johnson@duke.edu

K.T. Mullen, Ph.D.
Department of Ophthalmology, McGill Vision Research,
1650 Cedar Ave, L11.513, Montreal, QC, Canada, H3G 1A4
e-mail: kathy.mullen@mcgill.ca

J. Kremers et al. (eds.), *Human Color Vision*, Springer Series in Vision Research, DOI 10.1007/978-3-319-44978-4_7

can be divided into two broad categories: the perception of color differences (contrast) and color identification. The first role relies on the extraction of contrast from the visual scene to enhance visibility, but requires no knowledge or conceptualization of what the different colors are, whereas the second role, color identification, is based on an internal conceptualization of color and does not require the presence of color differences in the visual scene.

7.1.1 *The Role of Color Contrast Perception*

The vision of all sighted animals has the dimension of intensity, providing information about the black and white (achromatic) contrast in the image. Dichromatic mammals possess an additional dimension of contrast, based on the opponent combinations of short wavelength (S) cones with a middle wavelength (M) cone type. This produces a new dimension of contrast, one of color varying from "warm" to "cool," and passing through a neutral point so increasing the differences that can be perceived in the visual scene. In a further evolutionary step, catarrhines (Old World monkeys and apes) evolved into trichromats around 40 million years ago [1–3] and possess an additional dimension of contrast: one based on the combination of long wavelength (L) and M cones into a cone opponent system that enhances the perception of differences in the mid-long wavelength spectral regions. (These two dimensions of color contrast are frequently termed "blue-yellow" and "red-green" although these names used in this context are simply a matter of convenience and do not refer to the perception of unique hues.) Our understanding of the genetic basis of the evolution of the cone types that mediate these color contrast dimensions has developed rapidly over the past 3–4 decades [1, 2, 4–6] and is covered in other chapters of this book (Chaps. 1 and 11).

Given the rarity in the natural world of abutting surfaces of identical luminance (termed "isoluminance" or "equiluminance"), edges and contours with color contrast are likely to have an associated achromatic contrast; a color difference in the visual scene is typically generated from variations in the material properties of objects or surfaces, and so is most likely to have an associated change in luminance. On the other hand, there are many instances in which achromatic contrast may arise in the absence of color contrast; for example, by variations in the illumination of a surface causing shadows and shading, and inherent variations in surface reflectance of objects. Given the good sense that we can make of black and white images, it is self-evident that much information about shape and form is available based only on the achromatic components of the image.

The question is, what does an additional two dimensions of color contrast add to the vision of catarrhines that is so useful? Here we discuss three likely advantages of the enhanced contrast perception that color vision provides, illustrated with pictorial examples.

Fig. 7.1 Illustrations of two example roles of color contrast in natural scenes. *Top panels.* Detection of fruits and berries in achromatic and chromatic images of foliage (Photo courtesy of Martin LaBar). *Lower panels.* Use of color in image segmentation and as a linking feature: achromatic and chromatic images of fall foliage (from The Weather Channel)

7.1.1.1 Perception When "Lightness Is Varying Randomly"

Mollon [7] elegantly describes a fascinating series of historic accounts of color vision deficits, dating from as early as 1688 (by Robert Boyle), and the particular difficulty that ensues with searching for objects in "*dappled or variegated backgrounds where lightness is varying randomly*". As the colored versus black and white photograph pairs shown in Fig. 7.1 illustrate, in the natural environment, particularly amongst foliage or vegetation, the variation in achromatic contrast across the image is very high, creating a very complex scene. Multiple light sources are effectively created as the sun's rays penetrate through foliage or vegetation and the surfaces of the leaves reflect light at all possible angles with multiple occlusions. In this monochrome world, shapes perceived on the basis of boundaries and shading are masked by complex variations in shadows, reflected light, and multiple light sources. Figure 7.1 (top) shows how the shapes of the berries are all but invisible when viewed in monochrome, particularly at a distance. However, the addition of color contrast, reveals information in the scene that is absent or masked in the achromatic image, breaking the achromatic camouflage and enabling fruits or berries to be clearly detected against their leafy backgrounds. Color contrast enables the detection of objects that are otherwise invisible and is clearly advantageous when searching for objects in foliage, particularly at a distance [8]. This benefit of color vision is presumably also useful for birds, many of which have excellent color vision.

7.1.1.2 Color in Image Segmentation and as a Linking Feature

The visual scene can be segmented into common parts that belong together on the basis of a range of possible visual attributes, including stereoscopic depth, texture, motion or color. Once segmented, spatially distributed parts of the image with a common attribute may be perceptually linked together to extract an overall object, shape or contour. Color has long been recognized as an important basis for image segmentation and as a linking feature [7, 9, 10]. This important capacity of color vision is clearly exploited in the foliage example in Fig. 7.1 (lower). The achromatic image is too noisy for us to be able to extract any meaningful shapes or segregate the image into separate trees and distinct limbs. The addition of color contrast, however, causes the red, autumnal tree to dramatically pop out from the other still green ones, and enables us to link the common parts of the image based on their color. The Ishihara color vision tests also exploit this advantageous ability of color vision, in which circular disks of two contrasting colors, if distinguishable, can be grouped to form a figure against a background. Grouping and linking are by nature global tasks that allow shapes to be determined in the absence of contiguous boundaries or edges. The variable achromatic contrast of the individual elements masks any residual mean brightness differences between the two colors, effectively rendering them isoluminant.

Color contrast presumably combines with other spatial stimulus attributes, such as depth, texture and other higher order variations, to enhance image segmentation (Chap. 8). Segmentation by color and texture variations play a similar role as both are associated with changes in surface and specific object properties. On the other hand, color may also be manipulated to compete with image segmentation based on other visual attributes. Morgan et al. [11] show an example of competition between image segmentation based on texture and color: a small area of elements differently oriented from the others pops out in the achromatic version, but is much harder to see if random color variations are introduced. Image segmentation by color dominates over the conflicting segmentation by texture in this example. They argued that the reduced color vision of dichromats confers a visual advantage in this particular image; however, this conflict is probably uncommon in natural scenes in which texture boundaries and color boundaries are likely to coincide because both typically arise from common material variations. McIlhagga and Mullen [12] have shown that the linking of Gabor elements by common orientation into a winding contour is disrupted if the elements alternate between chromatic and achromatic contrast. An illustration of McIlhagga and Mullen's effect is shown by Shevell and Kingdom [13] (see their Fig. 14). This is another example of how competing image segmentation by color disrupts the spatial linking by orientation.

7.1.1.3 Color Vision and the Use of Edge Information

As already discussed, in a visual scene populated by surfaces and objects, variations in the material properties of surfaces produce color contrast that is typically associated with some level of achromatic contrast. Figure 7.2 illustrates the achromatic and chromatic (Red/Green (RG) and Blue/Yellow (BY)) edges present in an example image. Inspection shows that the edge plots are similar in each case, and the plots below show a strong correlation in this image between the chromatic and achromatic contrasts at the edges. A similar analysis averaging over multiple images, however, suggested that although most edges combine chromatic and achromatic contrast to varying degrees there is no overall correlation between the magnitudes of the two types of contrast [14]. In other words, achromatic edge contrast

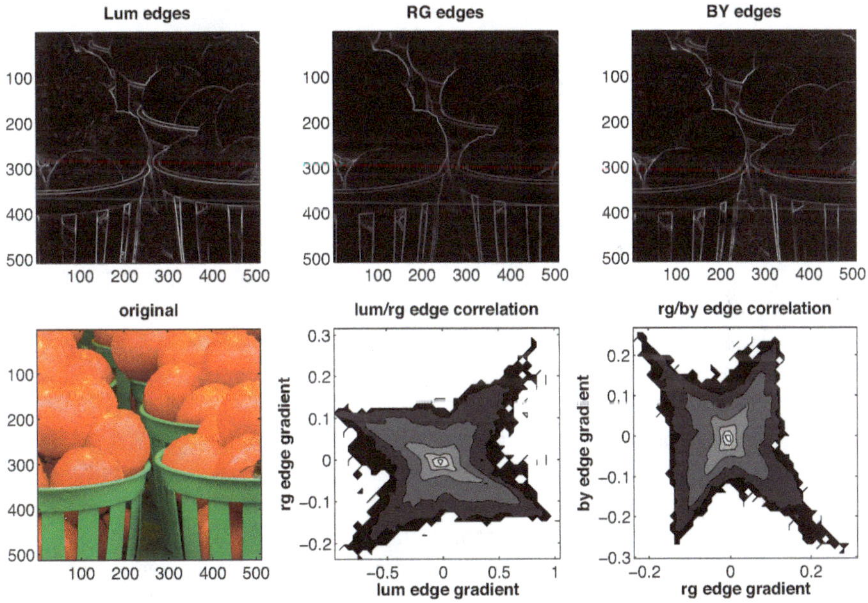

Fig. 7.2 An illustration of the color (red-green, rg, or blue-yellow, by) and achromatic (lum) edges of an image. The edges were computed as smoothed Gaussian derivatives in the x and y directions. The x and y edges were combined to form oriented edges at angles of 0°, 45°, 90°, and 135°. The edge values at each orientation can be positive (dark to light, red to green) or negative (light to dark, green to red). The edge values were divided by the local luminance as a rough normalization. The plot of luminance ("lum") versus chromatic edges in the *middle bottom row* suggests that for this image a combined color–luminance edge detector would be an effective way of encoding the edges (*Courtesy of William McIlhagga*)

does not predict chromatic edge contrast. How edges with both color and achromatic contrast might be encoded, either separately by purely chromatic or achromatic edge detectors or by cue invariant edge detectors that respond to both types of contrast, is discussed later in this chapter.

Shadows and shading in the visual scene produce achromatic edges and contrasts without associated color contrast. Thus, the presence or absence of color contrast in association with achromatic contrast can potentially provide crucial information for the disambiguation of shadows and shading from genuine changes in surface reflectance. Kingdom [15] elegantly demonstrated how simply the presence of color contrast can induce the perception of three-dimensional shape-from-shading from an overlaid achromatic grating. The presence of a chromatic grating, appearing as a continuous, striped colored pattern causes the overlaid orthogonal achromatic grating to appear as three-dimensional relief, giving rise to the perception of a corrugated surface [15, 16]. If a small amount of color contrast is added to the achromatic contrast, the 3D depth effect disappears, presumably because a spatially associated color and luminance variation cannot be interpreted as shading. This effect demonstrates a powerful role of color contrast in defining the continuity of a surface and in disambiguating shape from shading.

7.1.2 The Role of Color Identification

If asked to "find something red in the room," you would likely be able to quickly comply. This ability demonstrates "red" as an internalized color concept that can be used to search for and identify objects in the visual scene. Knowing what colors are is not the same as simply seeing colors as different from each other and requires a knowledge or cognition of color in addition to color perception [17]. That color cognition and color perception are distinct processes is clearly demonstrated by the neuropsychological deficit color agnosia. Color agnosic patients say they see "in color" but lack the ability to identify colors [18–21]. Color agnosic (J.T.) studied by one of the authors (KTM), who had acquired color agnosia after a brain hemorrhage, was in no doubt that he could see "in color." Psychophysical tests indicated that he had normal color contrast perception: his color contrast sensitivity function and suprathreshold color contrast discrimination were both normal [18]. He was also able to determine the isoluminant point using a minimum motion task and count the bars of a red-green isoluminant grating, which he reported to be made up of different colors, although he was incapable of identifying them. He had a profound deficit in color identification. He could verbally recite the colors of familiar objects with ease ("grass is green," etc.) suggesting that he did not suffer from color anomia, an aphasic condition in which colors cannot be connected to their names.

Color recognition relies on the existence of a cognitive or semantic color space. The color space is divided into fundamental response categories associated with different color sensations and names. Categories are not hard-edged boundaries but distributions around focal colors, the best examples of their categories. The presence of categorization means that the cognitive or semantic color space is of lower dimensionality than the perceptual color space. Color naming and categorization has a rich history that cannot be covered in this chapter, and is thought to be present in nonhuman [22] as well as human primates [20, 23–26]. That there is a fundamental distinction between color recognition and perception is also demonstrated by their different developmental trajectories. As reported by Petzold and Sharpe [27], in 1877 Charles Darwin wrote about his concerns that his children were color blind,

> "I attended carefully to the mental development of my young children, and with two or as I believe three of them, soon after they had come to the age when they knew the names of all common objects, I was startled by observing that they seemed quite incapable of affixing the right names to the colors in colored engravings, although I tried repeatedly to teach them. I distinctly remember declaring that they were color blind, but this afterwards proved a groundless fear. On communicating this fact to another person he told me that he had observed a nearly similar case. Therefore, the difficulty which young children experience either in distinguishing, or more probably in naming colors, seems to deserve further investigation".

That young children have difficulty with naming colors, or choosing named colors, has been borne out in many subsequent studies [27–30]. Throughout the twentieth century, the age at which children were reported to develop reliable naming of primary colors has consistently fallen, beginning at the age of 7 years in the early 1900s [29]. The most recent report shows that children rapidly acquire a reliable knowledge of 9 of the 11 basic colors within a 3-month period around the age of 36 months, after which there is a significant lag of 6–9 months before accurate knowledge of the final two colors, brown and gray, is acquired [30]. This delay in the acquisition of color knowledge is not explained by a delay in the development of color perception, which develops much earlier during infancy and in line with achromatic contrast perception [31–34].

The ability to identify color, as opposed to just seeing color differences, provides a huge set of advantages and provides much information about surfaces and objects. The sensation of color is based on the spectral reflectance of a surface and hence encodes a constant property of an object. Since the spectral composition of the light reflected from a surface is a confound of its spectral reflectance and the spectral composition of the illuminant, the effective, reliable use of color identification also requires color constancy, discussed elsewhere in this book (Chap. 6). In Fig. 7.3 we give some illustrated examples, although the list is far from exhaustive:

1. Food identification: Color identification plays an important role in the rapid identification of the type of food (fruit, leaves, etc.), especially at a distance, along with the judgment of its ripeness or off-ness.
2. Toxicity: Color may be used as a warning to signal the toxicity of animals by enhancing identification and so deterring predators.
3. Social signaling: This is used in countless examples of color identification. The socially dominant Gelada illustrated in Fig. 7.3 has vivid red chest markings, linked to testosterone levels (and resembling the markings on a hockey or football jersey). Color is also used to signal sexual availability, fertility, and estrous.
4. Identification of con-specifics and gender.

Fig. 7.3 Illustrations of examples roles of color identification in natural scenes. Wikimedia Commons Attributions: *"Fruit on display at La Boqueria market in Barcelona"* released into the public domain by copyright holder, Filip Maljkovic, via Wikimedia Commons. *"Poster frog"* released into the public domain by copyright holder, A. Amarnath, via Wikimedia Commons. *"Eclectus Parrot (Eclectus roratus)—pair"* by Flickr user Shiny Things used under Creative Commons Attribution 2.0 Generic license. Photos of "*Red salmon*" courtesy of Rachel A. Hovel and "*Male gelada*" courtesy of Noah Snyder-Mackler. (*Figure designed by Irem Onay*)

7.2 Color Vision in Form and Shape Perception

Our perception of color and achromatic contrasts are seen through very different but mutually complimentary spatiotemporal passbands, as illustrated in Fig. 7.4. In this figure, contrast sensitivity is compared in cone contrast units based on measurements of contrast detection thresholds. The use of the cone contrast metric to describe the stimuli allows visual sensitivity to two different physical properties, modulations in the spectral composition and the intensity of the stimulus, to be directly compared. From the *Principle of Univariance* we know that variations in the wavelength of light and variations in light intensity both have the same effect on the cones, causing a change in the amount of light absorbed and a consequent change in cone response. Hence, the visual response to color and luminance contrast can be directly compared based on their common effect on the cone photoreceptors response to absorbed light (Chap. 5).

Figure 7.4 (left panel) compares the spatial color and achromatic contrast sensitivity functions, and temporal contrast sensitivity is shown in the right panel. The lowpass shape of the color contrast sensitivity functions shows that we are highly sensitive to gradual changes in color occurring relatively slowly over large areas

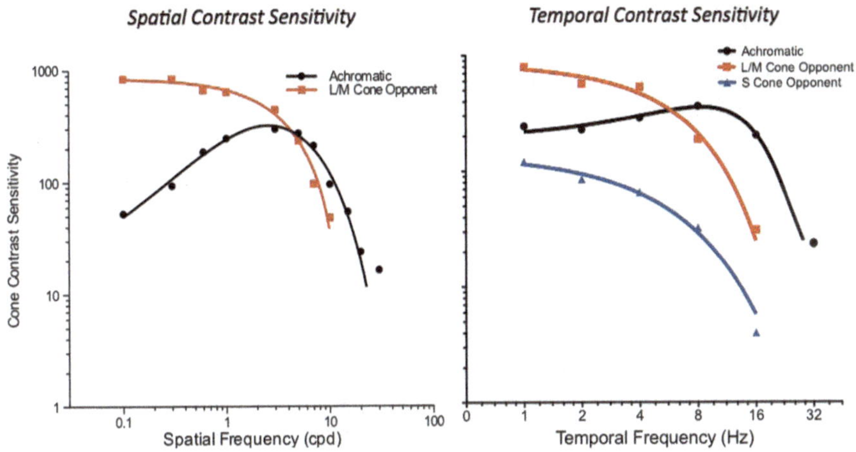

Fig. 7.4 Comparisons of chromatic and achromatic cone contrast sensitivity functions (CSF) for spatial frequency (*left panel*, in cycles per degree (cpd)) and temporal frequency (*right panel*, in Hz). *Red lines* and *data points* show L/M cone opponent responses, *blue* for S cone opponent, and *black* for achromatic. The use of cone contrast sensitivity allows chromatic and achromatic sensitivities to be directly compared. Cone contrast sensitivity for the spatial CSF is measured at 2 Hz and is based on data from Mullen [36], and for the temporal CSF is measured at 0.5 cpd, and is based on data from Mullen et al. [143]

[35–37]. In terms of cone contrast, color sensitivity is extremely high, achieving typical visual detection thresholds of around 0.3 %, and is significantly higher than the optimal achromatic contrast sensitivity, so winning a long-standing competition to answer the question, *what does the eye see best?* [38–40]. It remains unanswered whether color contrast sensitivity remains as high as spatial frequency is lowered to the point at which stimuli effectively become full fields of color. If so, it would reveal an absolute sensitivity to uniform color (dc) in the absence of spatial variation. (Clearly, some form of slow frequency temporal modulation will always remain arising from stimulus onset and offset.) While visual responses to achromatic contrast at low spatial frequencies (and mean luminance) are effectively removed by bandpass spatial filtering, these same attributes in terms of color are seen with the highest contrast sensitivity. The very high sensitivity of color vision to color changes over large areas is advantageous for tasks involving image segmentation and in defining surface colors, which typically operate over large spatial scales. It has also been specifically proposed that the spatial and chromatic properties of human red-green color vision has adapted to the detection of visual scenes of fruit against foliage; scenes that contained reddish objects (such as fruit) on a background of leaves, viewed within reaching distance, have been shown to correspond to the spatial properties of red-green human contrast sensitivity [40, 41].

As spatial frequencies increases above 0.5 cycles per degree (cpd), chromatic contrast sensitivity declines and color vision has poor visual resolution, with cut-off estimates ranging from around 12 cpd with chromatic aberrations optically corrected [36], to 20 cpd if the chromatic grating is created directly on the retina using laser interferometry to entirely by-pass the eye's optics [37]. Under natural viewing, however, visual resolution in color vision is limited by the chromatic aberration inherent in the eye's optics, which significantly degrades the color contrast of the image above 1–3 cpd and inserts achromatic contrast into the image [42]. Hence, the perception of fine spatial details and good visual resolution are supported entirely by achromatic vision.

During the 1980s the psychophysical study of color contrast at isoluminance became established as a tool for the investigation of color vision [43] and, at the same time, the idea that color vision played very little role in shape and form perception became popular [44, 45]. The lowpass, low-resolution nature of the color contrast sensitivity function initially encouraged the idea that color vision lacked the neural machinery for edge and contour detection. This view of color vision was termed the *coloring book* model by McIlhagga and Mullen [46] as it held that color vision played a subordinate role in shape and form perception and was poor at seeing form by itself, lacking the machinery for edge and contour detection. The main role of color vision was seen as one of filling-in between luminance-defined boundaries and linking areas of a similar color. An opposing model, named with the term *intrinsic images* borrowed from computer vision [46, 47], holds that both color and achromatic contrast can independently extract edges, performing the same computation, but on different data (color versus achromatic contrast). Color and luminance edges that coincide are due to object boundaries and unmatched luminance edges are due to changes in the intensity of the illuminant. A third model, more recently recognized, which we call the *integration* model, proposes that there are common, cue-invariant edge detectors that can respond to either color or achromatic contrast [48, 49]. This model has mainly been proposed on the basis of primate single cell data, particularly in V1, as described in further detail below [49–51]. How does the evidence stack up for these different models? Exploring their predictions reveals some inherent contradictions in the current psychophysical data.

Models 2 (*Intrinsic images*) and 3 (*Integrative*) both propose that color vision can support shape and form processing. Overwhelming support for this property has emerged in the psychophysical literature over the last three decades. Simple shape processing is hierarchical and begins with the encoding of local spatial frequency, edge, and orientation information in the primary visual cortex (V1). In the next stage (sometimes called "global") the distributed features such as local curvature, angles, etc. are integrated over space to extract an overall form, a process thought to occur at the level of cortical areas V2 to V4. At higher stages, in areas in human inferotemporal cortex, shape is encoded independently of its retinal position and spatial scale. Over several decades, the role that color contrast plays in these different hierarchical stages has been investigated, most commonly by comparing psychophysical performance using isoluminant chromatic and achromatic stimuli. The comparison is usually made using chromatic and achromatic stimuli of similar spatial frequency and matched in visibility. Stimuli have typically been matched in visibility by scaling their contrasts in multiples of their respective detection thresholds, which controls for the differences in chromatic and achromatic contrast sensitivity (see Fig. 7.4).

Because the differences in contrast sensitivity depend critically on the spatial frequency of the stimulus, this approach may only be used with stimuli that have a limited spatial or temporal frequency range (called spatiotemporally narrow band).

Tasks involving the discrimination or detection of local spatial frequencies, or the local orientation of edges or gratings are considered to be psychophysical tasks that are likely to be limited by neural performance at the level of V1. Accumulated evidence shows that such tasks are performed very similarly for color and achromatic contrast. Noise masking, sinewave masking, and adaptation experiments have revealed the presence of bandpass spatial filtering in color vision with similar bandwidths to achromatic contrast [52–56]. Presumably, the lowpass contrast sensitivity function is the upper envelope of these spatially tuned responses. A recent study using image classification has also indicated the presence of edge detectors in color vision similar to those of achromatic vision [57], supporting the presence of bandpass spatial tuning.

Psychophysical evidence definitively supports the presence of orientation tuning in color vision, based on orientation discrimination [58–60], sinewave masking [61], noise masking [62], adaptation [55, 63], and subthreshold summation experiments [64, 65], and is supported by fMRI results [66, 67]. Orientation discrimination is only slightly poorer in color than in achromatic contrast and orientation tuning has a similar or slightly broader bandwidth for color compared to achromatic contrast [61, 62]. A recent study of orientation tuning in color vision using subthreshold summation, however, found similar orientation bandwidths for color and achromatic contrast at mid spatial frequencies, but at low spatial frequencies orientation tuning dramatically broadened for the chromatic stimuli only, suggesting poor or absent orientation tuning for color [64, 65], as illustrated in Fig. 7.5. This result suggests two different types of orientation mechanisms may exist in color vision; an isotropic (untuned) detector at low spatial frequencies, which would be well-equipped for determining surface color, and orientation-tuned mechanisms at high spatial frequencies, which are configured for edge and contour detection. The presence of both isotropic and orientation tuned mechanisms in color vision is supported by the primate neurophysiological results in V1 discussed later in Sect. 7.3.2.

As color vision has the neural "apparatus" for form and shape coding, Model 2 (*Intrinsic images*), suggesting independent chromatic shape processing, and Model 3 (*Integration*) are both potential candidates to account for the role of color vision in shape perception. There is very little physiological support, however, for an independent, color-only shape processing system at this early cortical level (as proposed by Model 2). As detailed later in Sect. 7.3.2, color-only cells are found in primate V1 or V2 in only a limited number, and they do not typically have good spatial or orientation tuning. A more likely explanation is that neurons sensitive to both color and luminance contrast mediate shape processing in the early cortical stage. Thus circumstantial evidence supports the idea of the *Integration* model, in which edge detection is dependent on cue invariant mechanisms that can respond to either color or achromatic contrast. The presence of color–luminance neurons in significant numbers is well supported in primate V1, as described later in this chapter (Sect. 7.3.2).

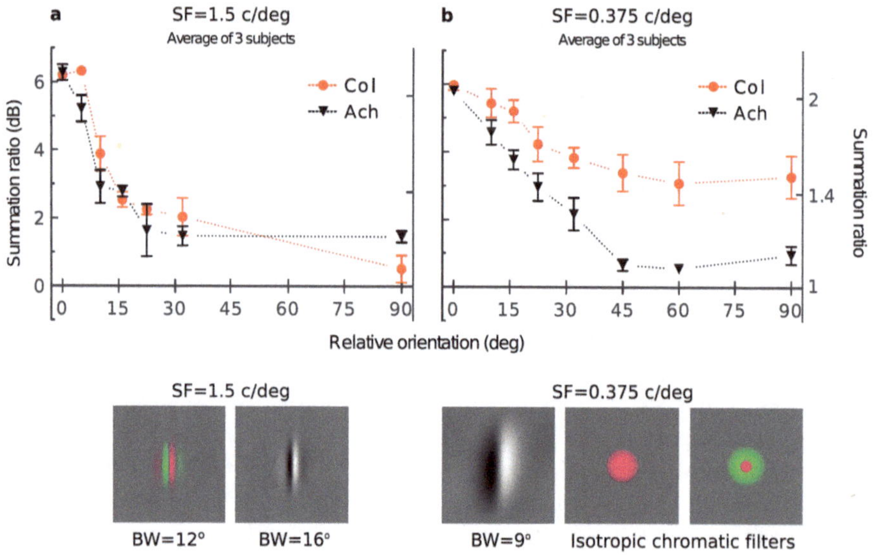

Fig. 7.5 *Top two panels* show orientation tuning curves for red-green (*red circles*) and achromatic (*black triangles*) stimuli at (**a**) 1.5 cpd and (**b**) 0.375 cpd for the average of three subjects (+/-se). The experiment measured the amount of subthreshold summation between two overlaid, cross-oriented gratings and is expressed as a summation ratio relative to the threshold for either grating presented alone, right axis, also expressed in dB, left axis). Summation ratio is plotted as a function of the relative orientation between the two gratings. Note the broad orientation tuning for low spatial frequency chromatic stimuli. Data are fitted with a model (not shown) to determine the orientation tuning of the underlying detection mechanisms. *Lower panels*. Representations of the optimal Log Gabor filters estimated from the model fits. At 1.5 cpd, similar narrow band orientation tuning is found for chromatic and achromatic stimuli. At low spatial frequencies (0.375 cpd), achromatic tuning remains essentially unchanged but chromatic orientation tuning is well fitted by an isotropic detector that may be spatially lowpass (e.g., red 'blob' detector) or bandpass (circularly symmetric color contrast detector). *Modified from Gheiratmand & Mullen* [65]

The abilities of color for form and shape processing typically extend to higher global tasks that depend on the integration of local information across space. The performance of color vision on contour integration [12, 68] and global shape perception using radial frequency patterns [69] and Glass patterns [70, 71] generally reveals relatively minor deficits in comparison to performance using achromatic contrast. These small deficits may arise from the marginally poorer orientation discrimination found at isoluminance [72].

The encoding of shape and form at an early cortical stage by cue-invariant, color–luminance mechanisms leads to the direct prediction that the processing of color and achromatic contrast do not occur independently. On the other hand, many different studies using a range of different approaches have tested for interactions between color and achromatic contrast at detection threshold, and these have mostly found independent mechanisms sensitive to color and achromatic contrast. A quantitatively rigorous approach has been to look for the effects of achromatic noise masking on chromatic detection. In general, when the test stimulus (signal) and masking noise both have the same contrast type (both achromatic or both chromatic), the energy of the test signal at detection threshold rises in proportion to the spectral density (power) of the noise mask—in other words the test becomes proportionally harder to see. If the test and masking noise are of different contrast types (chromatic vs. achromatic), however, little or no masking effect occurs [56, 73–75] indicating that the detection of the test and mask are mediated by separate color and achromatic neural responses. Interestingly, there is also little cross masking when test stimuli and masking noise have different polarities of response; for example, if the test stimulus is red (+L, −M) and the noise green (+M, −L), or vice versa. This suggests that the cone-opponent responses may be split into separable color poles (red vs. green) via a separation and rectification of the signals, in a similar fashion to the separation of the luminance system into ON and OFF pathways [73, 76].

A second rigorous psychophysical approach used to determine the nature of the responses at threshold is the measurement and modeling of detection threshold contours within a color space, pioneered by Stromeyer et al. [77] and King-Smith [78]. The shape of the threshold contours, particularly in the context of a biologically relevant space such as the cone contrast space, allows the cone contributions to separable post receptoral mechanisms to be determined. A series of studies has shown thresholds are determined by three separable mechanisms at detection threshold; two cone-opponents and one achromatic, each based on the linear combination of cone types [77, 79–82]. The absence of subthreshold summation between these mechanisms supports the presence of independent mechanisms that do not combine linearly at detection threshold [83, 84]. Other studies on overlay masking [85] and cross orientation masking [86] also suggest independent color and achromatic processes determine threshold.

There is a clear conundrum between the evidence pointing to the role of psychophysical cue invariant color–luminance responses in shape processing and the wide-ranging evidence for separable color and achromatic responses at detection threshold. Various explanations exist that might resolve this issue. First, by definition, measurement of simple detection thresholds will only reveal the most sensitive psychophysical mechanisms, a subset of all potential responses. It is quite likely that these most sensitive mechanisms are color-only or color dominated, and color–luminance mechanisms are only recruited when stimuli are presented at suprathreshold contrasts. In support of this, the responses of color-only neurons in V1 are reported to be among the most sensitive to cone contrast [51] (see Sect. 7.3.1). Furthermore, on the basis of one study of single cells in primate V1, around half of the neurons that responded to chromatic stimuli at the monkeys' detection threshold became responsive only to achromatic modulation at high contrast [87], hence switching their selectivity. So far, it is unknown how or where in the visual system detection thresholds are determined, with comparisons of neural sensitivities in V1 an educated guess.

Second, color–luminance responses may be mediated at the cortical level by the nonlinear combinations of cones or post-receptoral mechanisms. Such nonlinear combinations may only be revealed psychophysically under specific conditions. For example, it is well known that the presence of suprathreshold achromatic contrast facilitates chromatic detection, driving color thresholds down to extremely low levels (as low as 0.1 % cone contrast) [86, 88–91]. The psychophysical mechanism for this is under debate. However, a nonlinear model of single cell responses in V1 based on nonlinear combinations of chromatic and achromatic contrast has been proposed to explain the enhancement of color processing by achromatic contrast, at least for the S cone opponent response [92]. Hence color–luminance interactions at a suprathreshold level are still compatible with the presence of separable color and achromatic mechanisms under other conditions.

Third, color and luminance contrasts interact at suprathreshold levels. There are many wide ranging examples of interactions and cross effects between color and luminance contrast occurring at suprathreshold contrast levels. A full description of these is beyond the scope of this chapter, but some are described in reviews by Shapley and Hawken [49] and Rentzeperis et al. [48]. However, the presence of complex interactions at suprathreshold contrasts is entirely compatible with the existence of separable color and achromatic mechanisms at near-threshold low conditions. This issue remains to be explored further.

7.3 The Physiological Basis for Color Vision in Primate Cortex

Three questions continue to drive neurophysiological research into cortical color processing: Which neurons code color? How are color signals transformed in cortex? And how do these physiological responses impact human color perception?

7.3.1 Which Neurons Code Color?

Much debate has centered on what the requirements are for a neuron to be responsible for color signals. Are cells that contribute to color perception restricted to those that are highly specialized for color detection and highly selective for different colors? That is, are the only cells that contribute to color perception those that respond to color and not to achromatic stimuli? The most comprehensive study of color cortical neurophysiology is from macaque V1, with far fewer studies focusing on the processing of color signals in other cortical areas. Most of the projections from the lateral geniculate nucleus (LGN) to the cerebral cortex terminate in V1. Evidence that color-opponent neurons existed in the primate visual system and were likely homologous to the mechanisms involved in human cortical color processing was first provided by De Valois [93].

Early investigations found that many V1 neurons are tuned for the orientation and spatial frequency of black-and-white patterns, with few strongly selective for color [94, 95]. The small population of strongly color-selective V1 neurons lacked orientation selectivity and were spatially low-pass, which fit with the low-pass human color contrast sensitivity function [36] (see Sect. 7.2). Many subsequent studies have favored this viewpoint and have screened for color-responsive cells only if they respond best to isoluminant color [96–100]. This has had significant impact on the methods employed to determine the processing of cortical chromatic signals, namely that the localization and strength of cortical color signals are often determined (for example, with fMRI and optical imaging) by differencing the neural responses to isoluminant color stimuli (often full-field or low spatial frequency) and spatially defined black-white achromatic stimuli to determine a color "preference" (see Sect. 7.4).

However, there is growing evidence for the interaction of color and luminance, as described earlier in this chapter, as well as a prominent role for color in form processing (for recent reviews see Shapley et al. [101] and Rentzeperis et al. [48]). This makes it quite likely that neurons that respond to both chromatic and achromatic stimuli carry important information relevant for color perception. Single cell recordings in macaque V1 indicate that approximately 10 % respond exclusively to color (i.e., only to isoluminant color modulations), and these cells lack orientation tuning [50, 51, 96, 102, 103]. However, 40 % of all macaque V1 cells are color-responsive [50, 51, 103], and this percentage rises to 60 % of the cells in the superficial layers (layer 2/3), the main output layers of V1 [50, 51, 102, 103]. This larger fraction responds to both luminance and color contrast, and is selective for the orientation of both chromatic and luminance stimuli [50, 102, 104, 105].

As this chapter addressed above, there is mounting psychophysical evidence to suggest that color-preferring neurons, and neurons that respond to both color and luminance contrast, are likely involved in cortical color processing, perhaps at different contrast levels and in different contexts (for example, at edges versus color surfaces).

7.3.2 *How Are Color Signals Transformed in Early Visual Cortex?*

Much work has focused on the specific transformations of color signals as they progress from retina to LGN and then V1, and it is clear that color is not passively relayed even to early visual cortex. One very useful tool to examine how signals change from retina to thalamus and cortex is to use color-modulation stimuli that isolate signals from each of the cone types, which provides a direct way to study cone inputs to LGN and cortical neurons [96, 103, 106–108]. Such "cone-isolating" stimuli provide information about each of the cone inputs to a given neuron, the sign and weight of the inputs, and other spatiotemporal receptive field properties of the individual cone inputs when incorporated into tests of spatial frequency, temporal frequency, and subspace reverse correlation techniques (cf. Ringach et al. [109], see also Chap. 2). Studies using such cone-isolating stimuli indicate that the cone inputs to neurons in V1 are broadly distributed [103]. This is consistent with other findings of broad spectral tuning in various cortical areas, including V1, V2, and inferior temporal cortex (IT), when measured with cardinal stimuli derived from a color space based on the isolation of cone-opponent mechanisms (L-M, M-L, S-(M+L), L+M(+S)), often called "DKL space" [110], rather than L-, M-, and S-cone-isolating stimuli [111–114]. Many of the studies using cardinal stimuli derived from color-opponent mechanisms have suggested that the spectral diversity in V1 and subsequent visual areas is quite different from the narrow distribution in the LGN. However, the coordinate axes of color-opponent mechanisms in DKL space [110], or cardinal axes, are based on responses in the LGN to full-field stimuli. The majority of LGN color-sensitive neurons have receptive fields where the spatial extent of the center mechanism is smaller than that of the surround (Type I parvocellular cells [115]). These cells respond with the weighting L-M to full-field stimuli, but the weighting changes significantly when smaller spots fill just the receptive field center [108]. When spatial patterned stimuli are used, the interpretation of the axes of DKL space as the directions of stimuli that isolate LGN cell classes (often interpreted to be the input to cortex), is not correct because the spatiospectral response functions in the LGN as well as the cortex are not separable. Therefore, the ensemble of color spaces is also wider in the LGN than originally proposed by Derrington et al. [110], and the ensemble will shift for each different spatial pattern. Thus, the finding of a broader distribution of spectral responses in cortical neurons with spatially patterned stimuli is likely not a distinctively cortical transformation, but a by-product of emerging receptive field properties, such as orientation tuning, and thus the more common use of spatially defined stimuli. Cone contrast spaces do not pose the same problems because these axes represent retinal cone activation, not opponent color mechanisms.

There are significant transformations of color signals in early visual cortex. As noted above, many color-responsive V1 neurons are also selective for spatial patterns (orientation and specific spatial frequencies) [51, 105, 112, 116]. Although some V1 cells, like their retinal and LGN counterparts are "single-opponent," responding best to large, uniform areas of color, many more V1 cells are both chromatically and spatially opponent—thus termed "double-opponent" cells [103]. The single-opponent V1 cells are color-preferring. These cells are similar to their LGN counterparts, as they give maximal responses to red-green grating patterns of low spatial frequency, and low-pass spatial frequency responses for color overall. Unlike LGN parvocellular neurons, however, they respond extremely poorly to achromatic patterns of higher spatial frequency. When the receptive fields of V1 single-opponent cells are mapped with cone-isolating stimuli, the receptive fields have nearly equal and opposite inputs from L- and M-cones, with some receiving S-cone input [103], and they are roughly circularly symmetric—consistent with the finding that they have weak to nonexistent orientation selectivity [50, 51, 96, 97].

The double-opponent cells, however, often respond to both red-green grating patterns as well as achromatic patterns, with spatial specificity. Most of these cells are tuned for spatial frequency in the 1–3 cpd range, with similar spatial tuning for both chromatic and achromatic stimuli [51, 117]. When the receptive fields of V1 double-opponent cells are mapped with cone-isolating stimuli, the L- and M-cone inputs demonstrated spatial opponency for each cone and are of opposite sign at each location (as shown in Fig. 7.6). These cells respond poorly to extended areas of color or to color patterns of low spatial frequency, and are also orientation-

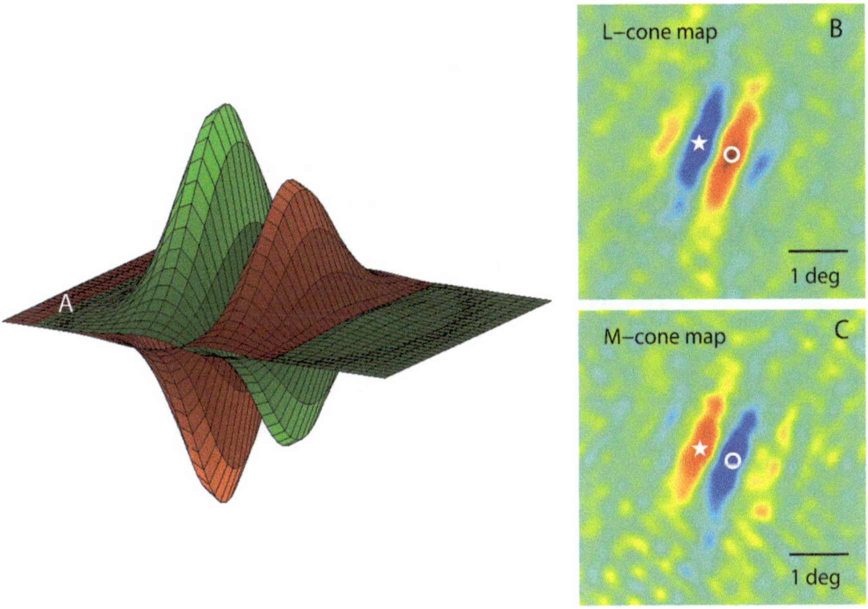

Fig. 7.6 Double-opponent cells in V1 (*from Johnson* et al. [50]). The spatial receptive field organization of an orientation-selective, spatial-frequency-bandpass, double-opponent neuron. (**a**) A schematic receptive field with side-by-side spatially antagonistic regions with opponent cone weights. The weighting above the horizontal plane is ON, where an increment of light will evoke an increase in response; the weighting below the line is OFF, where a decrement will result in a response. (**b**) Two-dimensional spatial map obtained from a neuron in V1 by means of subspace reverse correlation [109] with L-cone isolating grating stimuli. (**c**) Map obtained with M-cone isolating stimuli. At the starred location in **b**, the L-cone map is decrement excitatory. At the same location in (**c**), the M-cone map is increment excitatory, and vice versa for the locations marked by the *open circles*, demonstrating that the cell is indeed spatially and chromatically opponent (double-opponent). The schematic in (**a**) is a three-dimensional representation of the overlay of the two cone maps to give an overall profile. (**a**) is not to scale with respect to (**b**) and (**c**)

selective for both achromatic and chromatic stimuli, although a few double-opponent cells responded weakly to achromatic stimuli [50, 103]. The spatial maps of the receptive fields of the double-opponent cells were often elongated along the axis expected for their orientation preference, and most did not show even (concentric center/surround) symmetry, a departure from the classical double-opponent concept [118]. Since many double-opponent cells respond well to color differences across a border, they could play a crucial role in boundary detection in natural scenes, suggesting that color and form are inextricably bound together from the very earliest stage of cortical processing. Since most edges are defined by both color and luminance contrast with a wide variation in the proportions [14], the sensitivity of most double-opponent cells to both achromatic and isoluminant color boundaries is well-suited to natural image statistics and may be more efficient because they have the capacity of jointly detecting color and luminance-defined boundaries [119].

7.3.3 *Physiological Mechanisms for Color Appearance, Color Contrast, and Color Categorization and the Relevance to Human Color Perception*

A very important property of color appearance is its sensitivity to context. The presence of other colors (including black and white) in a specific spatial arrangement provides a powerful contextual influence on our color perception. The tendency for our perception of a color to change depending on its surrounding color context was discovered many centuries ago [120]. This phenomenon is especially intense at the edges where colors meet. An illustration of this kind of context effect is shown in Fig. 7.7, where a central square changes appearance depending on the color and brightness of the surrounding area.

There is increasing evidence that early visual cortical cells process chromatic context information [102, 113, 121] and that the presence of boundaries and extended regions (whether defined by color, luminance, or both) can significantly alter neuronal responses. These kinds of studies are likely to provide important information about the neurophysiological mechanisms underlying color appearance.

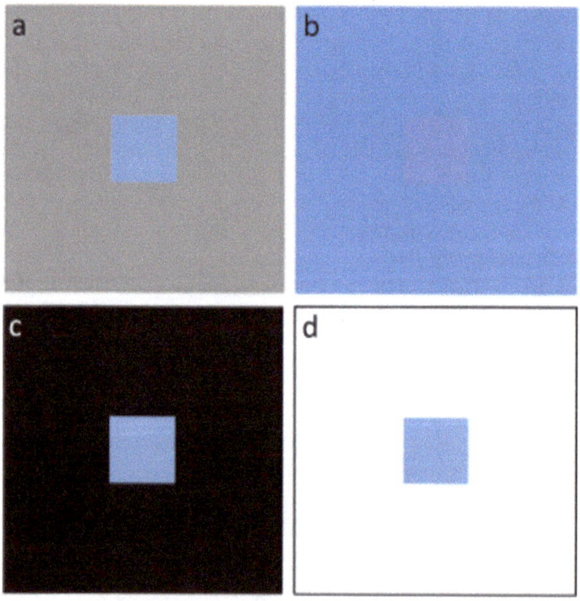

Fig. 7.7 Color contrast, brightness contrast, and color appearance. All four central squares have identical wavelength spectra, yet the color appearance of each central square may be strongly influenced by its surrounding context. (**a**) The central square surrounded by an isoluminant gray results in a saturated blue square. (**b**) The central square surrounded by a slightly different blue surround appears desaturated bluish purple, almost white. (**c**) The blue square on a black background results in a perception of reduced saturation and increased apparent brightness. (**d**) The blue square on a white background appears darker than in the other contexts

There is also a nascent, but growing body of work exploring the physiological interactions between color and luminance signals in V1 and other visual cortical areas, including V4, and inferior temporal cortex (IT) [121–125]. In the psychophysical literature, interactions between color and luminance are found to occur at suprathreshold contrasts and are often complex and nonlinear. Deviations from linearity in cortical neurophysiological responses to luminance and color are also likely, but have not yet been explored systematically.

But where do signals about color contrast emerge? The fact that edges seem to have a substantial influence on color appearance suggests that edge-sensitive color cells are likely important. This could place the neural basis of color contrast perception, which includes color constancy, as early as V1. Explicit tests to determine whether edge-sensitive V1 color cells respond to the color contrast of edges or boundaries are an obvious area for future research. Recent computational models suggest that double-opponent neurons with concentric center-surround receptive fields may contribute to color constancy by coding the external light source color [126, 127], while double-opponent cells with oriented receptive fields and responses to both achromatic and chromatic stimuli may be quite important for detecting salient boundaries in complex color scenes [119]. Nonoriented and spatially low-pass chromatic neurons, like the single-opponent neurons found in V1, could instead play a role in determining surface color and/or color identification.

Of course, color information is also represented in cortex beyond V1. Zeki discovered clusters of color-tuned cells within the macaque superior temporal sulcus in area V4, and proposed this area as a dedicated color center [99, 100]. This has been a source of controversy ever since, as has been its homologue in human cortex. With the mounting evidence for the interaction and mixing of chromatic and luminance signals as early as V1, along with the emergence of spatial specificity in many color-responsive neurons, one might argue that it is unlikely that subtracting responses like luminance, orientation, shape, etc. from pure color responses will give us much insight into the specific brain areas involved with color contrast perception, but these methods might instead provide insight into the areas involved with color identification, color categories, and/or color memory.

As noted in the introduction, discrimination and identification are different aspects of our visual perception, and both are apparent in human color vision. We can discriminate subtle differences in color contrast, and we categorize similar colors into groups, such as "red" and "green." The inferior temporal (IT) cortex is believed to play an important role in the recognition and memory of visual stimuli. Lesioning or inactivating cortical area IT in macaques degrades, but does not abolish their ability to discriminate colors [128, 129], making it likely that areas both within and outside IT cortex contribute to color discrimination. Cells in macaque anterior IT cortex (AITC) respond to hue [130], and response magnitudes in these neurons may depend on whether the behavioral task is discrimination or color categorization [131]. Figure 7.8 shows results for two representative AITC neurons that demonstrate task dependency. This allows for enhancement of the signal differentiating red from green colors during the categorization task and suppression during the discrimination task.

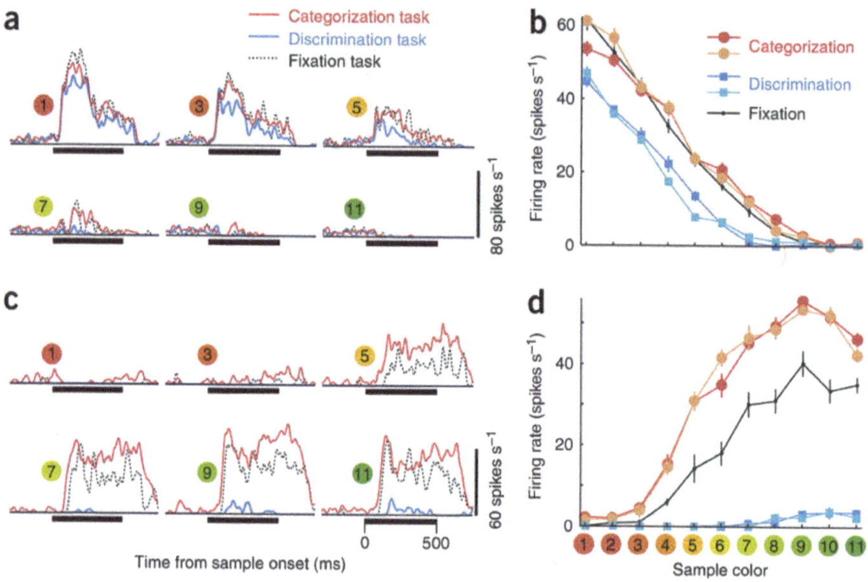

Fig. 7.8 Responses of two color-selective neurons in IT showing task dependency during a color categorization and a color discrimination task (*from Koida and Kamatsu* [131]). (**a** and **b**) Spike density functions showing the responses of each neuron to six sample colors (*filled circles* with *numbers*) during each task (*red line* shows the categorization task; *blue line* shows the discrimination task; *dotted black line* is a fixation task). (**c** and **d**) Color tuning curves of the same neurons as in (**a**) and (**b**) for the entire set of 11 sample colors recorded in separate task blocks (two blocks for each task). *Circles* indicate responses from the categorization task, *squares* are responses from the discrimination task, and *small dots* joined by the *black line* indicate those in a fixation task

Inferior temporal and ventral occipital cortical areas are also highly involved with object-related vision, so responses may be better understood in the future by utilizing stimuli that combine color and form, such as faces, shapes, scenes, etc., rather than simple, large fields of uniform color. That said, these kinds of stimuli are more difficult to control and the responses to them are more difficult to interpret.

7.4 Color Vision in Human Cortex: fMRI Studies

Until several decades ago insight into the physiological basis of color vision relied almost entirely on neurophysiological studies in nonhuman primates. The advent of fMRI, however, has allowed investigation of the visual cortex and LGN in the normal human brain to begin in earnest. fMRI is limited by its spatial and temporal resolution, with voxel size typically in the region of 3 mm^3 of cortex, although with improved processing methods and higher scanner strengths resolutions down to 1 mm^3 are possible. Even at this level, the method remains best at revealing responses in visual brain areas and cortical regions over a relatively large scale, and provides less insight into neural mechanisms or processes.

From an early stage it became clear that human area V1 responds very well to color contrast, at least as well as to achromatic contrast, with color contrast being a very effective driver of BOLD responses [117, 132–140]. Even the human LGN is driven very effectively by red-green color contrast [141]. It is worth noting, however, that the robust color response is typically elicited by using stimuli optimized for color contrast sensitivity (see the cone contrast sensitivity functions in Fig. 7.4) and may diminish relative to the achromatic response for stimuli with higher spatial frequencies, as color sensitivity is lost. As spatial contrast sensitivity functions of the BOLD response have not yet been systematically measured we do not yet know how well the psychophysical and BOLD functions match. In terms of the response to temporal frequency, psychophysical responses and BOLD responses in V1 do not correspond well since the BOLD responses in V1 are maintained with increasing temporal frequency whereas there is a steep decline in psychophysical sensitivity to higher temporal frequencies [142, 143]. In general, BOLD responses to color relative to achromatic contrast are also reduced as stimuli are presented further away from the fovea, particularly for RG stimuli [133], introducing a further source of variation in BOLD signals.

A number of fMRI studies have supported a reduced color response in the dorsal regions V3A and hMT+ of the occipital cortex when compared to early visual areas (V1, V2) and/or more ventral regions including V4 [136, 144, 145], although hMT+ retains some response to color contrast [146]. These studies support a division already well established in nonhuman primates based on single cell recordings and lesion studies.

Much interest has been focused on the role of the ventral pathway in color processing in human vision. The most common approach has been to search for areas that are highly responsive to color, and this has been done with two distinct methods. In one, a *Mondrian color localizer* is used, in which responses to a standard Mondrian stimulus with color and luminance contrast is differenced from the black and white (luminance-only) version [147–150]. In this case, the average luminance contrast is constant between display blocks but chromatic contrast is modulated. This localizer potentially reveals voxels with *any* response to color modulation. Another type of color localizer ("color preference") determines the voxels that show a significantly stronger activation to isoluminant chromatic gratings than to achromatic gratings (presented at the same cone contrast), and hence reveals regions with a color preference [144, 151, 152]. This localizer is more stringent and is likely to reveal fewer color selective regions as it determines voxels with a significant color preference rather than just any response to color. Differences between the two color localizers used may account for some of the variations in the reported color sensitive areas of the ventral visual pathway in human vision. As discussed above, variations in the spatial and temporal stimulus properties as well as their central/peripheral distributions across the visual field will also introduce variability into the determination of visual areas with color "preferences."

While it is clear that a retinotopically mapped area, thought to be human V4, responds well to color contrast, there is accumulating evidence that an area lying adjacent, but more anterior to it in the human brain shows a stronger and more robust color response. This area has been named variously V4a [147, 148], V8 [138] or VO [144, 152, 153] reflecting its controversial history and disputes over its precise location and naming [137]. The issue is particularly complicated given that there appears to be differences between human and nonhuman primates emerging in the IT cortex [151]. Nevertheless, this region can be retinotopically mapped as distinct from hV4 and probably represents a cluster of retinotopically organized areas (VO-1, VO-2) in human cortex [152, 153]. Taking a different approach, Mullen et al. defined an area in this region, also termed VO, which was localized functionally based on the color preference localizer, in which area VO comprised voxels with significant color preferences that were not part of hV4 [84]. Brewer et al. [152] suggested that stronger color responses are found in VO-1 rather than VO-2, but exactly how the region of color specialization defined by Mullen et al. [144] and the retinotopically mapped regions correspond is not yet well-established, and will be challenging to determine with confidence.

As raised in Sect. 7.3.2, single neurons in nonhuman primate V1 that are responsive to color differ in their color selectivity. Unselective neurons may respond similarly to both color and luminance contrast (called "color–luminance" neurons), whereas selective neurons will have their response dominated or exclusively driven by color contrast ("color-preferring" neurons). Both types will be responsive to color. Thus selectivity and responsivity to color are distinct properties. fMRI adaptation methods can be used to distinguish between these two properties in the human cortex [67, 154, 155]. If a BOLD response to color contrast can be significantly reduced (adapted) by previously viewing the same RG pattern, but is not adapted by viewing an achromatic pattern, it suggests the underlying neural response is selective for color and not driven by achromatic contrast. Conversely, if the BOLD response to color contrast can be significantly reduced by previously viewing both chromatic and achromatic patterns, it suggests an unselective response to both types of contrast. Using this approach Mullen et al. [155] found no sign of selectivity in the early visual areas (e.g., V1, V2), compatible with the dominant presence of color–luminance signals (although see Engel and Furmanski [154] for a different result). However, color selectivity appeared to increase along the ventral pathway with significant color selective adaptation found in area VO. This lends considerable support to a role for area VO in human color processing, although exactly what that role is remains to be determined.

A recent fMRI study on color memory suggests that responses from midlevel visual regions such as area V4 and lateral occipital cortex (LOC) may feed back onto V1 [156]. This finding raises issues with thinking strictly about color, and visual processing in general, as a hierarchical feed-forward process. This common assumption is likely to be as incorrect as classic conceptions of neural mechanisms in terms of single neurons rather than as neural populations working in concert. It is clear that much still remains unknown about cortical color processing, including how memory and sensory areas work together to recognize, discriminate, and remember colors and colored objects in both humans and nonhuman primates.

Acknowledgements We would like to acknowledge the contributions to figure design and preparation of Irem Onay (Figs. 7.1 and 7.3), William McIlhagga (Fig. 7.2), and Mina Gheiratmand (Fig. 7.5). We would like to thank Noah Snyder-Mackler for the gelada photograph, Rachel A. Hovel for the salmon photograph, both in Fig. 7.3, and Martin LaBar via Flickr for the berry photograph in Fig. 7.1.

References

1. Vorobyev M. Ecology and evolution of primate colour vision. Clin Exp Optom. 2004;87(4–5):230–8.
2. Nathans J. The evolution and physiology of human color vision: insights from molecular genetic studies of visual pigments. Neuron. 1999;24(2):299–312.
3. Nathans J, Thomas D, Hogness DS. Molecular genetics of human color vision: the genes encoding blue, green, and red pigments. Science. 1986;232(4747):193–202.
4. Mollon JD, Bowmaker JK, Jacobs GH. Variations of colour vision in a New World primate can be explained by polymorphism of retinal photopigments. Proc R Soc Lond B Biol Sci. 1984;222(1228):373–99.
5. Jacobs GH. Primate color vision: a comparative perspective. Vis Neurosci. 2008;25(5–6):619–33. doi:10.1017/S0952523808080760. S0952523808080760 [pii].
6. Deeb SS. Genetics of variation in human color vision and the retinal cone mosaic. Curr Opin Genet Dev. 2006;16(3):301–7. doi:10.1016/j.gde.2006.04.002.
7. Mollon JD. "Tho' she kneel'd in that place where they grew…". The uses and origins of primate colour vision. J Exp Biol. 1989;146:21–38.
8. Regan BC, Julliot C, Simmen B, Vienot F, Charles-Dominique P, Mollon JD. Fruits, foliage and the evolution of primate colour vision. Philos Trans R Soc Lond B Biol Sci. 2001;356(1407):229–83. doi:10.1098/rstb.2000.0773.
9. Barlow HB. The Ferrier Lecture, 1980. Critical limiting factors in the design of the eye and visual cortex. Proc R Soc Lond B Biol Sci. 1981;212(1186):1–34.
10. Koffka K. Principles of Gestalt psychology. New York: Harcourt, Brace and Company; 1935.
11. Morgan MJ, Adam A, Mollon JD. Dichromats detect colour-camouflaged objects that are not detected by trichromats. Proc Biol Sci. 1992;248(1323):291–5. doi:10.1098/rspb.1992.0074.
12. McIlhagga WH, Mullen KT. Contour integration with colour and luminance contrast. Vision Res. 1996;36(9):1265–79.
13. Shevell SK, Kingdom FA. Color in complex scenes. Annu Rev Psychol. 2008;59:143–66. doi:10.1146/annurev.psych.59.103006.093619.
14. Hansen T, Gegenfurtner KR. Independence of color and luminance edges in natural scenes. Vis Neurosci. 2009;26(1):35–49. doi:10.1017/S0952523808080796.
15. Kingdom FA. Color brings relief to human vision. Nat Neurosci. 2003;6(6):641–4. doi:10.1038/nn1060.
16. Kingdom FA, Beauce C, Hunter L. Colour vision brings clarity to shadows. Perception. 2004;33(8):907–14.
17. Davidoff J, Walsh V, Wagemans J. Higher-level cortical processing of colour. Acta Psychol (Amst). 1997;97(1):1–6.
18. Woodward TS, Dixon MJ, Mullen KT, Christensen KM, Bub DN. Analysis of errors in color agnosia: a single case study. Neurocase. 1999;5:95–108.
19. Victor JD, Maiese K, Shapley R, Sidtis J, Gazzaniga MS. Acquired central dyschromatopsia: analysis of a case with preservation of color discrimination. Clin Vis Sci. 1989;4(3):183–96.
20. Davidoff J. Cognition through color. Cambridge: The MIT Press; 1991.
21. Luzzatti C, Davidoff J. Impaired retrieval of object-colour knowledge with preserved colour naming. Neuropsychologia. 1994;32(8):933–50.

22. Sandell JH, Gross CG, Bornstein MH. Color categories in macaques. J Comp Physiol Psychol. 1979;93(4):626–35.
23. Berlin B, Kay P. Basic colour terms. Berkeley: University of California Press; 1969.
24. Uchikawa K, Boynton RM. Categorical color perception of Japanese observers: comparison with that of Americans. Vision Res. 1987;27(10):1825–33.
25. Guest S, Van Laar D. The structure of colour naming space. Vision Res. 2000;40(7):723–34.
26. Davidoff J, Davies I, Roberson D. Colour categories in a stone-age tribe. Nature. 1999;398: 203–4.
27. Petzold A, Sharpe LT. Hue memory and discrimination in young children. Vision Res. 1998;38(23):3759–72.
28. Raskin L, Maital S, Bornstein MH. Perceptual categorization of color: a life-span study. Psychol Res. 1983;45:135–45.
29. Bornstein MH. On the development of color naming in young children: data and theory. Brain Lang. 1985;26(1):72–93.
30. Pitchford NJ, Mullen KT. Is the acquisition of basic-colour terms in young children constrained? Perception. 2002;31(11):1349–70.
31. Morrone MC, Burr DC, Fiorentini A. Development of infant contrast sensitivity to chromatic stimuli. Vision Res. 1993;33(17):2535–52.
32. Pitchford NJ, Mullen KT. The role of perception, language, and preference in the developmental acquisition of basic color terms. J Exp Child Psychol. 2005;90(4):275–302. doi:10.1016/j.jecp.2004.12.005.
33. Allen D, Banks MS, Norcia AM. Does chromatic sensitivity develop more slowly than luminance sensitivity. Vision Res. 1993;33(17):2553–62.
34. Teller D. Spatial and temporal aspects of infant color vision. Vision Res. 1998;38:3275–82.
35. Kelly DH. Spatiotemporal variation of chromatic and achromatic contrast thresholds. J Opt Soc Am. 1983;73(6):742–50.
36. Mullen KT. The contrast sensitivity of human colour vision to red-green and blue-yellow chromatic gratings. J Physiol. 1985;359:381–400.
37. Sekiguchi N, Williams DR, Brainard DH. Aberration-free measurements of the visibility of isoluminant gratings. J Opt Soc Am A Opt Image Sci Vis. 1993;10(10):2105–17.
38. Chaparro A, Stromeyer III CF, Huang EP, Kronauer RE, Eskew Jr RT. Colour is what the eye sees best. Nature. 1993;361(6410):348–50. doi:10.1038/361348a0.
39. Watson AB, Barlow HB, Robson JG. What does the eye see best? Nature. 1983;302(5907):419–22.
40. Wolf K. Visual ecology: coloured fruit is what the eye sees best. Curr Biol. 2002;12(7):R253–5.
41. Parraga CA, Troscianko T, Tolhurst DJ. Spatiochromatic properties of natural images and human vision. Curr Biol. 2002;12(6):483–7.
42. Bradley A, Zang L, Thibos LN. Failures of isoluminance caused by ocular chromatic aberration. Appl Optics. 1992;31:3657–67.
43. Gregory RL, Heard PF. Color contrast without luminance contrast. J Physiol London. 1982;327:P12–P3.
44. Livingstone MS, Hubel DH. Anatomy and physiology of a color system in the primate visual cortex. J Neurosci. 1984;4(1):309–56.
45. Livingstone MS, Hubel DH. Psychophysical evidence for separate channels for the perception of form, color, movement, and depth. J Neurosci. 1987;7(11):3416–68.
46. McIlhagga W, Mullen KT. The contribution of colour to contour detection. In: Dickenson CM, editor. Colour Vision Research: Proceedings of the John Dalton Conference. London: Taylor & Francis; 1997. p. 187–96.
47. Barrow HG, Tenenbaum JM. Recovering intrinsic scene characteristics from images. In: Hanson AaR A, Riseman E, editors. Computer vision systems. New York: Academic; 1978.
48. Rentzeperis I, Nikolaev AR, Kiper DC, van Leeuwen C. Distributed processing of color and form in the visual cortex. Front Psychol. 2014;5:932. doi:10.3389/fpsyg.2014.00932.
49. Shapley R, Hawken MJ. Color in the cortex: single- and double-opponent cells. Vision Res. 2011;51(7):701–17. doi:10.1016/j.visres.2011.02.012.

50. Johnson EN, Hawken MJ, Shapley R. The orientation selectivity of color-responsive neurons in macaque V1. J Neurosci. 2008;28(32):8096–106. doi:10.1523/JNEUROSCI.1404-08.2008.
51. Johnson EN, Hawken MJ, Shapley R. The spatial transformation of color in the primary visual cortex of the macaque monkey. Nat Neurosci. 2001;4(4):409–16. doi:10.1038/86061.
52. Losada MA, Mullen KT. The spatial tuning of chromatic mechanisms identified by simultaneous masking. Vision Res. 1994;34(3):331–41.
53. Mullen KT, Losada MA. The spatial tuning of color and luminance peripheral vision measured with notch filtered noise masking. Vision Res. 1999;39(4):721–31.
54. Humanski RA, Wilson HR. Spatial frequency mechanisms with short-wavelength-sensitive cone inputs. Vision Res. 1992;32(3):549–60.
55. Bradley A, Switkes E, De Valois K. Orientation and spatial frequency selectivity of adaptation to color and luminance gratings. Vision Res. 1988;28(7):841–56.
56. Losada MA, Mullen KT. Color and luminance spatial tuning estimated by noise masking in the absence of off-frequency looking. J Opt Soc Am A. 1995;12(2):250–60.
57. McIlhagga W, Mullen KT. Classification images of chromatic edge detectors in human vision (abstract). Perception, 2015;44:289.
58. Reisbeck TE, Gegenfurtner KR. Effects of contrast and temporal frequency on orientation discrimination for luminance and isoluminant stimuli. Vision Res. 1998;38(8):1105–17.
59. Wuerger SM, Morgan MJ. Input of long- and middle-wavelength-sensitive cones to orientation discrimination. J Opt Soc Am A Opt Image Sci Vis. 1999;16(3):436–42. doi:10.1364/Josaa.16.000436.
60. Webster MA, De Valois KK, Switkes E. Orientation and spatial-frequency discrimination for luminance and chromatic gratings. J Opt Soc Am A. 1990;7(6):1034–49.
61. Pandey Vimal RL. Orientation tuning of the spatial-frequency-tuned mechanisms of the red-green channel. J Opt Soc Am A Opt Image Sci Vis. 1997;14(10):2622–32.
62. Beaudot WH, Mullen KT. Orientation selectivity in luminance and color vision assessed using 2-d band-pass filtered spatial noise. Vision Res. 2005;45(6):687–96. doi:10.1016/j.visres.2004.09.023. S0042-6989(04)00487-0 [pii].
63. Humanski RA, Wilson HR. Spatial-frequency adaptation: evidence for a multiple-channel model of short-wavelength-sensitive-cone spatial vision. Vision Res. 1993;33(5–6):665–75.
64. Gheiratmand M, Meese TS, Mullen KT. Blobs versus bars: psychophysical evidence supports two types of orientation response in human color vision. J Vis. 2013;13(1). doi:10.1167/13.1.2
65. Gheiratmand M, Mullen KT. Orientation tuning in human colour vision at detection threshold. Sci Rep. 2014;4:4285. doi:10.1038/srep04285.
66. Clifford CW, Spehar B, Solomon SG, Martin PR, Zaidi Q. Interactions between color and luminance in the perception of orientation. J Vis. 2003;3(2):106–15. doi:10.1167/3.2.1.
67. Engel SA. Adaptation of oriented and unoriented color-selective neurons in human visual areas. Neuron. 2005;45(4):613–23. doi:10.1016/j.neuron.2005.01.014.
68. Mullen KT, Beaudot WH, McIlhagga WH. Contour integration in color vision: a common process for the blue-yellow, red-green and luminance mechanisms? Vision Res. 2000;40(6):639–55.
69. Mullen KT, Beaudot WH. Comparison of color and luminance vision on a global shape discrimination task. Vision Res. 2002;42(5):565–75.
70. Mandelli MJ, Kiper DC. The local and global processing of chromatic Glass patterns. J Vis. 2005;5(5):405–16. doi:10.1167/5.5.2.
71. Cardinal KS, Kiper DC. The detection of colored Glass patterns. J Vis. 2003;3(3):199–208. doi:10.1167/3.3.2.
72. Mullen KT, Beaudot WH, Ivanov IV. Evidence that global processing does not limit thresholds for RF shape discrimination. J Vis. 2011;11(3):6.1–21. doi:10.1167/11.3.6. 11.3.6 [pii].
73. Gegenfurtner KR, Kiper DC. Contrast detection in luminance and chromatic noise. J Opt Soc Am A. 1992;9(11):1880–8.
74. Sankeralli MJ, Mullen KT. Postreceptoral chromatic detection mechanisms revealed by noise masking in three-dimensional cone contrast space. J Opt Soc Am A. 1997;14(10):2633–46.

75. Giulianini F, Eskew Jr RT. Chromatic masking in the (delta L/L, delta M/M) plane of cone-contrast space reveals only two detection mechanisms. Vision Res. 1998;38(24):3913–26.
76. Sankeralli MJ, Mullen KT. Bipolar or rectified chromatic detection mechanisms? Vis Neurosci. 2001;18(1):127–35.
77. Stromeyer III CF, Cole GR, Kronauer RE. Second-site adaptation in the red-green chromatic pathways. Vision Res. 1985;25(2):219–37.
78. King-Smith PE, Vingrys AJ, Benes SC. Visual thresholds measured with color video monitors. Color Res Appl. 1987;12(2):73–80.
79. Cole GR, Hine T, McIlhagga W. Detection mechanisms in L-, M-, and S-cone contrast space. J Opt Soc Am A. 1993;10(1):38–51.
80. Sankeralli MJ, Mullen KT. Estimation of the L-, M- and S-cone weights of the post-receptoral detection mechanisms. J Opt Soc Am A. 1996;13:906–15.
81. Chaparro A, Stromeyer III CF, Kronauer RE, Eskew Jr RT. Separable red-green and luminance detectors for small flashes. Vision Res. 1994;34(6):751–62.
82. Eskew RT, McLellan JS, Giulianini F. Chromatic detection and discrimination. In: Gegenfurtner KR, Sharpe LT, editors. Color vision: from molecular genetics to perception. Cambridge: Cambridge University Press; 1999. p. 345–68.
83. Mullen KT, Sankeralli MJ. Evidence for the stochastic independence of the blue-yellow, red-green and luminance detection mechanisms revealed by subthreshold summation. Vision Res. 1999;39(4):733–45.
84. Mullen KT, Cropper SJ, Losada MA. Absence of linear subthreshold summation between red-green and luminance mechanisms over a wide range of spatio-temporal conditions. Vision Res. 1997;37(9):1157–65.
85. Switkes E, Bradley A, Devalois KK. Contrast dependence and mechanisms of masking interactions among chromatic and luminance gratings. J Opt Soc Am A Opt Image Sci Vis. 1988;5(7):1149–62. doi:10.1364/Josaa.5.001149.
86. Mullen KT, Kim YJ, Gheiratmand M. Contrast normalization in colour vision: the effect of luminance contrast on colour contrast detection. Sci Rep. 2014;4:7350. doi:10.1038/srep07350.
87. Hass CA, Horwitz GD. V1 mechanisms underlying chromatic contrast detection. J Neurophysiol. 2013. doi:10.1152/jn.00671.2012. jn.00671.2012 [pii].
88. Cole GR, Stromeyer III CF, Kronauer RE. Visual interactions with luminance and chromatic stimuli. J Opt Soc Am A. 1990;7(1):128–40.
89. Eskew Jr RT, Stromeyer III CF, Picotte CJ, Kronauer RE. Detection uncertainty and the facilitation of chromatic detection by luminance contours. J Opt Soc Am A. 1991;8(2):394–403.
90. De Valois KK, Switkes E. Simultaneous masking interactions between chromatic and luminance gratings. J Opt Soc Am. 1983;73(1):11–8.
91. Mullen KT, Losada MA. Evidence for separate pathways for color and luminance detection mechanisms. J Opt Soc Am A. 1994;11(12):3136–51.
92. Horwitz GD, Chichilnisky EJ, Albright TD. Blue-yellow signals are enhanced by spatiotemporal luminance contrast in macaque V1. J Neurophysiol. 2005;93(4):2263–78. doi:10.1152/jn.00743.2004.
93. De Valois RL. Analysis and coding of color vision in the primate visual system. Cold Spring Harb Symp Quant Biol. 1965;30:567–79.
94. Hubel DH, Wiesel TN. Receptive fields and functional architecture of monkey striate cortex. J Physiol. 1968;195(1):215–43.
95. Schiller PH, Finlay BL, Volman SF. Quantitative studies of single-cell properties in monkey striate cortex. V. Multivariate statistical analyses and models. J Neurophysiol. 1976;39(6):1362–74.
96. Conway BR. Spatial structure of cone inputs to color cells in alert macaque primary visual cortex (V-1). J Neurosci. 2001;21(8):2768–83.
97. Conway BR, Hubel DH, Livingstone MS. Color contrast in macaque V1. Cereb Cortex. 2002;12(9):915–25.

98. Conway BR, Livingstone MS. Spatial and temporal properties of cone signals in alert macaque primary visual cortex. J Neurosci. 2006;26(42):10826–46. doi:10.1523/JNEUROSCI.2091-06.2006.
99. Zeki S. Colour coding in the cerebral cortex: the reaction of cells in monkey visual cortex to wavelengths and colours. Neuroscience. 1983;9(4):741–65.
100. Zeki S. Colour coding in the cerebral cortex: the responses of wavelength-selective and colour-coded cells in monkey visual cortex to changes in wavelength composition. Neuroscience. 1983;9(4):767–81.
101. Shapley R, Hawken M, Johnson E. Color in the primary visual cortex. In: Werner JS, Chalupa LM, editors. The new visual neurosciences. Cambridge: The MIT Press; 2014. p. 569–86.
102. Friedman HS, Zhou H, von der Heydt R. The coding of uniform colour figures in monkey visual cortex. J Physiol. 2003;548(Pt 2):593–613. doi:10.1113/jphysiol.2002.033555.
103. Johnson EN, Hawken MJ, Shapley R. Cone inputs in macaque primary visual cortex. J Neurophysiol. 2004;91(6):2501–14. doi:10.1152/jn.01043.2003.
104. Leventhal AG, Thompson KG, Liu D, Zhou Y, Ault SJ. Concomitant sensitivity to orientation, direction, and color of cells in layers 2, 3, and 4 of monkey striate cortex. J Neurosci. 1995;15(3 Pt 1):1808–18.
105. Thorell LG, De Valois RL, Albrecht DG. Spatial mapping of monkey V1 cells with pure color and luminance stimuli. Vision Res. 1984;24(7):751–69.
106. Estevez O, Spekreijse H. The "silent substitution" method in visual research. Vision Res. 1982;22(6):681–91.
107. Reid RC, Shapley RM. Spatial structure of cone inputs to receptive fields in primate lateral geniculate nucleus. Nature. 1992;356(6371):716–8. doi:10.1038/356716a0.
108. Reid RC, Shapley RM. Space and time maps of cone photoreceptor signals in macaque lateral geniculate nucleus. J Neurosci. 2002;22(14):6158–75. doi:20026444.
109. Ringach DL, Sapiro G, Shapley R. A subspace reverse-correlation technique for the study of visual neurons. Vision Res. 1997;37(17):2455–64.
110. Derrington AM, Krauskopf J, Lennie P. Chromatic mechanisms in lateral geniculate nucleus of macaque. J Physiol. 1984;357:241–65.
111. Hanazawa A, Komatsu H, Murakami I. Neural selectivity for hue and saturation of colour in the primary visual cortex of the monkey. Eur J Neurosci. 2000;12(5):1753–63.
112. Lennie P, Krauskopf J, Sclar G. Chromatic mechanisms in striate cortex of macaque. J Neurosci. 1990;10(2):649–69.
113. Wachtler T, Sejnowski TJ, Albright TD. Representation of color stimuli in awake macaque primary visual cortex. Neuron. 2003;37(4):681–91.
114. Horwitz GD, Hass CA. Nonlinear analysis of macaque V1 color tuning reveals cardinal directions for cortical color processing. Nat Neurosci. 2012;15(6):913–9. doi:10.1038/nn.3105.
115. Wiesel TN, Hubel DH. Spatial and chromatic interactions in the lateral geniculate body of the rhesus monkey. J Neurophysiol. 1966;29(6):1115–56.
116. Solomon SG, Lennie P. Chromatic gain controls in visual cortical neurons. J Neurosci. 2005;25(19):4779–92. doi:10.1523/JNEUROSCI.5316-04.2005.
117. Schluppeck D, Engel SA. Color opponent neurons in V1: a review and model reconciling results from imaging and single-unit recording. J Vis. 2002;2(6):480–92. doi:10.1167/2.6.5.
118. Daw NW. Goldfish retina: organization for simultaneous color contrast. Science. 1967;158(3803):942–4.
119. Yang KF, Gao SB, Guo CF, Li CY, Li YJ. Boundary detection using double-opponency and spatial sparseness constraint. IEEE Trans Image Process. 2015;24(8):2565–78. doi:10.1109/Tip.2015.2425538. Artn 2565.
120. Mollen J. Monge: the Verriest lecture, Lyon, July 2005. Vis Neurosci. 2006;23:297–309.
121. Xing D, Ouni A, Chen S, Sahmoud H, Gordon J, Shapley R. Brightness-color interactions in human early visual cortex. J Neurosci. 2015;35(5):2226–32. doi:10.1523/JNEUROSCI.3740-14.2015.
122. Bushnell BN, Harding PJ, Kosai Y, Bair W, Pasupathy A. Equiluminance cells in visual cortical area v4. J Neurosci. 2011;31(35):12398–412. doi:10.1523/JNEUROSCI.1890-11.2011.

123. Conway BR, Moeller S, Tsao DY. Specialized color modules in macaque extrastriate cortex. Neuron. 2007;56(3):560–73. doi:10.1016/j.neuron.2007.10.008.
124. Li M, Liu F, Juusola M, Tang S. Perceptual color map in macaque visual area V4. J Neurosci. 2014;34(1):202–17. doi:10.1523/JNEUROSCI.4549-12.2014.
125. Namima T, Yasuda M, Banno T, Okazawa G, Komatsu H. Effects of luminance contrast on the color selectivity of neurons in the macaque area v4 and inferior temporal cortex. J Neurosci. 2014;34(45):14934–47. doi:10.1523/JNEUROSCI.2289-14.2014.
126. Gao SB, Yang KF, Li CY, Li YJ. A color constancy model with double-opponency mechanisms. IEEE Int Conf Comput Vis. 2013;2013:929–36. doi:10.1109/Iccv.2013.119.
127. Gao SB, Yang KF, Li CY, Li YJ. Color constancy using double-opponency. IEEE Trans Pattern Anal Mach Intell. 2015;37(10):1973–85. doi:10.1109/TPAMI.2015.2396053.
128. Dean P. Visual cortex ablation and thresholds for successively presented stimuli in rhesus monkeys: II. Hue. Exp Brain Res. 1979;35(1):69–83.
129. Huxlin KR, Saunders RC, Marchionini D, Pham HA, Merigan WH. Perceptual deficits after lesions of inferotemporal cortex in macaques. Cereb Cortex. 2000;10(7):671–83.
130. Komatsu H, Ideura Y, Kaji S, Yamane S. Color selectivity of neurons in the inferior temporal cortex of the awake macaque monkey. J Neurosci. 1992;12(2):408–24.
131. Koida K, Komatsu H. Effects of task demands on the responses of color-selective neurons in the inferior temporal cortex. Nat Neurosci. 2007;10(1):108–16. doi:10.1038/nn1823.
132. Beauchamp MS, Haxby JV, Jennings JE, DeYoe EA. An fMRI version of the Farnsworth-Munsell 100-Hue test reveals multiple color-selective areas in human ventral occipitotemporal cortex. Cereb Cortex. 1999;9(3):257–63.
133. Vanni S, Henriksson L, Viikari M, James AC. Retinotopic distribution of chromatic responses in human primary visual cortex. Eur J Neurosci. 2006;24(6):1821–31. doi:10.1111/j.1460-9568.2006.05070.x.
134. Engel SA, Rumelhart DE, Wandell BA, Lee AT, Glover GH, Chichilnisky EJ, et al. fMRI of human visual cortex. Nature. 1994;369(6481):525. doi:10.1038/369525a0.
135. Wade A, Augath M, Logothetis N, Wandell B. fMRI measurements of color in macaque and human. J Vis. 2008;8(10):6.1–19. doi:10.1167/8.10.6.
136. Liu J, Wandell BA. Specializations for chromatic and temporal signals in human visual cortex. J Neurosci. 2005;25(13):3459–68.
137. Zeki S, McKeefry DJ, Bartels A, Frackowiak RS. Has a new color area been discovered? Nat Neurosci. 1998;1(5):335–6. doi:10.1038/1537.
138. Hadjikhani N, Liu AK, Dale AM, Cavanagh P, Tootell RB. Retinotopy and color sensitivity in human visual cortical area V8. Nat Neurosci. 1998;1(3):235–41. doi:10.1038/681.
139. Kleinschmidt A, Lee BB, Requardt M, Frahm J. Functional mapping of color processing by magnetic resonance imaging of responses to selective P- and M-pathway stimulation. Exp Brain Res. 1996;110(2):279–88.
140. Parkes LM, Marsman JB, Oxley DC, Goulermas JY, Wuerger SM. Multivoxel fMRI analysis of color tuning in human primary visual cortex. J Vis. 2009;9(1):1–13. doi:10.1167/9.1.1.
141. Mullen KT, Thompson B, Hess RF. Responses of the human LGN to different temporal frequencies for achromatic, L/M cone opponent and S-cone opponent stimuli measured with high field fMRI. J Vis. 2008;8(6):957a.
142. D'Souza DV, Auer T, Strasburger H, Frahm J, Lee BB. Temporal frequency and chromatic processing in humans: an fMRI study of the cortical visual areas. J Vis. 2011;11(8). doi:10.1167/11.8.8. 11.8.8 [pii]
143. Mullen KT, Thompson B, Hess RF. Responses of the human visual cortex and LGN to achromatic and chromatic temporal modulations: an fMRI study. J Vis. 2010;10(13):1–19. doi:10.1167/10.13.13. 10.13.13 [pii].
144. Mullen KT, Dumoulin SO, McMahon KL, de Zubicaray GI, Hess RF. Selectivity of human retinotopic visual cortex to S-cone-opponent, L/M-cone-opponent and achromatic stimulation. Eur J Neurosci. 2007;25(2):491–502. doi:10.1111/j.1460-9568.2007.05302.x

145. Tootell RB, Reppas JB, Kwong KK, Malach R, Born RT, Brady TJ, et al. Functional analysis of human MT and related visual cortical areas using magnetic resonance imaging. J Neurosci. 1995;15(4):3215–30.
146. Seidemann E, Poirson AB, Wandell BA, Newsome WT. Color signals in area MT of the macaque monkey. Neuron. 1999;24(4):911–7.
147. Bartels A, Zeki S. The architecture of the colour centre in the human visual brain: new results and a review. Eur J Neurosci. 2000;12(1):172–93.
148. McKeefry DJ, Zeki S. The position and topography of the human colour centre as revealed by functional magnetic resonance imaging. Brain. 1997;120(Pt 12):2229–42.
149. Zeki S, Watson JD, Lueck CJ, Friston KJ, Kennard C, Frackowiak RS. A direct demonstration of functional specialization in human visual cortex. J Neurosci. 1991;11(3):641–9.
150. Lueck CJ, Zeki S, Friston KJ, Deiber MP, Cope P, Cunningham VJ, et al. The colour centre in the cerebral cortex of man. Nature. 1989;340(6232):386–9. doi:10.1038/340386a0.
151. Lafer-Sousa R, Conway BR. Parallel, multi-stage processing of colors, faces and shapes in macaque inferior temporal cortex. Nat Neurosci. 2013;16(12):1870–8. doi:10.1038/nn.3555.
152. Brewer AA, Liu J, Wade AR, Wandell BA. Visual field maps and stimulus selectivity in human ventral occipital cortex. Nat Neurosci. 2005;8(8):1102–9. doi:10.1038/nn1507.
153. Wandell BA, Brewer AA, Dougherty RF. Visual field map clusters in human cortex. Philos Trans R Soc Lond B Biol Sci. 2005;360(1456):693–707. doi:10.1098/rstb.2005.1628.
154. Engel SA, Furmanski CS. Selective adaptation to color contrast in human primary visual cortex. J Neurosci. 2001;21(11):3949–54.
155. Mullen KT, Chang DH, Hess RF. The selectivity of responses to red-green color and achromatic contrast in the human visual cortex: an fMRI adaptation study. Eur J Neurosci. 2015;42(11):2923–33.
156. Bannert MM, Bartels A. Decoding the yellow of a gray banana. Curr Biol. 2013;23(22):2268–72. doi:10.1016/j.cub.2013.09.016.

Chapter 8
Interactions of Color Vision with Other Visual Modalities

Frederick A.A. Kingdom

Abstract Color vision is not only good for seeing hues but for seeing other visual dimensions, or "modalities," such as form, depth, material, and motion. The latter use of color vision relies in part on the exploitation of physical constraints that exist between the patterns of color and luminance in the natural visual world. Color vision on its own, however, that is, in the absence of luminance information, is in most cases less effective than luminance information for processing other modalities, often requiring more contrast relative to detection threshold to achieve commensurate levels of performance. Reasons for this are discussed.

Keywords Color vision • Equiluminance • Color and form • Color-is-material • Color and depth • Color and motion

8.1 Introduction

> "If he [Descartes] had examined that other, deeper opening upon things given us by the secondary qualities, especially color, then … [h]e would have been obliged to find out how the indecisive murmur of colors can present us with things, forests, storms—in short the world. … But for him it goes without saying that color is an ornament, mere coloring."
>
> Merleau-Ponty, *Eye and Mind,* p294 [1]; quoted in Chirimuuta & Kingdom [2].

Descartes' demotion of color to being merely ornamental undoubtedly pervades much of the popular imagination. It also anticipated, by some 300 years the ideas of Livingston and Hubel [3, 4], who argued that color vision fills in the gaps left after luminance-sensitive mechanisms have delineated the structural and motion properties of the scene. Their view is dubbed by some the "coloring book model" of color vision [5, 6]. A compelling challenge to the coloring book model is the experience of the cerebral achromatopsic, who though brain injury has lost the sensation of hue yet who

F.A.A. Kingdom (✉)
Department of Ophthalmology, McGill Vision Research, McGill University, Montreal General Hospital, 1650 Cedar Ave., Rm. L11.112, Montreal, QC, Canada, H3G 1A4
e-mail: fred.kingdom@mcgill.ca

J. Kremers et al. (eds.), *Human Color Vision*, Springer Series in Vision Research, DOI 10.1007/978-3-319-44978-4_8

often perceives the structure of purely chromatic patterns as well as normals do [7, 8] Achromatopsics appear to have a selective loss in the ability to use color vision to see hues, but not a loss in the ability to use color vision to see form. In short, achromotopsics display a dissociation between "form-from-color" and "hue-from-color".

This chapter examines how color vision interacts with other visual modalities, and not just form, but material perception, depth, and motion. In so doing the chapter endorses a more nuanced view about the role of color vision than the one proposed by Descartes: color vision is a way of seeing not only hues but *things* [2]. The remainder of the chapter is divided into four parts. Section 2 addresses some general issues concerning color vision's role in processing other visual modalities. Section 3 examines the spatial relationships between color and luminance in the natural visual world that are critical to how color vision interacts with other visual modalities. Section 4 examines how color vision subserves specific modalities, namely spatial resolution, contours, textures, materials, binocular function, and motion. The last section draws general conclusions.

8.2 General Issues

Given that color vision contributes at last some information to other visual modalities, a number of inter-related questions arise. First, do *purely* chromatic stimuli serve other modalities; second, if they do, is color vision *better or worse* than luminance vision at serving other modalities; third, does color and luminance *interact* when serving other visual modalities? While the first two of these questions have received considerable attention over the last 30 years [6, 9, 10], the question of how chromatic and luminance information interact when serving other visual modalities is a relatively recent line of investigation.

The first two of these questions have been addressed primarily through the use of equiluminant stimuli, that is, stimuli that have been behaviourally adjusted to produce a null response in the relevant luminance-sensitive mechanisms. Equiluminant stimuli are therefore designed to *isolate* the color vision system in order to probe its properties directly. The problem that inevitably arises when comparing color and luminance is that because performance usually improves with stimulus contrast, the absence of a common metric for equating color and luminance contrast means that there is no simple way to make the comparison. As is often said, it is like comparing "apples with oranges." One approach to this problem is to use a behavioral measure that is independent of how contrast is measured physically. One such measure is contrast given in multiples of the contrast needed to detect the stimulus [11–16]. With contrast defined in this way, form and motion tasks at equiluminance tend to require higher contrasts in order to achieve a level of performance comparable to that of their isochromatic (luminance contrast only) counterparts (see above references). The precise reason for this is not clear, but two possibilities spring to mind. The first is that chromatic form/motion mechanisms are subject to higher levels of internal noise; the second is that there are fewer neurons for processing chromatic form/motion signals than are available for chromatic detection [17, 18]. To this author's knowledge no studies to date have explicitly attempted to assess the contribution of these potential two factors to color-form and color-motion perception.

The third question, namely how color and luminance interact when subserving other modalities, is the least studied. Yet it is arguably the most important question. In natural scenes, for example, chromatic contrast may contribute very little to a given modality because enough luminance contrast is present to produce asymptotic performance. On the other hand, certain benefits of color vision might be revealed *only* when chromatic *and* luminance variations are present. This last point becomes obvious once one realizes that in the natural visual world color and luminance enjoy a spatial relationship that it would be surprising that vision did not exploit. In the following section I consider these relationships.

8.3 Spatial Relations Between Color and Luminance in the Natural Visual World

Figure 8.1 shows an image of a natural scene decomposed into its chromatic (sometimes termed here color) and luminance layers.[1] Although the color layer is unlikely to be equiluminant owing to the limitations of photographic reproduction, it nevertheless reveals a number of features about our chromatic visual experience. For example, the colors tend to form into relatively large clumps, or patches; in modern parlance color is "sparse." The fine detail in the image, such as the dense foliage in the background, is visible primarily in the luminance layer. It might be concluded from these observations that there is more high spatial frequency information in the luminance than in the color layer, but image analysis shows this not to be so, at least when measured across a range of natural scenes [19, 20]. The surprising parity in the amount of high spatial frequency information is because most high spatial frequency information originates from sharp edges, and as Fig. 8.1 testifies, edges are abundant in the color layer. Image analysis has also confirmed that clumpiness is a general feature of natural-scene color [20]. Luminance information is less clumpy because of dappling from shadows and shading and because many textures are defined by shading and shadows [21]. One must also bear in mind that humans are relatively insensitive to chromatic fine detail [22, 23]. Thus the clumpy appearance of the color layer in Fig. 8.1 is due to a combination of physical clumpiness and insensitivity to chromatic fine detail.

[1] To decompose the RGB camera image the image was first converted into the YUV color space, in which luminance information is represented in the Y layer and color information in the U and V layers. The luminance image was created by setting the U and V layers to zero then converting the YUV image back to RGB. The chromatic layer was created by setting the Y layer to 0.5 then converting the YUV image back to RGB. Note that the chromatic layer will not be isoluminant due to observer variability and the limitations of photographic reproduction.

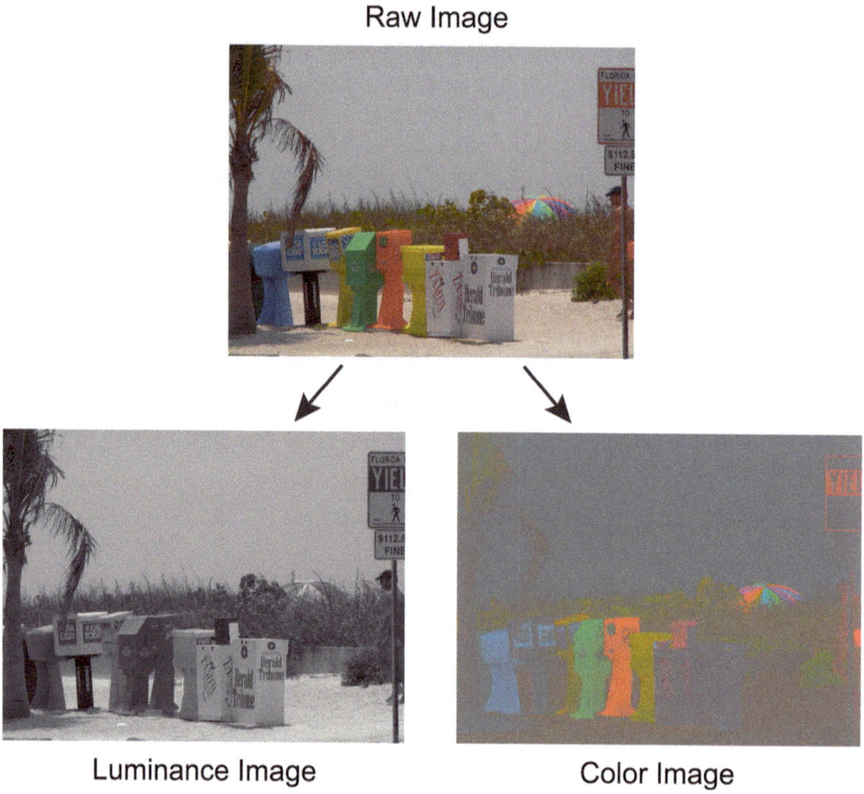

Fig. 8.1 Image of a natural scene decomposed into its chromatic and luminance layers. Based on Fig. 1 in Kingdom (2011), reproduced with permission

The figure also reveals that many edges are defined by both color and luminance (as with the edges of the bins in the foreground). In other words color and luminance changes are often spatially aligned [21, 24, 25]. Spatial alignment is a form of spatial correlation. One often reads how cone signals are decorrelated by the postreceptoral channels they feed into, resulting in independence among those postreceptoral channels. Yet the ubiquity of color–luminance alignment in natural scenes testifies to at least one form of non-independence. The advantage to vision of being sensitive to chromatic variations is of course premised on the fact that chromatic and luminance variations are not perfectly correlated, for if they were, one or other of color or luminance would be redundant. Therefore an understanding of the ways in which color and luminance are, and are not correlated, is critical to our understanding of the role of color vision in service to other visual modalities.

The initial stage of vision converts the L (long-wavelength-sensitive), M (middle-wavelength-sensitive), and S (short-wavelength-sensitive) cone signals into three postreceptoral channels: a luminance-sensitive channel which sums the signals from the L and M cones, and two chromatically selective channels, one which compares the responses of the L and M cones (the "L-M" response) and the other the response of the S cones with the summed responses from the L and M cones (the "S-(L+M)" response) [26–28] (see Chap. 2 on the retinal processing of photoreceptor signals and Chap. 5 on the psychophysical correlates of retinal processing.) These postreceptoral chromatic channels are often labelled "red-green" and "yellow-blue," respectively, though like the "red," "green," and "blue" labels often given to cones, these are strictly speaking misnomers. The postreceptoral channels decorrelate the cone signals by removing information that is redundant by virtue of their overlapping spectral sensitivities [21, 24, 29–31] (but see Lee et al. [32]; see also Chap. 4 on cone opponency). Shevell and Kingdom [10] reported the following Pearson R correlations for a typical image of a natural scene: 0.96 for L and M, 0.78 for M and S, and 0.73 for L and S. For the same scene on the other hand, the correlations between the postreceptoral channel responses were: 0.17 for L+M with L-M, 0.14 for L+M with S-(L+M) and –0.16 for L-M with S-(L+M), i.e., much smaller values. These low color–luminance correlations at the postreceptoral level however fail to embody important correlations at higher levels of processing. These higher-level correlations are in terms of color and luminance *changes* not *intensities*. Figure 8.2 illustrates the point. Shown are a sequence of colored squares—bright

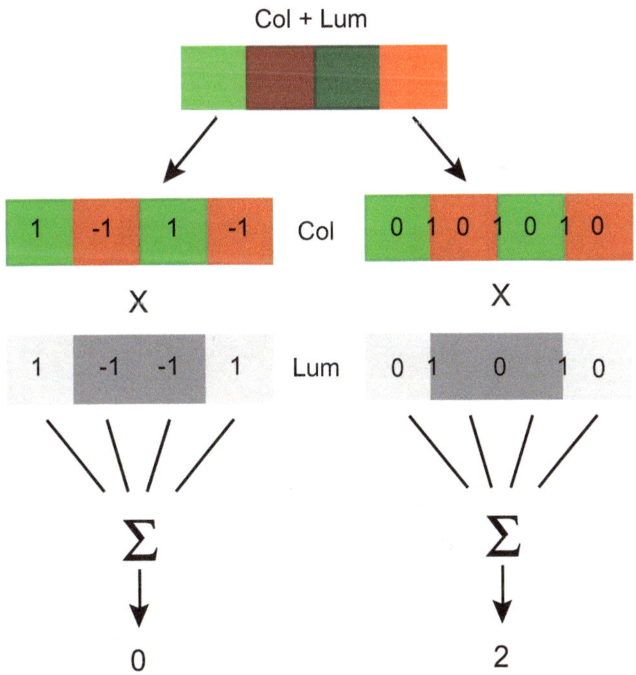

Fig. 8.2 Principle of correlation between color and luminance (*left*) and between color and luminance edges (*right*). See text for details

green, dark red, dark green, bright red—and their decomposition into color and luminance layers. On the left is the result of cross correlating the colors and luminances, using a simple code in which red equals "1" and green "−1", and with bright "1" and dark "−1". The integrated cross product of these values, in other words the overall correlation between the color and luminance layers, is zero. This zero correlation is a result of the presence of just as many red–green as green–red edges aligned with a given luminance edge. On the other hand on the right of the figure a different code is employed: "1" where there is a *change* in color or luminance, "0" elsewhere. Now the integrated cross product is non-zero, capturing the fact that color and luminance edges tend to be co-aligned.

Although color edges are typically aligned with luminance edges, it is less the case the other way round. Shadows, shading and other forms of nonuniform illumination, which are primarily luminance-defined, are not generally aligned with color, i.e., material borders. This property of natural color–luminance relations forms the basis for a number of special roles for color vision, some of which shall be considered later. For now it is sufficient to mention perhaps the best-known of these: the role of color vision in breaking camouflage [33, 34]. Camouflage-breaking by color is the basis of the well-known Ishihara Plates color vision test, and a naturalistic example is shown in Fig. 8.3a, b. Even for objects not fully camouflaged, color vision may augment figure-ground segmentation—compare the artificial textures in Fig. **8.3**c, d. Recently, Kingdom et al. [35] showed that suprathreshold chromatic variations have a small suppressive effect on suprathreshold luminance variations, in keeping with the role of color as camouflage breaker.

Fig. 8.3 The color in (**b**) helps to break the camouflage in (**a**). The color in (**c**) facilitates the segregation of the texture target in (**d**), but helps to mask it in (**e**)

The converse situation is that color can sometimes help to camouflage objects. Figure **8.2**e shows how the orientation-defined texture region in Fig. **8.3**d becomes harder to detect when irrelevant color variations are introduced. The figure was inspired by Morgan, Adam, and Mollon [36], who showed that than trichromats red-green dichromats were better than trichromats at detecting orientation-defined texture boundaries containing randomly arranged red and green colored bars (see also Pearson and Kingdom [37]). The flip-side of the coin is that Tritanopes will likely miss violet-colored flowers [33].

8.4 Color Vision and Other Visual Modalities

8.4.1 Color and Spatial Resolution

We noted in relation to Fig. **8.1** the relative lack of perceived fine detail in the color layer, and suggested that this was due to a combination of poor chromatic spatial resolution and the fact that natural textures tend to be primarily luminance-defined. Shevell and Kingdom [10] provide a striking demonstration of poor chromatic resolution (based on a figure in Moulden, Kingdom, and Wink [38]). When the checks in the center of a black and white checkerboard are made respectively dark blue and bright yellow, the colors appear vivid when viewed at close range, but fade to black and white when viewed from a distance.

One consequence of color vision's poor spatial resolution is that pure colored edges are blurred in the neural representation. In spite of this, thresholds for detecting blur in equiluminant features such as edges are surprisingly low [39, 40]. For example Wuerger et al. [39] found that when chromatic and luminance edges were equated in terms of cone contrasts, blur detection thresholds were similar for L–M ("red-green") and luminance edges. Blur detection thresholds for S cone ("blue-yellow") edges were however significantly worse than with luminance edges. There are a number of reasons why relative blur detection thresholds for L–M versus luminance edges might not be predictable from spatial resolution limits. These include differences in the color contrast metrics used in the two types of experiment and the possibility that blur detection recruits spatial frequencies below that of the spatial resolution limit.

On the other hand, if one separately blurs the chromatic and luminance content of an image of a natural scene, as in Fig. 8.4 (inspired by a similar figure in Wandell [41]: see his plate 7), one only obtains an impression of blur with the latter manipulation. Sharman, McGraw, and Peirce [42] measured blur detection thresholds in images of natural scenes whose color and luminance layers were separately blurred. They found higher thresholds for the color-only-blurred images, even when taking into account blur thresholds obtained for the color layer on its own. They suggested that this was due to the spreading of color into the areas between the luminance borders [43–45], which had the effect of reducing the gradients of color in the neural image. However, in Fig. **8.4** the chromatic blur *can* be seen spreading beyond the

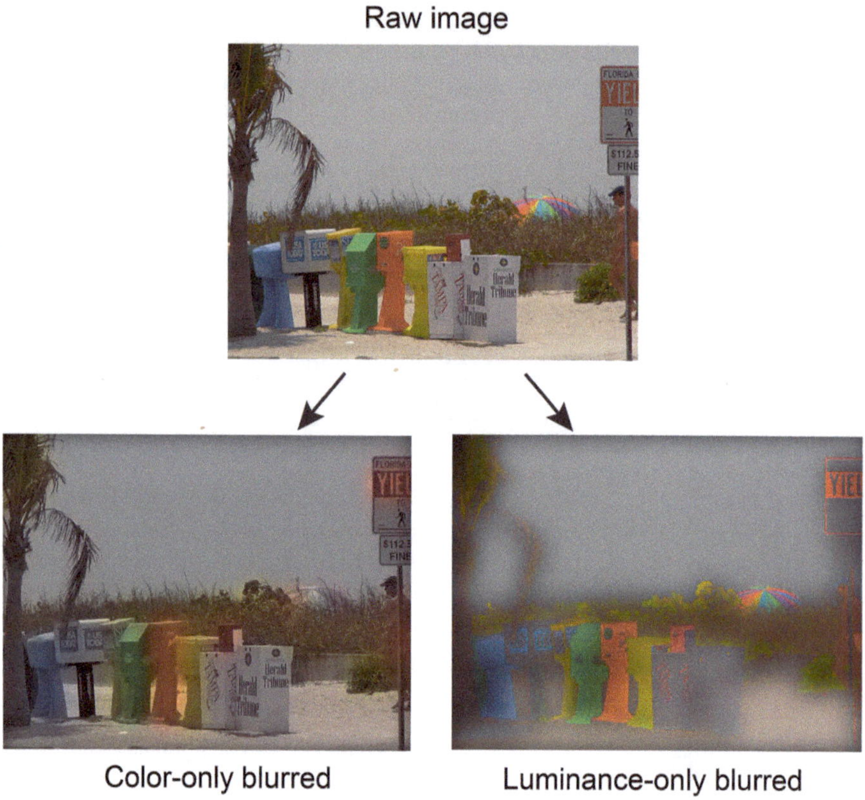

Fig. 8.4 Effect of separately blurring the color and luminance layers of an image of a natural scene

luminance edges, yet still there is little perceived blur. Kingdom et al. [46], based on measurements of perceived blur in mixed color-plus-luminance noise textures, suggested that in addition to the color spreading effect, color may contribute only minimally to our impression of blur. If so, the relative ease with which blur is detected in simple equiluminant edges may implicate the involvement of spatial mechanisms that are *sensitive* to color blur but which do not *signal* color blur.

8.4.2 Color and Contour

Provided that color features can be resolved, information about their orientations and positions provides the basis for processing higher-order features, such as contours and textures. Studies using simple equiluminant stimuli such as bars or Gabor patches have revealed that color can mediate position [11, 47] and orientation judgments [12, 48, 49], though typically more contrast is needed relative to detection

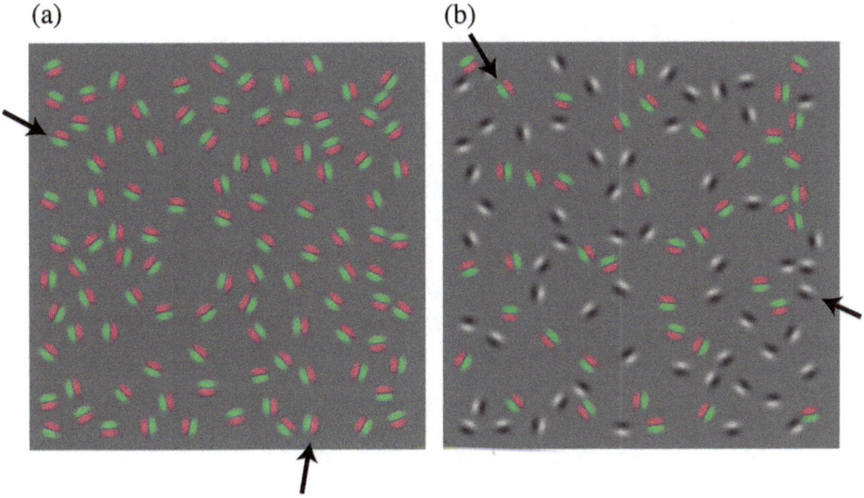

Fig. 8.5 Contour detection in a cluttered background. (**a**) all red–green elements; (**b**) alternating red–green and black–white elements. The arrows show the start and finish of each target path. From Fig. 14 in Shevell and Kingdom (2008), after McIlhagga and Mullen (1996) (images courtesy of K. Mullen)

threshold compared to luminance to produce commensurate levels of performance [12]. Color-defined features can be linked to form whole contours, as demonstrated by McIlhagga and Mullen [50] using the stimulus illustrated in Fig. **8.5**. The task is to find the string of elements that are collinearly arranged in the form of a "path." Paths constructed from equiluminant elements that appear red and green (Fig. **8.5**a) are detected just as easily as paths constructed from isochromatic elements that vary only in luminance [50]. However, when the path elements alternate between equiluminant and isochromatic (Fig. **8.5**b) or between L-M and S-(L+M), performance declines considerably [50, 51]. This shows not only that color can serve contour detection in cluttered surrounds, but that contour linking is selective along chromatic and luminance dimensions.

Moving up the processing hierarchy to contour shape processing, discriminating a perturbed circle from a circle requires, relative to detection threshold, about 1.7 times more contrast for red–green and about and two times more contrast for blue–yellow compared to luminance [16], showing that color vision is reasonably good at shape processing. However this leaves open the question as to whether there are shape mechanisms *selective* for color (see also Sect. 7.2 for a discussion of color vision in form and shape perception.) Using the shape-frequency after-effect as a tool, Gheorghiu and Kingdom [52] found a reduced after-effect when adaptation and test shapes were different in color (or when one was color-defined and the other luminance-defined), revealing color selectivity for contour-shape processing.

8.4.3 Color and Texture

Dense arrays of local orientations produce textures, and the detection of spatial variations in texture is important for segmenting a scene into surfaces and determining their three-dimensional shapes. At equiluminance, textural variations are detectable [53–55], but by mechanisms broadly tuned for color [55] and also sensitive to luminance contrast [54]. Equiluminant textures can elicit an impression of three-dimensional shape [56, 57], suggesting that chromatic signals contribute to "shape-from-texture."

One texture feature that continues to engage color vision researchers is mirror-symmetry. Figure **8.6** shows various mirror-symmetric arrangements involving three colors of dot: red, green, and blue. In Fig. **8.6**a all three colors are arranged in

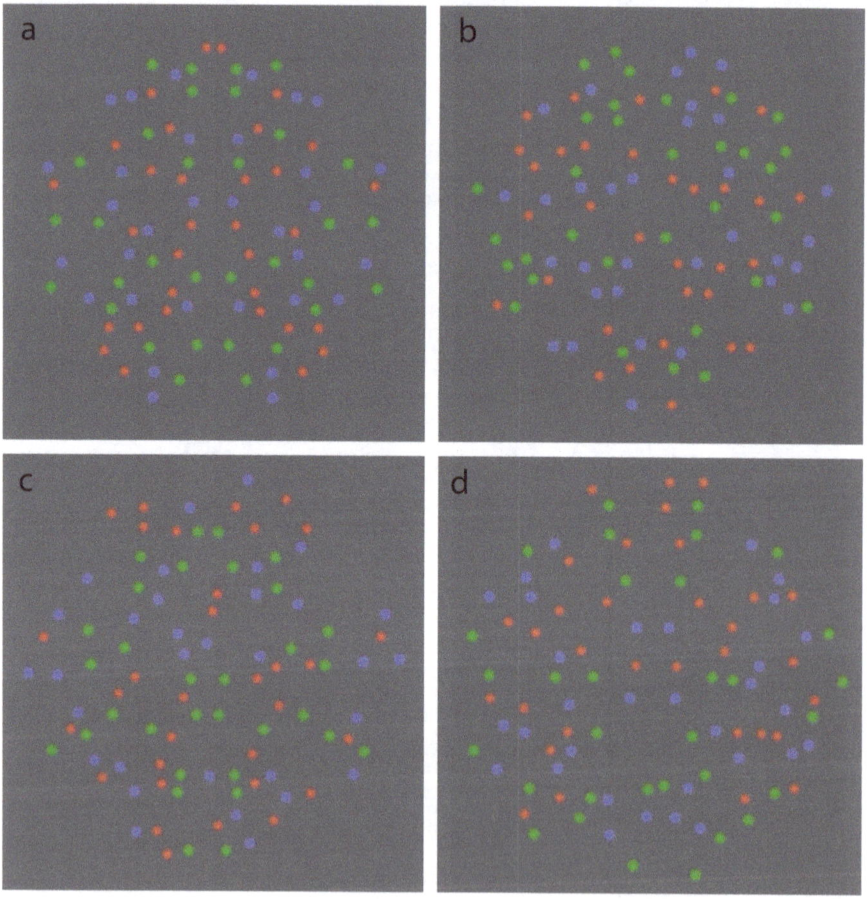

Fig. 8.6 Color and symmetry in stimuli made up from equal numbers of red, green, and blue dots. (**a**) all dots symmetric in position and color; (**b**) all dots positionally symmetric but anti-symmetric in color; (**c**) only the green dots symmetric, with the red and blue dots quasi-randomly positioned; (**d**) one third of the red, green, and blue dots symmetric, the other two-thirds of each color quasi-randomly positioned. Based on Figs. 1 and 2 in Gheorghiu et al. (2016), with permission

perfect mirror-symmetry, that is, each color dot on the left is mirror-paired with the same color dot on the right. In Fig. **8.6**b the dots are arranged in "anti-symmetry": green is paired with blue, red with green and blue with red. The symmetry pops out in Fig. **8.6**a but is hard to see in Fig. **8.6**b, showing that we are sensitive to the colors and not just positions of elements in mirror-symmetry. However, this begs the question: are there color-selective channels for mirror-symmetry? That is, do we have one channel that computes the degree of mirror-symmetry in the pattern of red dots, another in the pattern of green dots, another in the pattern of blue dots and so on? The observation that we are sensitive to color in mirror-symmetry patterns does not imply the existence of color-symmetry channels. For example, the visual system might assign to each dot a value for mirror-symmetry computed across a range of dimensions—color, orientation, size, position, etc.—and then combine those values across the set of dots, rather than first computing the amount of mirror-symmetry across the set of dots separately for each color. The issue of whether there are color-selective symmetry channels as so-defined has proved controversial, with Wu and Chen [58] in favor and Morales and Pashler [59] against. Gheorghiu et al. [60] have attempted to resolve the issue using stimuli such as those in Fig. **8.6**. In Fig. **8.6**c only one color – green - is mirror-symmetric (the other two colors are positioned quasi-randomly), and in Fig. **8.6**d one-third of all three colors are mirror-symmetric, with the other two-thirds of each of the three colors positioned randomly. Gheorghiu et al. found that unless observers were pre-cued as to which color was mirror-symmetric in Fig. **8.6**c such that they could selectively attend to it, they performed no better than with the stimuli in Fig. **8.6**d, and argued this went against the color-symmetry channel hypothesis. Gheorghiu et al. argue that their findings are consistent with Morales and Pashler's conclusion, that the benefits of color to symmetry perception of the symmetric pattern being in one color and the background in another is mediated through attention rather than through color-selective symmetry channels.

8.4.4 Color and Material

As noted earlier, most color edges in natural scenes are accompanied by luminance edges, but not the other way round. Spatially nonuniform illumination, such as shadows and shading, tends to be mostly luminance-defined [21, 61, 62]. It follows from this observation that color variations are more reliable indicators of material boundaries than are luminance variations, particularly in scenes where shadows and shading are prevalent. As Osorio and Vorobyev [63] write:

> "Chromaticity is probably relatively stable (constant) in natural illumination, so that it gives information about surface reflectance, pigmentation and other material properties..... Colour vision is therefore likely to be important for object detection or classification." (p. 1745)

Identifying whether a boundary is a material or an illumination change is not only important for object identification, but because nonuniform illumination plays a role in object perception, as with "shape-from-shading," "depth-from-shadows," etc. The

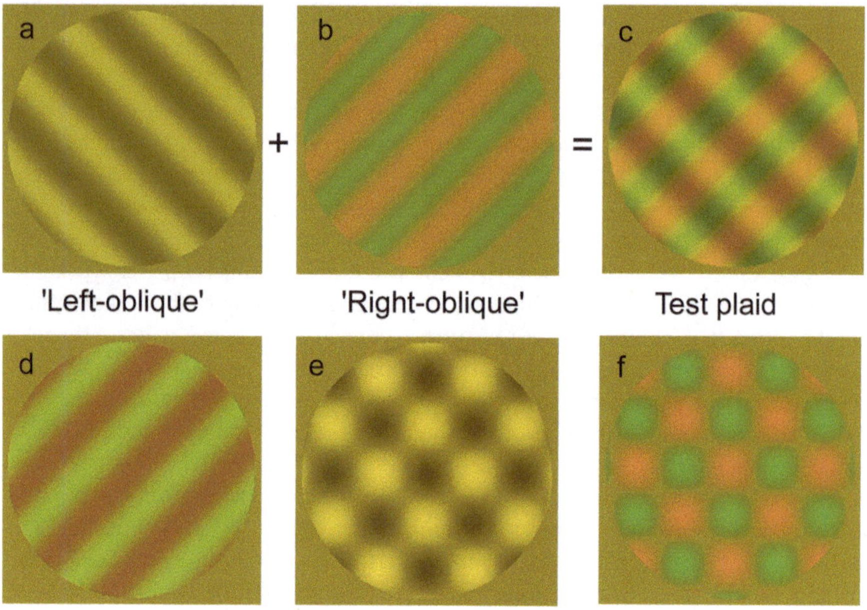

Fig. 8.7 The color-shading effect. (**a**) left-oblique luminance grating; (**b**) right-oblique red-green grating; (**c**) plaid formed by adding (**a**) and (**b**) together; (**d** (**a**) a and b are added together in spatial alignment; (**e**) luminance plaid; (**f**) red–green plaid. From Kingdom (2003), reproduced with permission

question remains however whether the physical relations between color and luminance in natural scenes are exploited by vision, i.e., does vision adopt the "colour-is-material" assumption? If it does, there must be behavioral evidence in support. Anecdotal evidence comes from a study of cerebral achromatopsia. Sacks [64] relates how an achromatopsic had trouble distinguishing shadows from changes in material structure: "A major problem occurred when he drove, in that he tended to misinterpret shadows as cracks or ruts in the road and would brake or swerve suddenly to avoid these." (p.7). Behavioral evidence in support of the color-is-material assumption comes from the "color-shading effect" [65, 66], of which an example is shown in Fig. 8.7. A purely luminance-varying grating pattern (Fig. 8.7a) elicits only a slight impression of rippling in depth. However, if mixed with an opposite-oriented pattern of red and green stripes (7b) to form a color–luminance plaid, the rippling becomes pronounced (7c). The addition of the color grating promotes the interpretation that the plaid is a material surface subject to a pattern of shading, characteristic of a corrugated material illuminated obliquely. No such effect is observed if the red-green grating is aligned with the black-white grating, as now the luminance grating is interpreted as a material not an illumination change. Moreover, it is not the plaid structure itself that generates the impression of a corrugated surface, as Fig. 8.2e, f show.

The color-is-material assumption is also evidenced in studies of shadows and transparency [62, 66, 67]. Achromatic transparencies such as dark glasses reduce

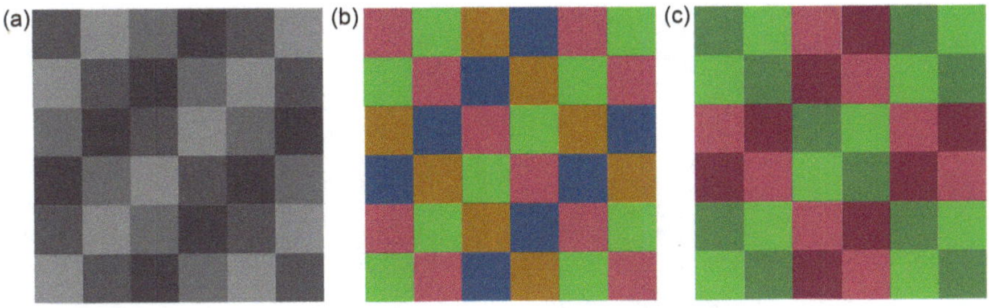

Fig. 8.8 Color and transparency. (**a**) achromatic figure simulating overlapping transparencies. The figure is formed by converting the color figure in (**b**) to grey-level. (**c**) simulated achromatic transparencies on a chromatic background

the level of light without altering chromatic content, akin to how a cast shadow affects primarily the luminance not color of the underlying surface. A sharp change in luminance that is unaccompanied by a change in color therefore implies the presence of an achromatic transparency or shadow, although compelling impressions of these image features typically require additional cues such as X-junctions and penumbra [67–69]. One might therefore expect that the introduction of chromatic changes across the borders of an achromatic transparency or shadow would reduce or eliminate the impression of transparency [62, 70]. Figure 8.8 shows the effect (the achromatic transparency on the left is in fact the grey-level version of the figure in the middle). The figure on the right reveals that it is not the introduction of color contrast per se that reduces perceived transparency. Provided that the changes in color across the transparency border are consistent along its length, a strong (and perhaps even stronger) impression of transparency remains [62, 70–74]. Thus the critical color requirement for perceiving transparency appears to be the consistency of the color change across the transparency border, not the uniformity of color. The consistency of color changes across borders is also an important factor in the closely related and very delicate phenomenon known as neon-color spreading [75, 76].

While the color-is-material assumption can affect perceived shape, the reverse is also true: perceived shape can affect perceived color. Bloj, Kersten and Hurlbert [77] painted a card magenta on its left half and white on its right half. When folded along the color boundary, such that it appeared like > when viewed from the left, the concave shape caused some of the light from the magenta half to fall onto the white half, resulting in a perceived pinkish glow on the white side. However when the perceived shape of the card was inverted using a pseudoscope, such that from the left the card appeared < (as if looking at a peaked roof from a helicopter), the white half of the card appeared a deep magenta color, even though physically the colors had not changed. The change in perceived color caused by a change in perceived shape shows that the visual system uses information about shape to counteract the effect of inter-reflections between surfaces. The light from the magenta half that falls on the white half tends to be discounted if the geometry is consistent with inter-reflection. However when the card is perceived with the center fold in front, the

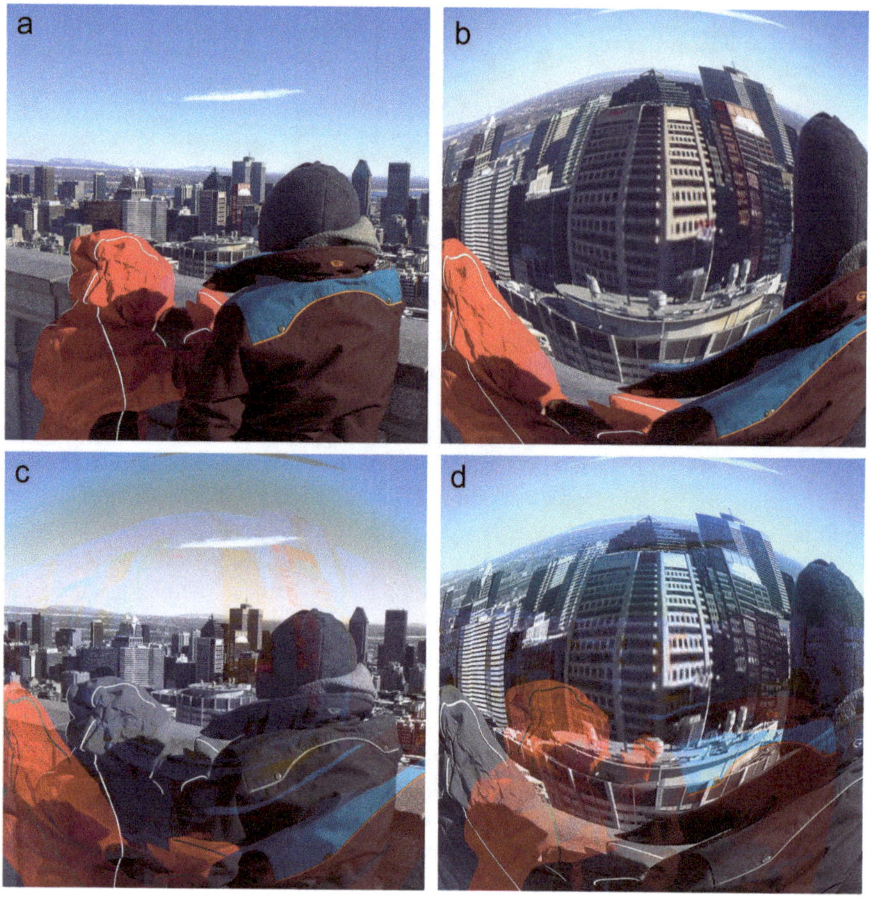

Fig. 8.9 Color and image distortion. (**a**) image of a natural scene; (**b**) image subjected to a barrel distortion such as from a fish-eye lens; (**c**) only the color layer has been subjected to the barrel distortion; (**d**) only the luminance layer has been subjected to the barrel distortion. Taken from Fig. 2 in Jennings et al. (2015)

geometry does not support inter-reflection, so the light reflected from the magenta half is attributed to the surface of the white half, causing it to appear magenta.

The color-is-material assumption may be relevant to the demonstration in Fig. 8.9, which is best thought of as the image-distortion analog of Fig. 8.4's color-blur demonstration. The figure, from Jennings et al. [78], shows the result of applying a barrel distortion, as produced for example by a fisheye lens, to an image (see also Hecht [79], Kingdom et al. [80]; Bex [81]). Figure 8.9c illustrates the same barrel distortion applied to just the color layer and Fig. 8.9d to just the luminance layer. When the distortion is applied just to the luminance layer, the image appears similar to the one in which the whole image is distorted, with both appearing like a reflection on a curved, shiny surface. On the other hand when just the color layer is distorted the image appears little different from the undistorted original, even though the disturbance to the colors is discernible: a "ghosting" can be observed whereby the color appears to have been smeared across the luminance boundaries. This demonstration suggests that our sense of image scale-distortion is largely mediated by the luminance not color layer. In addition, Jennings et al. [78], this time using periodic scale distortions, showed that observers were less *sensitive* to color-only compared to luminance-only distortions. They argued that there is more to the appearance of Fig. 8.9 however than mere lack of sensitivity to color distortion. In the natural visual world images are seen in specular reflections [82, 83]. An ideal specular surface is a mirror, with a shiny metallic surface or glossy paint a close approximation. Jennings et al. note that specular reflections from a number of materials appear to have reduced color spectra. Therefore the visual system might rely on luminance more than color when processing patterns of specular reflection in order to encode the shapes of curved, metallic objects, consistent with the appearance of Fig. 8.9.

8.4.5 Color and Binocular Function

Small differences in the two eyes' view of a scene provide the basis for stereoscopic depth perception [84, 85]. A compelling observation that led to the idea that color vision is stereoblind is that the depth-target in a random-dot-stereogram tends to disappear at equiluminance [3, 86–88]. However, there is evidence that stereoscopic depth perception is nevertheless possible at equiluminance, especially with relatively simple stimuli such as bars or grating patches [17, 89–92]. The poor depth quality in equiluminant random-dot-stereograms may be a result of a specific deficit in integrating local chromatic depth signals in order to generate a depth-defined surface [15].

A prerequisite for stereopsis is that corresponding features in the two eye's views must be matched. Spurious matches tend to occur in scenes containing dense arrays of similar elements lying in multiple depth planes. However, even if the elements have similar luminance contrasts, orientations and sizes, chromatic differences among them could serve to reduce the number of false matches. A number of studies have demonstrated that in complex stereo displays, the introduction of color contrasts that are matched between the eyes reduces the number of false matches, facilitating stereopsis [84, 93, 94].

Differences in the two eye's views of the world occur not only in terms of positional disparities, but in terms of disparities in contrast and hue. When different hues are presented to the two eyes the result is either fusion or rivalry depending on how far apart are the hues [95, 96], reinforcing the importance of color vision for binocular feature matching. Evidence has shown that the presence of matched features in the two eyes along one dimension tends to reduce rivalry among unmatched features along another dimension [97, 98]. A possible reason for this is that matched features reduce the interocular inhibition between the unmatched features that causes rivalry. In the chromatic domain a similar process has recently been observed in the context of dichoptic color mixing. Kingdom and Libenson [99] found that when two isoluminant colored patches with different saturations were presented to the two eyes, the perceived saturation of the mixture was close to the higher of the two saturations, as if the lower saturation did not contribute to the mixture. They attributed this to interocular inhibition between the saturations, resulting in a "winner-take-all" scenario. However, when binocularly matched luminance contrasts were added to the patches the percept shifted towards the average of the two saturations, presumably because interocular inhibition was reduced. Figure 8.10 shows one consequence of this process.

Fig. 8.10 Free fusion of the three stereo-pairs reveals a specifically dichoptic desaturation effect. The colors in the top and middle panels appear equally saturated when free-fused, whereas the colors in the bottom panel appear desaturated. From Figure 9 in Kingdom and Libenson (2015), reproduced with permission

The pairs of figures in the top and middle panels, when fused to form two single images, show strongly saturated colors. In the middle panel the presence of an achromatic blank field in one eye has no effect on the perceived color saturation in the fused image. However, when replaced by an achromatic luminance-varying image (bottom panel) the color appears desaturated. A possible explanation is that the matched luminance contrasts promote the interpretation that the two disparate saturations nevertheless emanate from the same object, and that under such circumstances vision tends to blend the saturations rather than subject them to competition. Thus a powerful perceptual interaction between color and luminance in binocular vision may be explained by the visual requirement of perceiving an unambiguous, unitary object world. A similar conclusion has emerged from other recent studies of dichoptic luminance [100] and chromatic [101, 102] vision.

8.4.6 Color and Motion

Evidence that color vision plays a minimal role in motion perception comes from the twin observations that an equiluminant moving grating appears to move more slowly than its luminance counterpart [13, 103–105] and that the target shapes in equiluminant random-dot kinematograms are difficult or impossible to discern [3, 106]. This second observation is analogous to the perception of equiluminant random-dot-stereograms discussed earlier. In addition, the motion of equiluminant chromatic L–M grating patches can be masked by the addition of randomly moving luminance "noise," even though the same noise fails to elevate L–M detection thresholds [107]. This suggests that the perceived motion of simple L–M chromatic stimuli is mediated by a luminance artifact generated within the visual system, possibly by a temporal difference in the L- and M-cone signals [108]. However this view runs counter to other evidence for a chromatic contribution to simple-stimulus motion perception [109–112] (reviewed by Cropper and Wuerger [113]). In reviewing color vision's role in motion perception, Shevell and Kingdom [10] conclude that even if there is not a dedicated input of color vision to the motion of simple stimuli, it is at best very weak.

For more complex objects, such as those defined by variations in color contrast rather than in color, the consensus is more favorable for a role in motion perception [107, 114]. Some have argued, however, that for both simple *and* complex objects, perceived motion at equiluminance is mediated by a general-purpose, attention-based, high-level mechanism that responds to the movement of any figure-ground relation, whether defined by color, luminance, or texture [105]. Such a mechanism is also implicated by the ability to attentionally track an equiluminant moving object even where the motion percept itself is severely degraded [115].

The importance of attention-to-color in motion perception also emerges from studies using the global-motion paradigm. In this paradigm the majority of elements are distractors that move in random directions, while a minority move coherently in one direction or another. Studies disagree as to whether coherent motion in global-motion displays is perceptible at equiluminance (e.g., against, Bilodeau and Faubert [116]; in favor, Ruppertsberg, Wuerger, and Bertami [117]). When the display elements are defined by both color and luminance contrast on the other hand, as in the color symmetry stimuli discussed above, it has been shown that the introduction of a color difference between the target and distractor elements reduces the proportion of target dots needed to identify the target motion direction, and this has been taken as evidence for a chromatic contribution to global motion [118]. However, subsequent studies have challenged this conclusion. Under peripheral viewing, in which selective attention to the target dot color becomes impossible, the chromatic-difference advantage disappears [119] as it does also in displays where the composition of elements is designed to render selective attention useless [120]. Rather than as evidence for a chromatic input to global motion, these studies point to color as a useful cue for picking out moving objects that would otherwise be camouflaged. An absence of chromatic input to global motion is also evidenced in studies of structure-from-motion [121]. There is however a caveat to this conclusion. Shevell and Kingdom [10], citing in support a number of physiological studies, argue that there may be a weak chromatic contribution to global motion but that in non-equiluminant displays the contribution of color does not manifest itself because the chromatic motion signal is swamped by the more effective luminance signal.

8.5 Conclusion

Color vision in conjunction with luminance vision can support the processing of form, material, depth, and motion. It achieves this in part by exploiting the natural constraints that define the spatial relationships between the patterns of color and luminance in the natural visual world. Attention-to-color is an important contributor to a range of form and motion tasks. For equiluminant patterns, that is, patterns that isolate color vision, color is able to contribute to the processing of some visual modalities, but invariably requires more contrast relative to detection compared to that required by luminance contrast. Future experiments will be needed to determine the precise reasons for this.

Acknowledgements Supported by a Canadian Institute of Health Research grant #MOP 123349 given to F.K.

References

1. Merleau-Ponty M. Eye and mind. In: Kearney R, Rasmussen D, editors. Continental aesthetics. Oxford: Blackwell; 1964/2001. Trans. Carleton Dallery.
2. Chirimuuta M, Kingdom FAA. The uses of colour vision: ornamental, practical and theoretical. Mind Machines. 2015;25(2):213–29.

3. Livingstone MS, Hubel DH. Psychophysical evidence for separate channels for the perception of form, color, movement and depth. J Neurosci. 1987;7(11):3416–68.
4. Livingstone MS, Hubel DH. Segregation of form, color, movement and depth: anatomy, physiology and perception. Science. 1988;240:740–9.
5. McIlhagga WH, Mullen KT. The contribution of colour to contour detection". In: Dickenson CM, Murray I, Carden D, editors. Colour vision research: proceedings of the John Dalton conference. London: Taylor & Francis; 1997. p. 187–97.
6. Gegenfurtner KR, Kiper DC. Color vision. Annu Rev Neurosci. 2003;26:181–206.
7. Heywood CA, Kentridge RW, Cowey A. From and motion from colour in cerebral achromatopsia. Exp Brain Res. 1998;123:145–53.
8. Cowey A, Alexander I, Heywood CA, Kentridge RW. Pupillary responses to coloured and contourless displays in total cerebral achromatopsia. Brain. 2008;131:2153–60.
9. Regan D. Human perception of objects. Massachussets: Sinauer Associates; 2000.
10. Shevell SK, Kingdom FAA. Color in complex scenes. Annu Rev Psychol. 2008;59:143–66.
11. Morgan MJ, Aiba TS. Positional acuity with chromatic stimuli. Vis Res. 1985;25:689–95.
12. Webster MA, DeValois KK, Switkes E. Orientation and spatial frequency discrimination for luminance and chromatic gratings. J Opt Soc Am A. 1990;7:1034–49.
13. Mullen KT, Boulton JC. Absence of smooth motion perception in color vision. Vis Res. 1992;32:483–8.
14. Simmons DR, Kingdom FAA. Contrast thresholds for stereoscopic depth identification with isoluminant and isochromatic stimuli. Vis Res. 1994;34:2971–82.
15. Kingdom FAA, Simmons DR, Rainville SJM. On the apparent collapse of stereopsis in random-dot-stereograms at isoluminance. Vis Res. 1999;39:2127–41.
16. Mullen KT, Beaudot WH. Comparison of color and luminance vision on a global shape discrimination task. Vis Res. 2002;42:565–75.
17. Kingdom FAA, Simmons DR. The relationship between colour vision and stereoscopic depth perception. J Soc 3D Broadcast Imag. 2000;1:10–9.
18. Solomon SG, Lennie P. The machinery of colour vision. Nat Rev Neurosci. 2007;8:276–86.
19. Parraga CA, Troscianko T, Tolhurst DJ. Spatio-chromatic properties of natural images and human vision. Curr Biol. 2002;12:483–7.
20. Yoonessi A, Kingdom FAA, Alqawlaq S. Is color patchy? J Opt Soc Am A. 2008;25:1330–8.
21. Johnson AP, Kingdom FAA, Baker Jr CL. Spatiochromatic statistics of natural scenes: first- and second-order information and their correlational structure. J Opt Soc Am A. 2005;22:2050–9.
22. Granger EM, Hurtley JC. Visual chromaticity-modulation transfer function. J Opt Soc Am A. 1973;63:1173–4.
23. Mullen KT. The contrast sensitivity of human colour vision to red-green and blue-yellow gratings. J Physiol. 1985;359:381–409.
24. Fine I, MacLeod DIA, Boynton GM. Surface segmentation based on the luminance and color statistics of natural scenes. J Opt Soc Am A. 2003;20:1283–91.
25. Hansen T, Gegenfurtner KR. Independence of color and luminance edges in natural scenes. Vis Neurosci. 2009;26:35–49.
26. DeValois RL. Analysis and coding of color vision in the primate visual system. Cold Spring Harb Symp Quant Biol. 1965;30:567–79.
27. Krauskopf J, Williams DR, Heeley DW. Cardinal directions of colour space. Vis Res. 1982;22:1123–31.
28. Derrington AM, Krauskopf J, Lennie P. Chromatic mechanisms in lateral geniculate nucleus of macaque. J Physiol. 1984;357:241–65.
29. Buchsbaum G, Gottschalk A. Trichromacy, opponent colour coding and optimum colour information transmission in the retina Proc. R Soc Lond B Biol Sci. 1983;220:89–113.
30. Zaidi Q. Decorrelation of L- and M-cone signals. J Opt Soc Am A. 1997;14:3420–31.
31. Ruderman DL, Cronin TW, Chiao C-C. Statistics of cone responses to natural images: implications for visual coding. J Opt Soc Am A. 1998;15:2036–45.

32. Lee T-W, Wachtler T, Sejnowski TJ. Color opponency is an efficient representation of spectral properties in natural scenes. Vis Res. 2002;42:2095–103.
33. Mollon JD. "Tho' she kneel'd in that place where they grew…": the uses and origins of primate color vision. J Exp Biol. 1989;146:21–38.
34. Mollon JD. Cherries among the leaves: the evolutionary origins of color vision. In: David S, editor. Color perception: philosophical, psychological, artistic and computational perspectives. New York: Oxford University Press; 2000. p. 10–30.
35. Kingdom FAA, Bell J, Gheorghiu E, Malkoc G. Chromatic variations suppress suprathreshold brightness variations. J Vis. 2010;10(10):13:1–13.
36. Morgan MJ, Adam A, Mollon JD. Dichromats detect colour-camouflaged objects that are not detected by trichromats. Proc R Soc Lond B Biol. 1992;248:291–5.
37. Pearson PM, Kingdom FAA. On the interference of task-irrelevant hue variation on texture segmentation. Perception. 2001;30:559–69.
38. Moulden B, Kingdom F, Wink B. Colour pools, brightness pools, assimilation, and the spatial resolving power of the human colour-vision system. Perception. 1993;22:343–51.
39. Wuerger SM, Owens H, Westland S. Blur tolerance for luminance and chromatic stimuli. J Opt Soc Am A. 2001;18:1231–9.
40. Webster MA, Mizokami Y, Svec LA, Elliott SL. Neural adjustments to chromatic blur. Spat Vis. 2006;19:111–32.
41. Wandell BA. Foundations of vision. Sunderland, MA: Sinauer; 1995.
42. Sharman RJ, McGraw PV, Peirce JW. Luminance cues constrain chromatic blur discrimination in natural scene stimuli. J Vis. 2013;13(4):14.
43. Boynton RM. Ten years of research with the minimally distinct border. In: Armington JC, Krauskopf J, Wooten BR, editors. Visual psychophysics and physiology. New York: Academic; 1978. p. 193–207.
44. Mollon J. Seeing Color. In: Lamb T, Bourriau J, editors. Colour, art and science (Ch 5). Cambridge, UK: Cambridge University Press; 1995.
45. Pinna B, Brelstaff G, Spillmann L. Surface color from boundaries: a new 'watercolor' illusion. Vis Res. 2001;41:2669–76.
46. Kingdom FAA, Bell J, Haddad C, Bartsch A. Perceptual scales for chromatic and luminance blur in noise textures. J Vis. 2015;15(9):6,1–10.
47. Krauskopf J, Forte JD. Influence of chromaticity on vernier and stereo acuity. J Vis. 2002;2:645–52.
48. Clifford CW, Spehar B, Solomon SG, Martin PR, Zaidi Q. Interactions between color and luminance in the perception of orientation. J Vis. 2003;3:106–15.
49. Beaudot WH, Mullen KT. Orientation selectivity in luminance and color vision assessed using 2-d band-pass filtered spatial noise. Vis Res. 2005;45:687–96.
50. McIlhagga WH, Mullen KT. Contour integration with color and luminance contrast. Vis Res. 1996;36:1265–79.
51. Mullen KT, Beaudot WH, McIlhagga WH. Contour integration in color vision: a common process for the blue-yellow, red-green and luminance mechanisms ? Vis Res. 2000;40:639–55.
52. Gheorghiu E, Kingdom FA. Chromatic tuning of contour-shape mechanisms revealed through the shape-frequency and shape-amplitude after-effects. Vis Res. 2007;47(14):1935–49.
53. McIlhagga W, Hine T, Cole GR, Snyder AW. Texture segregation with luminance and chromatic contrast. Vis Res. 1990;30:489–95.
54. Pearson PM, Kingdom FAA. Texture-orientation mechanisms pool colour and luminance. Vis Res. 2002;42:1547–58.
55. Cardinal KS, Kiper DC. The detection of colored glass patterns. J Vis. 2003;3:199–208.
56. Troscianko T, Montagnon R, Le Clerc J, Malbert E, Chanteau PL. The role of colour as a monocular depth cue. Vis Res. 1991;31:1923–9.
57. Zaidi Q, Li A. Three-dimensional shape from chromatic orientation flows. Vis Neurosci. 2006;23(3-4):323–30.
58. Wu C-C, Chen C-C. The symmetry detection mechanisms are color selective. Sci Rep. 2014;4:3893. doi:10.1038/srep03893.

59. Morales D, Pashler H. Attention and perception of color symmetry. Nature. 1999;399(6732):115–6.
60. Gheorghiu E, Kingdom FAA, Remkes A, Li HC-O, Rainville S. The role of color and attention-to-color in mirror-symmetry perception. Sci Rep. 2016;6:29287, doi:10.1038/srep29287.
61. Rubin JM, Richards WA. Color vision and image intensities: when are changes material? Biol Cybern. 1982;45:215–26.
62. Kingdom FAA, Beauce C, Hunter L. Colour vision brings clarity to shadows. Perception. 2004;33:907–14.
63. Osorio D, Vorobyev M. Photoreceptor spectral sensitivities in terrestrial animals: adaptations for luminance and colour vision. Proc R Soc Lond B. 2005;272:1745–52.
64. Sacks O. An anthropologist on Mars. London: Picador; 1995.
65. Kingdom FAA. Colour brings relief to human vision. Nat Neurosci. 2003;6:641–4.
66. Kingdom FAA. Illusions of colour and shadow. In: Biggam CP, Hough C, Kay CJ, Simmons DR, editors. New directions in colour studies. Amsterdam, The Netherlands: John Benjamins; 2011.
67. Kingdom FAA. Perceiving light versus material. Vis Res. 2008;48:2090–105.
68. Kanizsa G. Organization in vision. New York: Praeger; 1979.
69. Adelson EH. Perceptual organization and the judgement of brightness. Science. 1993;262:2042–4.
70. Ripamonti C, Westland S. Prediction of transparency perception based on cone– excitation ratios. J Opt Soc Am A. 2003;20:1673–80.
71. D'Zmura M, Colantoni P, Knoblauch K, Laget B. Color transparency. Perception. 1997;26:471–92.
72. Khang B-G, Zaidi Q. Accuracy of color scission for spectral transparencies. J Vis. 2002;2:451–66.
73. Fulvio JM, Singh M, Maloney LT. Combining achromatic and chromatic cues to transparency. J Vis. 2006;6:760–76.
74. Kingdom FAA, Kasrai R. Colour unmasks dark targets in complex displays. Vis Res. 2006;46:814–22.
75. Anderson BL. A theory of illusory lightness and transparency in monocular and binocular images: the role of contour junctions. Perception. 1997;26:419–53.
76. van Tujil HFJM. A new visual illusion: neonlike color spreading and complementary color induction between subjective contours. Acta Psychol. 1975;39:441–5.
77. Bloj MG, Kersten D, Hurlbert AC. Perception of three-dimensional shape influences colour perception through mutual illumination. Nature. 1999;402:877–9.
78. Jennings BJ, Wang K, Menzies S, Kingdom FAA. Detection of chromatic and luminance distortions in natural scenes. J Opt Soc Am A. 2015;32(9):1613–22.
79. Hecht E. Optics. 4th ed. London, UK: Addison-Wesley; 2001.
80. Kingdom FAA, Field DJ, Olmos A. Does spatial invariance result from insensitivity to change? J Vis. 2007;7(14):11:1–13.
81. Bex PJ. (In) sensitivity to spatial distortion in natural scenes. J Vis. 2010;10(2):23.1–15.
82. Mandelstam LI. Light scattering by inhomogeneous media. Zh Russ Fiz-Khim Ova. 1926;58:381.
83. Kerker K. The scattering of light and other electromagnetic radiation. New York: Academic; 1969.
84. Julesz B. Foundations of cyclopean perception. Chicago: University of Chicago Press; 1971.
85. Howard IP, Rogers BJ. Seeing in depth. 1, 2nd ed. Toronto, ON: Porteous; 2002.
86. Gregory RL. Vision with isoluminant colour contrast: 1. A projection technique and observations. Perception. 1977;6:113–9.
87. de Weert CMM. Colour contours and stereopsis. Vis Res. 1979;19:555–64.
88. Livingstone MS. Differences between stereopsis, interocular correlation, and binocularity. Vis Res. 1996;36:1127–40.

89. de Weert CMM, Sadza KJ. New data concerning the contribution of colour differences to stereopsis. In: Mollon JD, Sharpe LT, editors. Colour vision: physiology and psychophysics. London: Academic; 1983. p. 553–62.
90. Scharff LV, Geisler WS. Stereopsis at isoluminance in the absence of chromatic aberrations. J Opt Soc Am A. 1992;9:868–76.
91. Simmons DT, Kingdom FAA. On the independence of chromatic and achromatic stereopsis mechanisms. Vis Res. 1997;37:1271–80.
92. Ts'o DY, Roe AW, Gilbert CD. A hierarchy of the functional architecture for color, form and disparity in primate visual area V2. Vis Res. 2001;41:1333–49.
93. Jordan JR, Geisler WS, Bovik AC. Color as a source of information in the stereo correspondence process. Vis Res. 1990;30:1955–70.
94. den Ouden HEM, van Ee R, de Haan EHF. Colour helps to solve the binocular matching problem. J Physiol. 2005;567:665–71.
95. Hovis JK. Review of dichoptic color mixing. Optom Vis Sci. 1989;66:181–90.
96. Malkoc G, Kingdom FAA. Dichoptic difference thresholds for chromatic stimuli. Vis Res. 2012;62:75–83.
97. Blake R, Boothroyd K. The precedence of binocular fusion over binocular rivalry. Percept Psychophys. 1985;37:114–24.
98. Buckthought A, Wilson HR. Interaction between binocular rivalry and depth in plaid patterns. Vis Res. 2007;47:2543–56.
99. Kingdom FAA, Libenson L. Dichoptic colour saturation mixture: binocular luminance contrast promotes perceptual averaging. J Vis. 2015;in press.
100. Baker DH, Meese TS, Summers RJ. Psychophysical evidence for two routes to suppression before binocular summation of signals in human vision. Neuroscience. 2007;146:435–48.
101. Kingdom FAA, Wang D. Dichoptic colour-saturation masking is unmasked by binocular luminance contrast. Vis Res. 2015;116:45–52.
102. Jennings BJ, Kingdom FAA. Detection of between-eye differences in colour: Interactions with luminance. J Vis. 2016;In press.
103. Cavanagh P, Tyler CW, Favreau OE. Perceived velocity of moving chromatic gratings. J Opt Soc Am A. 1984;1:893–9.
104. Troscianko T, Fahle M. Why do isoluminant stimuli appear slower. J Opt Soc Am A. 1988;5:871–80.
105. Lu Z-L, Lesmes LA, Sperling G. Perceptual motion standstill in rapidly moving chromatic displays. Proc Natl Acad Sci U S A. 1999;96:15374–9.
106. Ramachandran VS, Gregory RL. Does colour provide an input to human motion perception? Nature. 1978;275:55–6.
107. Yoshizawa T, Mullen KT, Baker CL. Absence of a chromatic linear motion mechanism in human vision. Vis Res. 2000;40:1993–2010.
108. Mullen KT, Yoshizawa T, Baker CL. Luminance mechanisms mediate the motion of red-green isoluminant gratings: the role of "temporal chromatic aberration.". Vis Res. 2003;43:1235–47.
109. Gegenfurtner KR, Hawken MJ. Interaction of motion and color in the visual pathways. Trends Neurosci. 1996;19:394–401.
110. Cropper SJ, Derrington AM. Rapid colour-specific detection of motion in human vision. Nature. 1996;379:72–4.
111. Dobkins KR, Albright TD. What happens if it changes color when it moves?: Psychophysical experiments on the nature of chromatic input to motion detectors. Vis Res. 1993;33:1019–36.
112. Morgan MJ, Ingle G. What direction of motion do we see if luminance but not colour contrast is reversed during displacement? Psychophysical evidence for a signed-colour input to motion detection. Vis Res. 1994;34:2527–35.
113. Cropper SJ, Wuerger SM. The perception of motion in chromatic stimuli. Behav Cogn Neurosci Rev. 2005;4:192–217.
114. Cropper SJ, Derrington AM. Detection and motion-detection in chromatic and luminance beats. J Opt Soc Am A. 1995;13:401–7.

115. Cavanagh P. Attention-based motion perception. Science. 1992;257:1563–5.
116. Bilodeau L, Faubert J. Global motion cues and the chromatic motion system. J Opt Soc Am A. 1999;16:1–5.
117. Ruppertsberg AI, Wuerger SM, Bertami M. When S-cones contribute to chromatic global motion. Vis Neurosci. 2006;23:1–8.
118. Croner LJ, Albright TD. Image segmentation enhances discrimination of motion in visual distractors. Vis Res. 1997;37:1415–27.
119. Snowden RJ, Edmunds R. Colour and polarity contributions to global motion perception. Vis Res. 1999;39:1813–22.
120. Li H-CO, Kingdom FAA. Segregation by colour/luminance does not necessarily facilitate motion discrimination in noise. Percept Psychophys. 2001;63:660–75.
121. Li H-CO, Kingdom FAA. Motion-surface labeling by orientation, spatial frequency and luminance polarity in 3-D structure-from-motion. Vis Res. 2001;41:3873–82.

Chapter 9
Computational Modeling of Color Vision

Thomas Wachtler and Christian Wehrhahn

Abstract Modeling approaches have contributed in various ways to our understanding of the neural principles and mechanisms of color vision. We review computational models addressing aspects of color vision at different levels, from the photoreceptors to neural processing and color perception, and we consider the kinds of insights that have been enabled by different kinds of models.

Keywords Efficient coding • Decorrelation • Independent component analysis • Population coding • Natural scenes • Natural image statistics • Spatio-chromatic receptive fields • Cone mosaic • Opponency • Color constancy

9.1 Introduction

Color vision research, like all research in perception, has greatly benefited from computational models. Early studies were mostly concerned with mathematical descriptions of perceptual phenomena. Over the past decades, models at various levels of processing, from the sensory signals to neural coding principles and perception, have increasingly contributed to studying color vision. Progress in understanding of fundamental principles of sensory processing [1–3], together with development of computational concepts and methods and the rapid growth of data and knowledge on the physiological underpinnings of vision, have fostered the development of computational approaches to investigate color vision.

T. Wachtler (✉)
Computational Neuroscience, Department of Biology II, Ludwig-Maximilians-Universität München, Großhaderner Straße 2, D-82152 Planegg-Martinsried, Germany
e-mail: wachtler@bio.lmu.de

C. Wehrhahn
Vision Center Laboratory, The Salk Institute, La Jolla, CA, USA

J. Kremers et al. (eds.), *Human Color Vision*, Springer Series in Vision Research, DOI 10.1007/978-3-319-44978-4_9

9.2 Statistical and Information-Theoretical Approaches to Receptoral and Postreceptoral Processing

Theory-inspired approaches have advanced our understanding of sensory processing in general, and vision in particular. A particular focus in recent years was on the information processing aspects of sensory coding and attempts to explain properties of visual coding based on information-theoretical principles. A key concept in these studies is the notion of efficient coding of sensory information, which goes back to the work by Attneave, Barlow, and others [1, 2] (for review see Ref. [4]). Efficiency can be defined in many ways, but typical measures include minimal redundancy, sparseness, or other statistics that can be related to resource economy or information capacity [4]. Efficiency has also aspects related to optimal estimation or adaptation of processing to the structure of the signals to be represented [5]. Many studies have investigated how sensory signals should be coded to achieve such efficient representations, and these concepts have also been applied in studies on color vision.

9.2.1 Information in Cone Signals

Even at the earliest stages of color vision, important questions arise that may be answered by concepts of optimality and efficiency. For example, what is the benefit of trichromatic color vision, as in humans and some other primates, over dichromatic color vision as in most other mammals? Are the particular spectral sensitivities of human photoreceptors in any way optimal? The latter question has been addressed by simulating visual performance with hypothetical photoreceptors with different spectral peaks. Osorio and Vorobyev [6] modeled the discriminability of fruit spectra from leaf spectra. They calculated discriminability based on the distance in the space of cone responses, scaled by noise estimates for the different photoreceptor types according to their numerosity. Cone spectral sensitivities were varied to find the optimum peak sensitivities for discrimination. The results indicated a specific advantage of a trichromatic over a dichromatic retina for the detection of fruit against a background of leaves, a finding that has been corroborated by other studies [7]. Using a similar approach, Chiao et al. [8] analyzed the ability of dichromats to discriminate spectra from forest or aquatic environments. They found that spectral sensitivities of dichromatic species are consistent with an optimal spectral positioning of receptor sensitivities. Lewis and Zhaoping [9] analyzed the information content in cone signals by calculating the mutual information between input spectra and cone responses, using a number of different datasets of natural reflectances. The result from this analysis was that, when considering a wide range of possible peak wavelengths, human cone spectral sensitivities were not optimal. However, an increase in information capacity relative to the human cones would only be possible if the L cone spectral sensitivity would be shifted to much longer wavelengths. The feasibility of such a shift is probably limited by factors other than information transmission.

Information-theoretical analyses have also addressed the question how to best arrange the cones in a retinal mosaic. Garrigan et al. [10] analyzed the information capacity of cone mosaics using natural images. Interestingly, they found about equal information in all spectral channels, so that, based solely on spectral information, equal numbers of cones of each type would be predicted. However, taking into account optical blur, which is stronger for short-wavelength light, resulted in the prediction of a reduced number of S cones, consistent with the low S-cone density in the human retina. Interestingly, information capacity of the cone mosaic was found to be relatively insensitive to changes in L/M ratio, suggesting an explanation for the fairly high degree of interindividual variability of the relative numbers of L and M cones in humans (see Chaps. 2 and 3).

9.2.2 *Postreceptoral Processing*

Given the photoreceptors with their spectral sensitivities, how should their signals be encoded? The answer to this question certainly depends on what the goal of the visual system is, what costs and what constraints are imposed by neural processing, and which information is relevant and needs to be encoded. We do not know conclusive answers to these questions, and if we did it might still be difficult to relate them to corresponding coding mechanisms. Nevertheless, modeling can help explore hypotheses about the goals of visual coding and the underlying neural mechanisms.

9.2.2.1 Opponency

Spectral sensitivities of human cones show a high degree of overlap, especially for M and L cones, which results in highly correlated signals in these receptors. While this is not the only cause of redundancy in receptor signals, it is a substantial factor [11]. Buchsbaum and Gottschalk [12] investigated how to remove the correlation introduced by the overlap of spectral sensitivities. To find a decorrelating transformation for the cone signals, they performed a Principal Component Analysis (PCA) on the cone spectral sensitivities. PCA transforms the signals to an orthogonal basis that corresponds to the eigenvectors of the covariance matrix, resulting in a diagonal covariance matrix, i.e., zero correlations, in the transformed signals. The transformation is a rotation in signal space such that the directions of largest signal variance align with the coordinate axes. Decorrelating human cone spectral sensitivities in this way yielded one achromatic channel that summed all receptor signals with same sign, and two cone-opponent channels that combined cone signals with different signs, similar to the opponent channels in retina and LGN [12]. Considering the high amount of overlap of the cone spectral sensitivities, it is a plausible result that removing the strong correlations resulting from this overlap will yield an efficient representation. On the other hand, opponency is an inevitable result of orthogonal decorrelation as used by Buchsbaum and Gottschalk. However, further studies have confirmed these results without the constraint of orthogonality, and in addition have revealed the role of opponency for the encoding of spectral information in the receptor signals (see below).

Model calculations based on cone-opponent channels have successfully predicted spectral sensitivity, indicating that the limits to perception are set by the early stages of visual processing [13, 14]. Moreover, opponency predicts spectral sensitivity, wavelength discrimination, and choice behavior not only for trichromatic and dichromatic humans, but also for many other cases of dichromats, trichromats, and also tetrachromats [15–19]. This suggests that opponency is a fundamental principle underlying postreceptoral processing in animals capable of color vision.

9.2.2.2 Opponent Receptive Fields

Designing opponent channels from a retinal mosaic where each receptor has a unique position necessarily involves a spatial component. To derive predictions for the organization of postreceptoral processing, Atick and colleagues applied the concept of efficient coding to the spatio-chromatic organization of opponent receptive fields. The principle of reducing signal redundancy while at the same time minimizing noise led to a combination of low-pass filtering to reduce noise with subsequent decorrelation by flattening the frequency spectrum of natural scenes [20]. The corresponding spatio-chromatic filters for a retina of receptors with strongly overlapping spectral sensitivities, as in humans, showed spatially antagonistic center–surround organization with different weights for different photoreceptor types, implying cone-opponent receptive fields. Interestingly, for cone spectral sensitivities with less overlap the model predicted double-opponent receptive fields with center–surround organization of opposite signs for different photoreceptor types [21], as has been found in the goldfish retina [22].

9.2.2.3 Receptive Field Surround: Mixed or Specific?

Opponency is a fundamental coding principle at early stages of color vision, but how is color opponency constructed? S-cone opponency is established by a specialized circuitry in the retina [23], but L–M opponency is thought to arise from the center–surround organization of the midget pathway [24]. This pathway is specialized for high spatial resolution and likely acquired chromatic diversity only through the differentiation of an ancestral photoreceptor pigment into L and M cone pigments [25] (see Chaps. 1 and 11). The receptive field centers of foveal midget cells consist of single cones [26], and due to this exclusivity, introducing cones with different spectral sensitivities leads immediately to midget cells with different chromatic selectivities.

It has been a long-standing question whether there is cone-type specific wiring for chromatic signals in the midget pathway, and in particular, whether cone opponency is generated by cone-type specificity in the receptive field surround (see Chaps. 3 and 4). Interestingly, cone-type specificity of the receptive field surround is not necessary for opponency. Model calculations of retinal circuitry have shown that even when the surround is constructed from a random cone mosaic without cone-specific wiring, spectral opponency results from the purely spatial center–surround receptive field organization [27, 28]. This insight appears to be in line with the lack of conclusive evidence for differences between L and M cones other than their photopigments. However, physiological studies have found some degree of cone-type selectivity in the surround of midget retinal ganglion cells [29]. Lacking an anatomical basis for this specificity, it has been proposed that plasticity might underlie the chromatic receptive field organization of midget cells [25, 26, 30]. This view is supported by modeling studies that followed in spirit the work by Atick and colleagues considered above, but used spatio-chromatic data of natural scenes instead of making assumptions about their statistics. Decorrelation of cone mosaic responses to natural chromatic stimuli resulted in center–surround receptive fields with single-cone centers and mixed surrounds. The cone weights in the surround were biased towards the type of the center cone [11, 31]. This implied lower chromatic opponency as compared to a receptive field organization with pure surround or even unbiased mixed surround and may reflect the combined spatial and chromatic correlations in natural scenes [11], indicating that a mixed surround with functional bias is an efficient postreceptoral processing mechanism for natural spatio-chromatic signals.

9.2.2.4 Learning Cone Type Specificity

The finding that cone-type specific wiring of receptive fields can be learned without prior information about the type of each cone is remarkable because the high overlap of the spectral sensitivities and the mosaic arrangement, together with the spatial and chromatic correlation structure in natural scenes, strongly confound spatial and spectral variation in cone responses [11]. As a result, L and M cone responses are statistically so similar that they are indistinguishable for purely covariance-based methods like PCA [11, 31]. However, unsupervised methods that take into account higher-order statistics (see below) have yielded cone-type-specific receptive fields, showing that there is sufficient information in the cone responses to learn meaningful connectivity for color vision [11]. This in turn has implications for the ability of the visual system to organize during development and is particularly relevant if one considers the situation of the evolutionary step from dichromacy to trichromacy in primates, where it is assumed that a duplication of the X-chromosome opsin gene led to the expression of a third opsin in cone photoreceptors [25] (see Chap. 1). The ability to achieve in an unsupervised way a cone-type specific neural wiring that exploits the trichromatic retinal organization implies that such a gene duplication could potentially yield immediate benefits to the organism [11]. In this context, it is interesting that these models failed to predict consistent cone-type specificity for a fourth cone type in simulations of a tetrachromatic retina. Regarding the search for perceptual consequences of tetrachromacy in heterozygous carriers of color deficiencies [32], the conclusion of this finding was that if such a perceptual difference should be observed it would be in cases where the fourth cone opsin is maximally separated from the L and M opsins [11]. Such an observation has been made in a later experimental study [33].

The ability to determine the cone types from cone mosaic responses to natural scenes was confirmed using a different method [34]. The authors employed a multi-stage approach including multidimensional scaling and clustering, analyzing the full covariance matrix of cone responses based on a large amount of samples. With this method, the authors were able to achieve a perfect categorization of cones for the trichromatic case and a high degree of separation even for the tetrachromatic case.

9.2.2.5 Optimal Codes for Color

Data on the spectral content of natural scenes, notably multispectral images where for each pixel in an image the corresponding radiance or reflectance spectrum is available [35, 36] have enabled quantitative estimates of photoreceptor activations under natural conditions and thus have had strong impact on our understanding of the relation of color processing in the visual system to the properties of the natural environment. Párraga et al. [35] collected multispectral images of natural scenes from which they derived images of cone responses and analyzed their spatial frequency content. Somewhat surprisingly, in view of the common notion that color vision is biased towards low spatial frequencies, this study revealed that there is substantial high-frequency content in the L–M chromatic channel in response to natural scenes. Ruderman et al. [36] collected a dataset of reflectance images from forest scenes. They applied PCA to small patches of cone response images and obtained spatio-chromatic components with clearly separated chromatic dimensions, one non-opponent achromatic channel and two chromatic channels with opponent S–LM and L–M signals, respectively. This confirmed the results of Buchsbaum and Gottschalk [12], which had been based solely on cone spectral sensitivities, for the more realistic case of natural receptor stimulation. The study revealed two important aspects about the spatio-chromatic structure of natural scenes. First, the same spatial encoding patterns occurred in each of the three chromatic channels, indicating that spatial and chromatic information were separable in natural scenes. Second, the results showed that in natural scenes, variation along the opponent chromatic dimensions is uncorrelated, but not independent. For example, at object borders changes occur concomitantly in all channels, however their directions and amounts do not predict each other. This was an indication of higher-order statistical dependencies in the spatio-chromatic signals in natural scenes.

While PCA and related decorrelation methods are based on second-order correlations, methods like Independent Component Analysis (ICA) also take into account higher-order statistics [37]. The method derives from concepts of blind source separation and relates to information-theoretical concepts of efficient coding [5, 38]. The goal of ICA is to find a transformation such that the signals are not only uncorrelated but statistically independent. This includes taking into account higher-order statistics, exploiting the non-Gaussian properties of the signals. One approach to achieve this is to find the linear transformation that, together with a histogram equalization, maximizes the mutual information between input and output [38, 39]. Importantly, unlike PCA, ICA is not constrained to find an orthogonal solution.

Using methods for blind source separation in the context of sensory coding is motivated by the assumption that natural sensory signals are composed of independent sources, and that having filters representing these causes would be an efficient way of encoding the sensory signals [37]. For achromatic images of natural scenes, ICA yielded filters with receptive fields resembling those of simple cells in primary visual cortex [37], suggesting that the concept of efficiency can reveal insights into coding in the early visual system. Applied to natural spectra, ICA separated the spectra into components that resembled natural illumination and reflectance spectra [40, 41].

To investigate efficient spatio-chromatic codes for natural images, several studies, independently, but around the same time, analyzed natural color images using ICA [40, 42, 43]. In terms of the spatio-chromatic properties the results were largely similar, but in details the results varied, presumably because different kinds of image data were used. Some studies used RGB images [42] or even compressed JPEG images [43], while others used estimates of LMS cone signals derived from spectral image data [40, 41]. One consistent finding was a separation of chromatic information into three distinct channels, like in the study of Ruderman et al. [36], but in contrast to the PCA results the spatial structure revealed by ICA was different: Similar to the case of achromatic images, ICA showed mostly units with localized, oriented receptive fields that could be interpreted as chromatic edge detectors.

Studies that used cone response estimates based on the human cone spectral sensitivities [40, 41, 44, 45] enabled comparison of the model results to physiology. One interesting finding was that not all of the three opponency channels derived from natural scenes matched the cardinal axes of cone-opponent color space [41]. Two channels corresponded to the achromatic and S–LM cone axes, respectively, but the third channel represented a direction in color space intermediate to the S–LM and L–M opponency axes [41] (see also Chaps. 5, and 7 about issues related to cardinal axes). This oblique direction was a reflection of systematic statistical dependencies between cone-opponent signals under natural stimulation [46]. An implication of this finding is that the cardinal opponency axes might not be optimal in the sense of achieving an independent code, which seems to be supported by the absence of specificity for these axes in primate visual cortex [47–49].

Modeling receptor responses to natural scenes have also shed more light on the question of what underlies opponency as an efficient encoding. Buchsbaum and Gottschalk [12] had derived opponent codes by considering solely the correlations in cone signals introduced by the overlap of the cone spectral sensitivities. Similar results had been obtained by decorrelating cone responses to Munsell spectra by anti-Hebbian learning [50]. Using hyperspectral images of natural scenes, Lee et al. [44] in addition took into account the properties of natural spectra, which are known to be broadband and thus would imply spectral opponency to remove the correlations between wavelength bands. Indeed, it turned out that even when, instead of the human cone spectral sensitivities, hypothetical cone spectral sensitivities without overlap were simulated, opponency resulted as efficient code for the corresponding receptor signals. This result complemented the earlier findings and showed that opponency is efficient to encode chromatic information in natural scenes regardless of the spectral sensitivities of the photoreceptors.

These analyses were based on information-theoretical concepts and efficiency was defined in terms of information maximization. A related, but different rationale for opponency was given by von der Twer and MacLeod [51], based not on an information criterion but on minimizing the mean squared error in the presence of additive noise. The optimal nonlinear receptor response function derived for this case was predicted to be similar, though not identical, to the sigmoidal histogram equalization function achieving maximal output entropy [38]. Subsequently the authors elegantly showed that in the case of additive Gaussian noise, if more than one neuron is available to encode the signal, splitting the signal range by establishing an opponent code leads to better discriminability than having each neuron encode the entire signal range.

9.3 Cortical Processing and Perception

9.3.1 Spatio-chromatic Receptive Fields, Double Opponency

Do the spatio-chromatic receptive field properties predicted from efficient coding principles, as considered in Sect. 9.2.2. fit the physiology? Most chromatic units derived from Independent Component Analysis of cone response estimates to natural scenes had localized, oriented receptive fields resembling color edge detectors, which is in line with combined color and orientation selectivity of cells in primary visual cortex [52–56]. Indeed, the thorough analysis by Caywood et al. [45] revealed quantitative similarities between ICA filters and V1 spatio-chromatic tuning properties.

Considering the receptive field structure, the subfields of the ICA edge detectors encoded chromaticities in opposite directions in color space, indicating opponency. With coefficients of different signs, the ICA basis functions encoded similar features as orientation-selective double-opponent cells [52, 57]. However, other than double-opponent cells, regions of positive and negative weights for different cone types tended to be out of phase, which points more to single opponency with anisotropic receptive fields [44] (but see Ref. [45]). However, regarding the significance of such results, it should be kept in mind that, while ICA is less restricted than PCA, it nevertheless has certain constraints and assumptions, in particular the assumption of a linear decomposition, that may influence the results. Thus, overall, the predictions of efficient coding models regarding the distinction of single and double opponent cells so far are inconclusive.

Double opponent cells, i.e., cells showing spatially opponent receptive field organization with chromatically opponent subfields [58, 59], have been of interest in color vision research for a long time, and have been intensively studied in recent years [52, 57, 60, 61], see also Chap. 7. Interestingly though, there have not been many studies to investigate what information double-opponent responses provide under natural conditions, and their functional relevance has not been conclusively established. Double-opponent cells have been discussed to contribute to color contrast perception and color constancy [61], or, due to their sensitivity to chromatic context, to encode hue [56]. A model by Spitzer and Barkan [62] showed that double-opponent cells could have a role in color contrast induction. Rubin and Richards [63] presented a model that assigned a very specific function to double-opponent cells, namely the signaling of object borders. The key assumption behind this model was that a chromatic edge with a change in spectral slope, which doubleopponent cells would be suitable to detect, would be an indication of a change in reflectance as opposed to a change in illumination occurring at a shadow border. A later model proposed that for this task a combination of single and double opponent cells would be more appropriate [64], which would be consistent with studies reporting both types of cells in visual cortex [57].

9.3.2 *Color tuning and population coding*

While the spatio-chromatic receptive fields predicted by analysis of natural scenes seemed to be biased towards three color space axes, the analysis by Caywood et al [45] has shown that if the color tuning of these model receptive fields is tested using similar methods as in the physiological studies, the distribution of chromatic preferences is quite similar to color tuning in primary visual cortex. In visual cortex, color preferences do not strongly cluster around the cardinal axes of cone opponency, but are distributed continuously [47–49], and many neurons show a color tuning that results from a nonlinear combination of cone signals [48, 49, 65]. Recent results indicate that such a distributed representation emerges from an efficient coding objective if the center-surround receptive field processing and the nonlinear split in ON and OFF channels in the retina are taken into account [66]. Thus, coding of color in visual cortex seems largely consistent with the notion that sensory processing in the visual system achieves an efficient representation.

The finding of distributed chromatic preferences in cortical neurons point towards a representation of color by a population code, that is, a representation where neurons have relatively broad and overlapping selectivities, such that a given value is encoded in the pattern of responses in a population of neurons. Population coding, also called ensemble coding or distributed coding, is ubiquitous in the nervous system, and models based on population coding of visual stimuli reflect the common notion about the coding of visual features in early visual cortex [67, 68]. Population coding enables robust representation of stimulus features and the implementation of higher-level mechanisms that enhance coding efficiency (see below). However, so far this type of coding has rarely been considered in the context of color representation in early visual cortex [61] (see Chap. 7).

9.3.3 Adaptation and Induction

A population coding scheme is particular suitable for neural implementation of modulations of processing like those that occur from spatial or temporal context, in adaptation and induction [69]. Modulation of the gain of neurons in a population code has been shown to explain perceptual effects of color adaptation, such as shifts in color appearance or changes of perceived contrast [70–73]. Color tuning curves in primary visual cortex are consistent with a population code for color, and analysis of physiological data on chromatic tuning in primary visual cortex using a population coding model has reproduced effects that were consistent with color induction [49]. These results have been obtained from the entire population responsive to the stimuli without any selection based on receptive field properties, and in particular included both opponent and non-opponent neurons. Population coding models that have been proposed for the neural mechanisms underlying perceptual effects of lateral interactions in orientation vision [74–76] similarly appear suitable to describe the effects of color induction [77, 78]. However, quantitative modeling of color induction effects using a population coding model has not been performed.

Induction has been investigated quantitatively by modeling interactions of cone responses, see also Chap. 6. With a set of 16 equally detectable, isoluminant stimuli [79] the chromatic induction of a small white test-field surrounded by a large chromatic inducer was investigated [80]. Three different approaches with observers viewing the stimulus binocularly were applied to measure the shift in color of the test field evoked by the 16 surrounds. The experimental results were consistent with the hypothesis that the shift in color of the test field is due to a postreceptoral, opponent process. The contributions of the respective cone contrasts are known quantitatively and are reliably predicted by a model [81]. Validity of this mechanism is restricted to the isoluminant plane and cone contrast amplitudes of inducing colors up to fivefold detection threshold. When the small white test field was presented on one eye and the surround colors to the other eye the test field color was shifted in a similar way [81].

9.3.4 Color Constancy

Color constancy is an important perceptual ability in humans and other species [82–85]. It is one of the most prominent topics in color vision [86] and has been the target of many modeling studies. Color constancy refers to the ability of assigning color descriptors—or percepts—to object surfaces in a visual scene such that these descriptors are invariant under changes in the spectral composition of the illumination (see Chap. 6). This amounts to recovering some information about the reflectance properties of the surfaces from the reflected light. Achieving color constancy poses a problem because, if the illuminant is neither known nor directly measurable, determining the surface reflectances unambiguously is not possible. The spectrum of light reflected from a surface is given by the product of the illuminant spectrum and the reflectance spectrum of the surface, so an observed spectrum of reflected light can arise from an arbitrary number of combinations of illuminants and reflectances. The problem becomes even more difficult by the visual system's limited ability to encode spectral variation, which in case of trichromatic humans with three types of cone photoreceptors is limited to three degrees of freedom. Solving this underdetermined problem requires additional information to constrain the space of possible solutions. Models of color constancy have proposed different kinds of constraints, and we will consider some of them below.

All models necessarily make simplifying assumptions about the conditions under which the color constancy problem has to be solved. Common assumptions are that surfaces are flat and diffusely reflecting, and that there is only a single illuminant which illuminates the scene uniformly or at most varies smoothly across the scene. This so-called Mondrian world resembles the simplified situation often found in experimental studies on color constancy, but a number of phenomena that occur under natural conditions, like specular highlights, mutual illumination, multiple illuminants in varying mixtures, shadows, are not taken into account under these conditions. While the simplifying assumptions reduce the mathematical complexity, the price paid is that some cues are neglected that, under realistic conditions, might be helpful in solving the problem.

Modeling studies have approached color constancy from different directions. A number of studies have focused on solving the color constancy problem exactly, analyzing the problem mathematically and determining under which conditions a solution would be possible and how such a solution could be calculated. Those models typically gave little regard to possible implementations in the visual system. Another direction was to propose algorithms or processing mechanisms, motivated more or less strongly by processing in the visual system, and to investigate their ability to achieve color constancy. Some of these models were specifically designed to model neural processing in the visual system.

9.3.4.1 Mathematical Solutions: Linear Basis Models

Under the simple Mondrian-world conditions the mathematical description of the reflection of illuminant light at object surfaces and the subsequent encoding by the photoreceptors is given by a simple integral equation [87]. The problem can be further simplified by assuming that the shapes of both illuminant and reflectance spectra are not arbitrary, but that their variation is restricted to a relatively low number of degrees of freedom, and that they can be described as linear combinations of correspondingly low numbers of basis functions [88, 89]. This turns the initially infinite-dimensional problem into a discrete set of linear equations with limited dimensionality, which greatly facilitates the analysis. The linear basis approach has been used by a large number of studies [90–97]. It puts the problem in a form that is mathematically simple, but still the problem is underdetermined. In the simplest case, the unknowns correspond to one reflectance spectrum for each surface in the scene plus one illumination spectrum, while the available information corresponds to one observed spectrum from each surface in the scene. To account for the missing information, further constraints are needed.

Spectral Constraints

Buchsbaum [90] presented a model to recover reflectances under the constraint that the average of the reflectance spectra in the scene is known. This is a slightly generalized version of the "gray world" assumption. Buchsbaum further assumed that the subspaces of illuminant and reflectance spectra each are three-dimensional, which ensured that sampling with three photoreceptor types yielded sufficient information to solve the problem. The model consisted of two processing stages. As a first step, an estimate of the illumination spectrum was derived from comparing the average of the observed receptor signals to the known average. This illumination spectrum was then used to correct the observed surface spectra, yielding the actual reflectance spectra. Buchsbaum's approach to solving the linear system by deriving an explicit estimate of the illuminant and using that to calculate the reflectances has been prototypical for a large number of models. The models of this type differed mainly in how they constrained the mathematical problem.

Dimensionality Constraints

Maloney and Wandell [91] explored the possibility that the dimensionalities of illuminations and reflectances together would be low enough for the set of equations to be solved exactly. They showed that, given the number of degrees of freedom of surface reflectances is lower than the number of photoreceptor types, the illuminant can be estimated unambiguously if the number of different reflectances in the scene is sufficiently large. A similar study was conducted at the same time by Yuille [92], with essentially the same conclusions. An obvious problem with this constraint on dimensionality is that the small number of photoreceptor types poses severe limits on the spectral variation that could be accounted for. For a trichromatic visual system, the number of degrees of freedom of surface reflectances could not be more than two. Even if this number would be increased to three, by allowing for rods to contribute to color vision [92], or for multiple observations under different illuminants [96], it seems hardly plausible that this would be sufficient for color constancy under natural conditions. From an analysis of natural reflectance spectra, supported by considerations on the physical processes underlying light reflectance, Maloney [98] concluded that the dimensionality of natural reflectance spectra is five to seven.

Based on psychophysical experiments using hyperspectral images of natural scenes approximated by different numbers of principal components, Nascimento and colleagues [99] argued that this number may be even as high as eight.

General Approaches

Regularization theory provides a general framework specifically to solve underdetermined problems [100]. The approach is to introduce additional terms to a cost function that constrain the solution. For the color constancy problem, the treatment as an optimization problem also avoids the two-step process. Rubner and Schulten [101] used the linear basis model with three additional terms, one term penalizing the deviation of the illuminant from a neutral daylight, one term penalizing the deviation of the average of the reflectances from neutral gray, and one term minimizing the spatial gradient of the illuminant. Similarly, in the model by Takebe et al. [102] a cost function that included constraints on reflectances (gray world) and illumination (neutral illuminant with minimal energy) was minimized by a network of modules implementing the corresponding terms. An advantage of the regularization approach is that different constraints can be combined with the possibility to control their relative importance by giving different weights to the regularizing terms. Therefore, such models could be used in combination with psychophysical experiments to investigate which constraints are relevant for human color constancy [101]. However, such studies have not been performed so far.

Color constancy amounts to an estimation problem, which makes the Bayesian framework for optimal inference [103] an attractive approach. To apply the Bayesian method to the problem of color constancy, Brainard and Freeman [97] started from the linear basis model with three degrees of freedom each for reflectances and illuminant. In this case, applying the Bayesian method to achieve color constancy means to determine the probabilities of all possible combinations of illuminants and reflectances that could have caused the observations, and to determine the one with the highest probability. Those (posterior) probabilities need to be calculated based on the conditional probabilities (likelihood) of each possible observation given luminances and reflectances, and the (prior) probabilities of occurrences of illuminants and reflectances. In a Mondrian world, the observed spectra result in a deterministic way from the combinations of luminances and reflectances. Therefore, the probabilistic component arises solely from the uncertainties about illuminants and reflectances, that is, the priors. Thus, the difference between the Bayesian method used by Brainard and Freeman and the traditional linear basis models was that the constraints on reflectances and illuminants were expressed not just in terms of their averages, but in a more general way by their entire probability distributions.

In comparison with other models, including a gray-world model according to Buchsbaum's model [90], the estimates by the Bayesian method were closest to the actual illuminant, although the performance of the gray-world model was almost as good. When scenes with sets of reflectances were used that were chromatically biased, both models yielded systematic deviations from the correct illuminant, which were slightly smaller for the Bayesian model. That both models suffered from a similar inability to account for biases in the reflectance spectra is not so surprising, since both relied on the assumption of unbiased ensembles of reflectances. However, the Bayesian model additionally incorporated assumptions about the illuminant spectra in the form of prior probabilities, which led to a tendency of the illuminant estimate towards the center of the illuminant distribution and thus reduced the effect of the bias in the reflectances. It is not clear whether the gray world model would have performed similarly if such an additional constraint (see for example [102]) would have been included. Therefore, it is hard to say how much improvement in color constancy could be gained by taking into account entire probability distributions in a Bayesian way instead of just making assumptions about averages.

Brainard et al. [104] performed an interesting follow-up study, in which the predictions of their Bayesian model were compared with psychophysical data from human subjects. The results of the Bayesian algorithm showed some similarity to the psychophysical data, but systematic differences were apparent. Two modifications of the model led to a better match to the psychophysical data. One modification was to make the distribution for the illumination prior much broader than the originally assumed distribution, which was close to that of natural daylights. The second modification was to restrict the surfaces taken into account in the calculations to a limited viewing region around the test field. For the first manipulation, the Bayesian approach was particularly suitable and could potentially elucidate whether human color perception is consistent with an estimation based on prior assumptions about illuminants. The second modification concerned a generic model assumption that would affect deterministic models in the same way. It would be interesting to systematically investigate whether the chosen modifications were optimal or whether other priors and viewing regions, or other modifications of the model, might lead to similar results.

In principle the Bayesian approach provides a framework for taking into account all sorts of constraints and assumptions, but those must be known and defined in terms of probability distributions. The simplistic conditions that have been considered in computational models of color constancy so far almost certainly do not represent the variety of cues the visual system is using, and in this case the benefits of a Bayesian model seems limited. It is also unlikely that experiments under correspondingly simple conditions would be informative about whether humans use Bayesian strategies in cases like color constancy. Maybe Bayesian modeling would be able to play out its full potential when applied under more complex conditions, for example in the case of a more comprehensive model of cue combination [105, 106].

Other constraints to the mathematical problem of color constancy have been proposed that apply to more realistic situations than the Mondrian world. In particular, solutions based on specular highlights [94, 107] and mutual illumination [108] have been considered. Both effects create systematically different mixtures of reflectance and illuminations spectra in a scene, which provides information to infer the illuminant. Golz and MacLeod [109] proposed another cue to estimate the chromaticity of the illuminant, which exists even in the Mondrian world. It had previously been suggested that color perception is affected not only by the mean chromaticity of a scene, but also by second-order statistics [110], and in particular that the amount of variance might be a cue to a chromatically biased illuminant [111]. Golz and MacLeod noted that a chromatic illuminant has the consequence that surfaces with a color similar to the illuminant tend to have a higher luminance. This correlation between luminance and surface color depends on the illuminant color—"when the lighting gets red, the reds get lighter" [112]—and therefore can be used to distinguish whether a chromatic bias in a scene is due to a bias in the reflectances or a chromatic illuminant. Golz and MacLeod showed that the predictions from such a model closely match psychophysical results on color appearance. However, although analysis of natural scenes suggests that such higher-order dependencies could be used to infer the color of the illuminant [109], experimental evidence is inconclusive about whether this kind of information is actually used for color constancy [113, 114].

Under natural conditions, a number of cues to the illuminant are available, and it is fair to assume that the visual system will use most of them to achieve color constancy. Thus, a realistic model of human color vision would have to account for a process of cue combination [105, 115, 116].

9.3.4.2 Processing Mechanisms

While the goal of the linear basis models was to find an exact solution to the mathematical problem, other models focused on processing mechanisms. Those models typically did not involve an explicit estimate of the illuminant. Modeling approaches of this kind include the lightness algorithms, of which Land's retinex algorithm [117] is probably the most famous, and network models of neural processing in the visual system.

Lightness Algorithms

Lightness algorithms try to achieve color constancy by estimating surface reflectance—lightness—in different spectral channels. The channels are treated as being independent, and lightness estimates are achieved by some kind of normalization process taking into account the signals within the respective spectral channel. The basic idea behind lightness algorithms can be related to the physiologically motivated adaptation rule of von Kries [118, 119]. Hurlbert [87, 93] provides comprehensive reviews of these models with analyses on their mutual relationships as well as their relationships to other models.

Lightness algorithms arise from a different approach than the linear basis models discussed above, but the assumptions and constraints they incorporate, often implicitly, are similar. In particular, they assume a Mondrian world with flat, matte surfaces and illumination that is uniform or has a shallow gradient. Lightness algorithms typically include a spatial processing component to disentangle spatial changes in reflectance from changes in illumination. This involves suppression of small gradients, so that the weak variation due to illumination is discarded. Reintegration recovers variation due to surface reflectances without the variation due to the illuminant. These signals then are normalized for each spectral channel to derive lightness values. The normalization used for a given lightness algorithm imposes a constraint that reflects an implicit assumption about the set of reflectances. Depending on the specific kind of normalization, this can be similar to the assumption that the average reflectance in each channel is the same ("gray world" assumption), or that the highest value in each channel is close to a perfect reflectance ("brightest is white" assumption) [87]. The main differences between the various lightness algorithms [93, 117, 120–123] lie in the exact form of the processing steps and how they are implemented. For example, in the early retinex algorithms the gradient calculation and thresholding was done by summing differences along a random path of discrete steps across the scene. Other algorithms performed the calculations in the continuous case using the gradient or Laplacian operators and spatial integration across the scene.

A special approach was taken by Hurlbert and Poggio [124]. Instead of defining an algorithm, they determined the (linear) transformation that would best approximate the desired input–output mapping as defined by a large number of examples, using optimal linear estimation. Simulated Mondrians under illuminations with linear gradients were used as inputs, and the outputs were the respective reflectances of the Mondrians. The solution was found by inverting the system of linear equations. The interesting outcome was that the resulting linear operator showed a center–surround structure with a sharp positive peak and very broad negative surround—the same kind of spatial processing that had been assumed by other lightness algorithms [93, 121, 123].

Although they represent quite different approaches, the results of linear basis models and lightness algorithms have been similar. Lightness algorithms tend to yield color constancy under conditions under which also the basis algorithms lead to good color constancy [87]. Implicitly they rely on a kind of "gray world" assumption due to the normalization of signals in each spectral channel. Lightness algorithms are typically designed to be insensitive to shallow illumination gradients, but are also misled by biased distributions of reflectances [87]. However, without the requirement of a strict mathematical solution as in linear basis models, lightness algorithms may be successful in maintaining approximately invariant surface descriptors in a given scene under changing illuminant—which is the minimal requirement for color constancy—although they may assign different descriptors to the same surface in different scenes [91, 95]. In contrast to linear basis models, lightness algorithms do not derive an explicit estimate of the illuminant. They achieve color constancy by transforming the local receptor signals in relation to some global or semi-global measure of the receptor activity in each receptor type. The result is an encoding of the relative variation in receptor signals. In this sense, lightness algorithms are related to the notion of relational color constancy [125], a concept based on the finding that cone ratios are almost invariant under changes of natural illuminations.

Multi-stage Network Models

Attempts to uncover the neural mechanisms playing a role in human color constancy have included models that incorporate aspects of the multi-stage visual processing in the human visual system. Dufort and Lumsden [126] constructed a network model for color in area V4 from basic opponent cell types by defining the circuitry such that the cells would have desired properties to encode basic color categories. Color constant cells were obtained when the connections were chosen such that a cell received local excitatory input and inhibitory input from a wide region, which essentially resulted in double-opponent type receptive fields.

The study by Courtney et al. [127] focused on the multi-stage aspect, with a network model that comprised a hierarchy of levels of visual processing, including cone adaptation, opponent neurons with different receptive field structures and sizes, and an antagonistic circuitry at the highest stage in the model hierarchy, which was associated with area V4. The network demonstrated moderate levels of color constancy and showed some sensitivity to the specific spatio-chromatic properties of the stimulus, similar to human color perception. For example, effects of stimulus geometry on color judgments in humans, like changing the size of the surround or the size of a gap between stimulus and surround, could be reproduced by the model. The main result of the study was that, when either the adaptation stage or the V4 stage were removed, the remaining network still showed some color constancy. The authors concluded that both kinds of processing, receptor adaptation and antagonistic lateral interactions, could contribute to color constancy, but both together would achieve the goal more reliably.

The model by Wray and Edelman [128] had a multi-stage architecture as well, but modeled a wider hierarchy of cortical visual areas. The main emphasis was on intra- and inter-areal recurrent connections, which were assumed to produce modulatory lateral interactions, so-called non-classical receptive field surrounds, and antagonistic feedback. Like the model by Courtney et al, the network showed results that qualitatively matched the psychophysical effects of stimulus-surround geometry on human color constancy. Testing the role of feedback by removing different kinds of connections in the model, the authors reported that within-area long-range horizontal connections had strong effects on color constancy, while lesioning inter-areal feedback, which was modeled with little color specificity, had only a mild effect.

The results of these network studies have largely confirmed the implications of the studies on lightness algorithms. A key element to achieve color constancy is suppression or normalization of local signals by an antagonistic mechanism with a large integration range. Variants of implementations of this fundamental processing scheme produce variants in results, but the relevance of these specifics for our understanding of the visual system remains unclear. In the visual system, lateral interactions supporting color constancy mechanisms happen probably at various stages and on different spatial and temporal scales. Photoreceptor adaptation may be a component, as well as receptive fields with spatio-chromatic opponency or large silent surrounds, which could be mediated by long-range interactions within a cortical area, or by feedback between areas in the cortical hierarchy. The models suggest that possibly a variety of such neural mechanisms could contribute to color constancy.

9.3.4.3 Conclusions

Modeling studies of color constancy have provided insights into the problem the visual system is facing in estimating object surface properties based on limited sensory information, and the possible cues and mechanisms the visual system might use to solve the problem. Mathematical models have determined the information carried by different cues. Algorithmic approaches have investigated fundamental mechanisms to approximately achieve color constancy. Interestingly, the statistical modeling and machine learning approaches, together with quantitative analysis of the natural sensory conditions, which have been successful in providing insights in other aspects of color vision, so far have not much been applied to the problem of color constancy.

A vast variety of different models, implying different ways of processing, seem to be able to qualitatively explain color constancy. However, the notorious variability of color constancy when tested in experiments and the strong dependence of the results on the particular experimental or simulated conditions have so far prevented relating the models more closely to experimental results. In the Mondrian world the complexity and dimensionality of the problem is so much reduced that, both in experiments and in models, due to the limited degrees of freedom it is difficult to distinguish between alternative mechanisms that could lead to the same results. This makes it hard to identify the mechanisms of color constant perception in humans. Understanding which mechanisms play a role in the human visual system will require models that reach beyond the Mondrian world, and experiments specifically designed to test particular hypotheses.

9.4 Color Categories

One of the most intriguing questions in color vision is why and how, in a continuous color space, human observers consistently identify certain regions of different color categories [129]. It is debated to which extent cultural factors play a role in categorical color judgment, but it is hard to dispute that the four so-called unique hues—pure red, green, blue, yellow—have a special status in that they appear unmixed with other hues [130, 131] (but see Ref. [132] for a competing view). For example, in experimental settings color-normal trichromatic observers are typically able to adjust a stimulus such that it elicits the percept of one of these unique hues without any contribution of any of the other hues [130], and it is reasonable to assume that these perceptual points have a specific neural correspondence. The unique hues represent perceptually opponent pairs, but it is evident that these hues do not coincide with the cardinal axes of opponency established at precortical stages [131, 133–135]. To account for this discrepancy, De Valois and De Valois [136] proposed a model with an additional processing stage after the opponent combination of receptor signals. This stage introduced a rotation to align the coordinate axes with the axes of perceptual opponency. However, the difference between cardinal axes and perceptual opponency is not a simple rotation [131]. So far, there is no conclusive understanding of why we have these particular color categories. In general, there is no compelling reason for having one set of categories at certain locations in color space as opposed to another, or having categories at all [137]. Categorical coding requires nonlinear processing, but there are fairly simple, neurobiologically plausible, nonlinear mechanisms that would likely be sufficient [138, 139]. Lehky and Sejnowski [138] trained a four-layer backpropagation network receiving input from three cone types with human spectral sensitivities to learn color selectivities for various regions in color space. This shows that with a multi-stage architecture any kind of mapping from the space of receptor signals to some categorical color representation would be possible to achieve for the visual system.

Since color categories cannot be explained by neural coding properties, a possibility could be that properties of the sensory environment underlie the emergence of color categories. Shepard [129] suggested that the structure of our color percepts, including color categories, is a result of an adaptation to the visual environment. If this is the case, it should be possible to deduce color categories from the statistics of natural spectral stimuli. Some indication towards such a relation comes from analyses of color images using a clustering technique, which resulted in a number of clusters that were close to the foci of basic color terms [140].

A quite different explanation for how basic color categories might derive from the properties of the natural visual world was proposed by Philipona and O'Regan who tested the hypothesis that the visual system would preferentially encode those surface colors that under changes in illumination yielded invariant stimulation of a subset of photoreceptor types. For each of a set of surfaces they determined the transformation describing the variation in photoreceptor responses under different illuminants, and determined the surface spectra for which the transformation was close to singular in the sense of leaving a subset of the photoreceptor responses invariant. Those surface colors turned out to lie close to the four focal colors red, green, blue, and yellow [141]. This remarkable match is suggestive of a relation of categorical perception to sensory invariances and supports the notion that perceptual organization is adapted to the structure of the sensory environment [4, 142]. A caveat is however that for the analysis the authors had used Munsell reflectance spectra [143] or natural spectra with artificially enhanced saturation. The use of artificially constructed stimuli like the Munsell colors, which have also been used in other models on color categories [144], always bears the danger of circularity. In the analysis by Philipona and O'Regan, the identified singularities corresponded to the stimuli with highest saturation [145], so further investigations would be necessary to conclusively determine what was responsible for the results.

9.5 Future Perspectives

Modeling approaches have contributed to our understanding of the neural principles and mechanisms of color vision. We have reviewed some of the computational work on the various aspects of processing and perception of color, notably studies on efficient coding and on color constancy. Interestingly, these two aspects do not seem to have been considered together. Finding out whether color constancy could be understood as a result of an adaptation to properties of the visual environment might be a rewarding challenge for the future. This approach may also help to gain better insights into the neural mechanisms of coding and perception of color. A complementary contribution could come from the development of advanced experimental paradigms and models that go beyond the simplistic Mondrian world and account for more complex, realistic conditions. Computational studies will certainly continue to play an important role in providing unifying concepts for the variety of empirical findings, and in turn will motivate informative experimental studies, but achieving fruitful interactions of modeling and experiments in this field will remain a challenge.

References

1. Attneave F. Some informational aspects of visual perception. Psychol Rev. 1954;61(3):183.
2. Barlow HB. Possible principles underlying the transformations of sensory messages. In: Rosenblith WA, editor. Sensory communication. Cambridge, MA: MIT press; 1961. p. 217–34.
3. Marr D. Vision. San Francisco: W.H. Freeman and Company; 1982.
4. Barlow H. The exploitation of regularities in the environment by the brain. Behav Brain Sci. 2001;24(04):602–7.
5. Barlow H. Redundancy reduction revisited. Netw Comput Neural Syst. 2001;12(3):241–53.
6. Osorio D, Vorobyev M. Colour vision as an adaptation to frugivory in primates. Proc R Soc Lond B Biol Sci. 1996;263(1370):593–9.
7. Sumner P, Mollon J. Catarrhine photopigments are optimized for detecting targets against a foliage background. J Exp Biol. 2000;203(13):1963–86.
8. Chiao CC, Vorobyev M, Cronin TW, Osorio D. Spectral tuning of dichromats to natural scenes. Vis Res. 2000;40(23):3257–71.
9. Lewis A, Zhaoping L. Are cone sensitivities determined by natural color statistics? J Vis. 2006;6(3):8.
10. Garrigan P, Ratliff CP, Klein JM, Sterling P, Brainard DH, Balasubramanian V. Design of a trichromatic cone array. PLoS Comput Biol. 2010;6(2):e1000677.
11. Wachtler T, Doi E, Lee TW, Sejnowski TJ. Cone selectivity derived from the responses of the retinal cone mosaic to natural scenes. J Vis. 2007;7(8):6.
12. Buchsbaum G, Gottschalk A. Trichromacy, opponent colours coding and optimum colour information transmission in the retina. Proc R Soc Lond B Biol Sci. 1983;220(1218):89–113.
13. Ingling CR, Tsou BHP. Orthogonal combination of the three visual channels. Vis Res. 1977;17(9):1075–82.
14. Guth SL. Model for color vision and light adaptation. J Opt Soc Am A. 1991;8(6):976–93.
15. Vorobyev M, Osorio D. Receptor noise as a determinant of colour thresholds. Proc R Soc Lond B Biol Sci. 1998;265(1394):351–8.

16. Kelber A. Receptor based models for spontaneous colour choices in flies and butterflies. Entomol Exp Appl. 2001;99(2):231–44.
17. Kelber A, Vorobyev M, Osorio D. Animal colour vision–behavioural tests and physiological concepts. Biol Rev. 2003;78(01):81–118.
18. Koshitaka H, Kinoshita M, Vorobyev M, Arikawa K. Tetrachromacy in a butterfly that has eight varieties of spectral receptors. Proc R Soc B Biol Sci. 2008;275(1637):947–54.
19. Schnaitmann C, Garbers C, Wachtler T, Tanimoto H. Color discrimination with broadband photoreceptors. Curr Biol. 2013;23(23):2375–82.
20. Atick JJ, Redlich AN. What does the retina know about natural scenes? Neural Comput. 1992;4(2):196–210.
21. Atick JJ, Li Z, Redlich AN. Understanding retinal color coding from first principles. Neural Comput. 1992;4(4):559–72.
22. Daw NW. Colour-coded ganglion cells in the goldfish retina: extension of their receptive fields by means of new stimuli. J Physiol. 1968;197(3):567.
23. Dacey DM, Lee BB. The 'blue-on' opponent pathway in primate retina originates from a distinct bistratified ganglion cell type. Nature. 1994;367(6465):731–5.
24. Mollon JD. "Tho' she kneel'd in that place where they grew …": the uses and origins of primate colour vision. J Exp Biol. 1989;146(1):21–38.
25. Nathans J. The evolution and physiology of human color vision: insights from molecular genetic studies of visual pigments. Neuron. 1999;24(2):299–312.
26. Boycott B, Wässle H. Parallel processing in the mammalian retina: the Proctor Lecture. Invest Ophthalmol Vis Sci. 1999;40(7):1313–27.
27. Paulus W, Kröger-Paulus A. A new concept of retinal colour coding. Vis Res. 1983;23(5):529.
28. Lennie P, Haake PW, Williams DR. The design of chromatically opponent receptive fields. In: Computational models of visual processing, Vol 71. Cambridge, MA: MIT Press; 1991, p. 82.
29. Buzás P, Blessing EM, Szmajda BA, Martin PR. Specificity of M and L cone inputs to receptive fields in the parvocellular pathway: random wiring with functional bias. J Neurosci. 2006;26(43):11148–61.
30. Martin PR, Lee BB, White AJ, Solomon SG, Rüttiger L. Chromatic sensitivity of ganglion cells in the peripheral primate retina. Nature. 2001;410(6831):933–6.
31. Doi E, Inui T, Lee TW, Wachtler T, Sejnowski TJ. Spatiochromatic receptive field properties derived from information-theoretic analyses of cone mosaic responses to natural scenes. Neural Comput. 2003;15(2):397–417.
32. Jordan G, Mollon JD. A study of women heterozygous for colour deficiencies. Vis Res. 1993;33(11):1495–508.
33. Jordan G, Deeb SS, Bosten JM, Mollon JD. The dimensionality of color vision in carriers of anomalous trichromacy. J Vis. 2010;10(8):12.
34. Benson NC, Manning JR, Brainard DH. Unsupervised learning of cone spectral classes from natural images. PLoS Comput Biol. 2014;10(6):e1003652.
35. Párraga C, Brelstaff G, Troscianko T, Moorehead I. Color and luminance information in natural scenes. J Opt Soc Am A. 1998;15(3):563–9.
36. Ruderman DL, Cronin TW, Chiao CC. Statistics of cone responses to natural images: implications for visual coding. J Opt Soc Am A. 1998;15(8):2036–45.
37. Bell AJ, Sejnowski TJ. The "independent components" of natural scenes are edge filters. Vis Res. 1997;37(23):3327–38.
38. Laughlin S. A simple coding procedure enhances a neuron's information capacity. Z Naturforsch. 1981;36(9-10):910–2.
39. Bell AJ, Sejnowski TJ. An information-maximization approach to blind separation and blind deconvolution. Neural Comput. 1995;7(6):1129–59.
40. Lee TW, Wachtler T, Sejnowski TJ. The spectral independent components of natural scenes. In: Biologically motivated computer vision. Springer; 2000. p. 527–34
41. Wachtler T, Lee TW, Sejnowski TJ. Chromatic structure of natural scenes. J Opt Soc Am A. 2001;18(1):65–77.

42. Hoyer PO, Hyvärinen A. Independent component analysis applied to feature extraction from colour and stereo images. Netw Comput Neural Syst. 2000;11(3):191–210.
43. Tailor DR, Finkel LH, Buchsbaum G. Color-opponent receptive fields derived from independent component analysis of natural images. Vis Res. 2000;40(19):2671–6.
44. Lee TW, Wachtler T, Sejnowski TJ. Color opponency is an efficient representation of spectral properties in natural scenes. Vis Res. 2002;42(17):2095–103.
45. Caywood MS, Willmore B, Tolhurst DJ. Independent components of color natural scenes resemble V1 neurons in their spatial and color tuning. J Neurophysiol. 2004;91(6):2859–73.
46. Webster MA, Mollon J. Adaptation and the color statistics of natural images. Vis Res. 1997;37(23):3283–98.
47. Thorell LG, de Valois RL, Albrecht DG. Spatial mapping of monkey VI cells with pure color and luminance stimuli. Vis Res. 1984;24(7):751–69.
48. Lennie P, Krauskopf J, Sclar G. Chromatic mechanisms in striate cortex of macaque. J Neurosci. 1990;10(2):649–69.
49. Wachtler T, Sejnowski TJ, Albright TD. Representation of color stimuli in awake macaque primary visual cortex. Neuron. 2003;37(4):681–91.
50. Usui S, Nakauchi S, Miyamoto Y. A decorrelating neural network for color constancy. In: 14th Annual International Conference of the IEEE Engineering in Medicine and Biology Society; 1992. p. 1030–1.
51. von der Twer T, MacLeod DIA. Optimal nonlinear codes for the perception of natural colours. Netw Comput Neural Syst. 2001;12(3):395–407.
52. Michael CR. Color vision mechanisms in monkey striate cortex: dual-opponent cells with concentric receptive fields. J Neurophysiol. 1978;41(3):572–88.
53. Leventhal AG, Thompson KG, Liu D, Zhou Y, Ault SJ. Concomitant sensitivity to orientation, direction, and color of cells in layers 2, 3, and 4 of monkey striate cortex. J Neurosci. 1995;15(3):1808–18.
54. Friedman HS, Zhou H, Heydt R. The coding of uniform colour figures in monkey visual cortex. J Physiol. 2003;548(2):593–613.
55. Johnson EN, Hawken MJ, Shapley R. The spatial transformation of color in the primary visual cortex of the macaque monkey. Nat Neurosci. 2001;4(4):409–16.
56. Conway BR, Hubel DH, Livingstone MS. Color contrast in macaque V1. Cereb Cortex. 2002;12(9):915–25.
57. Johnson EN, Hawken MJ, Shapley R. The orientation selectivity of color-responsive neurons in macaque V1. J Neurosci. 2008;28(32):8096–106.
58. Daw NW. Goldfish retina: organization for simultaneous color contrast. Science. 1967;158:942–4.
59. Livingstone MS, Hubel DH. Anatomy and physiology of a color system in the primate visual cortex. J Neurosci. 1984;4:309–56.
60. Conway BR. Spatial structure of cone inputs to color cells in alert macaque primary visual cortex (V-1). J Neurosci. 2001;21(8):2768–83.
61. Shapley R, Hawken MJ. Color in the cortex: single-and double-opponent cells. Vis Res. 2011;51(7):701–17.
62. Spitzer H, Barkan Y. Computational adaptation model and its predictions for color induction of first and second orders. Vis Res. 2005;45(27):3323–42.
63. Rubin JM, Richards W. Color vision and image intensities: when are changes material? Biol Cybern. 1982;45(3):215–26.
64. Gershon R, Jepson AD, Tsotsos JK. Ambient illumination and the determination of material changes. J Opt Soc Am A. 1986;3(10):1700–7.
65. De Valois RL, Cottaris NP, Elfar SD, Mahon LE, Wilson JA. Some transformations of color information from lateral geniculate nucleus to striate cortex. Proc Natl Acad Sci. 2000;97(9):4997–5002.
66. Kellner CJ, Wachtler T. A distributed code for color in natural scenes derived from center-surround filtered cone signals. Front Psychol. 2013;4.

67. Lehky SR, Pouget A, Sejnowski T. Neural models of binocular depth perception. In: Cold Spring Harbor symposia on quantitative biology, Vol. 55. Cold Spring Harbor Laboratory Press; 1990. p. 765–77.
68. Paradiso M. A theory for the use of visual orientation information which exploits the columnar structure of striate cortex. Biol Cybern. 1988;58(1):35–49.
69. Schwartz O, Hsu A, Dayan P. Space and time in visual context. Nat Rev Neurosci. 2007;8(7):522–35.
70. Webster MA, Mollon JD. Changes in colour appearance following post-receptoral adaptation. Nature. 1991;349(6306):235–8.
71. Webster MA. Human colour perception and its adaptation. Netw Comput Neural Syst. 1996;7(4):587–634.
72. Juricevic I, Webster MA. Variations in normal color vision. V. Simulations of adaptation to natural color environments. Vis Neurosci. 2009;26(01):133–45.
73. Webster MA. Adaptation and visual coding. J Vis. 2011;11(5):3.
74. Gilbert CD, Wiesel TN. The influence of contextual stimuli on the orientation selectivity of cells in primary visual cortex of the cat. Vis Res. 1990;30(11):1689–701.
75. Clifford CWG, Wenderoth P, Spehar B. A functional angle on some after-effects in cortical vision. Proc R Soc Lond B Biol Sci. 2000;267(1454):1705–10.
76. Schwartz O, Sejnowski TJ, Dayan P. Perceptual organization in the tilt illusion. J Vis. 2009;9(4):19.
77. Kellner CJ, Haas O, Obando M, Wachtler T. A model of lateral interactions in color vision. Bernstein Conference 2013. 2013;doi:10.12751/nncn.bc2013.0230.
78. Klauke S, Wachtler T. "Tilt" in color space: hue changes induced by chromatic surrounds. J Vis. 2015;15(13):7.
79. Teufel HJ, Wehrhahn C. Evidence for the contribution of S cones to the detection of flicker brightness and red–green. J Opt Soc Am A. 2000;17(6):994–1006.
80. Teufel HJ, Wehrhahn C. Chromatic induction in humans: How are the cone signals combined to provide opponent processing? Vis Res. 2004;44(20):2425–35.
81. Wehrhahn C, Teufel HJ. How are lateral chromatic interactions computed from cone signals? Neural Comput. 2010;22(11):2763–84.
82. Locke NM. Color constancy in the rhesus monkey and in man. Columbia University; 1935.
83. Dörr S, Neumeyer C. The goldfish—a colour-constant animal. Perception. 1996;25(2):243–50.
84. Werner A, Menzel R, Wehrhahn C. Color constancy in the honeybee. J Neurosci. 1988;8(1):156–9.
85. Garbers C, Henke J, Leibold C, Wachtler T, Thurley K. Contextual processing of brightness and color in Mongolian gerbils. J Vis. 2015;15(1):13.
86. Foster DH. Color constancy. Vis Res. 2011;51(7):674–700.
87. Hurlbert AC. Computational models of color constancy. In: Walsh V, Kulikowski J, editors. Perceptual constancy: why things look as they do. Cambridge University Press: Cambridge; 1998. p. 283–322.
88. Yilmaz H. Color vision and an new approach to general perception. In: Bernard EE, editor. Biological prototypes and synthetic systems. New York: Plenum; 1973. p. 126–41.
89. Sällström P. Colour and physics: Some remarks concerning the physical aspects of human colour vision. USIP Report 73-09. University of Stockholm; 1973.
90. Buchsbaum G. A spatial processor model for object colour perception. J Franklin Inst. 1980;310(1):1–26.
91. Maloney LT, Wandell BA. Color constancy: a method for recovering surface spectral reflectance. J Opt Soc Am A. 1986;3(1):29–33.
92. Yuille A. A method for computing spectral reflectance. Biol Cybern. 1987;56(2-3):195–201.
93. Hurlbert A. Formal connections between lightness algorithms. J Opt Soc Am A. 1986;3(10):1684–93.
94. D'Zmura M, Lennie P. Mechanisms of color constancy. J Opt Soc Am A. 1986;3(10):1662–72.
95. Forsyth DA. A novel algorithm for color constancy. Int J Comput Vis. 1990;5(1):5–35.

96. D'Zmura M. Color constancy: surface color from changing illumination. J Opt Soc Am A. 1992;9(3):490–3.
97. Brainard DH, Freeman WT. Bayesian color constancy. J Opt Soc Am A. 1997;14(7):1393–411.
98. Maloney LT. Evaluation of linear models of surface spectral reflectance with small numbers of parameters. J Opt Soc Am A. 1986;3(10):1673–83.
99. Nascimento SMC, Foster DH, Amano K. Psychophysical estimates of the number of spectral-reflectance basis functions needed to reproduce natural scenes. J Opt Soc Am A. 2005;22(6):1017–22.
100. Tikhonov AN, Arsenin VY. Solutions of ill-posed problems. Washington, DC: VH Winston & Sons; 1977.
101. Rubner J, Schulten K. A regularized approach to color constancy. Biol Cybern. 1989;61(1):29–36.
102. Takebe K, Nakauchi S, Usui S. A computational model for color constancy by separating reflectance and illuminant edges within a scene. Neural Netw. 1996;9(8):1405–15.
103. Knill DC, Richards W, editors. Perception as Bayesian inference. New York: Cambridge University Press; 1996.
104. Brainard DH, Longere P, Delahunt PB, Freeman WT, Kraft JM, Xiao B. Bayesian model of human color constancy. J Vis. 2006;6(11):10.
105. Maloney LT. Illuminant estimation as cue combination. J Vis. 2002;2(6):6.
106. Brainard DH, Maloney LT. Surface color perception and equivalent illumination models. J Vis. 2011;11(5):1.
107. Lee HC. Method for computing the scene-illuminant chromaticity from specular highlights. J Opt Soc Am A. 1986;3(10):1694–9.
108. Funt BV, Drew MS, Ho J. Color constancy from mutual reflection. Int J Comput Vis. 1991;6(1):5–24.
109. Golz J, MacLeod DI. Influence of scene statistics on colour constancy. Nature. 2002;415(6872):637–40.
110. Brown RO, MacLeod DIA. Color appearance depends on the variance of surround colors. Curr Biol. 1997;7(11):844–9.
111. Mausfeld R, Andres J. Second-order statistics of colour codes modulate transformations that effectuate varying degrees of scene invariance and illumination invariance. Perception. 2002;31(2):209–24.
112. MacLeod DIA, Golz J. A computational analysis of color constancy. In: Heyer D, Mausfeld R, editors. Colour perception: mind and the physical world. Chichester, UK: Wiley; 2004. p. 205–46.
113. Granzier JJM, Brenner E, Cornelissen FW, Smeets JBJ. Luminance—color correlation is not used to estimate the color of the illumination. J Vis. 2005;5(1):2.
114. Golz J. The role of chromatic scene statistics in color constancy: spatial integration. J Vis. 2008;8(13):6.
115. Maloney LT. Physics-based approaches to modeling surface color perception. In: Gegenfurtner KR, Sharpe LT, editors. Color vision: from genes to perception. Cambridge, UK: Cambridge University Press; 1999. p. 387–416.
116. Yang JN, Maloney LT. Illuminant cues in surface color perception: tests of three candidate cues. Vis Res. 2001;41(20):2581–600.
117. Land EH. Experiments in color vision. Sci Am. 1959;200(5):286–98.
118. von Kries J. Die Gesichtsempfindungen. In: Nagel WA, editor. Handbuch der Physiologie des Menschen. Vieweg; 1905. p. 109–282.
119. Jameson D, Hurvich LM. Essay concerning color constancy. Annu Rev Psychol. 1989;40(1):1–24.
120. Land EH, McCann JJ. Lightness and retinex theory. J Opt Soc Am. 1971;61(1):1–11.
121. Horn BKP. Determining lightness from an image. Comput Graphics Image Process. 1974;3(4):277–99.

122. Blake A. On lightness computation in Mondrian world. In: Ottoson D, Zeki S, editors. Central and peripheral mechanisms of colour vision. New York: Springer; 1985. p. 45–59.
123. Land EH. An alternative technique for the computation of the designator in the retinex theory of color vision. Proc Natl Acad Sci U S A. 1986;83(10):3078–80.
124. Hurlbert A, Poggio T. Synthesizing a color algorithm from examples. Science. 1988; 239(4839):482.
125. Foster DH, Nascimento SMC. Relational colour constancy from invariant cone-excitation ratios. Proc R Soc Lond B Biol Sci. 1994;257(1349):115–21.
126. Dufort PA, Lumsden CJ. Color categorization and color constancy in a neural network model of V4. Biol Cybern. 1991;65(4):293–303.
127. Courtney SM, Finkel LH, Buchsbaum G. Network simulations of retinal and cortical contributions to color constancy. Vis Res. 1995;35(3):413–34.
128. Wray J, Edelman GM. A model of color vision based on cortical reentry. Cereb Cortex. 1996;6(5):701–16.
129. Shepard RN. The perceptual organization of colors: an adaptation to regularities of the terrestrial world? In: Barkow JH, Cosmides L, Tooby J, editors. The adapted mind: evolutionary psychology and the generation of culture. Oxford, UK: Oxford University Press; 1992. p. 495–532.
130. Mollon J, Jordan G. On the nature of unique hues. In: Dickinson C, Murray I, Carden D, editors. John Dalton's colour vision legacy. London: Taylor & Francis; 1997. p. 381–92.
131. Valberg A. Unique hues: an old problem for a new generation. Vis Res. 2001; 41(13):1645–57.
132. Bosten MJ, Boehm AE. Empirical evidence for unique hues. J Opt Soc Am A. 2014;31(4):385–93.
133. De Valois RL, De Valois KK, Switkes E, Mahon L. Hue scaling of isoluminant and cone-specific lights. Vis Res. 1997;37(7):885–97.
134. Wuerger SM, Atkinson P, Cropper S. The cone inputs to the unique-hue mechanisms. Vis Res. 2005;45(25):3210–23.
135. Webster MA, Wilson JA. Interactions between chromatic adaptation and contrast adaptation in color appearance. Vis Res. 2000;40(28):3801–16.
136. De Valois RL, De Valois KK. A multi-stage color model. Vis Res. 1993;33(8):1053–65.
137. MacLeod D. Receptoral constraints on colour appearance. In: Ottoson D, Zeki S, editors. Central and peripheral mechanisms of colour vision. New York: Springer; 1985. p. 103–16.
138. Lehky SR, Sejnowski TJ. Seeing white: Qualia in the context of decoding population codes. Neural Comput. 1999;11(6):1261–80.
139. Wachtler T, Dohrmann U, Hertel R. Modeling color percepts of dichromats. Vis Res. 2004;44(24):2843–55.
140. Yendrikhovskij SN. A computational model of colour categorization. Color Res Appl. 2001;26(S1):S235–8.
141. Philipona DL, O'Regan JK. Color naming, unique hues, and hue cancellation predicted from singularities in reflection properties. Vis Neurosci. 2006;23(3-4):331–9.
142. Shepard RN. Perceptual-cognitive universals as reflections of the world. Psychon Bull Rev. 1994;1(1):2–28.
143. Parkkinen JPS, Hallikainen J, Jaaskelainen T. Characteristic spectra of Munsell colors. J Opt Soc Am A. 1989;6(2):318–22.
144. Buchsbaum G, Bloch O. Color categories revealed by non-negative matrix factorization of Munsell color spectra. Vis Res. 2002;42(5):559–63.
145. Witzel C, Cinotti F, O'Regan JK. What determines the relationship between color naming, unique hues, and sensory singularities: illuminations, surfaces, or photoreceptors? J Vis. 2015;15(8):19.

Chapter 10
Color Vision in Clinical Practice

Cord Huchzermeyer, Jan Kremers, and John Barbur

Abstract The principal aim of this chapter is to introduce the reader to clinical applications of color assessment and to describe how knowledge of color vision loss is used in clinical practice. The chapter starts with an overview of important aspects of normal, trichromatic color vision. Congenital deficiencies are then reviewed briefly and the factors that cause the large variability in congenital deficiency are discussed. This is then followed by a detailed classification of the retinal, optic nerve and systemic diseases that can cause loss of color vision. The principal color vision tests that are used routinely in clinical practice are then introduced. Emphasis is placed on detection of small changes in color vision that can precede the earliest, clinical signs of a disease, the classification of the patient's class of color vision (i.e., normal trichromacy, deutan, protan, tritan, or acquired deficiency) and the test's ability to quantify the severity of color vision loss. Important diseases of the eye and the visual system are reviewed and the associated loss of red/green and yellow/blue chromatic sensitivity discussed and illustrated with emphasis on new findings and future developments. Finally, a typical approach for the use of color vision tests in clinical practice is explained with case reports that illustrate key findings.

Keywords Acquired dyschromatopsia • Clinical ophthalmology • Retinal dystrophies • Optic neuropathies • Glaucoma • Clinical color vision testing • Color Vision Assessment and Diagnosis (CAD) test

---

C. Huchzermeyer (✉) • J. Kremers
Department of Ophthalmology, University Hospital Erlangen, Schwabachanlage 6, 91054 Erlangen, Germany
e-mail: Cord.Huchzermeyer@uk-erlangen.de; Jan.Kremers@uk-erlangen.de

J. Barbur
Applied Vision Research Centre, School of Health Sciences, City University London, Northampton Square, London EC1V 0HB, UK
e-mail: J.L.Barbur@city.ac.uk

J. Kremers et al. (eds.), *Human Color Vision*, Springer Series in Vision Research, DOI 10.1007/978-3-319-44978-4_10

10.1 Normal Color Vision

Normal human color vision is trichromatic and relies on two independent post-receptoral (red–green and yellow–blue) chromatic mechanisms to signal changes in the spectral composition of the light in addition to achromatic channels that are more concerned with detection of edges and contours defined by luminance contrast (see Chaps. 2 and 4); the neural substrate for trichromacy relies on three classes of cones that map the short (S), middle (M), and long (L) wavelength regions of the visual spectrum. These wavelength regions are perceptually associated with "blue," "green," and "red" colors. The post-receptoral neural stages, in particular the retinal ganglion cells belonging to the retinogenicular pathways, provide the chromatic opponency needed to generate red/green (RG) and yellow/blue (YB) chromatic pathways. An extensive discussion of the origins and processing of photoreceptor signals can be found in Chaps. 2 and 4. Briefly, light perception starts with the photoisomerization of the light sensitive pigments in the cones. This triggers the phototransduction cascade, leading to changes in the membrane potential of the photoreceptors. For convenience the three different cone classes in the outer retina have been labelled as L-, M-, and S-cones [1]. Lights of different spectral composition that cause identical rates of photoisomerization in all three cone types appear identical to the observer and are said to be metameric. This important property follows as a consequence of the broad and overlapping spectral responsivity functions of the cones. In contrast, the neural substrate of color opponency lies in the retinal ganglion cells (RGCs) [2], who are located in the inner retina. Their axons form the optic nerve which connects the eye with the brain. The midget ganglion cells receive input from L- and M-cones via on- and off- midget bipolar cells. Their output is determined by opponent signals from L- and M-cones. The small bistratified RGCs receive input from S-cones that is opponent to the combined input from L- and M-cones. These two populations of ganglion cells are the origin of the parvocellular and part of the koniocellular retinogeniculate channels. In the standard model the signals derived from these two channels combine in central visual pathways as needed to generate the perception of many suprathreshold hues [1]. Interestingly, this is not the case at threshold when the vastly different RG and YB cone contrast signals contribute in isolation to color perception and produce either RG or YB hues [3]. Rod signals are fed into both pathways under certain conditions and may influence the appearance of colored objects at mesopic light levels [4], but do not necessarily contribute to either RG or YB chromatic mechanisms [5].

Recently, a modification of this standard model has been proposed by Schmidt, Neitz, and Neitz [7], to account for unexpected experimental findings in patients with a certain type of congenital stationary night blindness. These patients retain relatively normal hue perception in central vision as shown using D15 tests in spite of large field absence of S-cones [6]. The authors have suggested that the outer retina forms the basis for hue perception [7]. In their model, inputs from S-cones are carried to a subpopulation of midget ganglion cells via H2-horizontal cells. More studies are needed to understand how such patients manage to carry out color-related tasks, whether they rely entirely on color signals and the implications of this new theory for clinical practice.

10.2 Categories of Disease

10.2.1 Etiology

Etiology is one of the categories used in medicine to structure knowledge about different diseases; it denotes the ultimate cause of disease. Diseases can be caused or contributed to by endogenous factors like inherited alterations in one or several genes or by exogenous factors, including physical, chemical, or biological noxious agents. Many diseases are multifactorial and involve an endogenous predisposition and exogenous factors that trigger the disease.

10.2.2 Pathogenesis

Pathogenesis is the mechanism by which the etiological agent causes the observed morphological and functional changes. Common mechanisms include inflammation, ischemia, genetic alterations, tumor development, and tissue degeneration. Inflammation is an unspecific tissue reaction to noxious agents aimed at eliminating this agent. It is characterized by dilatation and increased permeability of the blood vessels, leakage of serous fluid with protein components of the blood and infiltration of the tissue by immune cells. Inflammation may cause significant "collateral" damage to the body or specific tissues. Ischemia is the state when an organ or tissue is not receiving sufficient blood supply, resulting in a lack of oxygen and other nutrients at the cellular level. Ischemia plays an important role for example in the advanced stages of diabetic retinopathy. Tissue degeneration is the progressive loss of a tissue or of important constituents that are indispensable for its function, ultimately leading to atrophy of the tissue. Necrosis is the violent death of a cell with uncontrolled release of cell components and a consecutive inflammatory reaction. In contrast, apoptosis is an ordered cell death, where a physiological program is activated and endogenous endonucleases digest nuclear DNA. In the end, the cell turns into a membrane-bound body that is finally cleared via phagocytosis. Several of the mechanisms may be involved in the pathogenesis specific diseases and to color vision deficits. For example, advanced stages diabetes mellitus are associated with vascular changes, inflammation, and neurodegeneration [8].

10.2.3 Structural and Functional Consequences

The events occurring during the pathogenesis of a disease have observable consequences on the structure and the function of an organ system. These may be noted by the patient (symptoms), be demonstrated during the basic clinical examination (signs) or measured with advanced techniques. Reduced chromatic sensitivity can often be either a sign or a symptom of a disease. When this loss is restricted to RG color vision as the only abnormality (as in deuteranomaly), the condition normally remains stable throughout the lifespan. Loss of chromatic sensitivity that often affects both the RG and YB chromatic mechanisms may be associated with acquired functional defects (such as those observed in optic neuritis). In ophthalmology, the most important indicator of visual function is best-corrected, high contrast visual acuity. This measure describes the "clarity" of vision under optimal conditions in the absence of refractive errors, and although the task is one of recognizing specific optotypes, the outcome is strongly linked to the resolving power of the optics of the eye. When visual acuity is reduced, other measures of spatial vision such as the ability to see fine detail in low contrast will be severely affected. The latter is often described as functional contrast sensitivity and is strongly affected by light scatter, in addition to residual refractive errors and imperfections in the image forming properties of the eye known as aberrations. Perimetry is normally used to measure the ability of the visual system to detect the presence of briefly presented visual stimuli or flicker at different locations in the visual field.

Structural changes of the eye can often be seen by the ophthalmologist using relatively simple optical instruments. Opacification of the optic media can be seen with the slit lamp, whereas changes of the retina and the optic nerve require imaging of the back of the eye, the so-called fundus, usually carried out with an ophthalmoscope. In clinical practice, additional imaging techniques are used, including fundus photography, fluorescein angiography, fundus autofluorescence, optical coherence tomography, and ultrasound of the eye. Some of the images are shown in Fig. *10.1*. Changes along the visual pathway that are located downstream from the eye can sometimes be demonstrated using magnetic resonance imaging (MRI) or computer tomography (CAT), but these techniques are unable to resolve fine spatial detail.

10.2.4 Prognosis

The prognosis is a prediction for the expected course of a disease. A disease may resolve completely, become chronic, progress steadily, or exhibit alternative remissions and relapses. For example, X-linked congenital color vision deficits are often stationary, whereas cone dystrophies show a progressive course with extensive deterioration of color discrimination and other visual functions over time. After an episode of optic neuritis, visual function may recover (almost) completely, but subsequent episodes may lead to progressive damage of the optic nerve with structural and functional deficits.

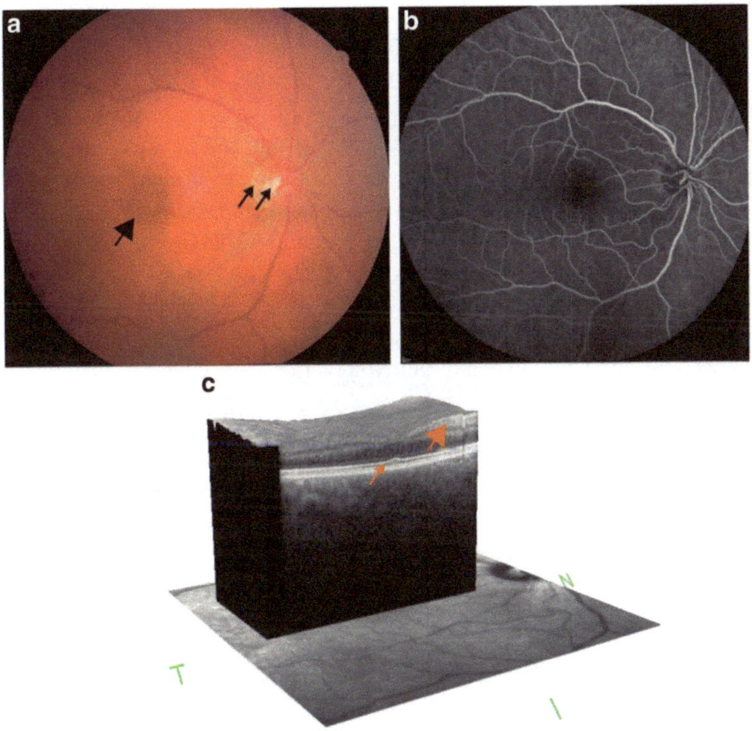

Fig. 10.1 Clinical examination of the fundus (back of the eye), which includes the retina and the optic disc, i.e., the intraocular portion of the optic nerve. Panel (**a**) shows a color fundus photograph. The optic nerve head (*small arrows*) consists of a neuroretinal rim (*orange*) and a central excavation (*yellow*). The macula lutea (*large arrow*) is the portion of the retina responsible for the clear central vision and is located on the temporal side of the optic disc. Important qualities of the macula are the absence of retinal blood vessels and the presence of a pigment (the macular pigment) in the inner retina that absorbs short-wavelength light. Panel (**b**) shows a similar photograph obtained during fluorescein angiography. Fluorescein is a fluorescent dye that can be injected into a vein. It emits a green light after being excited with blue light. A filter blocks the reflected blue light and allows visualization of the circulation of the dye in the fundus. Panel (**c**) shows an OCT (ocular coherence tomography) image. High-resolution three-dimensional data can be obtained by comparing phase shifts between coherent light entering the eye and the reflected light exiting the eye. The small red arrow shows the inner/outer photoreceptor junction (changed in hereditary retinal disease) and the large red arrow shows the retinal nerve fiber layer (changed in diseases of the optic nerve)

10.3 Assessment of Color Vision in Clinical Practice

Color vision tests have three primary purposes in clinical practice:

- Detection of a significant reduction in either RG or YB chromatic sensitivity.
- Classification of the patient's class of color vision (i.e., normal trichromacy, deutan, protan, or tritan congenital deficiency or acquired loss of color vision).
- Quantifying the severity of color vision loss.

When tests are used to screen for color vision deficiency, the aim is to quickly detect significant reductions in chromatic sensitivity without aiming to classify the patient's type of color vision or to quantify the severity of loss. More tests are usually needed and modern tests often provide a rapid "screening" option and full RG and YB tests to be administered to those patients that fail the screening test [3, 9]. Categories for classification must be precise and indicate clearly whether one is dealing with congenital or acquired loss. An accurate measure of the severity of RG color vision loss is important in many occupational environments where some degree of color vision loss is considered safe [10]. This section reviews briefly the most commonly used, conventional color vision tests and provides some information on recently developed, display based tests that are finding important applications in occupational medicine and clinical practice.

10.3.1 Pseudoisochromatic Plates

A large number of pseudoisochromatic plate tests have been designed following some common principles. They are based on the recognition of simple patterns—a task that is easily performed by untrained observers. Pseudoisochromatic plates consist of a set of images that are made up of differently sized dots. Some of the dots form a foreground image, usually a shape, a number or a letter. These dots differ from those forming the background only in color, but not in size, luminance, or saturation. Size, luminance contrast, and saturation are also slightly varied (noise added) so that small individual differences in sensitivity in any of these dimensions cannot be used to carry out the task. The test stimulus shape and background colors are chosen to be easily confused by subjects with specific color vision defects, because their chromaticities fall on or near the typical confusion lines in color space when the plates are used with the appropriate illuminant. An important requirement in any design is to ensure that the patient can only make use of e.g. RG color signals to see the shape. Random variation in dot luminance and the addition of YB color noise are often employed to isolate and test for RG color vision (as in the Ishihara test). Usually, the object or image of interest can be recognized by subjects with normal trichromatic color vision, but not by those with congenital and/or acquired deficiencies (vanishing design). The Ishihara test plates are extremely sensitive when no errors are allowed. All protans fail and only 0.6 % of deutans pass. The disadvantage of having such high sensitivity is that ~19 % of normal trichromats also fail (see Fig. *10.2*) when no errors are allowed [10, 11]. Other designs that are less sensitive do not employ YB color masking with the advantage of passing almost all normal trichromats and the disadvantage of also passing some deutan subjects. Plates have also been designed so that the shapes of interest can only be seen by color-deficient subjects (hidden digit design), those where color-normal and color-deficient designs see different images (transforming design) and those that are designed specifically to differentiate protan- from deutan-observers (qualitatively diagnostic design). In spite of these efforts, pseudoisochromatic test plates produce poor specificity and the number of plates at which the subject fails is only a very poor indicator of the severity of color vision loss [11].

a

Classification of class of colour vision using Ishihara				
	Normals (n=340	Deutans (n=343)	Protans (n=169)	Tritans (n=3)
% classified as normal	81	1	0	100
% classified with CVD	19	99	100	0
% classified as deutan	0	62	6	0
% classified as protan	0	0	48	0
% classified as indeterminate	19	38	46	0
% classified as tritan	0	0	0	0

b

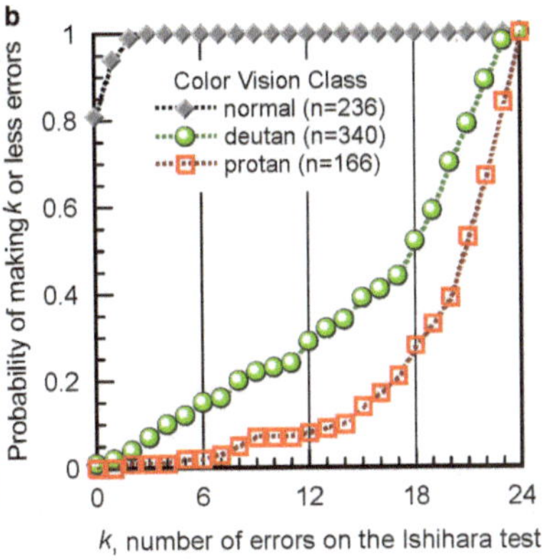

Fig. 10.2 Classification outcome for the Ishihara test plates (38 plate edition). Panel (**a**) 855 subjects were investigated and a criterion of zero errors is used as a pass. Note that no YB assessment is carried out. Panel (**b**) shows the probability of making (k) or less errors when reading the numerals on the Ishihara test plates plotted for a group of normal trichromats and for subjects with congenital, deutan, and protan color deficiency. Note that if four or less errors are allowed, all normal trichromats pass but so do 10 % of deutans and just under 1 % of protans. After learning the first plate which all subjects passed, the order of presentation was random for the remaining 24 plates of the Ishihara 38-plate test (From Rodriguez-Camora 2012 [3])

One advantage of pseudoisochromatic plate tests is that they are relatively cheap, fast, and easy to use, but good results require the use of standardized illumination. The most commonly used sets are the Ishihara plates and the Richmond HRR (Hardy, Rand, and Rittler) plates [12], although the Dvorine test plates (which is similar in design to Ishihara, but slightly easier to pass) are used extensively in the USA. The Ishihara and Dvorine plates detect only RG deficits, whereas the HRR plates also include cards for YB deficiencies. Although somewhat less sensitive than the Ishihara test plates, the Richmond HRR plates allow better classification of protan- vs. deutan-defects and an estimation of the severity of color vision defects [13]. Pseudochromatic plates are well suited as a screening tool especially for congenital red-green-defects (Ishihara plates), but also for acquired color vision defects (HRR plates). If few or no errors are allowed, so as to increase the sensitivity of these tests, the specificity is poor with some normal subjects failing the tests. The ability to classify the class of color vision involved and to grade the severity of these defects with pseudoisochromatic test plates remains very limited [11, 14].

10.3.2 *Color Arrangement Tests*

Color arrangement tests are better suited to detect color discrimination deficits in acquired disorders. The observer has to arrange colored caps of similar brightness and saturation so that they form a gradient. When plotted in the CIE chromaticity diagram, the caps form a contour that can be approximated by a circle. The order of the caps is noted in a diagram that is evaluated qualitatively, but quantitative analyses are also possible. The Farnsworth-Munsell 100 Hue test is the prototype of a test that measures color discrimination. There are 85 caps that are subdivided into four series with similar colors. Caps are confused where the characteristic lines of confusion are tangential to the circle in color space. In acquired dyschromatopsias, i.e., color vision defects that are caused by acquired eye diseases, this test can also demonstrate unspecific confusions without any clear axis. Its clinical use as a screening tool is limited, because it is very time-consuming, but in the past it has been the gold-standard to classify acquired dyschromatopsia. In contrast, the Panel-D15 test is designed to rapidly identify subjects with large, congenital color vision defects when the errors follow the typical axes of confusion. The differences between the Fransworth Munsell 100 Hue and the Panel-D15 are illustrated in Fig. 10.3. Observers with such defects confound caps that are on opposite sides of the color circle. The test is not suitable for assessing acquired deficiencies since the adjacent caps vary significantly in the rod and S cone signals generated, even when the L and M differences remain small. Another problem of the Panel-D15 is its poor sensitivity. Both problems are illustrated in Fig. 10.4. Desaturated versions of this test are more sensitive, but also less specific.

10.3.3 Metameric Matches

To quantify color vision defects that are caused by quantitative or qualitative changes in the visual pigments and/or spectrally selective prereceptoral absorption of light, color matching tests are performed. A detailed introduction to metameric matches in clinical practice can be found elsewhere [15]. Two lights of different spectral composition are said to be metameric, if they cause the same rate of photoisomerizations (quantum catches) in all photoreceptor subtypes and, therefore, appear to be identical to the observer. Simplified forms of metameric matches are made with an anomaloscope, i.e., an apparatus where a mixture of two test lights and a reference light, presented in the two halves of a circular field, are adjusted so that the whole field appears uniform. Although the anomaloscope employs spectral lights that are not absorbed by S-cones, this reduced color match is extremely effective at revealing shifts in the wavelengths of peak spectral responsivity of L and M cones, changes in their optical density and even differences in the relative numbers of L and M cones in the retina [16].

The subject's task is to adjust the relative amounts of red and green light in the mixture field and the luminance of the monochromatic yellow reference field. The Rayleigh equation is applied to detect abnormalities of the red/green system (see Fig. *10.5*). In the standard configuration (see diagram), the reference light has a wavelength of 589 nm and the test light consists of a mixture of 545 and 671 nm light, presented separated in the two halves of a 2° field. Under normal conditions, only L- and M-cones are involved in the match [15]. For dichromats all mixture ratios are metameric if the luminance of the monochromatic yellow field is adjusted

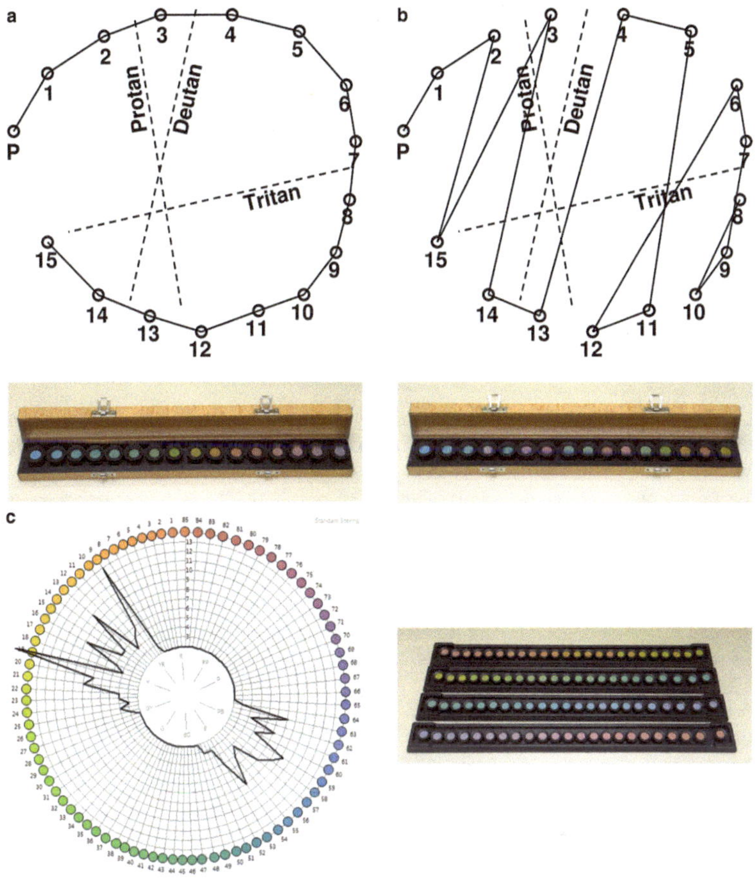

Fig. 10.3 Comparison of the Panel D15 and the Farnsworth-Munsell 100 hue test. Panel (**a**) shows a normal result from the Panel D15. The colored caps form a gradient. The sequence of the caps as sorted by the observer is shown in the diagram, indicating correct order. Panel (**b**) shows the result from a deuteranope. The subject has confused colors that lie on opposite sites of the color circle. The axes of confusion are parallel to the deutan axis. Panel (**c**) demonstrates that the FM 100 test consists of four sets that are sorted separately. Therefore, the deuteranope observer confused colors that appear very similar even to a normal observer. Confusions occur where the typical axes of confusion are tangential to the color circle

appropriately. In achromats (subjects with no functioning cones and only rods) and patients with cone disorders, rod signals may also be involved in the match. The rod curve has a very steep negative slope. Although the anomaloscope is eminently suitable at separating subjects with deutan- and protan-like deficiencies (see Fig. *10.6*), neither the matching range nor the midpoint describe the subject's severity of color vision loss. Some subjects with variant L and M cone pigment genes can produce anomaloscope parameters that are completely within normal limits, but exhibit RG thresholds that do not fall within the normal range [16]. The anomaloscope is also not suitable for investigating acquired loss of color vision, particularly when the cause does not link uniquely to photoreceptor damage.

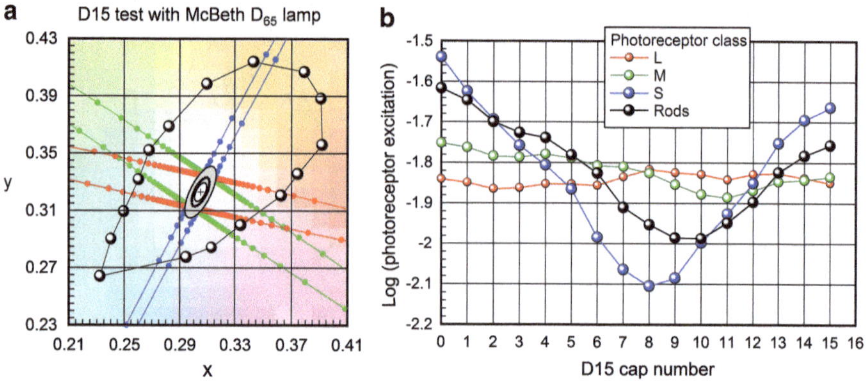

Fig. 10.4 CIE 1931 chromaticity plot showing the measured chromaticities of the reference and the 15 caps employed in the D15 test (Panel **a**). The center *black*, *ellipse* plots the mean thresholds measured 333 young normal trichromats. The *red*, *green*, and *blue* lines show the color confusion bands for protanopes, deuteranopes, and tritanopes, respectively. The plot shows that chromaticity differences between adjacent caps are suprathreshold for normal trichromats. Panel (**b**) shows the photoreceptor excitations generated by the D15 caps when illuminated with daylight (CIE illuminant D65). It is of interest to note that although the L and M cone excitation changes between adjacent caps are small, S cones and also rods exhibit very large changes. This may well explain why some subjects with poor or absent RG discrimination can carry out this cap arrangement test with close to zero errors. A comparison of CAD and D15 results in 406 participants (81 normal trichromats and 325 subjects with congenital color deficiency) shows that according to the traditional criterion for a D15 pass (i.e., no major crossings and two or less adjacent errors) ~57 % of color deficient subjects pass. Needless to say, all normal trichromats also pass. This finding illustrates the poor sensitivity of the D15 test. Although the majority of color deficients that pass tend to have relatively small color thresholds, some severe deutan and protan subjects with thresholds above 15 SN CAD units also pass

To examine the S-cones, other methods, the Moreland equation being the most prominent, have been developed [17, 18]. In this test one compares two mixture lights, one consisting of 480 and 580 nm, the other of 430 and 500 nm light. The matches are more difficult for the observer and residual differences between the hemifields may remain. All four photoreceptor types (L-, M-, S-cones and rods) can in principle be involved in the match. Like normal observers, X-linked anomalous subjects will make a match at the intersection of the LM- and S-cone lines; therefore, these abnormalities are not detected. The Moreland equation is also significantly influenced by prereceptoral filtering caused by the lens and the macular pigment.

10.3.4 Display-Based Tests

Visual displays offer great advantages over conventional pseudoisochromatic test plates both in terms of color threshold measurements and by making it easier to isolate the use of RG or YB color signals [19]. In addition, the strength of the color

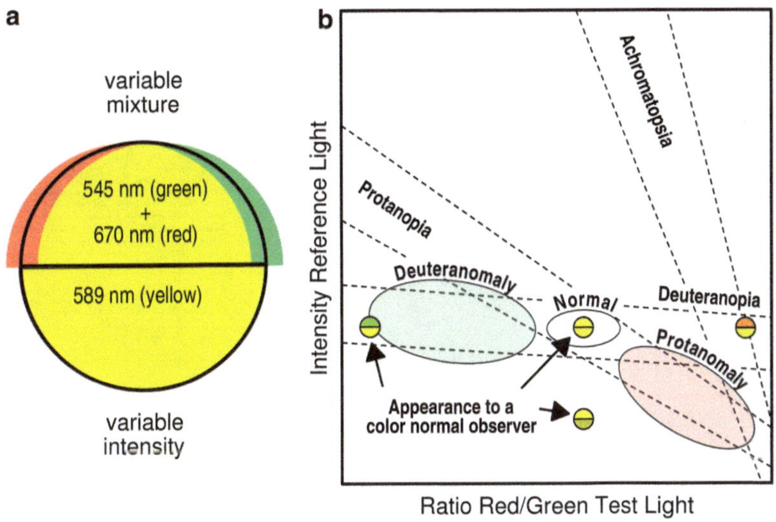

Fig. 10.5 The anomaloscope and the Rayleigh equation. Panel (**a**) shows the composition of the test field and the reference field in the anomaloscope when using the Rayleigh equation. The examiner regulates the ratio between the 545 and the 670 nm lights in the test field, and the observer adjusts the intensity of the reference light. Thus, matches where both field appear identical to the observer are sought. Panel (**b**) shows a diagram of the Rayleigh equation. The four circles demonstrate the appearance of the stimulus at different settings for a trichromat subject. Dichromats, S-cone monochromats and achromatopsia patients accept a wide variety of different setting as equal (*areas between dashed lines*). Normal trichromats and abnormal trichromats only accept a much smaller range of settings (*ellipses*)

signals can be made directly proportional to the cone contrasts generated and this is useful when one wishes to quantify the severity of color vision loss by means of threshold responses [20].

The only remaining problem is the need to ensure that the colored stimulus generated cannot be detected by means of luminance contrast and that the test isolates one of the two chromatic mechanisms and not both! Although stimuli that are "isoluminant" for the CIE 1931 standard observer can be generated easily using modern visual displays, this approach only works best in those normal trichromats that follow closely the standard observer, both in terms of peak spectral responsivities and the relative numbers of L and M cones (i.e., the L:M ratio of 2:1 which describes the standard CIE observer). Subjects with anomalous cone pigments, or those with significant differences in L:M ratio end up with spectral luminous efficiency functions that differs significantly from the standard, V(λ) function. This is illustrated in Fig. *10.7* which shows how the V(λ) function changes from one extreme L:M ratio to another. Protanopes and deuteranopes flank the two extremes which correspond to only M and L cones, respectively.

Colored stimuli that are isoluminant for the standard normal CIE observer will generate small luminance contrast signals in most normal trichromats. For certain colors the differences in luminance contrast can be quite large, particularly in deuteranopes and protanopes.

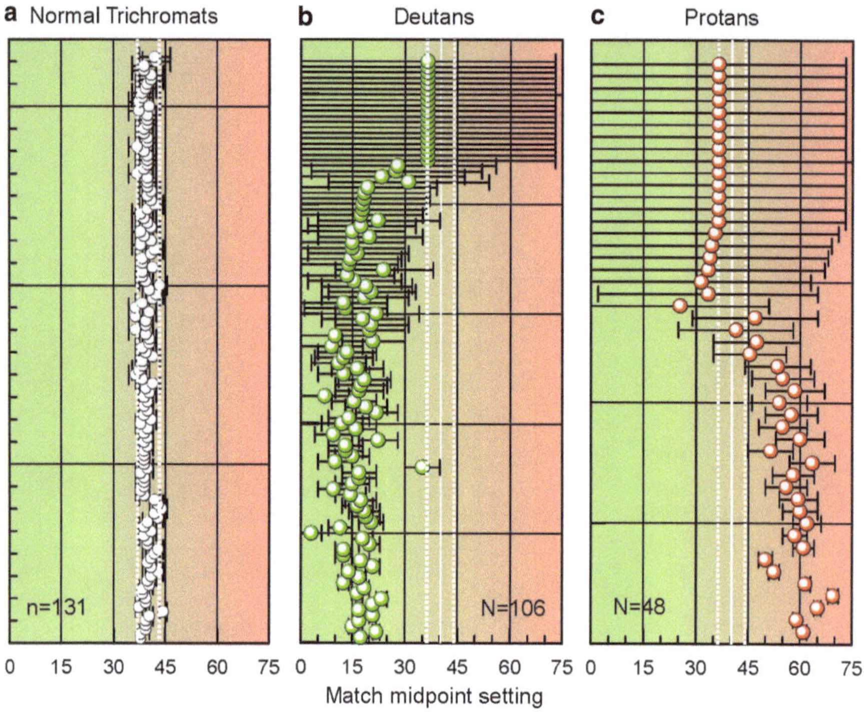

Fig. 10.6 Match midpoint and range values ranked in order on increasing matching range measured in normal trichromats (Panel **a**) and in subjects with deutan- (Panel **b**) and protan-like deficiency (Panel **c**). Some of the deutan and protan subjects can discriminate red/green differences in the match as well if not better than some normal trichromats. They do require either more green or more red in the match when compared to normal trichromats [11]

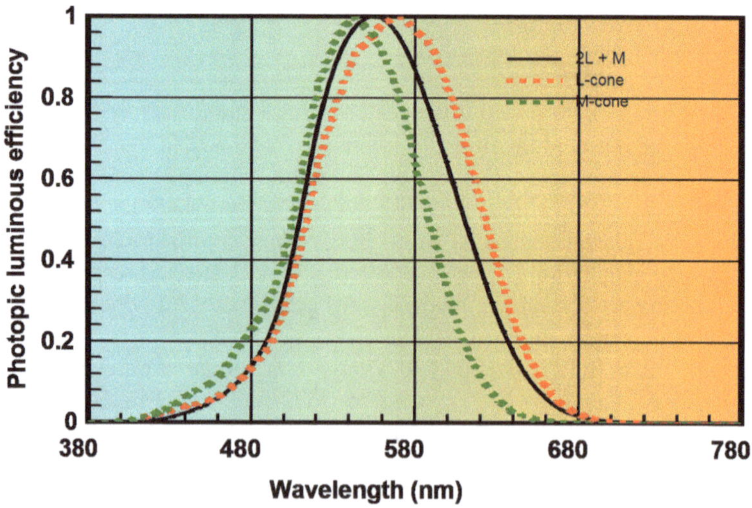

Fig. 10.7 Changes in the spectral luminous efficiency of the eye as a result of congenital deficiency and/or variation in the relative numbers of L and M cones in the retina. Although the exact form of the spectral luminance efficiency function does not affect the contrast of an achromatic object (i.e., an object defined by light of the same spectral composition as the background), this is not the case when colored objects are involved

To overcome the variable L:M ratio, a cone specific stimulation technique (see also Chap. 2) has been developed and used in some tests [21, 22]. The spectral composition of the stimulus is modulated to ensure that only one class on cone is stimulated. The test consists of letters drawn in a uniform color on a gray background starting with high chromatic contrast and then reducing this contrast until the letters are no longer identified correctly. This is done using chromatic displacements that "isolate" L, M, and S cones. The approach works well in normal trichromats with cone spectral responsivities that are similar to the standard CIE observer and in dichromats, but less well in anomalous trichromats since the variant cone function is unknown. The selective cone contrast technique has been implemented in the "Cone Contrast Test" which consists of columns of ten letters with the cone specific contrast increasing with letter height along each column [20].

The most common approach used to minimize the contribution of luminance contrast to the detection of a colored stimulus is to use the principles of pseudoisochromatic test plates by varying the luminance of the elements that make up the background texture (e.g., solid discs, squares, and diamonds) so as to mask any local luminance contrast signals generated when color signals are added to some of the background checks. The use of a dark background also helps to minimize detection of luminance differences since the visual system is relatively insensitive at detecting small changes in the luminance of a "pedestal" of light. The only disadvantage of this technique is that the chromatic sensitivity is also reduced when using a dark background field which results in higher color thresholds [11].

The most common, computer-based color vision test that employs the principles of the pseudo-isochromatic plates is the Cambridge Color Test (CCT). Discs of varying size and luminance are added to a dark background thus generating a large textured field. A chromatic signal is then added to some of the background discs to form a large Landolt ring with the color-defined gap placed randomly along the horizontal or vertical meridians. The subject's task is to press a button following the presentation of each stimulus to indicate the position of the gap [23]. The CCT employs two different procedures known as the "*trivector*" and the "*ellipse*" tests. The former yields unidirectional color detection thresholds along the protan, deutan, and tritan color confusion lines while the latter measures full chromatic threshold contours. In the case of subjects with congenital deficiency the orientation and the size of the corresponding "ellipses" yields useful information on the chromatic mechanism involved and the severity of color vision loss. Changes in color vision as a result of normal aging have been investigated with the CCT test [24] and compared to results obtained in similar studies using the CAD test [25].

The Color Assessment and Diagnosis (CAD) test employs a new technique based on findings from studies on camouflage [26]. More effective isolation of color signals is achieved by generating outline, color-defined targets in coherent motion which are buried in dynamic luminance contrast noise (see Fig. *10.8* for screen dumps of the stimuli that isolate RG and YB chromatic mechanisms). The rationale behind the use of dynamic luminance contrast noise (that causes the perception of random motion and flicker detection) is to mask the transient visual mechanisms that are most sensitive to luminance contrast. The "color-defined target" is isoluminant for the "standard" observer and moves diagonally across the $5° \times 5°$

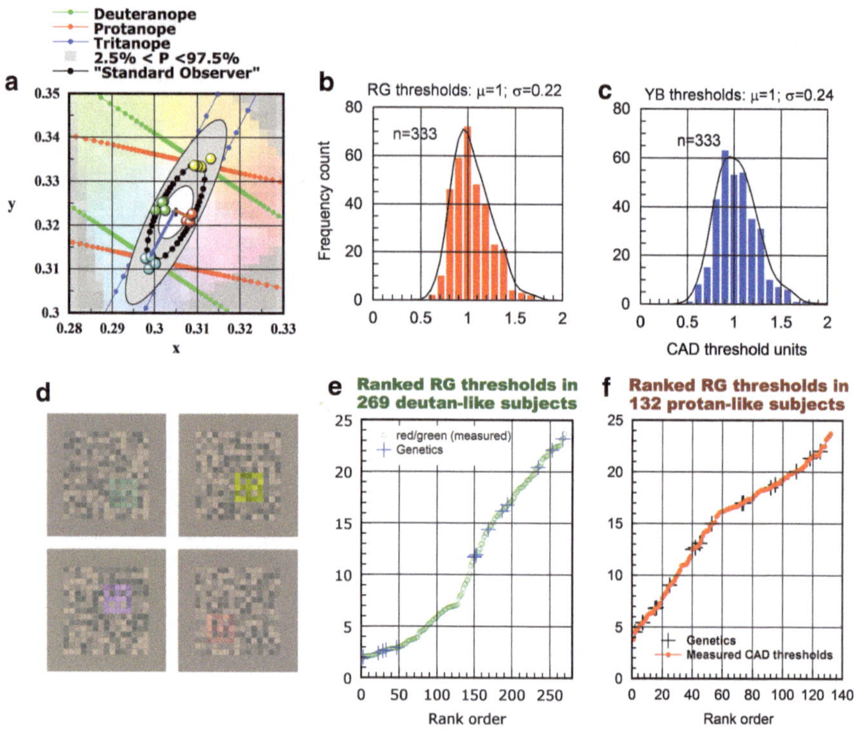

Fig. 10.8 Panel (**a**) shows the mean thresholds measured for 333 young subjects with normal trichromatic color vision. Panels (**b**) and (**c**) show the frequency distributions of RG and YB thresholds normalized with respect to the mean. The minor and major axes of the center ellipse define the Standard Normal (SN) CAD thresholds for RG and YB color vision. The screen dumps in Panel (**d**) show the colored stimuli that isolate the YB and RG chromatic mechanisms. All thresholds are expressed in SN CAD units by dividing every measurement by the corresponding mean. Panels (**e**) and (**f**) show the ranked RG thresholds for 269 subjects with deutan-like deficiency and 132 subjects with protan-like deficiency. The ranks show that the loss of RG chromatic sensitivity forms a continuum although the shapes of the deutan and protan ranks reflect the number and likelihood of occurrence of the variant genes in each class [11]

central square of rapidly flickering checks. The dynamic luminance contrast noise reduces preferentially the sensitivity of transient ganglion cells that respond best to flicker and motion defined by luminance contrast [26]. When using these techniques, dichromats are unable to detect anything moving over the square background of flickering checks for color directions close to their corresponding color confusion lines even for chromatic saturations that are limited only by the phosphors of the display. During the test, the subject's task is simply to indicate the direction of motion of the colored stimulus. A four-alternative, forced-response procedure is used to measure the subject's threshold for each of several interleaved color directions. As a result of some intersubject variability in the orientation of the tritan color confusion line, a number of color directions that favor the YB and the

RG chromatic mechanism have to be investigated. The CAD test employs four directions for YB thresholds and 12 directions for RG thresholds. The larger number of RG directions is needed to obtain enough information to classify accurately the subject's class of color vision.

Figure *10.8* shows the frequency distribution of RG and YB thresholds measured in 333 young subjects, mostly students and pilots, all with normal trichromatic color vision. Each subject had their color vision assessed using a batch of conventional color vision tests, including the Nagel anomaloscope. All subjects with congenital deficiency or diagnosed medical conditions such as retinal or systemic diseases were excluded. The measured RG and YB thresholds were normalized with respect to the mean threshold. The standard deviations are shown in Fig. *10.8*. The mean RG and YB thresholds as measured in young, healthy subjects define the Standard Normal CAD units for RG and YB color vision, respectively. The results for every subject are then expressed in CAD units. For example, a subject with a RG threshold of 0.7 CAD units is more sensitive than the average normal young trichromat, while another subject with a threshold of 3 CAD units would require three times as large RG color signal at threshold when compared to the mean. The results of Fig. *10.8* also reveal ~2.3-fold increase in threshold between the least and most sensitive young normal trichromats. Subjects with congenital deficiency form a continuum with thresholds ranging from just above normal limits to the maximum chromatic saturation that can be achieved on visual displays (see Fig. *10.8*).

The CAD test has close to 100 % sensitivity and specificity in detecting congenital color vision deficiency (see Fig. *10.9*) and in classifying the subject's class of color vision (i.e., normal trichromacy, deutan, protan, tritan, or acquired deficiency). A fast screening option which takes only 1–2 min to carry out can be used to determine whether the subject has normal trichromatic color vision. If the subject fails the fast screening test, a full YB test (which lasts

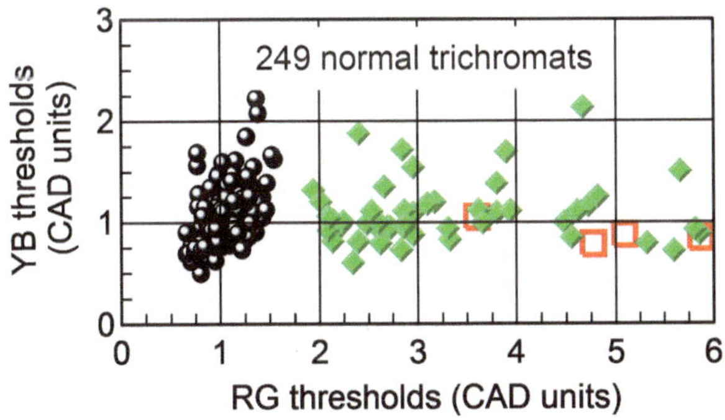

Fig. 10.9 YB and RG CAD thresholds measured in 249 normal trichromats (*black symbols*), 140 deutans (*green diamonds*), and 63 protans (*red squares*). Only subjects with RG thresholds less than 6 standard normal CAD units are shown. The separation between the least sensitive normal trichromats and the least affected deutan subject is sufficiently large to screen for normal trichromacy. Protan-like subjects start with higher thresholds which extend up to the maximal saturation limit of the display (~34 CAD units for the RG directions)

~3 min) and a RG test (which lasts ~10 min) are carried out to determine the subject's class of color deficiency and the severity of loss. The test also detects and diagnoses automatically the less common case of acquired deficiency in a subject with congenital RG color loss. Upper thresholds limits for normal healthy aging for binocular and monocular RG and YB color vision have been measured and incorporated in the program [25]. The CAD test results are always expressed in standard normal CAD units together with the normal limits for the subject's age. The results show that chromatic sensitivity for binocular vision is highest ~ 20 years of age and that color thresholds remain remarkably stable in healthy aging with ~1 % increase per year for RG and ~1.6 % per year for YB over the remaining life span. The establishment of reliable upper normal threshold limits for RG and YB color vision as a function of age has important clinical applications by enabling the detection of the earliest signs of chromatic sensitivity loss that cannot be attributed to healthy aging.

Another important advantage of the CAD test is that by controlling the amount of luminance contrast masking one can establish whether the subject is using color or luminance contrast signals to detect the moving, color-defined stimulus. The strength of luminance contrast masking is controlled by its amplitude which is simply specified as a percentage of background luminance. The largest strength corresponds to 100 % modulation when the luminance of each check could have any value from zero to twice the background luminance. When the color signal strength is above the subject's color threshold any further increase in the amplitude of luminance contrast masking noise has no significant effect on the subject's threshold for detection of coherent movement [22]. If the subject detects the moving stimulus by means of its luminance contrast, an increase in luminance contrast noise amplitude causes a monotonic increase in the corresponding color thresholds. Subjects with complete absence of color vision (e.g., rod monochromats) require different CAD test parameters to reveal the full extent of their color vision loss. Other subjects with genetic mutations that affect the normal functioning of the neural retina, but are still capable of some residual color vision do not show the pattern of results obtained in rod monochromats. The color thresholds do not continue to increase above a certain level of luminance contrast masking and are no longer limited by the phosphors of the display. Interestingly, some of these subjects fail all the Ishihara test plates, but can carry out the D15 test with few or zero errors. The CAD test is therefore ideal for studies of acquired, partial loss of chromatic sensitivity since by manipulating the effectiveness of luminance contrast masking one can reveal the true extent of color vision loss.

A number of subjects examined for congenital deficiency in the AVOT clinics at City University London (http://www.city.ac.uk/avot) were diagnosed with acquired loss of color vision. Some had clinically diagnosed retinal and optic nerve diseases. Although some patients, particularly those diagnosed with AMD showed greater YB losses, in general the YB loss was also accompanied by significant loss of RG color vision.

10.3.5 Color Matches and Color Discrimination

Subjects with dyschromatopsia often make abnormal color matches or/and have reduced chromatic sensitivity and confuse similar colors. Metameric matches are determined at the photoreceptor level. Therefore, altered color matches may either be caused by prereceptoral filtering (causing changes of the spectral composition of the lights at the photoreceptor level) or by qualitative or quantitative changes in the visual photopigments. Many patients with dyschromatopsia have impaired ability for color discrimination. This can be demonstrated by an increased number of errors in the color sorting tests, especially in the FM 100 Hue test, where caps with very similar hues have to be sorted. When making metameric matches, these observers will accept a wider range of matches, but the point where both lights are metameric will always be included in the matching range. As a general rule, it can be said that diseases of the optic nerve and the visual pathway affect mostly color discrimination, whereas changes of the optic media and the photoreceptors can also cause abnormal color matches, sometimes as an isolated phenomenon, but often also associated with both YB and RG color thresholds well above the normal limits that describe healthy aging [27, 28].

10.4 Classification of Color Vision Defects

Color vision defects are classified into congenital and acquired. The term "congenital" means literally "present at birth," regardless of etiology. However, the term congenital color vision defects often refers to the stationary color vision defects caused by mutations in the cone opsin genes. These are inherited following a X-chromosomal mode for L- and M-opsins and an autosomal dominant mode for the S-opsins. The classification of these defects is summarized in Table 10.1.

In acquired dyschromatopsia, deficits are usually more complex than in congenital forms. When both the RG and the YB chromatic mechanisms are damaged in different ways, there are no normal axes of confusion and the distribution of the

Table 10.1 Classification of congenital color vision defects

Anomalous trichromacy:	Shift in λ_{max} for either L- or M-cones
Protanomaly	L-cones
Deuteranomaly	M-cones
Dichromacy:	Lack/loss of function of one photoreceptor type
Protanopia	L-cones
Deuteranopia	M-cones
Tritanopia	S-cones
S-cone monochromatism:	Lack/loss of function of L- and M-cones
Achromatopsia:	Lack/loss of function of all three cone types

Table 10.2 Verriest classification of acquired dyschromatopsia

Type	Axis	Defect	Visual acuity loss	Disorder
I	Red–green + scotopization	Protan	Moderate to severe	Cone degeneration
II	Red–green	Deutan	Mild to severe	Optic neuropathies
IIIa	Blue + pseudoprotanomaly	Tritan	Mild	Chorioretinal disorders with sensory elevation
IIIb	Blue	Tritan	Mild to moderate	Glaucoma, autosomal-dominant optic atrophy

thresholds reflects the loss of sensitivity in both chromatic mechanisms. Koellner's rule states that diseases of the retina are associated with blue-yellow defects and those of the optic nerve with red–green defects, but recent findings suggest that this is not always the case. In the course of the chapter, notable exceptions, like autosomal-dominant optic atrophy, are discussed. Verriest published extensive work on acquired dyschromatopsia and proposed a classification that is still in use today, albeit with modifications [29, 30]. Based primarily on the axes in the Farnsworth-Munsell 100 Hue test, he divided acquired dyschromatopsia into three categories that are shown in Table 10.2.

Protan-like defects (Verriest Type I) are usually associated with severe impairment of visual acuity in retinal disease, for example in cone dystrophies. They are characterized by confusion along a red-green axis and a marked decrease in the sensitivity to long wavelength light. In advanced stages rods will play an important part in the Rayleigh matches, resulting in a steep negative slope in the Rayleigh function (scotopization). One typical example is juvenile macular dystrophy (Stargardt disease).

Type II defects are associated with a decreased color discrimination but normal spectral sensitivity, resulting in a deutan-like defect. It is typical for optic neuropathies (except glaucoma and autosomal-dominant optic atrophy).

Type III defects are associated with tritan-like defects and can be subdivided in two groups, depending on the presence of pseudoprotanomaly, that is a red-shift in the Rayleigh match similar to congenital protanomaly. Pseudoprotanomaly points toward a retinal disorder with disruption of the photoreceptor alignment (Type IIIa). In optic neuropathies like glaucoma and autosomal-dominant optic atrophy, the midpoint of the Rayleigh match often remains unaffected (Type IIIb). BY color vision losses are first observed in the earliest stages of glaucoma, significant loss of RG color vision is observed in intermediate and late stages [31].

In reality, the association between a certain disease entity and color vision defects may well be test dependent and is by no means straightforward. In many diseases, the degree to which the RG or YB chromatic mechanisms exhibit the greatest loss of sensitivity may change throughout the course of the disease. It remains, however, important to be able to detect the presence of acquired loss of chromatic sensitivity in subjects with congenital RG color deficiency (see Figs. *10.10* and *10.11*).

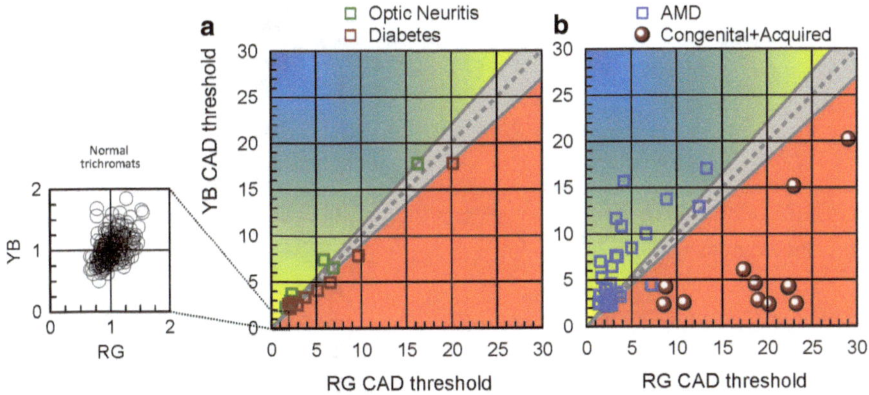

Fig. 10.10 Preliminary findings intended to illustrate the potential use of color assessment in ophthalmology. YB and RG color thresholds measured in patients with different retinal and optic nerve diseases. The AMD patients (*blue squares*) exhibit a greater loss for YB chromatic sensitivity compared to RG. Other findings (not shown), suggest that significant YB loss of chromatic sensitivity precedes clinical signs of retinopathy. For diabetic patients (*black squares*) the gradient is significantly less than unity which indicated greater loss of RG sensitivity. For optic neuritis patients (*green squares*) a good correlation between RG and YB chromatic thresholds is observed in the affected eyes implying that both RG and YB axes are affected to a similar extent after an episode of optic neuritis. Also shown are a number of patients that have not been diagnosed clinically, but have both acquired and congenital loss of color vision (*discs*). These patients have a much larger loss of RG sensitivity due to the underlying congenital deficiency

10.5 Retinal Dystrophies

Retinal *dystrophies* are marked by progressive retinal degeneration that is caused by monogenetic mutations (in contrast to retinal *degenerations*, like age-related macular degeneration, that are caused by an interaction between multiple genetic and environmental factors). The characteristics of retinal dystrophies are listed in Table 10.3. Because these disease classes have a very direct impact on the photoreceptors, color vision tests can play an important part in the diagnosis and management of these conditions. More than 200 genes that may cause retinal dystrophies have been identified (see https://sph.uth.edu/RetNet/disease.htm), and in some retinal dystrophies no causative genetic defect has yet been discovered. A systematic approach to classify the genetic defects in the cone dystrophies is shown in Fig. *10.12* [32]. The retinal dystrophies that may be involved can have a predilection for specific photoreceptor types, for example the rods in retinitis pigmentosa, or they may cause unspecific damage to the outer retina (photoreceptors and retinal pigment epithelium). Some forms of retinal dystrophies are exceedingly rare, so that only a few families are affected worldwide. However, other forms are more frequent and collectively retinal dystrophies are among the most important causes of visual

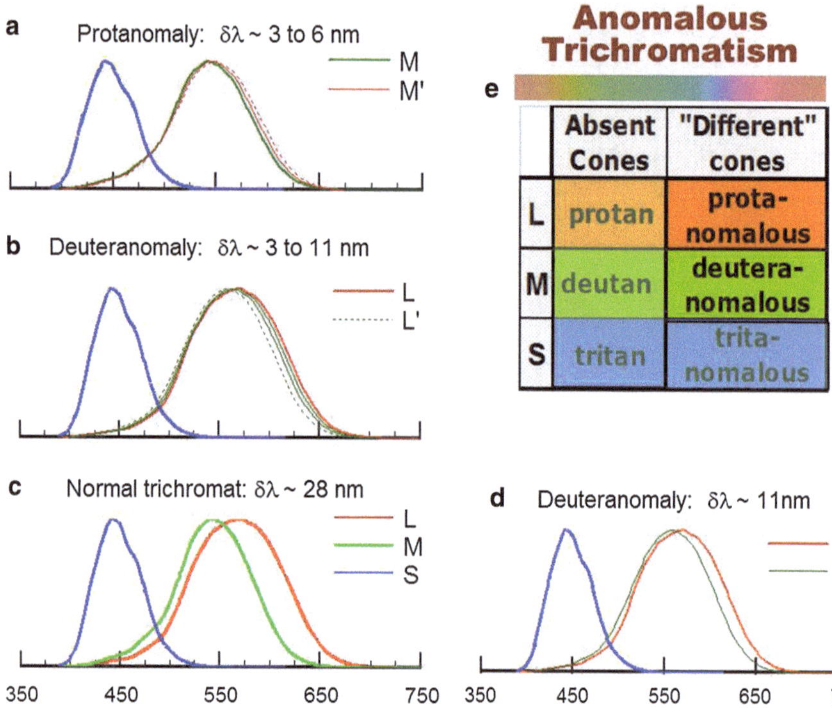

Fig. 10.11 Subjects with normal trichromatic color vision rely on L and M cones separated by ~ 28 nm (**c**). The least affected deuteranomalous subjects rely on a normal L cone pigment and a variant L cone pigment separated by ~11–12 nm (**d**). Deuteranomalous subjects have smaller separations (**b**). Protanomalous subjects rely on normal and variant M cone pigments (with a maximum separation of 6 nm) (**a**). In dichromats one cone class is either absent or completely nonfunctional (**e**)

Table 10.3 Characteristics of retinal dystrophies

Etiology	Inherited monogenetic defect
Pathophysiology	Loss of function or dysfunction of proteins relevant for the function of the photoreceptor/retinal pigment epithelium (RPE) functional unit; apoptosis of photoreceptors and RPE cells
Structural and functional defects	Atrophy of the choriocapillaris, RPE and the outer retina; loss of visual acuity, color vision defects, glare (cones); visual field loss, night blindness (rods)
Prognosis	Slowly progressive course; currently no therapy

impairment and blindness in children and young adults [33]. The degeneration of the retina occurs in a slowly progressive fashion. The rate of progression may be quite variable, even within families that share one specific mutation, indicating the influence of other endogenous or exogenous factors [34]. Currently, there is no established therapy for the majority of retinal dystrophies. However, advances in the

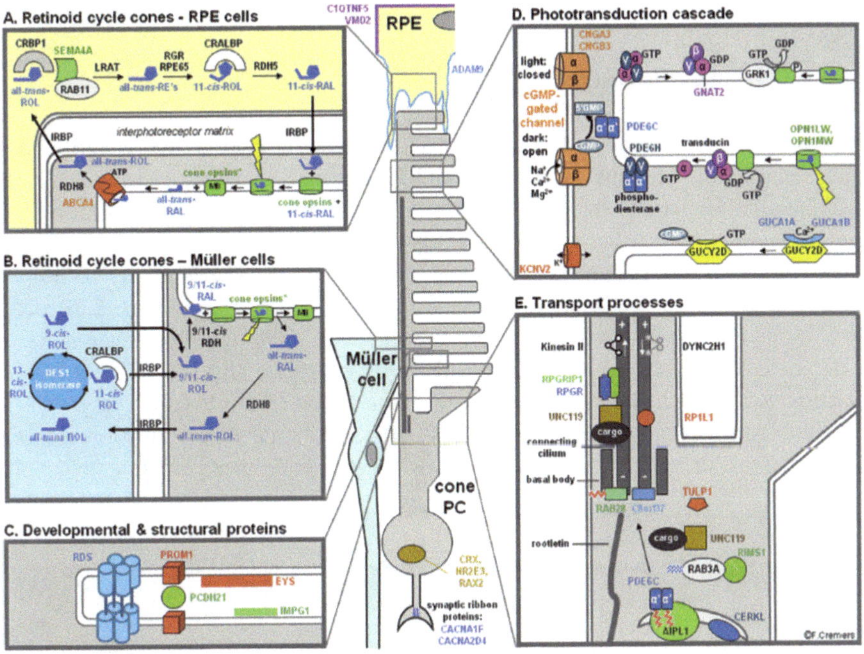

Fig. 10.12 The diagram shows different proteins that are important for the function of the outer retina. Mutations in the genes that code for the proteins can lead to retinal dystrophies. These proteins belong to important functional units: the retinoid cycle (Panels **a** and **b**), development and structural proteins (Panel **c**), the phototransduction cascade (Panel **d**) and transport processes (Panel **e**) [Courtesy of Roosing et al.]

knowledge of the disease mechanisms have led to the development of a number of therapeutic approaches that are currently evaluated in the laboratory or in clinical studies [35].

The distinction between congenital, stationary color vision defects, on the one hand, and retinal dystrophies, on the other hand, is somewhat vague. However, in congenital color vision defects, the number of cones often is similar to those in normal subjects indicating that the all cones are functional and that the missing pigment is replaced (see Chap. 1). There are notable exceptions in which defects in the pigment genes not only lead to congenital color vision defects but also to cone degeneration [36–38]. In contrast to the red-green deficiencies, the mechanism underlying tritanopia seems to be similar to an S-cone dystrophy, resulting in progressive loss of these photoreceptors. S-cone monochromacy and achromatopsia are associated with severe changes of other visual function losses such as poor visual acuity, nystagmus, and photophobia and with structural changes that can be demonstrated using modern imaging techniques like OCT or adaptive optics.

10.5.1 Macular Dystrophies

10.5.1.1 Stargardt Disease

Stargardt disease is the most common macular dystrophy, often associated with a considerable loss of visual acuity early in life. Because its pathophysiology is well understood, it is a good example of a retinal dystrophy affecting the central retina. Figure *10.13* compares Stargardt disease and retinitis pigmentosa—a predominantly peripheral retinal dystrophy (see below). Stargardt disease is commonly accompanied by a central scotoma, but there are also forms with ring scotomata and spared foveal function (Bull's eye maculopathy) [39]. The morphological correlate of these changes is an atrophy of the RPE and the photoreceptors (Fig. *10.13*). Currently, there is no established therapy, but a number of clinical trials are in progress [40]. The etiology is a mutation of the *ABCA4* gene, which codes for an ATP-binding cassette transporter that transports retinoids across membranes [41]. It is present in

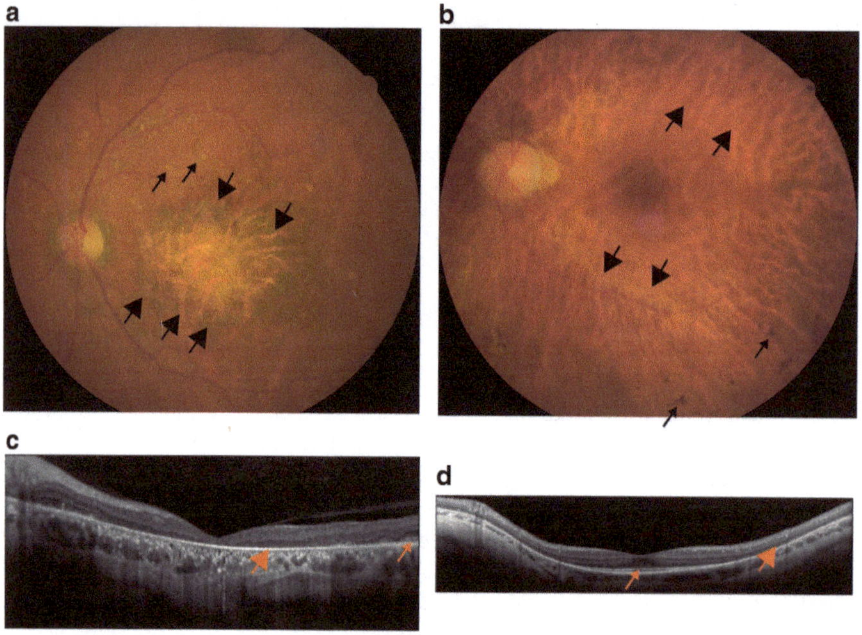

Fig. 10.13 Two important groups of retinal dystrophies. Panels (**a**) and (**c**) show the fundus photograph and the OCT scan of a patient with Stargardt disease. On the photograph the typical yellow flecks of fundus flavimaculatus can be seen (*small arrows*). There is marked atrophy of the retinal pigment epithelium and the photoreceptors in the central retina (*large arrows*). This can also be seen on the OCT scan. Panels (**b**) and (**d**) show the fundus and OCT of a patient with retinitis pigmentosa. Here, there is an atrophy of the RPE and photoreceptors in the periphery (*large arrows*), because the rods are predominantly affected. Toward the lower border of the photograph, there are typical bone spicules (*small arrows*). The *red arrow* in the OCT scans show zones with regular structure (*small arrows*) and with severely damaged retinal structure (*large arrows*)

rods [41] and in cone photoreceptors [42]. The mode of inheritance is autosomal recessive. Mutations in the *ABCA4* gene can also cause other ocular diseases including cone–rod dystrophy, retinitis pigmentosa, and a predilection for age-related macular degeneration. The defect transporter leads to the deposition of *N*-retinylidene-*N*-retinyl-ethanolamine (A2E), a metabolite of all-trans retinal, in the retinal pigment epithelium [43]. Funduscopically these deposits may be visible as yellow flecks (Fundus flavimaculatus). They are autofluorescent and can therefore be imaged using a scanning-laser opthalmoscope (fundus autofluorescence) [44]. The A2E deposits enhance the phototoxicity of short wavelength light ("blue light hazard") [45]. Fundus flavimaculatus may be present without central atrophy: in these cases, the visual impairment is only mild.

The changes in color vision observed in Stargardt disease/fundus flavimaculatus may depend on the severity of the disease, especially on the best corrected visual acuity [46]. Typically, advanced stages are accompanied by red-green defects of Verriest type I [29]. In these cases, visual acuity is typically very low [46]. The defects are commonly associated with scotopization, i.e., a sensitivity loss to long wavelengths, resulting in a pattern resembling achromatopsia in the Rayleigh matches [47].

In subjects with preserved visual function and in asymptomatic relatives (heterozygous carriers), YB defects may be present. The Cambridge Color Test revealed decreased sensitivity along all three axes [48]. The presence of pseudoprotanomaly in these milder cases may point towards a disruption of the photoreceptor mosaic [49].

10.5.1.2 Vitelliform Macular Dystrophy (VMD)

Vitelliform macular dystrophy (VMD, also Best disease) is another important juvenile macular dystrophy. It has an autosomal dominant mode of inheritance. There is an early dysfunction of the retinal pigment epithelium that can be demonstrated with electro-oculography [50]. Funduscopically there is a characteristic focal accumulation of lipofuscin-like material below and within the retinal pigment epithelium. This results in a very characteristic morphological appearance. In the early stages, there is a vitelliform lesion consisting in an egg-yolk-like yellow lesion. Here, visual function is relatively well preserved. Subsequently, the yellow lipofuscin-like material may displace downward to form a visible level and finally get a scrambled-egg-like appearance with marked retinal atrophies and deterioration of visual function. Mutations in the *VMD2* gene are responsible. *VMD2* codes for the Bestrophin protein, a chloride channel with a regulatory influence on Calcium channels [51]. The exact mechanism how dysfunction of bestrophin leads to accumulation of the lipofuscin-like material is not known [52].

In the early stages, there are typical macular color vision defects due to a disturbance of the photoreceptor arrangement [49, 50]. There are color discrimination defects predominantly on the *YB* axis and a pseudoprotanomaly in the anomaloscope. In the late stages with limited visual acuity, Verriest type I defects are present due to a marked atrophy of the retinal pigment epithelium of the macula [53].

10.5.1.3 Progressive Cone Dystrophy/Cone–Rod Dystrophy

In contrast to Stargardt and Best disease, the progressive cone and cone–rod dystrophies are genetically very heterogeneous [32]. Only a small fraction of the responsible genes are known. These lead to atrophy of the photoreceptors or the retinal pigment epithelium [32]. Cone dystrophies are manifest by a progressive decrease of visual acuity, accompanied by photophobia, at a variable age ranging from early youth to adulthood. Photophobia is caused by saturation of rods at photopic light levels that play an increasingly important role in mesopic vision when a majority of cone photoreceptors are lost. Color vision defects occur early and may be an important sign because morphologic changes in the retina are often absent or minimal in the early stages.

In contrast to cone dystrophies, cone–rod dystrophies also show signs of progressive rod degeneration. This manifests itself by the development of night blindness and visual field constriction in the course of the disease.

Due to the genetic heterogeneity, different color vision defects may occur. Only two special cases are discussed here. First, mutations of the opsin genes, that usually lead to stationary color vision defects described elsewhere in the book (Chap. 1), may also lead to progressive photoreceptor degeneration. Some of these forms can also be considered as cone dystrophies. They are associated with characteristic congenital color vision defects and progressive decrease of visual acuity [54, 55]. Secondly, there is a form of mild autosomal-dominant cone dystrophy with only mild impairment of visual function that is associated with disturbances along the YB axis [56, 57].

10.5.2 Retinitis Pigmentosa

Retinitis pigmentosa is a heterogeneous group of diseases characterized by a progressive rod-cone dystrophy [58]. In contrast to cone–rod dystrophies, early symptoms are caused by rod degeneraton with concentric restriction of the visual fields and night blindness. Visual acuity is often preserved for a long time, and in some forms, the central retina may be normal even at late stages of the disease [59]. However, there are forms with early central involvement. In the late stages, retinitis pigmentosa and cone–rod dystrophy cannot be differentiated anymore. Retinitis pigmentosa has its name from the characteristic morphological appearance with "bone spicules," i.e., pigment deposits in the inner retina.

Color vision defects occur in later stages and are predominantly along the YB axis. One reason may be a reduced opsin density [60], but a disruption of the cone mosaic may also be present, resulting from cystoid macular edema. Red-green defects (Verriest type I) may also be present [29, 61].

10.5.3 Future Directions in Retinal Dystrophies

Knowledge about changes in color vision in many of the discussed pathologies may be enhanced by studies of functional consequences using stimuli that are specific for one class of cone photoreceptor. The silent substitution paradigm (see Chap. 2) can be used to generate spectral modulations that cause output changes in only one class of cone photoreceptor. Scholl, Kremers, and coworkers have found a general decrease in the amplitude of L- and M-cone driven ERGs in patients with cone dystrophies, cone–rod dystrophies and retinitis pigmentosa [62, 63]. They observed a phase-delay in the L-cone and a phase-advance in the M-cone responses that may be characteristic for all retinal dystrophies. Although there was increased interindividual variability, on average, amplitudes were not reduced in patients with macular dystrophies (Stargardt disease/Fundus flavimaculatus and Best disease), but there was increased variability in Stargardt patients [64]. In patients with Best disease, L- cone amplitudes were paradoxically large [65]. The characteristic phase changes in the ERG responses were present in both types of macular dystrophy. Important future directions are the development of S-cone isolating stimuli [66] and the development of psychophysical methods that allow the application of the silent substitution paradigm in patients with hereditary retinal disease.

Imaging techniques of the outer retina have undergone tremendous improvements with the development of two methods: ocular coherence tomography (OCT) and adaptive optics (AO; see Chap. 3). These techniques have major implications for color vision research in retinal disease. For example, Carroll et al. were able to demonstrate that some forms of X-linked defects of the photopigments associated with dichromacy resulted in a significant loss of cone photoreceptors in contrast to the other forms of X-linked dichromacy that are characterized purely by a substitution of the photopigment [37]. McClement and coworkers were able to demonstrate morphological changes in the photoreceptors of patients with a novel mutation in a hybrid L/M opsin gene using OCT and adaptive optics [54]. Furthermore, there are efforts to combine adaptive optics and eye tracking systems to ensure the stimulation of single photoreceptors in psychophysical experiments [67].

10.6 Macular Color Vision Defect

The "macular color vision defect" is a change in color vision caused by macular diseases in which best-corrected visual acuity is relatively preserved [68]. There is a loss in color discrimination that is either nonspecific or along the YB axis (Verriest type III). In some patients there is a type I defect. A hallmark of the macular color vision defect is pseudoprotanomaly: a red-shift in the Rayleigh match resembling that observed in protanomaly.

Several hypotheses have been brought forward to explain these changes. Some have postulated an increased vulnerability of S-cones to damaging agents as the cause of the very frequent YB deficits in a variety of diseases associated with mild defects in photoreceptor disease, but also in disease of the optic nerve like glaucoma. Alternatively, the loss of S-cones may have a stronger impact their function because of their relative scarcity. However, Hood, Benimoff, and Greenstein have demonstrated that a limited response-range of the S-cone photoreceptors may account for the observed dominance of YB defects in mild retinal defects [69].

The frequent presence of pseudoprotanomaly cannot be explained by this. Possible explanations for pseudoprotanomaly include changes in preretinal filtering, reduced effective optical density of the photopigments, photoreceptor distortion, a kinetic disorder of the regeneration of photopigment, and alterations of the photopigments. Pokorny and coworkers have used several psychophysical techniques to examine the macular color vision defect. They have made extended Rayleigh series with additional reference lights in addition to the usual 589 nm, Rayleigh matches with different field sizes and brightness matching with the test light entering the pupil at different locations. The latter is used to identify a disruption of the photoreceptor layer through an abnormal Stiles–Crawford effect. Because an abnormality of this effect was present in most patients with a macular color vision defect, photoreceptor disruption seems to be an important factor in this type of defect [49]. This is illustrated in Fig. *10.14*. Other factors may involve a reduced optical density of the photopigment [60] which can cause a shift in the midpoint of anamolascope matches similar to protanomaly [16].

10.7 Diabetes Mellitus

Diabetes mellitus is a very common systemic disease and it is one of the leading causes of blindness worldwide [70]. In this chronic condition, many organ systems are damaged by the effects of enduringly elevated blood glucose levels (hyperglycemia), which is caused by absolute or relative lack of the hormone insulin. The etiology is an autoimmune destruction of the insulin secreting beta cells of the pancreas (Type I) or an insulin resistance in the context of a generalized metabolic syndrome, most commonly associated with obesity (Type II).

End organ damage in diabetes is mediated by several mechanisms (see review by Antonetti et al. [71]): neurodegeneration, inflammation, and ischemia. Changes in visual function, including color vision, can be demonstrated prior to the presence of diabetic vascular changes in the retina. This is a hallmark of neurodegeneration of neural and glial cells (reviewed by Lierth and coworkers [72]). In spite of such findings, there is no established neuroprotective therapy to treat these early manifestations (see review by Zhang et al. [73]). When monitoring patients with diabetes, physicians screen for the appearance of diabetic retinopathy, defined as the presence of vascular changes in the retina, and then treat complications of this vasculopathy according to severity (see AAO preferred practice patterns, http://www.aao.org/preferred-practice-pattern/diabetic-retinopathy-ppp--2014). Sequelae leading to

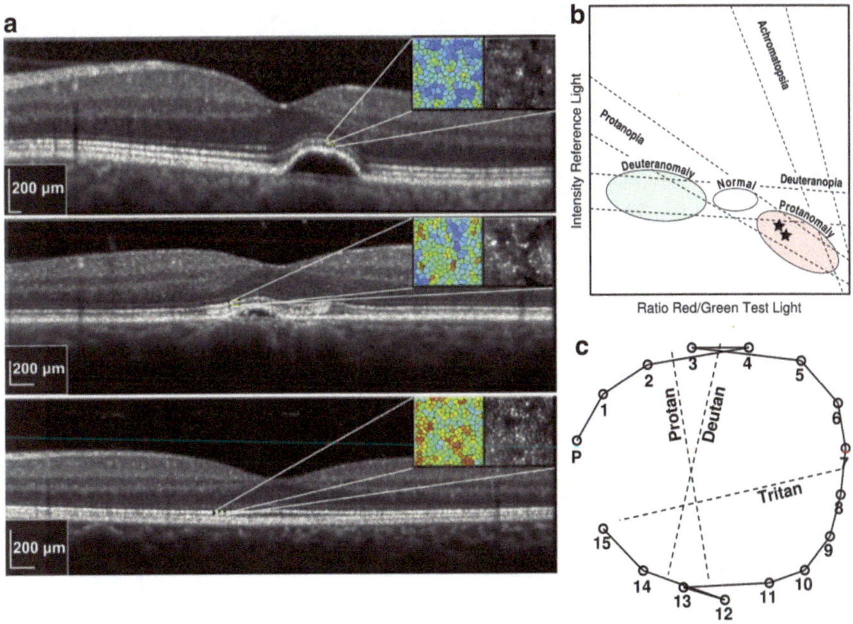

Fig. 10.14 Example of a macular color vision defect. Central serous chorioretinopathy (CSCR) is a rare disease of the retina that is characterized by serous detachment of the retinal pigment epithelium and the sensory retina. It is commonly self-limiting with complete resolution of symptoms in less than 3 months, although it may recur. Impairment of visual function is mild consisting of central scotomata and color vision defects. Panel (**a**) shows the impact of CSCR on the cone mosaic imagined with OCT and adaptive optics. [Courtesy of Flores et al., AJO 2015 160:588–95, Copyright Elsevier Limited] Panel (**b**) shows the presence of pseudoprotanomaly with the anomaloscope and Panel (**c**) demonstrates several confusions along the tritan axis with the Panel D15 test in a subject with CSCR. This combination is characteristic of a macular color vision defect

Table 10.4 Characteristics of diabetic retinopathy

Etiology	Multifactorial: genetic predisposition + either viral infection (Type I) or diet (Type II)
Pathophysiology	Neurodegeneration with loss of glial cells and neurons; inflammation; microvascular changes with ischemia; proliferative retinopathy caused by secretion of vascular endothelial growth factor
Structural and functional defects	Early functional defects with reduced color vision and contrast sensitivity, possibly as a sign of neurodegeneration; vascular changes: microaneurysms, capillary dropout, intraretinal hemorrhage, lipid deposits in the retina, retinal neovascularization, vitreous hemorrhage
Prognosis	Control of blood glucose levels is an important factor; vision can be preserved by treatment in some patients; remains very common cause of blindness

blindness include chronic macular edema and proliferation of vascular tissue in the eye [70], both mediated predominantly by vascular endothelial growth factor (VEGF) which is secreted in response to tissue malperfusion [74]. The characteristics of diabetic retinopathy are summarized in Table 10.4.

Due to the direct implications for treatment, diagnostic methods, that are used routinely nowadays, focus mostly on morphological changes (see AAO preferred practice patterns): funduscopy, color fundus photographs, OCT, and fluorescein angiography. Currently, color vision is not routinely assessed in diabetes. Improving screening methods for early diabetic retinopathy is an important goal for the future [75], and color vision tests might play a role. The key issue is the relevance of early functional changes in color vision for treating patients. It seems convincing that the demonstration of neurodegeneration in the retina would warrant modification of metabolic control, but its value has not been proven in large trials. Equally importantly, the more reliable and accurate color vision tests developed in last decade may gain clinical relevance as neuroprotective treatments have become available and may halt neurodegeneration in the early stages of diabetes [73]. It remains to be shown how effective and informative color assessment becomes in diabetes when compared to other functional, morphological, or even biochemical markers of neurodegeneration [76].

In clinical research, color vision tests can be used to quantify small changes and compare different treatments. Although the most common outcome measure in clinical trials is visual acuity, other aspects of vision may show differential changes and are important for the patients' quality of life [31, 77].

10.7.1 Changes in Color Vision Caused by Diabetes Mellitus

It is well established that diabetes can cause significant loss of color vision. When compared with the occurrence of impaired color vision in older subjects within the normal population, a much larger number of patients investigated in the Early Treatment Diabetic Retinopathy Study (ETDRS) exhibited abnormal color discrimination performance. Poor color discrimination also correlated with the presence and severity of macular oedema. Earlier findings using conventional color assessment tests revealed mostly tritan-like defects in diabetic patients with macular oedema. The patients included in the ETDRS study were selected because they had retinopathy. It remains of great interest to establish the degree of color vision loss in patients with diabetes before signs of retinopathy can be detected using OCT and stereo photography. Several studies have suggested that loss of yellow-blue color vision precedes the first clinical signs of retinopathy [78–82]. It is also of equal interest to establish whether RG loss of color vision correlates with the YB loss and whether this correlation remains unchanged during the progress of the disease.

10.7.2 Signs of Color Vision Alterations in Diabetes Assessed with CAD

More recent results using the CAD system for color assessment yield a number of interesting findings.

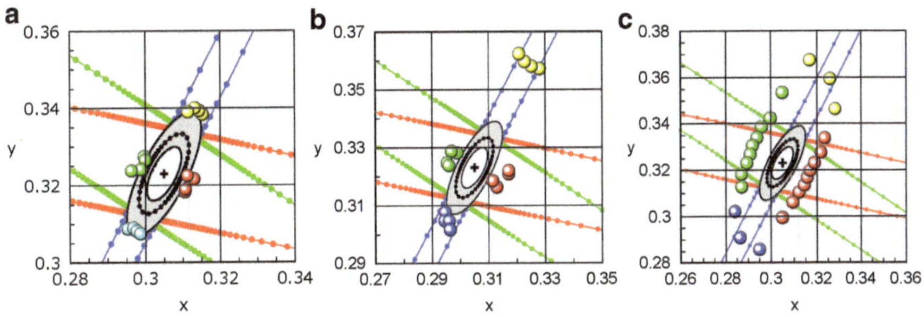

Fig. 10.15 Minimal color vision loss in two patients with diabetes (**a**, **b**) in the absence of any signs of retinopathy. More severe loss of both RG and YB color vision in a patient with retinopathy but normal visual acuity (**c**)

Figure *10.15* shows that diabetic patients can exhibit early losses of both RG and YB color vision in the absence of a retinopathy. The results show CAD test thresholds measured with the stimulus centered at the fovea (i.e., ~ 4° around the point of regard). The gray area bounded by the inner and outer ellipses shows the region associated with normal trichromatic color vision. The red-, green-, and blue-dotted lines show the color confusion bands for protan-, deutan-, and tritan-like subjects. When the color thresholds are expressed in standard normal CAD units the patient with the most severe loss (C) shows marginally greater RG thresholds.

However, the loss of color vision is not limited to the foveal region. The loss of both RG and YB color vision was present at every location tested suggesting that neural changes that cause loss of color vision affect the whole retina.

A large number of patients diagnosed with diabetic retinopathy were investigated in another study which attempted to establish the severity of color vision loss and the extent to which diabetes affects the YB chromatic system. In general, there is a significant correlation between RG and YB sensitivity loss. There is, however, little doubt that when the RG chromatic sensitivity loss is small, this is accompanied by a greater YB loss. When the patient exhibits RG thresholds larger than some 15 standard normal CAD units, comparing the severity of RG and YB losses becomes more difficult since chromatic displacements along the tritan axis become limited by the output of the display (see Fig. *10.16*).

The use of color assessment in the management of diabetes has been very limited, largely because of variability associated with conventional color assessment tests and lack of direct implications for treatment. Advances in color vision assessment and progress in understanding the underlying neurodegenerative changes in diabetes may increase future interest in the regular use of color tests to detect the subclinical effects of diabetes, to monitor its progression and the effectiveness of medication. In addition, the knowledge derived from wider use of color vision assessment in diabetes may also improve our understanding of the effects of diabetic retinopathy on neural retinal changes in addition to improved understanding of the correlation with glucose-induced microvascular damage.

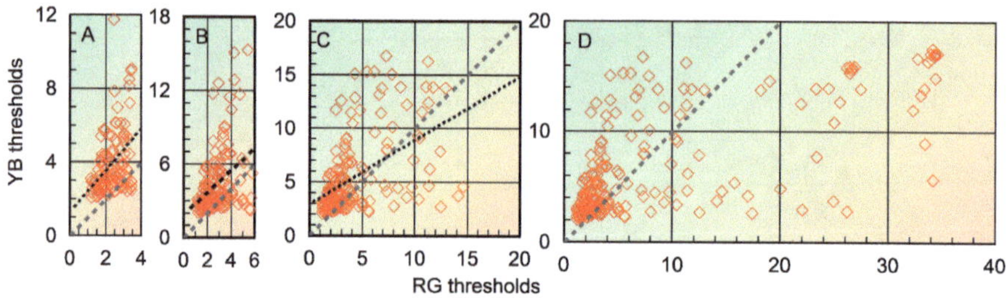

Fig. 10.16 YB thresholds plotted against the corresponding RG thresholds measured in 219 patients with diabetes, without retinopathy. The *dashed, gray, straight lines* are used to indicate the case of perfect correlation. The *dotted black lines* show the linear regressions to the data points shown in each graph. The data shown in Section *D* cannot be fitted with a straight line since the visual display limits the maximum YB thresholds that can be generated to ~18 standard normal CAD units. A chromatic displacement of just over 34 CAD units is possible in the RG direction. Section *A* shows patients with RG thresholds ≤3.5 CAD units. 52 % of the patients fall into this category. The YB thresholds are significantly higher than the corresponding RG thresholds. The linear regression with $r^2=0.15$. Section *B* (RG ≤ 6 CAD units) captures 66 % of the sample, $r^2=0.14$. Section *C* (RG ≤15 CAD units) captures 82 % of the sample with an r^2 value of 0.26. The data shown in Section *D* cannot be fitted with a straight line since the visual display limits the maximum YB thresholds that can be generated to ~18 standard normal CAD units. A chromatic displacement of just over 34 CAD units is possible in the RG direction

10.8 Glaucoma

Glaucoma is the most frequent optic neuropathy with 60.5 million people worldwide affected in 2010 [83]. It is a very frequent ocular condition and it is among the most common causes of blindness. Therefore, a whole subspecialty of ophthalmology is dedicated to it. Glaucoma is a heterogeneous group of diseases ("the glaucomas"), all having in common a progressive atrophy of the optic nerve accompanied by a characteristic cupping of the optic nerve head (the intraocular portion of optic nerve) that is absent in other forms of optic nerve atrophy. Cupping occurs when not only retinal ganglion cell axons but also microglial cells (especially astrocytes) undergo apoptosis. Examples of the appearance of the optic disc are shown and compared with changes in other optic neuropathies in Fig. *10.17*. Although an elevated intraocular pressure (IOP) is often involved in the pathogenesis, it is no longer part of the definition of glaucoma. In clinical practice, exact diagnosis of functional changes, including color vision, is most important in primary open-angle glaucoma and some related forms like pseudoexfoliation glaucoma and pigment dispersion glaucoma, which are managed in a similar fashion. Therefore, this section is concerned only with these conditions. Information about etiology and pathogenesis is shown in Table 10.5.

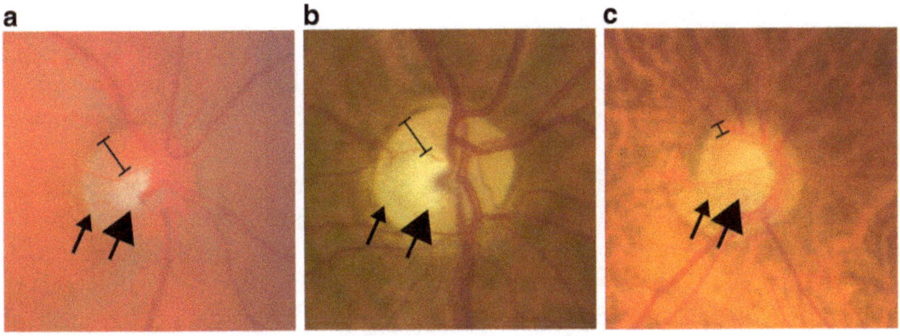

Fig. 10.17 Morphology of the optic disc in different diseases discussed in this section. Panel (**a**) shows a normal disc. In patient with optic neuritis, the optic disc often appears completely normal. Affection of the optic nerve can only be demonstrated with functional tests, including a longer latency between a visual stimulus and the elicited visual evoked potentials (VEPs) in the occipital cortex evoked by this stimulus, impaired color vision with characteristic red-green desaturation and a relative afferent pupillary deficit. Panel (**b**) shows optic atrophy, characterized by pallor of the neuroretinal rim. Panel (**c**) shows glaucomatous optic atrophy with typical enlargement of the central excavation of the optic disc (cupping). The *bars* indicate the width of the neuroretinal rim, the *small arrows* the border of the optic disc and the *large arrows* the border of cup

Table 10.5 Characteristics of glaucoma

Etiology	Unknown; some hereditary forms; multiple genetic risk factors (largely unknown)
Pathophysiology	Largely unknown; mechanical stress, neurodegeneration, inflammation, autoimmune disease, mitochondrial dysfunction all have been implicated in the pathogenesis
Structural and functional defects	Loss of retinal ganglion cells with optic atrophy, loss of glia in the optic disc (cupping); typical visual field defects; blindness
Prognosis	Progression can be stopped/slowed down by lowering the intraocular pressure; poor prognosis in advanced cases when patients present with advanced optic atrophy

10.8.1 Clinical Management of Primary Open-Angle Glaucoma

The diagnosis of glaucoma is based on the identification of characteristic morphological and functional defects of the optic nerve. Several tools are used for diagnosing glaucoma:

Morphological techniques measure the retinal nerve fiber layer (RNFL) thickness and the three-dimensional structure of the optic disc. Measurements can be performed by funduscopy, scanning laser ophthalmoscopy using the Heidelberg Retina Tomograph (HRT), and ocular coherence tomography (OCT). Glaucoma is accompanied by a decrease of the RFNL thickness and, as mentioned, by a cupping of the optic disc.

Functional (psychophysical) techniques mainly include those that establish visual performance with perimetric techniques (e.g., white-on-white perimetry). Glaucoma is often accompanied by localized losses in the visual field. In advanced stages of glaucoma, these losses increase in size and can ultimately lead to complete blindness.

Although the intra-ocular pressure is still an important risk factor for glaucoma, intraocular hypertension is nowadays mainly used to support one of the two previous methods.

The diagnosis of glaucoma is not always straightforward and a relatively large number of glaucoma suspects display only one of the above mentioned features of glaucoma (either a suspect morphology of the optic disc, suspect visual fields or an elevated IOP) [84] and therefore cannot be definitely diagnosed for having glaucoma. Particularly, for these subjects, improved measurements of very early functional deficits are of importance, so that effective treatment can start at an early stage.

Once glaucoma is diagnosed, IOP is lowered either pharmacologically or surgically below a certain target pressure. This target pressure is the IOP level below which a deterioration of vision that will affect the vision-related quality of life in the lifetime of the patients is unlikely to occur [85]. The target pressure is estimated from statistical data based on IOP before treatment and stage of optic nerve damage among other factors. Therapy has to be monitored by repeated IOP measurements to assure that the target IOP is indeed reached.

It is absolutely important to monitor morphological and functional measures to exclude a progression, even when the target IOP has been reached. If morphological or functional progression of glaucoma is observed, then the target IOP is further decreased [85]. In late stages of the disease, glaucomatous damage may progress even at very low IOP levels, suggesting that other pathophysiological mechanisms dominate in these late stages.

The ultimate goal for diagnosing early glaucoma as well as for monitoring progression is achieved by demonstrating a correlation between morphological and functional changes. However, this correlation is often poor. Among other factors, the limited sensitivity of current psychophysical techniques may be the cause for this. Therefore, an improvement in the understanding of functional changes and the methods to measure these is important. Some recent studies have shown that the parafoveal loss of color vision often precedes the foveal loss and that changes in color vision can be more sensitive than either perimetry or functional contrast sensitivity [31].

10.8.2 Pathophysiology of Retinal Ganglion Cell Damage

The etiology of glaucomatous optic neuropathy is multifactorial and its pathophysiology is poorly understood. Topographical considerations indicate that the damage occurs at the level of the lamina cribrosa, where the axons of the retinal ganglion cells (RGCs) are arranged in bundles to penetrate the sclera and leave the eye.

According to the classical theory, RGC damage occurs due to mechanical stress caused by elevated IOP levels, but it may also occur in the absence of elevated IOP. A more recent hypothesis states that mechanical damage may actually be mediated by the pressure gradient over the lamina cribrosa; therefore, a normal intraocular and a low intracranial pressure may result in a similar pressure gradient as with elevated IOP and normal intracranial pressure [86]. However, additional, non-mechanical, factors are likely to be involved in the pathogenesis of glaucoma. These factors may include ischemia, the presence of reactive oxygen species [87], glutamate excitotoxicity [87], mitochondrial dysfunction [88], autoimmunity [89], and neurodegenerative changes similar to Alzheimer's disease [90]. This section does not consider these factors in detail. The interested reader is referred to the review by Almasieh and coworkers [87].

10.8.3 Retinal Ganglion Cells: Loss vs. Damage

It is not clear whether malfunctioning RGCs play a role in the functional deficits observed in glaucoma, or if only a loss of cells is functionally relevant. On the one hand, there is evidence that RGCs are damaged before their axons are lost. Changes in the polarization behavior of axons at the levels of the RNFL are thought to indicate disruption of the cellular ultrastructure [91] and functional deficits in the inner retina can be demonstrated using pattern and multifocal ERGs [92]. On the other hand, at the time when visual field defects can be reproducibly measured in the standard automated perimetry, already around 20–40 % of RGC axons are lost [93]. Axon loss may either be focal, reflecting the arrangement of the axons while passing the lamina cribrosa, or diffuse affecting the majority of the visual field [94]. Generally, the relationship between morphological changes caused by axon loss and functional defects are still largely unknown. Furthermore, possible RGC dysfunction, that may occur prior to their loss and may be psychophysically measurable, needs to be studied more intensively. Color vision may play an important role in these studies.

10.8.3.1 Selective Damage of RGC Subtypes in Glaucoma

It is not clear whether certain subpopulations of RGCs are affected earlier or more profoundly than others. It has been postulated that the large parasol RGCs that form the magnocellular pathway are more severely affected [95]. Others have provided evidence that larger cells within different subpopulations may be more vulnerable than smaller ones. However, this view has been questioned by more recent studies [96]. The foveal or parafoveal location of the RGCs may also be a relevant factor [95, 97]. Driven by such theories, advanced perimetric tests have been developed that are trying to favor or even isolate one of the three reticulogeniculate pathways (MC, PC, KC). These are summarized in Table 10.6. However, the fact that there are techniques focusing on every one of those pathways suggests that all three may be significantly affected early in glaucoma. A review can be found elsewhere [98].

Table 10.6 Perimetric methods for detecting dysfunction of isolated retinogeniculate pathways

Method	Mechanism
Short-wavelength automated perimetry (SWAP)	Increment threshold of blue flashes on a yellow background favoring of short wavelength (SW)-mechanisms and thus the, "blue-on" RGCs of the koniocellular (KC) pathway
Frequency-doubling technology (FDT) and matrix perimety	Is based on the frequency-doubling technique where the spatial frequency of a flickering grating is observed to be doubled. This perimetry is thought to favor the magnocellular (MC) pathway. But see Ref. [129] for challenges of this idea.
Flicker perimetry	Based on the threshold of luminance flicker and therefore considered to be based on the luminance channel, MC pathway.
Flicker defined form	Detection of a shape formed from dots flickering in counterphase to the other dots; MC pathway
Motion perimetry	Measuring either minimum position displacement, minimum proportion of moving dots in a field of stationary dots, or minimum size of clustered moving dots; motion channel, MC and cortical
High pass resolution perimetry (HPRP)	Spatial resolution; parvocellular (PC) pathway
Rarebit perimetry	Spatial resolution; PC pathway

10.8.4 Foveal Color Vision in Glaucoma Patients

Foveal color vision as measured with the standard tests like the FM-100 or the CAD test is altered in glaucoma [31, 99]. A deterioration of color discrimination has been demonstrated by many authors showing an increased error score with the FM-100 test [29, 99–101] and enlarged matching range with the anomaloscope [100–103].

Results concerning the main axis of color vision deficits were conflicting. Koellner's rule suggests RG defects, because glaucoma is an optic neuropathy [104], and these have been demonstrated many times, but other authors have found only BY defects, or initial BY defects early followed by RG defects [99]. As an explanation, some authors have postulated differences between subtypes of glaucomas—suggesting different pathophysiological mechanisms for distinct color pathways. Studies have found differences between normal-tension and high-tension glaucomas [105] and between pigment dispersion and primary open-angle glaucoma [106], where patients with high-tension glaucoma had lower chromatic and achromatic sensitivities than normal-tension glaucoma patients with comparable visual fields defect and patients with pigmentary glaucoma had less RG defects than those with primary glaucoma.

Often, color vision defects may be found at an early stage, even before the presence of visual field defects [103]. On the other hand, subjects with advanced glaucoma may have normal color vision [102]. The correlation of color vision deficits with morphology is poor, especially in eyes with predominantly local defects [102]. However, recent data are lacking.

Many older studies have not accounted for differences in pupil size, which were often caused by the glaucoma medications used at that time, and changes in the optic media have not been considered either [99, 107].

10.8.5 Perimetric Procedures and Color Vision

Functional defects in glaucoma often start in the periphery. Thus, tests have been developed that test color vision at different retinal locations to account for the topography of damage to the retinal ganglion cells (color perimetry) [99]. Short-wavelength automated perimetry uses a yellow stimulus on a blue background attempting to favor the YB sensitive RGCs of the koniocellular system [108]. A problem in many of the earlier tests has been the fact that the stimulus also contained a luminance component so that a complete response isolation of the S-cones or the YB-RGCs is not warranted. More recent studies [109] have not confirmed earlier evidence that SWAP was able to detect functional defects before the appearance of visual field defects in the standard white-one-white perimetry.

The parvocellular system as the substrate of the red-green (L–M-cone) opponent system is currently not used in combination with perimetry as a diagnostic tool in glaucoma. However, there are perimetric techniques that test for very small spatial discrimination tasks that are thought to rely on the parvocellular system as well (according to the "double duty model"; see Chap. 4). The value of these tests indicates that RG color vision defects may be detected locally early in the course of glaucoma. One problem with the isolation of different channels in the past was that the different techniques for the different pathways were not comparable. Therefore, advantages or disadvantages of a given test may not have been due to a predominant damage to this system, but rather to characteristics of the specific test.

10.8.6 Newer Developments in Glaucoma

One example for such characteristics of perimetric tests is Ricco's area in SAP and SWAP. Redmond and colleagues suggest that problems with the variability and reproducibility of perimetric tests are possibly related to spatial summation of the stimulus [110, 111]. When the stimulus area is smaller than a certain criterion area, called Ricco's area, there is complete spatial summation and the sensitivity is directly proportional to the stimuli area (i.e., doubling the area would also lead to a doubling of the sensitivity) thus following Ricco's law [112]. Once the stimulus is larger than Ricco's area, the increase of sensitivity with increasing stimulus size is less than proportional [110]. To achieve good correlation between structure and function, stimulus size should ideally not exceed Ricco's area, so that there is complete spatial summation. However, the latter may vary with retinal eccentricity and with state of adaptation (for example if blue is added to the background) [110].

Ricco's area is found to be larger in glaucoma patients [110]. The typical stimulus size in SWAP is larger than Ricco's area. The authors suggest that it may be advantageous to change the stimulus sizes in SWAP to match Ricco's area.

Furthermore, efforts have been made to test different post-receptoral channels with similar tests in glaucoma. Antón and coworkers [113] have created a method based on sinusoidal gratings with varying spatial and temporal frequency using stimulus conditions that favor the PC, MC, and KC pathway respectively. To achieve this, the gratings are modulated along the cardinal directions of the opponent modulation spaces. In contrast, Rauscher et al. [31] have used first-order motion perception with static luminance contrast noise to compare RG and YB color detection. Interestingly, both studies have found decreased sensitivity to chromatic RG stimuli to be the best diagnostic criterion in glaucoma patients compared with the sensitivities of the other channels.

There is increasing evidence that glaucomatous changes also affect cortical visual areas [114, 115]. The research concerning the role of these changes for psychophysical techniques, including color vision, has only started. Functional magnetic resonance imaging may play an important role for this.

10.9 Optic Neuritis

Optic neuritis is the most common optic neuropathy in younger patients. It is characterized by inflammation and demyelinization of the optic nerve fibers, as can be demonstrated by a prolonged latency in the visually evoked potentials or the Pulfrich phenomenon, where linear movements of a pendulum in 2D plane are perceived to be elliptically in a 3D plane. Sometimes it is associated with visible swelling of the intraocular portion of the optic nerve (papilledema). There are idiopathic forms and those associated with multiple sclerosis. Because optic neuritis is frequently the first manifestation of multiple sclerosis, they are initially indistinguishable. More information on optic neuritis and multiple sclerosis can be found in Table 10.7.

Table 10.7 Characteristics of optic neuritis

Etiology	Unknown environmental factors (larger incidence in certain temperate regions, migrants have lower risk after migrating to regions with low incidence, but risk increases only in the next generation after migrating to regions with high incidence), also some hereditary risk
Pathophysiology	Inflammation; autoimmune reaction against myelin sheath of the RGC axons in the optic nerve; therefore, loss of saltatory signal conduction
Structural and functional defects	Loss of visual acuity, central scotoma, red-green desaturation, in late stages optic atrophy
Prognosis	Usually recovery of visual acuity, but incomplete recovery of contrast sensitivity and color vision; recurrences frequent; later optic atrophy; a number of patients will develop multiple sclerosis

Patients note an acute unilateral loss of vision with central scotoma, red-green desaturation and loss of contrast sensitivity, frequently accompanied by ocular pain that is aggravated by eye movements. In the course of a view weeks, it resolves spontaneously in most cases with visual acuity returning to 20/20. However, color vision and contrast sensitivity often remain impaired when compared with the other non-affected fellow eye.

Patients are usually treated with a high dose of intravenous steroids. This accelerates recovery, but does not lead to an improved outcome. In the Optic Neuritis Treatment Trial (ONTT), the incidence of multiple sclerosis was lower in the group treated this way at year 2, but not at 3 years or later.

Color vision abnormalities were already noted in the first case reports on optic neuritis. Red and green colors appear desaturated, i.e., washed out. This is an important sign in clinical practice, although not very objective. In contrast to glaucoma, where selective damage of RGC subpopulations is still debated, there is convincing evidence that the small fibers of the parvocellular and koniocellular systems are affected more strongly by demyelinization than the thicker ones of the magnocellular pathway. Objects may also appear darker, when the magnocellular system is also affected. It has to be noted, however, that the reduced function of the chromatic and luminance pathways may interact to result in a perceived desaturation.

When examined with color sorting tests, color vision changes in optic neuritis are thought to be red-green deficits compatible with an optic neuropathy following Koellner's rule or Verriest type II. However, in the ONTT patients, no characteristic color vision defects were associated with optic neuritis [116]. Actually, the main axis of the deficits in the FM-100 test changed during the course of the disease in a considerable number of patients. Generally, YB defects were more frequent at onset, whereas RG defects were more common at a later stage. Patients with better visual acuities tended to have relatively larger YB than RG defects.

Furthermore, the main axis of dyschromatopsia may depend on type of visual field defect (diffuse defects, central scotoma, or altitudinal scotoma) [116, 117]. When foveal visual field defects were present, RG defects dominated, whereas parafoveal scotomata tended to have YB defects. This is possibly related to the fact that S-cones are not present in the fovea and have the largest density in the parafovea.

10.10 Hereditary Neuropathies

Hereditary optic neuropathies are frequently associated with dysfunction of proteins of the respiratory chain, either directly by mutations in the mitochondrial DNA (in Leber's Hereditary Optic Neuropathy; LHON) or indirectly in Autosomal Dominant Optic Atrophy (ADOA) [118]. Both forms may be triggered or aggravated by exposition to toxins, like ethambutol, tobacco or alcohol.

The comparison of these two forms is interesting for color vision scientists, because LHON often follows Koellner's rules, being associated with RG defects, whereas ADOA is a prominent exception. Important characteristics of hereditary optic neuropathies are presented in Table 10.8.

Table 10.8 Characteristics of hereditary optic neuropathies

Etiology	Hereditary defect of single genes associated with mitochondrial function; may be triggered/aggravated by toxins
Pathophysiology	Apoptosis of RGC axons
Structural and functional defects	Loss of visual acuity, different color vision defects, optic atrophy
Prognosis	Visual acuity relatively preserved in ADOA, often poor visual acuity in LHON

10.10.1 Leber's Hereditary Optic Neuropathy (LHON)

LHON is caused by mutations in the mitochondrial DNA. Therefore, it follows a maternal pattern of inheritance: men do not transmit the defect, whereas all children of an affected mother inherit the mutation. However, as a result of incomplete penetrance only 50 % of male and 15 % of female carriers of the mutation develop the disease. It is not known, why these mutations do not seem to manifest in other organ systems than the retina or why patients remain asymptomatic up to early adulthood before there is a severe subacute bilateral loss of vision (visual acuity is frequently 20/200 or worse). Morphologically the subacute stage is associated with disc hyperemia, abnormally dilated peripapillary vessels and edema of the retinal nerve fibers, whereas optic atrophy with pallor of the optic disc develops later on.

Color vision defects are predominantly RG defects of Verriest type II [29], but YB discrimination may also be affected [119]. There is evidence for a preferred damage to small RGC fibers, especially in the papillomacular bundle [120]. Of note, color vision defects were also demonstrated in asymptomatic carriers [121, 122].

10.10.2 Autosomal Dominant Optic Atrophy (ADOA)

ADOA is the most common form of inheritable optic atrophy. In contrast to LHON, it starts slowly at an early age of 4–8 years, often without being noticed. Visual acuity is only mildly reduced. Morphologically, a pallor of the optic disc can be observed. The mutation is located in the *OPA1* gene on chromosome 3, but this gene seems to play an important role for the correct mitochondrial function.

ADOA is the example par excellence of an exception to Koellner's rule because it is characterized by YB defects, in contrast to LHON and other optic neuropathies [123]. However, RG defects have also been found [124]. Possibly, the prominently affected color axis is also determined by the retinal location of the visual field defects [125]. It seems that defects of the papillomacular bundle which are usually accompanied by severe impairment of visual acuity (as in LHON) are associated with RG defects, whereas YB defects seem to be more common in subjects with relatively preserved vision and superotemporal or inferotemporal defects of the retinal nerve fiber layer (as in glaucoma and ADOA).

10.11 Central Achromatopsia

Central defects of the visual system can also cause defective color vision. However, color vision defects are not so characteristic for lesions of the chiasma, the optic tract, or the optic radiation. Central achromatopsia is commonly caused by ischemic insults of occipitotemporal cortex, affecting the area V4, V8, and V4α [126–128]. If the defect is unilateral, there will be hemiachromatopsia, i.e., defects are restricted to homonymous (the same side of the visual field for the two eyes) hemifields. Central achromatopsia is usually associated with homonymous superior visual field defects (characteristic for damage of the optic radiation in the temporal lobe) or prosopagnosia, where subjects can see and describe faces but not name familiar faces although language and intellect is intact. This is caused by interruption of pathways that connect the visual cortex (V1) with higher centers of vision.

10.12 Using Color Vision Test in Clinical Practice

A typical approach when a patient visits the clinic for the first time is initially asking for a detailed history of current and past disease (anamnesis), including the symptoms, after which a basic clinical exam is performed. In ophthalmology this includes measuring refraction and visual acuity, slit-lamp examination, and inspection of the fundus. Depending on the history of the patients, a screening for color vision defects may be performed. Most commonly, this is performed with pseudoisochromatic plates. The Ishihara plates are designed specifically to screen for congenital X-linked defects (see above). The Hardy–Rand–Rittler plates include plates for blue-yellow defects and are more useful to identify acquired color vision defects. Alternatively, a desaturated Panel D15 test can be used to find color vision defects in acquired disease.

If defects are identifies, further color vision tests may be initiated to further classify and grade the abnormality. This may include the Farnsworth-Munsell 100 Hue test for acquired deficiencies and the anomaloscope with the Rayleigh equation to classify the X-linked defects. By including upper normal threshold age limits for healthy color vision, the CAD test may have enough sensitivity and specificity to detect changes in color vision that precede recognizable clinical signs. The anomaloscope can also facilitate distinguishing photoreceptor disease from optic neuropathies in unexplained visual loss with blue-yellow defects, by demonstrating pseudoprotanomaly. Scotopization can also be useful in indicating progressive degeneration of cones.

10.13 Case Reports

10.13.1 Case Report 1

A 20-year-old man visits the eye clinic for further evaluation. He applied for a job in the police force, but failed to identify numbers on the pseudoisochromatic plate test used to screen for color vision abnormalities during the required medical examination. He strongly rejects the notion of having a disease and wants the ophthalmologist to attest normal color vision, so that he can pursue his career. When examined with the anomaloscope, the Rayleigh match was shifted towards the green, but the matching range was narrow, so that the diagnosis of a deuteranomaly was made. He leaves the clinic angrily, saying that he will find a more competent doctor who will recognize that he can see perfectly well.

10.13.2 Case Report 2

A 26-year-old woman is referred for a neuroophthalmologic examination by a general ophthalmologist who had performed a routine examination. The reason for the referral was a slightly subnormal visual acuity and pallor of the temporal parts of the optic disc in both eyes. The patient has not noted any symptoms but reports that her sister has been diagnosed to have a color vision defect in early childhood. Her mother has normal vision, but they do not know their father, who has left the family. Visual acuity is 20/40 and the Panel D15 shows several errors along the tritan axis. Examination with the anomaloscope and the Rayleigh equation are unremarkable.

An extensive work-up for optic atrophy is performed showing normal brain scans, neurologic examinations and cerebrospinal fluid. Genetic examination shows pathogenic mutations in the *OPA1* gene, so that the diagnosis of autosomal-dominant optic atrophy is made.

10.13.3 Case Report 3

A 55-year-old woman is seen by an ophthalmologist because her mother has gone blind from glaucoma at the age of 75 and she wants to know whether this may also happen to her. Visual acuity and white-on-white perimetry are normal. The anterior portion of the eye appears unremarkable with slit-lamp examination. The intraocular pressure is slightly elevated at 24 mmHg. The retina is normal. The optic disc is tilted and has an area that is larger than average, so that it cannot be judged whether there is glaucomatous damage. The patient has two factors indicating an elevated risk for glaucoma: the family history and the elevated intraocular pressure. The ophthalmologist explains that she may choose to receive treatment against the elevated intraocular pressure to reduce her risk of developing glaucoma, but treatment may have harmful side effects and one cannot be sure if she will develop glaucoma if she is not treated. Therefore, she could also wait and be monitored frequently for visual field defects. She finds both options unsatisfactory and asks if there is another test to determine if she needs treatment. The ophthalmologists offers making a decision based on the detections of early functional defects in short-wavelength automated perimetry.

10.13.4 Case Report 4

A 14-year-old boy and his mother are referred for further examination because he has problems seeing the blackboard at school and an attempt to prescribe glasses resulted in little improvement. His best corrected visual acuity is markedly reduced so that he cannot read regular newspaper print at a normal reading distance. Everybody else in the family has normal vision, with the exception of an uncle who has to wear strong glasses. Funduscopic examination shows a slight irregularity of the macular retinal pigment epithelium and a temporal pallor of the optic disc. The Panel D-15 test shows very mild abnormalities along the tritan axis and the Rayleigh match is shifted towards red. This is an indication for a retinal disease rather than an optic neuropathy, so that a multifocal ERG is performed. Here, a marked decrease of the central amplitudes is diagnostic of a macular dystrophy. Two mutations of the *ABCA4* gene that are known to be associated with Stargardt disease can be demonstrated after genetic examination. During a follow-up examination at age 25, there is a zone of atrophy of the retinal pigment epithelium involving the fovea centralis that can be easily identified on funduscopy. Furthermore, there are some yellow flecks surrounding the central atrophy (fundus flavimaculatus).

References

1. Jacobs GH. The discovery of spectral opponency in visual systems and its impact on understanding the neurobiology of color vision. J Hist Neurosci. 2014;23:287–314.
2. Hurvich LM, Jameson D. An opponent-process theory of color vision. Psychol Rev. 1957;64(Part 1):384–404.
3. Rodriguez-Carmona M, O'Neill-Biba M, Barbur JL. Assessing the severity of color vision loss with implications for aviation and other occupational environments. Aviat Space Environ Med. 2012;83:19–29.
4. Zele AJ, Cao D. Vision under mesopic and scotopic illumination. Percept Sci. 2015;5:1594.
5. Walkey HC, Barbur JL, Harlow JA, Makous W. Measurements of chromatic sensitivity in the mesopic range. Color Res Appl. 2001;26:S36–42.
6. Terasaki H, Miyake Y, Nomura R, Horiguchi M, Suzuki S, Kondo M. Blue-on-yellow perimetry in the complete type of congenital stationary night blindness. Invest Ophthalmol Vis Sci. 1999;40:2761–4.
7. Schmidt BP, Neitz M, Neitz J. Neurobiological hypothesis of color appearance and hue perception. J Opt Soc Am A Opt Image Sci Vis. 2014;31:A195–207.

8. Gardner TW, Antonetti DA, Barber AJ, LaNoue KF, Levison SW. Diabetic retinopathy: more than meets the eye. Surv Ophthalmol. 2002;47 Suppl 2:253–62.
9. Squire TJ, Rodriguez-Carmona M, Evans ADB, Barbur JL. Color vision tests for aviation: comparison of the anomaloscope and three lantern types. Aviat Space Environ Med. 2005;76:421–9.
10. Ballard J. Colour-vision safety on track. Occup Health Work. 2013;10:20–3.
11. Barbur JL, Rodriguez-Carmona ML. Variability in normal and defective colour vision: consequences for occupational environments. In: Best J, editor. Colour des. Cambridge: Woodhead; 2012. p. 24–82.
12. Hardy LH, Rand G, Rittler MC. A screening test for defective red-green vision; test based on eighteen pseudoisochromatic plates from the American Optical Company's compilation. Arch Ophthalmol (Chic Ill 1929). 1947;38:442–9.
13. Cole BL, Lian K-Y, Lakkis C. The new Richmond HRR pseudoisochromatic test for colour vision is better than the Ishihara test. Clin Exp Optom. 2006;89:73–80.
14. Cole BL. Assessment of inherited colour vision defects in clinical practice. Clin Exp Optom. 2007;90:157–75.
15. Pokorny J, Smith VC. Effect of field size on red–green color mixture equations. J Opt Soc Am. 1976;66:705–8.
16. Barbur JL, Rodriguez-Carmona M, Harlow JA, Mancuso K, Neitz J, Neitz M. A study of unusual Rayleigh matches in deutan deficiency. Vis Neurosci. 2008;25:507–16.
17. Moreland JD, Kerr J. Optimization of stimuli for trit-anomaloscopy. Mod Probl Ophthalmol. 1978;19:162–6.
18. Moreland JD, Kerr J. Optimization of a Rayleigh-type equation for the detection of tritanomaly. Vision Res. 1979;19:1369–75.
19. Parry NRA. Color vision deficiencies. In: Elliot AJ, Fairchild MD, Franklin A, editors. Handbook of color psychology. Cambridge: Cambridge University Press; 2015. p. 216–42.
20. Rabin J, Gooch J, Ivan D. Rapid quantification of color vision: the cone contrast test. Invest Ophthalmol Vis Sci. 2011;52:816–20.
21. Rabin J. Cone-specific measures of human color vision. Invest Ophthalmol Vis Sci. 1996;37:2771–4.
22. Rabin J. Quantification of color vision with cone contrast sensitivity. Vis Neurosci. 2004;21:483–5.
23. Regan BC, Reffin JP, Mollon JD. Luminance noise and the rapid determination of discrimination ellipses in colour deficiency. Vision Res. 1994;34:1279–99.
24. Paramei GV. Color discrimination across four life decades assessed by the Cambridge Colour Test. J Opt Soc Am A Opt Image Sci Vis. 2012;29:A290–7.
25. Barbur JL, Rodriguez-Carmona ML. Color vision changes in normal aging. In: Elliot AJ, Fairchild MD, Franklin A, editors. Handbook of color psychology. Cambridge: Cambridge University Press; 2015. p. 180–96.
26. Barbur JL. "Double-blindsight" revealed through the processing of color and luminance contrast defined motion signals. Prog Brain Res. 2004;144:243–59.
27. Barbur JL, Moro S, Harlow JA, Lam BL, Liu M. Comparison of pupil responses to luminance and colour in severe optic neuritis. Clin Neurophysiol. 2004;115:2650–8.
28. Moro SI, Rodriguez-Carmona ML, Frost EC, Plant GT, Barbur JL. Recovery of vision and pupil responses in optic neuritis and multiple sclerosis. Ophthalmic Physiol Opt. 2007;27:451–60.
29. Verriest G. Further studies on acquired deficiency of color discrimination. JOSA. 1963;53:185–97.
30. Pokorny J, Smith VC, Verriest G, Pinckers A. Congenital and acquired color vision defects. New York: Grune & Stratton; 1979.
31. Rauscher FG, Chisholm CM, Edgar DF, Barbur JL. Assessment of novel binocular colour, motion and contrast tests in glaucoma. Cell Tissue Res. 2013;353:297–310.
32. Roosing S, Thiadens AAHJ, Hoyng CB, Klaver CCW, den Hollander AI, Cremers FPM. Causes and consequences of inherited cone disorders. Prog Retin Eye Res. 2014;42:1–26.

33. Kocur I, Resnikoff S. Visual impairment and blindness in Europe and their prevention. Br J Ophthalmol. 2002;86:716–22.
34. Lois N, Holder GE, Fitzke FW, Plant C, Bird AC. Intrafamilial variation of phenotype in Stargardt macular dystrophy-Fundus flavimaculatus. Invest Ophthalmol Vis Sci. 1999;40: 2668–75.
35. McClements ME, MacLaren RE. Gene therapy for retinal disease. Transl Res. 2013;161:241–54.
36. Carroll J, Dubra A, Gardner JC, Mizrahi-Meissonnier L, Cooper RF, Dubis AM, et al. The effect of cone opsin mutations on retinal structure and the integrity of the photoreceptor mosaic. Invest Ophthalmol Vis Sci. 2012;53:8006–15.
37. Carroll J, Neitz M, Hofer H, Neitz J, Williams DR. Functional photoreceptor loss revealed with adaptive optics: an alternate cause of color blindness. Proc Natl Acad Sci U S A. 2004;101:8461–6.
38. Carroll J, Rossi EA, Porter J, Neitz J, Roorda A, Williams DR, et al. Deletion of the X-linked opsin gene array locus control region (LCR) results in disruption of the cone mosaic. Vision Res. 2010;50:1989–99.
39. Fujinami K, Sergouniotis PI, Davidson AE, Wright G, Chana RK, Tsunoda K, et al. Clinical and molecular analysis of Stargardt disease with preserved foveal structure and function. Am J Ophthalmol. 2013;156:487–501.e1.
40. Auricchio A, Trapani I, Allikmets R. Gene therapy of ABCA4-associated diseases. Cold Spring Harb Perspect Med. 2015;5:a017301.
41. Allikmets R, Singh N, Sun H, Shroyer NF, Hutchinson A, Chidambaram A, et al. A photoreceptor cell-specific ATP-binding transporter gene (ABCR) is mutated in recessive Starqardt macular dystrophy. Nat Genet. 1997;15:236–46.
42. Molday LL, Rabin AR, Molday RS. ABCR expression in foveal cone photoreceptors and its role in Stargardt macular dystrophy. Nat Genet. 2000;25:257–8.
43. Sparrow JR, Fishkin N, Zhou J, Cai B, Jang YP, Krane S, et al. A2E, a byproduct of the visual cycle. Vision Res. 2003;43:2983–90.
44. Lois N, Halfyard AS, Bird AC, Holder GE, Fitzke FW. Fundus autofluorescence in Stargardt macular dystrophy-fundus flavimaculatus. Am J Ophthalmol. 2004;138:55–63.
45. Sparrow JR, Nakanishi K, Parish CA. The lipofuscin fluorophore A2E mediates blue light-induced damage to retinal pigmented epithelial cells. Invest Ophthalmol Vis Sci. 2000;41:1981–9.
46. Vandenbroucke T, Buyl R, De Zaeytijd J, Bauwens M, Uvijls A, De Baere E, et al. Colour vision in Stargardt disease. Ophthalmic Res. 2015;54:181–94.
47. Pinckers A, Uvijls A. The luminance fall in anomaloscope examination: clinical examples. Acta Ophthalmol Scand. 2000;78:465–7.
48. Maia-Lopes S, Silva ED, Silva MF, Reis A, Faria P, Castelo-Branco M. Evidence of widespread retinal dysfunction in patients with stargardt disease and morphologically unaffected carrier relatives. Invest Ophthalmol Vis Sci. 2008;49:1191–9.
49. Pokorny J, Smith VC, Ernest JT. Macular color vision defects: specialized psychophysical testing in acquired and hereditary chorioretinal diseases. Int Ophthalmol Clin. 1980;20:53–81.
50. Marmorstein AD, Marmorstein LY, Rayborn M, Wang X, Hollyfield JG, Petrukhin K. Bestrophin, the product of the Best vitelliform macular dystrophy gene (VMD2), localizes to the basolateral plasma membrane of the retinal pigment epithelium. Proc Natl Acad Sci. 2000;97:12758–63.
51. Marquardt A, Stöhr H, Passmore LA, Krämer F, Rivera A, Weber BH. Mutations in a novel gene, VMD2; encoding a protein of unknown properties cause juvenile-onset vitelliform macular dystrophy (Best's disease). Hum Mol Genet. 1998;7:1517–25.
52. Hartzell HC, Qu Z, Yu K, Xiao Q, Chien L-T. Molecular physiology of bestrophins: multifunctional membrane proteins linked to best disease and other retinopathies. Physiol Rev. 2008;88:639–72.
53. Campos SH, Forjaz V, Kozak LR, Silva E, Castelo-Branco M. Quantitative phenotyping of chromatic dysfunction in best macular dystrophy. Arch Ophthalmol (Chic Ill 1960). 2005;123:944–9.

54. McClements M, Davies WIL, Michaelides M, Carroll J, Rha J, Mollon JD, et al. X-linked cone dystrophy and colour vision deficiency arising from a missense mutation in a hybrid L/M cone opsin gene. Vision Res. 2013;80:41–50.
55. Baraas RC, Carroll J, Gunther KL, Chung M, Williams DR, Foster DH, et al. Adaptive optics retinal imaging reveals S-cone dystrophy in tritan color-vision deficiency. J Opt Soc Am A Opt Image Sci Vis. 2007;24:1438–47.
56. Van Schooneveld MJ, Went LN, Oosterhuis JA. Dominant cone dystrophy starting with blue cone involvement. Br J Ophthalmol. 1991;75:332–6.
57. Bresnick GH, Smith VC, Pokorny J. Autosomal dominantly inherited macular dystrophy with preferential short-wavelength sensitive cone involvement. Am J Ophthalmol. 1989;108:265–76.
58. Hamel C. Retinitis pigmentosa. Orphanet J Rare Dis. 2006;1:40.
59. Jacobson SG, Roman AJ, Aleman TS, Sumaroka A, Herrera W, Windsor EAM, et al. Normal central retinal function and structure preserved in retinitis pigmentosa. Invest Ophthalmol Vis Sci. 2010;51:1079–85.
60. Swanson WH, Fish GE. Color matches in diseased eyes with good acuity: detection of deficits in cone optical density and in chromatic discrimination. J Opt Soc Am A Opt Image Sci Vis. 1995;12:2230–6.
61. Pinckers A, van Aarem A, Keunen JE. Colour vision in retinitis pigmentosa. Influence of cystoid macular edema. Int Ophthalmol. 1993;17:143–6.
62. Scholl HPN, Kremers J. Large phase differences between L-cone– and M-cone–driven electroretinograms in retinitis pigmentosa. Invest Ophthalmol Vis Sci. 2000;41:3225–33.
63. Scholl HPN, Kremers J. Alterations of L- and M-cone driven ERGs in cone and cone–rod dystrophies. Vision Res. 2003;43:2333–44.
64. Scholl HP, Kremers J, Vonthein R, White K, Weber BH. L- and M-cone-driven electroretinograms in Stargardt's macular dystrophy-fundus flavimaculatus. Invest Ophthalmol Vis Sci. 2001;42:1380–9.
65. Scholl HPN, Kremers J, Apfelstedt-Sylla E, Zrenner E. L- and M-cone driven ERGs are differently altered in Best's macular dystrophy. Vision Res. 2000;40:3159–68.
66. Kuchenbecker JA, Greenwald SH, Neitz M, Neitz J. Cone-isolating ON-OFF electroretinogram for studying chromatic pathways in the retina. J Opt Soc Am A Opt Image Sci Vis. 2014;31:A208–13.
67. Harmening WM, Tuten WS, Roorda A, Sincich LC. Mapping the perceptual grain of the human retina. J Neurosci. 2014;34:5667–77.
68. Krill AE, Archer DB, editors. Krill's hereditary retinal and choroidal disease. Hagerstown: Harper & Row; 1977.
69. Hood DC, Benimoff NI, Greenstein VC. The response range of the blue-cone pathways: a source of vulnerability to disease. Invest Ophthalmol Vis Sci. 1984;25:864–7.
70. Yau JWY, Rogers SL, Kawasaki R, Lamoureux EL, Kowalski JW, Bek T, et al. Global prevalence and major risk factors of diabetic retinopathy. Diabetes Care. 2012;35:556–64.
71. Antonetti DA, Barber AJ, Bronson SK, Freeman WM, Gardner TW, Jefferson LS, et al. Diabetic retinopathy seeing beyond glucose-induced microvascular disease. Diabetes. 2006;55:2401–11.
72. Lieth E, Gardner TW, Barber AJ, Antonetti DA. Retinal neurodegeneration: early pathology in diabetes. Clin Exp Ophthalmol. 2000;28:3–8.
73. Zhang X, Wang N, Barile GR, Bao S, Gillies M. Diabetic retinopathy: neuron protection as a therapeutic target. Int J Biochem Cell Biol. 2013;45:1525–9.
74. Tolentino MJ, Miller JW, Gragoudas ES, Jakobiec FA, Flynn E, Chatzistefanou K, et al. Intravitreous injections of vascular endothelial growth factor produce retinal ischemia and microangiopathy in an adult primate. Ophthalmology. 1996;103:1820–8.
75. Rodgers M, Hodges R, Hawkins J, Hollingworth W, Duffy S, McKibbin M, et al. Colour vision testing for diabetic retinopathy: a systematic review of diagnostic accuracy and economic evaluation. Health Technol Assess. 2009;13. Available from: http://www.journalslibrary.nihr.ac.uk/hta/volume-13/issue-60. Accessed 7Oct 7, 2015.

76. Kurtenbach A, Kernstock C, Zrenner E, Langrová H. Electrophysiology and colour: a comparison of methods to evaluate inner retinal function. Doc Ophthalmol Adv Ophthalmol. 2015;131:159–67.
77. Mangione CM, Lee PP, Gutierrez PR, et al. Development of the 25-list-item national eye institute visual function questionnaire. Arch Ophthalmol. 2001;119:1050–8.
78. Bearse MA, Han Y, Schneck ME, Barez S, Jacobsen C, Adams AJ. Local multifocal oscillatory potential abnormalities in diabetes and early diabetic retinopathy. Invest Ophthalmol Vis Sci. 2004;45:3259–65.
79. Bresnick GH. Diabetic retinopathy viewed as a neurosensory disorder. Arch Ophthalmol (Chic Ill 1960). 1986;104:989–90.
80. Greenstein VC, Shapiro A, Zaidi Q, Hood DC. Psychophysical evidence for post-receptoral sensitivity loss in diabetics. Invest Ophthalmol Vis Sci. 1992;33:2781–90.
81. Parisi V, Uccioli L. Visual electrophysiological responses in persons with type 1 diabetes. Diabetes Metab Res Rev. 2001;17:12–8.
82. O'Neill-Biba M, Sivaprasad S, Rodriguez-Carmona M, Wolf JE, Barbur JL. Loss of chromatic sensitivity in AMD and diabetes: a comparative study. Ophthalmic Physiol Opt J Br Coll Ophthalmic Opt Optom. 2010;30:705–16.
83. Quigley HA, Broman AT. The number of people with glaucoma worldwide in 2010 and 2020. Br J Ophthalmol. 2006;90:262–7.
84. Prum Jr. BE, Lim MC, Mansberger SL, Stein JD, Moroi SE, Gedde SJ, et al. Primary open-angle glaucoma suspect preferred practice pattern® guidelines. Ophthalmology. http://www.sciencedirect.com/science/article/pii/S0161642015012786. Accessed 19 Nov, 2015.
85. Prum BE, Rosenberg LF, Gedde SJ, Mansberger SL, Stein JD, Moroi SE, et al. Primary open-angle glaucoma preferred practice pattern® guidelines. Ophthalmology. 2015 http://linkinghub.elsevier.com/retrieve/pii/S0161642015012762. Accessed 19 Nov, 2015.
86. Morgan WH, Balaratnasingam C, Lind CRP, Colley S, Kang MH, House PH, et al. Cerebrospinal fluid pressure and the eye. Br J Ophthalmol. 2016;100(1):71–7. bjophthalmol – 2015–306705.
87. Almasieh M, Wilson AM, Morquette B, Cueva Vargas JL, Di Polo A. The molecular basis of retinal ganglion cell death in glaucoma. Prog Retin Eye Res. 2012;31:152–81.
88. Lee S, Van Bergen NJ, Kong GY, Chrysostomou V, Waugh HS, O'Neill EC, et al. Mitochondrial dysfunction in glaucoma and emerging bioenergetic therapies. Exp Eye Res. 2011;93:204–12.
89. Bell K, Gramlich OW, Von Thun Und Hohenstein-Blaul N, Beck NS, Funke S, Wilding C, et al. Does autoimmunity play a part in the pathogenesis of glaucoma? Prog Retin Eye Res. 2013;36:199–216.
90. Doucette LP, Rasnitsyn A, Seifi M, Walter MA. The interactions of genes, age, and environment in glaucoma pathogenesis. Surv Ophthalmol. 2015;60:310–26.
91. Fortune B, Cull G, Reynaud J, Wang L, Burgoyne CF. Relating retinal ganglion cell function and retinal nerve fiber layer (RNFL) retardance to progressive loss of RNFL thickness and optic nerve axons in experimental glaucoma. Investig OpthalmolVis Sci. 2015;56:3936.
92. Porciatti V. Electrophysiological assessment of retinal ganglion cell function. Exp Eye Res. 2015;141:164–70.
93. Quigley HA, Dunkelberger GR, Green WR. Retinal ganglion cell atrophy correlated with automated perimetry in human eyes with glaucoma. Am J Ophthalmol. 1989;107:453–64.
94. Jonas JB, Dichtl A. Evaluation of the retinal nerve fiber layer. Surv Ophthalmol. 1996;40:369–78.
95. Quigley HA, Dunkelberger GR, Green WR. Chronic human glaucoma causing selectively greater loss of large optic nerve fibers. Ophthalmology. 1988;95:357–63.
96. Morgan JE. Selective cell death in glaucoma: does it really occur? Br J Ophthalmol. 1994;78:875–9.
97. Dandona L, Hendrickson A, Quigley HA. Selective effects of experimental glaucoma on axonal transport by retinal ganglion cells to the dorsal lateral geniculate nucleus. Invest Ophthalmol Vis Sci. 1991;32:1593–9.

98. Anderson RS. The psychophysics of glaucoma: improving the structure/function relationship. Prog Retin Eye Res. 2006;25:79–97.
99. Sample PA, Weinreb RN, Boynton RM. Acquired dyschromatopsia in glaucoma. Surv Ophthalmol. 1986;31:54–64.
100. Drance SM, Lakowski R, Schulzer M, Douglas GR. Acquired color vision changes in glaucoma. Use of 100-hue test and Pickford anomaloscope as predictors of glaucomatous field change. Arch Ophthalmol (Chic Ill 1960). 1981;99:829–31.
101. Lakowski R. The Pickford-Nicolson anomaloscope as a test for acquired dyschromatopsias. Mod Probl Ophthalmol. 1972;11:25–33.
102. Airaksinen PJ, Lakowski R, Drance SM, Price M. Color vision and retinal nerve fiber layer in early glaucoma. Am J Ophthalmol. 1986;101:208–13.
103. Hamill TR, Post RB, Johnson CA, Keltner JL. Correlation of color vision deficits and observable changes in the optic disc in a population of ocular hypertensives. Arch Ophthalmol (Chic Ill 1960). 1984;102:1637–9.
104. Köllner H. Die Störungen des Farbsinnes. Berlin: Karger; 1912.
105. Yamazaki Y, Lakowski R, Drance SM. A comparison of the blue color mechanism in high- and low-tension glaucoma. Ophthalmology. 1989;96:12–5.
106. Greenstein VC, Halevy D, Zaidi Q, Koenig KL, Ritch RH. Chromatic and luminance systems deficits in glaucoma. Vision Res. 1996;36:621–9.
107. Sample PA, Boynton RM, Weinreb RN. Isolating the color vision loss in primary open-angle glaucoma. Am J Ophthalmol. 1988;106:686–91.
108. Sample PA. Short-wavelength automated perimetry: its role in the clinic and for understanding ganglion cell function. Prog Retin Eye Res. 2000;19:369–83.
109. Sample PA, Medeiros FA, Racette L, Pascual JP, Boden C, Zangwill LM, et al. Identifying glaucomatous vision loss with visual-function-specific perimetry in the diagnostic innovations in glaucoma study. Investig Opthalmol Vis Sci. 2006;47:3381.
110. Redmond T, Garway-Heath DF, Zlatkova MB, Anderson RS. Sensitivity loss in early glaucoma can be mapped to an enlargement of the area of complete spatial summation. Investig Opthalmol Vis Sci. 2010;51:6540.
111. Redmond T, Zlatkova MB, Vassilev A, Garway-Heath DF, Anderson RS. Changes in Ricco's area with background luminance in the S-cone pathway: optom. Vis Sci. 2013;90:66–74.
112. Ricco A. Relazione fra il minimo angolo visuale e l'intensità luminosa. Mem Della Regia Acad Sci Lett Ed Arti Modena. 1877;17:47–160.
113. Antón A, Capilla P, Morilla-Grasa A, Luque MJ, Artigas JM, Felipe A. Multichannel functional testing in normal subjects, glaucoma suspects, and glaucoma patients. Investig Opthalmol Vis Sci. 2012;53:8386.
114. Weber AJ, Chen H, Hubbard WC, Kaufman PL. Experimental glaucoma and cell size, density, and number in the primate lateral geniculate nucleus. Invest Ophthalmol Vis Sci. 2000;41:1370–9.
115. Yücel YH, Zhang Q, Weinreb RN, Kaufman PL, Gupta N. Atrophy of relay neurons in magno- and parvocellular layers in the lateral geniculate nucleus in experimental glaucoma. Invest Ophthalmol Vis Sci. 2001;42:3216–22.
116. Katz B. The dyschromatopsia of optic neuritis: a descriptive analysis of data from the optic neuritis treatment trial. Trans Am Ophthalmol Soc. 1995;93:685–708.
117. Silverman SE, Hart WM, Gordon MO, Kilo C. The dyschromatopsia of optic neuritis is determined in part by the foveal/perifoveal distribution of visual field damage. Invest Ophthalmol Vis Sci. 1990;31:1895–902.
118. Skidd PM, Lessell S, Cestari DM. Autosomal dominant hereditary optic neuropathy (ADOA): a review of the genetics and clinical manifestations of ADOA and ADOA+. Semin Ophthalmol. 2013;28:422–6.
119. Effects of idebenone on color vision in patients with Leber hereditary optic neuropathy. http://pubmedcentralcanada.ca/pmcc/articles/PMC3658961/;jsessionid=849539C2104C5431E02A2F5A49B0A573.eider?lang=en-ca. Accessed 27 Nov, 2015.

120. Sadun AA, Win PH, Ross-Cisneros FN, Walker SO, Carelli V. Leber's hereditary optic neuropathy differentially affects smaller axons in the optic nerve. Trans Am Ophthalmol Soc. 2000;98:223.
121. Ventura DF, Gualtieri M, Oliveira AGF, Costa MF, Quiros P, Sadun F, et al. Male prevalence of acquired color vision defects in asymptomatic carriers of Leber's hereditary optic neuropathy. Invest Ophthalmol Vis Sci. 2007;48:2362–70.
122. Quiros PA, Torres RJ, Salomao S, Berezovsky A, Carelli V, Sherman J, et al. Colour vision defects in asymptomatic carriers of the Leber's hereditary optic neuropathy (LHON) mtDNA 11778 mutation from a large Brazilian LHON pedigree: a case-control study. Br J Ophthalmol. 2006;90:150–3.
123. Jaeger W. [Hereditary optic atrophy with dominant transmission; with special reference to the associated color-sense disorder]. Albrecht Von Graefes Arch Für Ophthalmol. 1954;155:457–84.
124. Votruba M, Fitzke FW, Holder GE, Carter A, Bhattacharya SS, Moore AT. Clinical features in affected individuals from 21 pedigrees with dominant optic atrophy. Arch Ophthalmol (Chic Ill 1960). 1998;116:351–8.
125. Hart Jr WM. Acquired dyschromatopsias. Surv Ophthalmol. 1987;32:10–31.
126. Adachi-Usami E, Tsukamoto M, Shimada Y. Color vision and color pattern visual evoked cortical potentials in a patient with acquired cerebral dyschromatopsia. Doc Ophthalmol Adv Ophthalmol. 1995;90:259–69.
127. Bouvier SE, Engel SA. Behavioral deficits and cortical damage loci in cerebral achromatopsia. Cereb Cortex. 2006;16:183–91.
128. Crognale MA, Duncan CS, Shoenhard H, Peterson DJ, Berryhill ME. The locus of color sensation: cortical color loss and the chromatic visual evoked potential. J Vis. 2013;13. http://www.journalofvision.org/content/13/10/15. Accessed 31 Aug, 2013.
129. Swanson WH, Sun H, Lee BB, Cao D. Responses of primate retinal ganglion cells to perimetric stimuli. Invest Ophthalmol Vis Sci. 2011;52:764–71.
130. Gale CR, Martyn CN. Migrant studies in multiple sclerosis. Prog Neurobiol. 1995;47:425–48.

Chapter 11
Evolution of Color Vision

Almut Kelber and Gerald H. Jacobs

Abstract Research conducted in recent years has documented the widespread presence of various forms of color vision in species from across the animal kingdom, helped to develop an understanding of the basic biological mechanisms that underlie this sensory capacity, and provided some insights into the utility of color vision in the natural world. This chapter draws information from all these various sources to reveal what has been learned about the evolution of color vision. In doing so, we examine the extent and nature of animal color vision, describe various structural features of eyes and photoreceptors that condition color vision, look at photopigment opsin genes and their color vision concomitants, and comment on specific cases where progress has been made toward understanding the functional utility of color vision. Although results from a wide range of different species are considered, we focus particular attention on issues relevant to the evolution of vertebrate color vision.

Keywords Photopigment evolution • Chromatic vision • Color vision diversity • Retinal mosaic • Opsin expression patterns • Vertebrate color vision • Monochromacy • Dichromacy • Trichromacy • Tetrachromacy

11.1 Introduction

Experimental studies of human color vision started in earnest around two hundred years ago [1], and by the end of the eighteenth century studies of nonhuman species also began to appear. Results from many of these early animal studies were flawed because experimenters assumed erroneously that the visual sensitivities of other animals were necessarily the same as those of humans. But there were some

A. Kelber (✉)
Department of Biology, Lund University, Sölvegatan 35, 22362 Lund, Sweden
e-mail: almut.kelber@biol.lu.se

G.H. Jacobs
Department of Psychological and Brain Sciences, University of California, Santa Barbara, CA 93106, USA
e-mail: jacobs@psych.ucsb.edu

J. Kremers et al. (eds.), *Human Color Vision*, Springer Series in Vision Research, DOI 10.1007/978-3-319-44978-4_11

significant advances; for example, the work of von Frisch [2, 3] showed clearly that species as disparate as honeybees and fish both have a capacity for color vision. In the years that followed behavioral proof of color vision was reported for a number of other species. Despite the accumulation of such results, Gordon Walls [4], the great comparative vision specialist, was led to conclude that, "the decades of study of the distribution of color vision in vertebrates has thus yielded no play-by-play account of the course of human color vision evolution."

In the years since Walls' pessimistic conclusion we have learned that color vision is widespread across the animal kingdom and common in most clades where advanced eyes occur. These facts imply that color vision must have evolved early and often. Color vision has been most intensively studied in arthropods and chordates, but color vision—or at least the preconditions for this sensory capability—has also evolved independently in molluscs and polychaetes, and it may still await recognition in additional phyla. All of this work has provided significant insights into many of the puzzling questions surrounding the evolution of color vision. This chapter reviews what has been learned. In doing so, we consider the distribution and utility of color vision across the animal kingdom, comment on the mechanisms that support and condition color vision, and describe various scenarios that have been advanced to explain the evolution of color vision.

11.2 How and Where did Color Vision Evolve?

11.2.1 Preconditions for Color Vision

To support color vision—that is the ability of an animal to discriminate spectral differences irrespective of variations in intensity—two basic conditions have to be met: An animal has to (1) possess photoreceptor cells with different spectral sensitivities, and (2) have some means for comparing the signals from the different types of photoreceptors.

The first condition evolved repeatedly because the visual photopigments that all animals use embed a chromophore in an opsin protein that, depending on the polarity of a small number of amino acids in the binding pocket, adjusts the peak spectral sensitivity of the photopigment to values lying between 310 and 630 nm. Single substitutions of amino acids, and thus point mutations in the opsin genes, can shift the sensitivity of the resulting visual pigment by up to 30 nm and so satisfy the first precondition for color vision. There has been major progress in our understanding of the evolution of opsin-based pigments in recent years (for an excellent review of this topic see Lamb [5]). That is particularly true for the C-opsins found in the ciliary photoreceptors that vertebrates use for vision [6] while many steps in the evolution of the R-opsins—those characteristic of invertebrate rhabdomeral photoreceptors — still need to be resolved [7].

Such a mutation and shift in the spectral sensitivity had to occur in one of two copies of an opsin gene, thus both gene and entire genome duplications have played a crucial role in the evolution of color vision [8]. An additional way to adjust spectral sensitivity is by switching the chromophore; for example, compared to retinal the presence of the chromophore 11-cis didehydroretinal broadens photopigment sensitivity and shifts it toward longer wavelengths by up to 30 nm (e.g., [6]) while 3-hydroxyretinal shifts the sensitivity to shorter wavelengths by up to 12 nm [9]. (See Chap. 1 for more about genetics.) Spectral sensitivity can also be altered through the variable presence of screening pigments. In such cases, receptors containing identical photopigments, some screened and some not screened, will have different spectral sensitivities. While this is known to occur in some taxa, as for example in some butterflies [10] and in stomatopod crustaceans [11], there are no documented cases where a change like this has led to the emergence of novel color vision.

The second condition necessary for color vision is that light absorbed by pigments with differing sensitivities must be compared neuronally. In the parietal eye of a lizard, chromatic antagonism is established within a single receptor cell, and this may well have been the case in early color vision [12]. Most often, however, pigments with different spectral sensitivities are expressed in different receptor cells, and the signals from these cells are then compared by downstream neuronal opponent mechanisms. To date, the evolution of differential opsin expression is less well understood than the evolution of opsin pigments (e.g., [13]). The evolution of spectrally opponent mechanisms is also only vaguely understood. One likely possibility is that the basic nervous system wiring that underlies spatial discrimination — a comparison of photoreceptor signals originating at neighboring locations in the receptor mosaic — was simply co-opted to generate comparisons from photoreceptors having different spectral sensitivities. That such a step can work was established by adding a novel photopigment to the retinas of mice using genetic engineering. From that single change alone new color vision emerged. [14]. A similar outcome occurred in genetic engineering experiments done on monkeys [15]. In each instance, it seems possible that new spectral opponency was set up through the use preexisting neural circuits and, arguably, this is what happened during the evolution of color vision.

Finally, and importantly, for color vision to evolve, a duplication and mutation of an opsin gene — or any other change leading to the chromatic signals discussed above—has to spread through the population. Why would animals with such mutations, and thus the potential for novel color vision, have an adaptive advantage over conspecifics that are not so equipped; in short, what causes new color vision to evolve as a stable trait?

11.2.2 What Are the Advantages of Color Vision?

This is an important question given that chromatic or color signals are generally weaker, thus containing considerably less information than achromatic or intensity signals [16]. Moreover, color vision usually comes at a cost: An animal that expresses different pigments in subsets of its photoreceptors usually sacrifices higher spatial resolution or lower absolute sensitivity (e.g., [17]). In addition, comparing signals in an opponent fashion rather than adding them reduces the signal-to-noise ratio. Evolution of sensory capabilities is driven by selection for new behavioral tasks that a new ability makes possible [12, 18]. Given that color vision competes for space and energy with achromatic vision, what are the new tasks that color vision supports? (See Chap. 7 for discussion about color–luminance interactions).

As color vision most likely evolved independently in several taxa, the original selective advantages may have been quite varied, and it is possible that several of them were of importance. A salient fact is that color vision — as vision in general — evolved in shallow water environments [18], and thus the initial selective advantage of adding new photoreceptor types must have had to do with the utilization of information contained in the spectral composition of light. Animals use color vision for two general purposes: the determination of the illumination spectrum, and the identification and recognition of objects and object properties [19]. The first of these seems likely to have been the original use of color vision as it does not require highly advanced eyes with good optical properties, high spatial resolution, and high absolute sensitivity. Since color vision evolved initially in aquatic animals, its original use may have been for the assessment of depth from the spectral composition of the light reaching the eye [18].

Another, independent advantage of possessing two rather than one photopigment could have been for UV protection. In that case, a pigment with maximal sensitivity to light of 450–500 nm would have been used for vision while a second, UV-sensitive pigment, initially employed to monitor this high-energy and dangerous light close to the surface, would then have been converted to play a role in vision as a secondary step [18].

The second use of color information — object vision — requires higher absolute sensitivity and some spatial resolution. While color vision in animals with highly evolved eyes is likely used mainly to enhance the contrast between objects and backgrounds, and to identify objects of relevance, this was probably not the reason for the initial evolution of color vision. Campenhausen [20] pointed out that under changing illumination the chromatic properties of objects are more stable than their lightness properties, and suggested that chromatic vision may originally have evolved to support lightness constancy. This may be specifically true for aquatic animals that need to determine object color at different depths under the water surface, but could also apply to animals occupying terrestrial habitats. Johnsen and colleagues [21] have given examples of this for the illumination conditions present at dawn and dusk when the color of the illuminating light changes from sunlight, rich in long wavelengths, to the blue skylight present when the sun is below the horizon, and then back to moonlight or to the more red-shifted starlight. A blue flower that is darker than the green foliage background in sunlight or moonlight may well be brighter than the background in late twilight making the achromatic signal unreliable while the chromatic signal remains relatively stable. The situation is similar when an animal sees an object in full sunlight or in the shade (see Fig. 11.1). While relative luminance and achromatic contrast to the background change dramatically, the chromatic signal remains the same. On a different time scale, a similar idea was put forward by

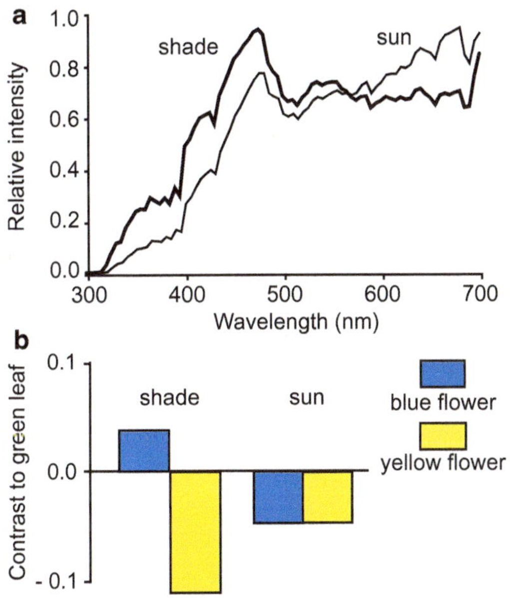

Fig. 11.1 (**a**) Spectral composition of illumination in bright sunlight and skylight in the shade. (**b**) The achromatic (Michelson) contrast of a typical blue and yellow flower, against a green leaf background, for a receptor with λ_{max} of 520 nm, under both illuminations. For this achromatic channel, the blue flower has a higher luminance than the green leaf and the yellow flower in the shade, but a lower than the green leaf and equal in luminance to the yellow flower in direct sunlight. Viewing situations like this can render achromatic vision unreliable, and may have been an important reason for color vision to evolve

Maximov [22]. He suggested that the processing of chromatic signals could reduce the effects of the high frequency luminance modulation caused by wave action in shallow water where color vision most likely evolved. Contrary to the expectations of Campenhausen [20], no animal has been described so far that uses chromatic signals exclusively to achieve luminance constancy, although it must be admitted that for many species the utility of color vision remains unknown.

11.2.3 Dimensionality of Color Vision in Animals

The dimensionality of color vision is an important aspect for its evolution. During the evolution of color vision, achromatic vision must have first been replaced by dichromatic color vision, based on the presence of two receptor types and some means of comparing their relative photon capture. Beyond this, receptor types were further added in some clades, leading to trichromatic, tetrachromatic and, potentially, even higher-dimensional color vision. What were the driving forces for increasing the dimensionality of color vision and where and why has such diversity evolved?

11.2.3.1 Dichromacy

As outlined above, dichromacy allows an animal to disentangle intensity and the spectral composition of light and distinguish between shorter and longer wavelengths. Many dichromats can discriminate ultraviolet from regions of human-visible light (e.g., rodents [23], decapod crustaceans [24]). Some marine animals have two receptor types with very similar spectral sensitivities that allow them to discriminate between downwelling light and bioluminescence or bicolored bioluminescence (e.g., the firefly squid [25]). Dichromacy is fully sufficient for the control of phototaxis (see Sect. 11.2.4), and it enables animals to show preferences for short- or long-wavelength light, such as a preference for the solar versus the anti-solar half of the sky, or vegetation versus sky [26]. It also allows for basic color-constancy, but it does not permit an animal to distinguish broad-band stimuli from a medium wavelength close to a "neutral point" which is perceived as "white" or from various mixtures of short and long wavelengths (e.g., [27]). In other words, dichromatic animals can discriminate light of shorter wavelengths from light of longer wavelengths, but many colors are metameric for them.

Dichromacy was likely the first type of color vision to evolve. It probably evolved in ancestral arthropods [28] and it occurs in contemporary polychaetes [29], mesopelagic squid [25], and in a majority of the crustaceans [24]. Among vertebrates, and probably also among insects, dichromacy always reflects a reduction from a previous higher dimensionality (see Sects. 11.3 and 11.4).

11.2.3.2 Trichromacy

Trichromatic color vision allows for disentangling the three basic dimensions that for trichromatic humans are called hue, brightness, and saturation, and it allows for the determination of the material properties of objects such as pigmentation, glare, and texture. At the same time, given the broad spectral sensitivity of opsin-based visual pigments and the broad spectra of most natural colors, trichromacy is sufficient for the discrimination of most terrestrial spectra, and thus good enough for "general purpose color vision" [19, 30]. With the exception of primates, most studied trichromats have the sensitivity peaks of the three receptors evenly spaced across the spectral range. Bees and moths, as typical examples, have photoreceptors peaking at 340–360, 420–440, and 520–560 nm ([31, 32], and see Fig. 11.2). By contrast, among primates that possess three cone pigments, the middle (M) and long-wavelength sensitive (L) pigments are positioned relatively close together, most often peak separated by ~30 nm [33].

11.2.3.3 Tetrachromacy and Beyond

In recent years the application of molecular techniques has revealed that an unexpectedly wide range of animals express more than three opsin types in their eyes, and thus they have the potential for still higher-order color vision. Here we discuss

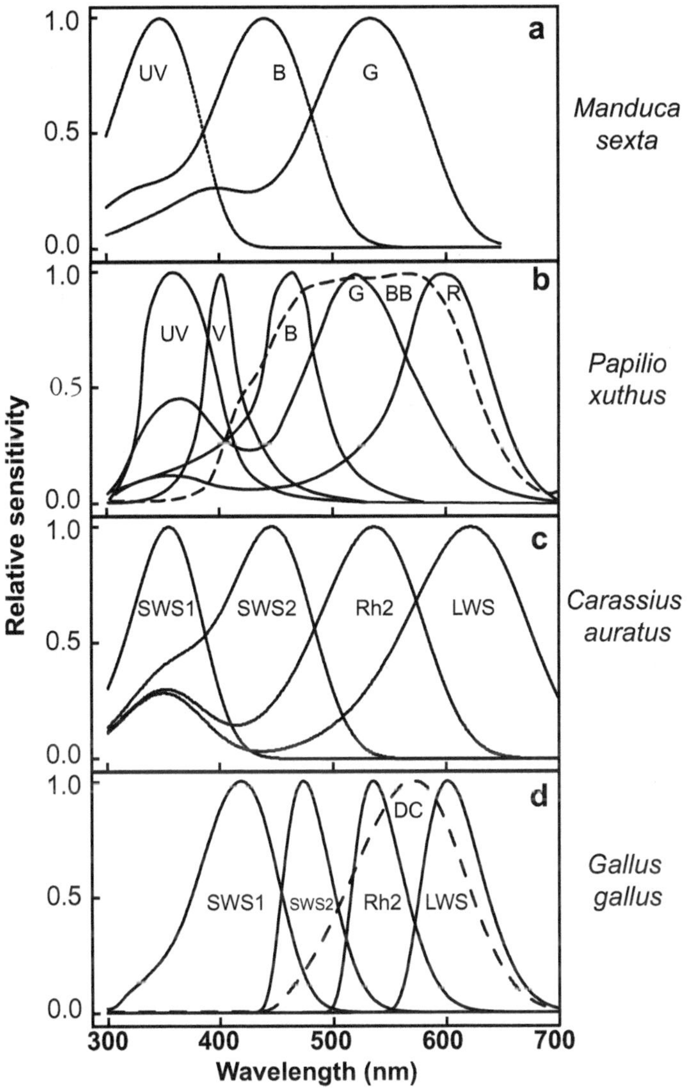

Fig. 11.2 Typical spectral sensitivities of receptors with and without filter pigments. (**a**) *Manduca sexta*, a moth with three spectral receptor types respectively sensitive to light of ultraviolet (UV), short (*blue*, *B*) and medium (*green*, *G*) wavelengths. (**b**) *Papilio xuthus*, a butterfly with filter and fluorescent pigments narrowing the sensitivity of violet-sensitive (V) and long-wavelength sensitive (*red*, *R*) receptors. BB receptors are broadband receptors expressing two opsin genes. The sensitivity curves of two additional receptor types expressing the same opsins as B and G but affected by different filters are omitted for clarity. (**c**) Goldfish expressing four cone opsin genes with no known filters. (**d**) Chicken expressing four cone opsins and carotenoid filters in oil droplets in four types of single cone and Rh2 in double cones (DC). *Dashed curves* are specialized broadband receptors assumed not to contribute to color vision

cases where four or more receptor types with different spectral sensitivities are present in the same retinal region at the same time, and likely to be used for color vision. Three advantages may have favored the evolution of more than three receptor types and higher dimensions of color vision: the need to extend the visible spectrum, the need to handle the complex illumination conditions often characteristic of aquatic environments (see Sect. 11.2.2), and specific behavioral tasks that require high spectral resolution.

Compared to humans (See Chap. 5 for more about spectral sensitivity in humans), many animals enjoy an expanded visual range in the shorter wavelengths, often down to 300 nm, as well as to longer wavelengths. To cover this extended range, an additional receptor is well suited, especially if receptor sensitivities are narrowed by filtering pigments, as they are in butterflies, birds, and reptiles ([30], and see Fig. 11.2b, d). Narrowed receptor sensitivities have been argued to be beneficial for color discrimination and color constancy [34, 35].

The complex light environment in aquatic habitats and the need for good lightness and color constancy have been considered the most important reason for the early evolution of tetrachromacy in basal vertebrates [22, 36]. This is also reflected by the large number of opsin pigments that have evolved in several groups of crustaceans, such as the water flea *Daphnia* and some stomatopods [24], as well as in fishes [6], and in the odonates (dragonflies and damselflies), where aquatic larvae and terrestrial adults can express different subsets of up to 20 different opsin genes [37].

Among terrestrial arthropods, the colorful butterflies deserve special attention because they have evolved the highest degree of color vision diversity. Trichromacy, sufficient for general color vision tasks in terrestrial environments, is found in moths and basal butterfly families, but in Lycaenids, Pierids, and Papilionids almost species-specific color vision systems have been described, sometimes comprised of eight or more separate receptor types [10]. In *Papilio xuthus*, a species having eight spectrally discrete receptor types, the presence of tetrachromacy has been strongly suggested, although not definitively proven, by behavioral tests [38]. In this species the additional receptor types include a broadband receptor, which may support achromatic vision, and a narrowband violet (λ_{max} 390 nm) receptor, which potentially has a function in a specific behavioral context (Fig. 11.2b). Behavioral needs, such as the choice of suitable mates, food sources, and oviposition sites, in conjunction with the requirement of parallel information processing in small insect brains may have driven the evolution of this apparent complexity. Among butterflies, sexual dimorphism in the expression of opsin pigments and the resulting spectral types of receptors is common. The only documented case of a likely co-evolution between coloration and visual pigments has been reported for *Heliconius* butterflies where a gene for a second UV-sensitive opsin, expressed in a subset of ommatidia, is thought to have co-evolved with UV-reflective wing colors in part of the genus [39].

11.2.4 *Four Grades of Color Vision*

Animals use chromatic signals in different ways, which Kelber and Osorio [19] have suggested comprise four separate grades of color vision. We describe these using examples drawn from various animal taxa, but they can be thought of as possibly representing four different stages in the evolution of color vision.

11.2.4.1 Grade 1: Color-Guided Photokinesis and Phototaxis

As pointed out above, the most basic form of chromatic vision is possible in animals that have eyes with low absolute sensitivity and low or absent spatial and temporal resolution. While many terrestrial animals with such primitive eyes use achromatic information to find suitable habitats, only the spectral composition of light can be used to inform aquatic animals about depth in the water column, independent of the time of day [40]. Such information also allows terrestrial animals to discriminate between the solar and anti-solar halves of the sky, or to find other habitats that are defined by their spectral composition. Spectrally guided phototaxis has been described in several unicellular organisms: nematodes, the water-flea *Daphnia* and a number of amphibians (reviewed in [41]). More advanced color vision that extracts chromatic information for object detection and recognition may well have evolved from this basic form [12, 18].

11.2.4.2 Grade 2: Innate Color Preferences

The simple eyes of butterfly caterpillars allow their owners to find food sources by using light reflected from green leaves. Likewise, caterpillars and many other herbivorous insects have innate color preferences based on a simple chromatic mechanism that guides them to their food sources [42, 43] while many flower-visiting insects such as bees and butterflies innately prefer blue in their search for a first nectar source (e.g., [26]). Jumping spiders prefer to attack blue prey [44], and even tsetse flies searching for a host have a preference for blue. Animals with small brains also use innate color preferences to guide them to mates. For instance, the attraction of male glow-worms by the color of female bioluminescence is based on color opponency while female fiddler crabs use color to choose yellow-clawed males over males both with brighter or darker claws of another color ([19] and references therein).

In principal, innate color preferences could be controlled through hard-wired neuronal pathways that send a processed color signal directly to a motor center without the need for any neuronal modification of the color signal. Peripheral information processing or "matched filtering" reduces the need for elaborate processing in the brain, thus innate color preferences allow animals with a small brain capacity to react efficiently to relevant objects of constant and predictable color. Despite relying on very simple neuronal mechanisms these animals can take advantage of the reliability of the chromatic signal as compared to lightness signals that change as a function of illumination conditions [19].

11.2.4.3 Grade 3: Color, Learning, and Cognition

The ability of animals to learn and freely choose any color does not require any new sensory capabilities. It does, however, place higher demands on central neuronal processing, and it has been argued that "learning gives evidence of a perceptual dimension of color" [45]. If we include the degree of central processing in the concept, color learning can be regarded as indication for a third grade of color vision. Animals that have learned a color often then approach all objects of that color, a behavior which seems to imply the acquisition of an internal representation. To date, color representation in the brain is poorly understood in primates, and even less so in other animals.

From an ecological point of view, the ability to change color preferences through experience must have given animals an evolutionary advantage as it allowed them to modify their behavior in response to changing food sources and so avoid bad choices. Many flower visitors and herbivores learn to recognize the color of profitable food sources. And even though predatory animals most often employ colorblind motion vision for the detection of prey, they do learn to avoid brightly colored aposematic prey [46]. At the same time, animals that have evolved an internal representation of color and are capable of color learning may still show innate preferences for objects of predictable colors.

11.2.4.4 Grade 4: Color Categories and Color Appearance

The classical experiments used in studies of animal color vision measure discrimination thresholds, but don't allow asking animals how they perceive large color differences or decipher color appearance. Humans can group colors with respect to hue, saturation, and brightness, and assign easily discriminable colors to a small number of color categories [47]. Although still debated, it appears likely that nonhuman primates form color categories [48]. Beyond primates, questions about categorization or generalization of colors have rarely been asked. Birds, with their excellent color discrimination abilities, large brain capacities, and high levels of cognitive abilities, can form color categories based on experience [49]. Clearly, more work is needed on this topic and, specifically, the question of whether the scaling of color differences can be extrapolated linearly from thresholds to large color differences is highly relevant for studies on the evolution of coloration, aposematism, and visual communication in animals.

11.2.5 Evolution of Retinal Mosaics

Although the number of receptor types expressing different opsins often determines the dimensionality of color vision, two factors that are often ignored in studies of the evolution of color vision are the total numbers of various receptor types and their distributions across the retinal mosaic [48]. Both of these features vary widely across species. Moreover, many eyes have regionalized mosaics, organized mostly along a dorsoventral axis. In general, two or more types of receptors can be organized into a regular or a random mosaic or anything in between these arrangements [50]. Theoretically, in a retina having two receptor types, a regular array with the same number of both types should be optimal for coding a scene with naturalistic spatial and spectral statistics and, in fact, even with unequal numbers of both types, a regular mosaic remains optimal [51]. Humans and other primates have a regular spatial array of S cones, but a randomly clumped arrangement M and L cones (Fig. 11.3f). The human retina also has high cone density in the rod-free fovea, and relatively low cone densities in the periphery where rods outnumber cones with ratios of ~10:1 (e.g., [55]). (See Chaps. 3 and 5). Among mammals, diurnal squirrels have heavily (~85 %) cone-dominated retinas, while nocturnal fruit bats have cone–rod ratios of only 1:150 to 1:400 [48] making it likely that color vision holds a very different significance for these two groups of animals. Chickens have 41 % double cones that are thought to be used for achromatic vision only (see below), with 21 % M cones (expressing Rh2 opsin), 17 % L cones (LWS), 13 % S cones (SWS2), and 8 % very-short-wavelength sensitive (VS) cones (SWS1) in a mosaic as regular as the unequal numbers of cone types allow ([50], and see Fig. 11.3d). The retinas of most fish are arranged as crystalline mosaics with 50 % L/M double cones, and 25 % S and 25 % UV-sensitive single cones (e.g., [52], and see Fig. 11.3c).

In camera-type eyes the nervous system compares signals from adjacent arrays of receptors with slightly different visual fields. It is one of the few benefits of the compound eyes possessed by many invertebrates that each ommatidium in the hexagonal or square pattern houses a set of usually 8 or 9 receptor sharing the same visual field [56]. In many crustaceans, each ommatidium has a dichromatic system consisting of one UV and 7 blue-green-sensitive receptors, but a majority of insects have random mosaics, composed of two or three ommatidial types [57] with different receptor compositions. Flies have a random mosaic of two ommatidial types each containing different sets of spectral receptors (Fig. 11.3a). Bees, moths, and butterflies have three ommatidial types: approximately 25 % of the ommatidia have UV- and green-sensitive, 25 % have blue- and green-sensitive receptors, and 50 % have all three receptors types (Fig. 11.3b). In the eye of the butterfly *Papilio xuthus*, additional red-sensitive receptors are found in 50 % of ommatidia, broadband receptors in 25 % of ommatidia, and narrow-band violet-sensitive receptors in 25 % of ommatidia. In other species of butterflies, males and females have different ommatidial types with different subsets of receptor types [10].

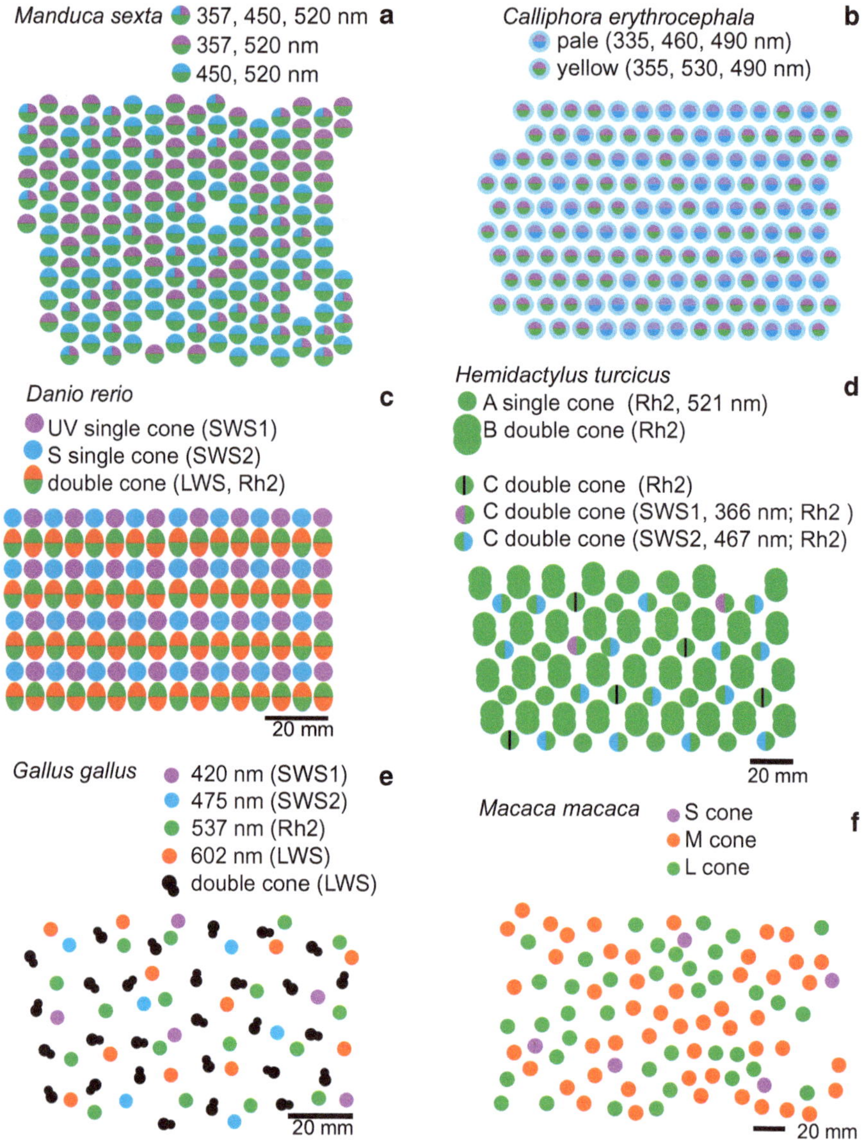

Fig. 11.3 Retinal mosaics of insects, (**a**) a fly (*C. erythrocephala*) and (**b**) a moth (*M. sexta*), as well as the vertebrates (**c**) zebra fish (*D. rerio*), (d) gecko (*H. turcicus*;), (**e**) chicken (*G.gallus*;), and (**f**) macaque (*M. macaca*). For the insects, each disk indicates an ommatidium, and the λ_{max} values of all receptor types in each ommatidial type are given. For the vertebrates, each color-coded disk indicates a cone, and sensitivity peaks and/or opsin types are given. Geckos have two morphological types of double cones, one of which can express different opsins in both members. (**a**, **b**) adapted from [26]; (**c**) adapted from [52]; (**d**) data from [53]; (**e**) mosaic data from [50] but double cone orientations adjusted from unpublished data; (**f**) data from [54]

Regionalisation is most commonly found along the dorsoventral axis. For instance, in the Apollo butterfly *Parnassius apollo*, only the ventral half of the compound eye—the part looking downwards—contains specific red-sensitive receptors while the dorsal half—looking at the sky—is trichromatic [10]. In mammals, the relative proportion of S and L cones varies widely [58]; in birds both the relative frequency of cone types and the coloration of the oil droplets differ across the retina [59]. Even more extreme, the dorsal part of the eye of the dragonfly *Sympetrum*, or the males of honeybees and many flies, as well as the ventral part of the eye of the mesopelagic squid *Watasenia*—all looking upwards—are color-blind while remaining parts of these eyes support varying degrees of color vision [26]. Dorso-ventral regionalization is assumed to be an adaptation to the common spectral properties of backgrounds seen in the dorsal and ventral parts of the visual fields. These properties differ across habitats and thus retinal mosaics appear to vary accordingly (e.g., [60, 61].

The determination of opsin expression in different photoreceptors is at least partly understood in the fly, *Drosophila,* but it remains to be determined how other random arrays, the regular receptor array found in fishes, or receptor arrays showing a dorso-ventral gradient are controlled [13]. In mice, S cones express opsin before the L cones, and the presence of thyroid hormone has been found to inhibit the expression of SWS1 opsin. (See Chap. 1 for more about thyroid function and opsin expression in humans). Thyroid hormone is also known to control opsin expression in salmonid fishes, but the control of thyroid hormone expression patterns in the retina is still under investigation (reviewed by [58]). As long as the control mechanisms remain largely unknown, it will be difficult to gain an understanding of how they evolved. We discuss selected examples of opsin gene expression patterns in vertebrates in a later section (Sect. 11.3.2).

11.2.6 Constraints on Color Vision Evolution

Naturally not all features of visual systems that resulted from evolutionary processes were optimized to support color vision. As most animals use the same eyes and, with some exceptions, the same photoreceptors, to support both color vision and other visual capacities [57], additional constraints on visual systems have to be taken into account. For instance, the spectral range over which color vision is possible is limited by the transmittance of optical media. The corneas or lenses of many animals block ultraviolet light thereby preventing this potentially dangerous radiation from reaching the retina. Such an arrangement is seen in many fishes [62], birds [63], and mammals [64].

It was noted above that adding new spectral types of photoreceptor often comes at the cost of higher spatial resolution or lower absolute sensitivity, a fact which is particularly relevant for nocturnal animals [65]. Moreover, the spectral sensitivity of receptors is under selection for high achromatic contrast as well as chromatic contrast [16]. In vertebrates, low cone–rod ratios and a diminished potential for color vision are often closely related to the need of a species to have highly sensitive vision at night. In the following section we discuss the evolution of cone-based vision in vertebrates.

11.3 Origin and Evolution of Color Vision in Vertebrates

Since Karl von Frisch's classic work on fish vision [2] behavioral proof for color vision has been achieved in all the major vertebrate lines. To understand the evolution of vertebrate color vision, we first look at the evolution of visual pigments, and then discuss a number of adaptations specific to vertebrates. Mammals are considered separately below.

11.3.1 Early Vertebrate Opsin Evolution

Having evolved around 550 mya (million years ago), hagfishes are the most basal chordates. They apparently use their rather primitive eyes as circadian organs rather than for purposes of vision [66], while lampreys have eyes that are similar to those of jawed vertebrates. Lampreys express five different opsins (Fig. 11.4), probably one in each of the five spectral types of photoreceptors. Four of these opsins evolved prior to the time lampreys and gnathostomes separated around 500 mya. These are believed to be homologous to the four classes of vertebrate cone opsins (SWS1, SWS2, Rh2, LWS; see Chap. 1). Behavioral evidence for color vision is still lacking, but the Australian lamprey *Geotria australis* has the potential for tetrachromatic color vision [67].

It is not known exactly how vertebrate color vision evolved from monochromacy. Very likely the five opsins that appeared first in the lampreys were derived from two whole genome duplications that occurred early in vertebrate evolution, although two opsins, including the ancestor of the LWS opsin, must have existed prior to that event [8]. As no extant species can serve as a "missing link" between the monochromatic hagfishes and the potentially tetrachromatic lampreys, the reconstruction of the early evolution of vertebrate color vision remains challenging.

Four cone opsins (SWS1, SWS2, Rh2, LWS), and with those the potential for tetrachromatic color vision, are the basal condition for all vertebrates, including the teleost fishes and tetrapods that evolved from the lineage whose extant representative is the lungfish ([68], and see Fig. 11.4). All cases of lower color vision dimensionality, including that characteristic of the mammals (see below), are thus presumably reductions from this basal state.

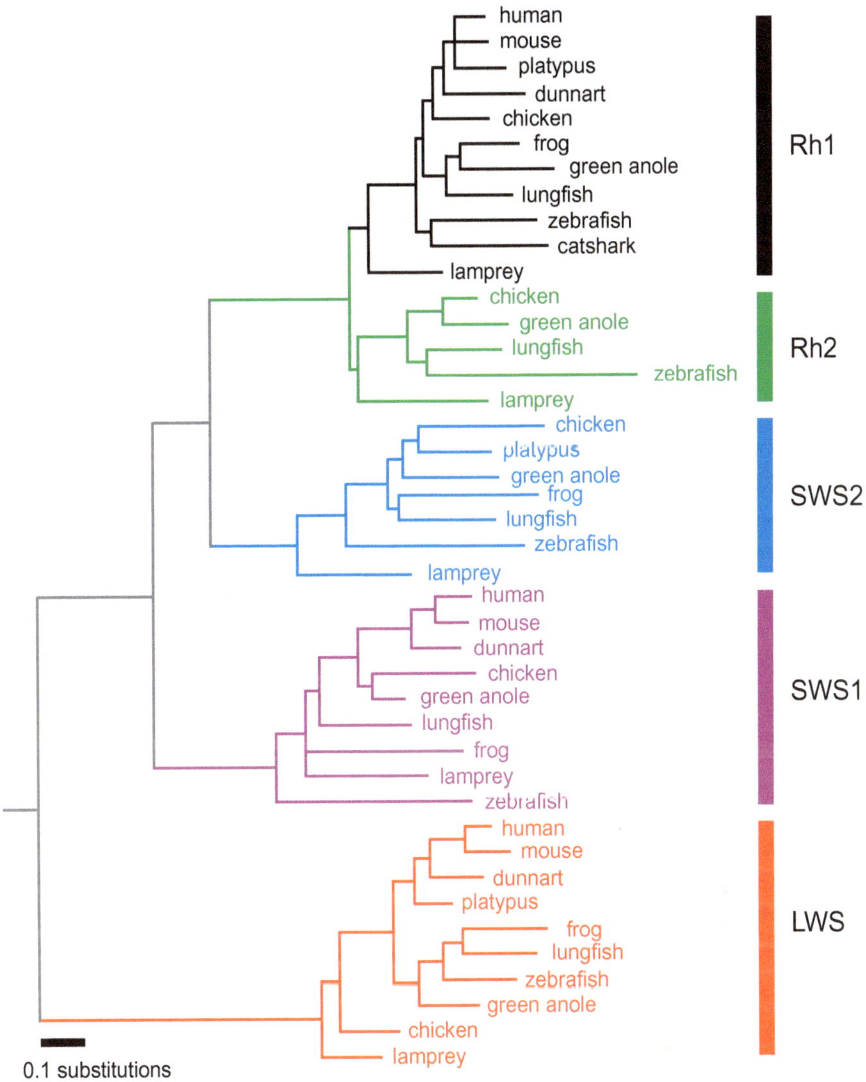

Fig. 11.4 Phylogeny of vertebrate visual pigment (opsin) gene families. The scale bar gives nucleotide substitutions per site. Adapted from [67]

11.3.2 Opsin Gene Expression Patterns in Vertebrates

As pointed out earlier, a basic requirement for color vision is the presence of photoreceptors with differing spectral sensitivities. Variations in photopigment absorption spectra typically result from differences in opsin gene structure (see Chap. 1), but to translate appropriately those gene variations into color vision abilities requires

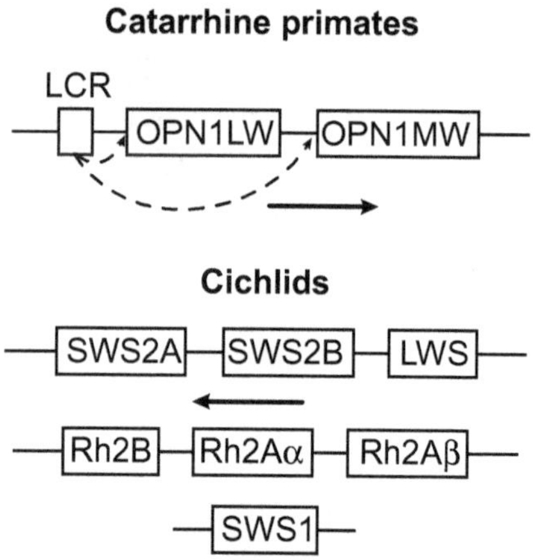

Fig. 11.5 Schematic illustrating the arrangement of the cone-opsin genes in catarrhine primates and cichlid fishes. *Top*: In catarrhines, two versions of the LWS-derived opsin genes are positioned in a tandem array on the X-chromosome. An upstream locus control region (LCR) pairs stochastically (dashed lines) with the promoter sequences of the two genes, thus differentially expressing the pigments that make up the L and M cones. *Bottom*: Cichlids typically possess single versions of the LWS and SWSS1 opsin genes along with multiple versions of the SWS2 and Rh2 genes. This arrangement allows for the production of a total of seven spectrally discrete pigments. Generally, only three of these genes are expressed at any one time

a mechanism that selectively places gene products into different photoreceptors. In most cases this means expressing a single type of cone opsin gene into each photoreceptor. Further, that expression pattern varies across the retinal mosaic in a way that is at least roughly consistent across members of a species and across the lifespan of an individual, even though in some species the expression pattern is subject to programmed alteration in response to developmental state or to environmental conditions. So far, the mechanisms that drive these processes are only partially understood. We highlight some of these issues by referring to select cases.

Early insight into the issue of opsin gene expression came from work by Nathans and his colleagues [69]. They discovered that multiple LWS-derived genes in humans are positioned in a head-to-tail tandem array on the X-chromosome (shown schematically in Fig. 11.5) By virtue of a small number of sequence differences these genes (OPN1LW and OPN1MW) specify, respectively, the cone photopigments that absorb maximally at about 560 nm (L pigment) and 530 nm (M pigment). In most individuals, there is a single copy of the former gene and multiple copies of the latter. Chromosome 7 contains an SWS1 gene (OPN1SW) that specifies a pigment with a peak at ~420 nm (S pigment). In each case, flanking upstream regions are responsible for directing the expression of these genes [70, 71]. For the X-chromosome genes, an upstream locus control region (LCR) pairs with promoter sequences for either the OPN1LW or OPN1MW gene, thus determining which opsin gene is expressed in a given photoreceptor [72]. That pairing process is assumed to be stochastic in nature thus resulting in a variable mixture of M and L cones across the retinal mosaic. A similar arrangement appears to direct the expression of the two LWS opsin genes in zebrafish and, taken together, these observation suggests that such a mechanism represents a general way of facilitating differential expression in cases where there are duplicated visual opsin genes [73].

In recent years, studies of opsin genes in the hundreds of species of cichlids that inhabit the East African Rift Lakes have yielded dramatic evidence of the ways in which color vision may be influenced by differential gene expression [74, 75]. In cichlids, as in other teleost fishes, there are numerous instances of duplications within the four vertebrate cone-opsin gene families. As illustrated in Fig. 11.5, cichlids usually have only single versions of the LWS and SWS1 genes while having multiple versions of both SWS2 and Rh2 genes. The result is that they are capable of expressing a total of seven spectrally discrete types of cone pigment with peaks spanning the spectrum from ~360 to ~565 nm [74]. Although there seem to be some exceptions [76], individual species typically express only three photopigments from this gene set, but the identity of the three varies from species to species. The level of expression of the various opsins can also vary dramatically; for instance, low expression of the SWS1 gene can occur together with much higher levels of expression of an Rh2 gene. Finally, the gene-expression patterns may change over the course of development and thus, as a consequence, potentially alter both spectral sensitivity and color vision. The mechanism underlying all these variations is differential gene expression [77].

Attempts to link the patterns of opsin gene expression to visual behavior have thus far yielded mixed results. For example, during the breeding season male cichlids become intensely colored, an adaptation believed to attract the more drably colored females. If sexual selection is the only force driving the evolution of male color patterns it follows that the intensely colored males should appear more conspicuous to their conspecifics. Dalton and colleagues [78] examined models of the inferred color spaces of cichlids to compute how conspicuous these fish should appear to other fish within their photic habitats. In some of the several species of cichlids examined males were in fact found to be more conspicuous than females, but in others the males were not judged more conspicuous than the females. Similarly, the photic environments in which these cichlids live vary considerably, from clear to turbid waters, and it might be expected that there would be corresponding photopigment variations. In some cases that prediction proves to be true; for example, in species that inhabit waters that greatly attenuate short wavelength light the photopigments maximally sensitivity to short wavelengths were expressed at low levels [79], but in other photic environments the expression patterns did not appear to maximize visual sensitivity [75]. In summary, although there is now abundant evidence for variations in opsin gene expression patterns in various fishes (and, apparently, also in some birds — [80]) there is still much to be learned about how these are or are not adaptively matched to visual behavior. For that task, the colorful cichlids continue to provide an attractive target.

11.3.3 Adaptations in Vertebrate Cones

Cone photoreceptors form the basic cellular adaptation for color vision in most vertebrates. They have evolved a number of interesting adaptations.

11.3.3.1 Co-Expression of Opsin-Based Visual Pigments and Chromophore Changes

It was long believed that each photoreceptor expresses a single visual pigment, but in the past 20 years co-expression of cone pigments, as two are in mice [81], or three cone pigments, are in tiger salamanders [82], has been observed in an increasing number of vertebrates (see Ref. [58]). Co-expression could result from the incomplete separation of the expression of newly duplicated pigments, an effect related to a developmental sequence in pigment expression, spatial gradients that do not separate expression completely, or a genetic program that simply facilitates the co-expression as a means of broadening the spectral sensitivity of the recipient cones (see Ref. [82]). Among fish, amphibians, and some reptiles, many species additionally tune cone sensitivity by switching between A1 (11-cis retinal) or A2 (11-cis 3, 4 didehydroxyretinal) based chromophores. Such a switch may occur seasonally, at different life stages, or even in different populations of the same species (e.g., in sticklebacks, [83]).

11.3.3.2 Ellipsoids and Oil Droplets

Spectral sensitivity is conditioned by optical filters in the cornea and lens; in addition, many vertebrates have optical filters in the cones themselves: the ellipsoid or ellipsosome—which consists of densely packed mitochondria—and the oil droplet. These cell structures have a much higher refractive index than the surrounding cytoplasm, but their optical role is poorly understood [84]. Many of these optical filters are pigmented and thus they can have direct effects on color vision. Pigmented oil droplets are found in lungfishes, reptiles, and birds. These droplets are filled with carotenoids that act as long-pass filters, absorbing light of shorter wavelengths and thus narrowing and shifting the spectral sensitivity of single cones types to longer wavelengths [84]. Narrowing of the spectral sensitivity function has been argued to improve both color discrimination and color constancy [34, 35], but it comes at the cost of greatly decreased photon capture and thus lower sensitivity. Because of this potential loss of sensitivity, the cones of nocturnal species typically contain only unpigmented oil droplets.

11.3.3.3 Single and Double Cones

Besides rods and single cones, most vertebrates—with the notable exceptions of lungfishes and placental mammals—have double cones (in some teleost fishes, such as anchovies, even triple cones are found [85]). The two members of a double cone are closely attached to each other, but make contacts to different bipolar cells [86]. In fish, both members are often morphologically very similar ("twin cones"), but in most other cases a principal member and an accessory member differ in size and structure; for instance, in the retinas of birds the accessory member often lacks an oil droplet. The double cones of birds, in which both members express LWS visual pigment, may not contribute to color vision (reviewed in Ref. [87]). Double cones of fish most often express LWS and MWS pigments in separate photoreceptors and probably do contribute to color vision [88], while geckos have both, i.e., double cones with the two members expressing the same pigment and others expressing different pigments ([53], and see Fig. 11.3). In general, the functions and roles of double cones are still largely unresolved.

11.3.4 Dimensionality of Color Vision in Different Vertebrate Lines

11.3.4.1 Elasmobranchs

Elasmobranchs have rod-dominated retinas [89] with low cone–rod ratios, between 1:50 (spiny dogfish, *Squalus acanthias*) and 1:3 (Atlantic stingray *Dasyatis sabina*). Sharks are cone monochromats: the grey Bamboo shark (*Chiloscyllium griseum*) has found to be color-blind in behavioral tests [90]. Rays, by contrast, have up to three types of cones expressing different pigments thus giving them the potential for trichromatic color vision. Giant shovelnose rays (*Glaucostegus typus*) are the only rays that have been shown to be able to discriminate colors [91].

11.3.4.2 Teleosts

Ray-finned fishes went through a third genome duplication in the Devonian [8] and thus have potentially twice as many opsin genes as compared to the sister group that includes coelacanths, lungfishes and all tetrapods. Additional gene duplications have occurred in many lineages, and extant bony fish have preserved different numbers of functional opsin genes: the SWS1 pigment is rarely duplicated, SWS2 occurs maximally in two copies, while Rh2 and LWS are found in up to four copies (4 LWS in guppies, 4 Rh2 in zebra fish; [79]. Many species express subsets of these opsin genes, either at different life stages [92], in different habitats [75], or in different regions of the retina [61]. Their color vision ranges from rod monochromacy (in some deep sea fish, [93]) to trichromatic (e.g., most cichlids) or tetrachromatic color vision (e.g., goldfish, [94], Fig. 11.2; zebrafish [52], Fig. 11.3c; Fig. 11.4). As in the crustaceans, the diversity and highly complex color vision systems of teleost fish likely mirror the complex illumination conditions present in the aquatic world that can result from the absorption and scattering properties of water (see Sect. 11.2.2).

11.3.4.3 Amphibians

Many amphibians (anurans and caudates) have lost the Rh2 opsin gene (Fig. 11.4) and are probable trichromats expressing SWS1, SWS2, and LWS visual pigments in their double cones and in two types of single cones [95, 96] while caecilians have only rods and the rod opsin visual pigment Rh1 [97]. Behaviorally, color vision has been demonstrated in frogs, toads, salamanders, and newts (see Ref. [41] for references) but only few taxa have been studied so far.

Unlike all other vertebrates, anurans and some caudates have two spectral classes of rods. In frogs and toads, around 90 % of rods express Rh1; the remaining 10 % express the same SWS2 pigment as the SWS2 cones but now coupled to proteins of the rod phototransduction cascade, an arrangement characterized by very low levels of dark noise. Theoretically, this creates the conditions for frogs and toads to use their highly sensitive rods for color vision in very dim light, although so far there is no proof that such capability exists.

11.3.4.4 Reptiles and Birds

Many reptiles and probably most birds have retained the tetrachromacy that they inherited from their common ancestor with lungfishes ([95], and see Fig. 11.4). Reptiles are rather diverse: crocodiles have rod-dominated retinas with double cones expressing two different pigments along with two classes of single cones. Many nocturnal snakes are likely to be dichromats, but turtles and lizards have one class of double cones and four classes of single cones, and there is behavioral proof for the presence of color vision in the latter groups (reviewed in Ref. [41]). There has also been considerable work on the neuronal correlates of color vision in turtles ([98] and references therein). As lizards have lost their rods, evolution has changed the three classes of cones of nocturnal geckos into highly sensitive rod-like receptors that allow them to discriminate colors under illumination conditions that are scotopic for humans [65].

Most birds, like the turtles and lizards, have cone-dominated retinas with approximately 50 % double cones, 20 % L (LWS opsin) and M (Rh2) cones, and 10 % S (SWS2) and VS (SWS1) single cones (Fig. 11.3e, Fig. 11.4). Both members of the double cones contain LWS pigment (with the single known exception seen in the retina of the bobolink) and differ only in their spectral filter: the principal member contains yellow carotenoids in an oil droplet, the accessory member contains a lower density of carotenoids that are not always located in an oil droplet.

While the spectral sensitivities of their SWS2, MWS and LWS pigments are fairly conserved, birds fall into two groups with respect to the SWS1 pigment. It is currently assumed that the ancestral situation featured an SWS1 pigment sensitive to violet light (λ_{max} ~410 nm), from which a UV-sensitive pigment (λ_{max} ~370 nm) has evolved several times, in passerines, parrots, trogons and seagulls [58]. Three classes of single cones have spectral sensitivities that have been narrowed (see Fig. 11.4d) by filtering with colored oil droplets containing astaxanthin (R droplet, L cone), zeaxanthin (Y, M cone), or galloxanthin (C, S cone) [99]. The violet sensitive (VS) cones have transparent oil droplets, but their spectral properties are modulated by the UV transmittance of the lens and cornea. While ocular media of some passerines and parrots transmit light with wavelengths down to 310 nm, the ocular media of raptors, swifts, and some other birds absorb almost as much UV as do the ocular media of humans [63]. These differences are partly correlated with the occurrence of violet-sensitive (VS) or ultraviolet-sensitive (UVS) cones. Birds with UVS cones also have a different composition of oil droplet carotenoids that serves to shift S-cone sensitivity to shorter wavelengths [100] resulting in a more even spacing of the spectral sensitivities of all the cone types.

Otherwise, the sensitivities of single cones are so similar among bird that they likely evolved to support good general color vision abilities. Birds with UVS cones can discriminate plumage colors better than birds with VS cones [101], but there is no evidence for an influence of bird spectral sensitivities on visual communication in birds [74]. Although theoretical modeling predicts that narrowed spectral sensitivities can potentially improve color discrimination, it has been recently shown that chickens cannot make finer color discriminations than humans, probably because bird cones have higher noise levels [35].

11.3.4.5 Rod Influences on Vertebrate Color Vision

Classical duplicity theory, stating that vision is supported by rod signals at low light levels and by cone signals at high light levels, linked vertebrate color vision to the operation of cone photoreceptors [102]. Since then, it has often been implicitly assumed that the number of cone photopigment types and their spectral properties both set the dimensionality of color vision and can be used to predict fundamental features of color discrimination. One complication in that picture is the fact that rods and cones operate simultaneously over a large intermediate range of illuminance levels – covering some four log units for human observers (see Chap. 5). There is evidence from numerous studies showing that across this so-called mesopic interval rod and cone signals can interact. These interactions have been shown to modify post-receptoral signals in several ways and to impact human vision, for example, by biasing hue perceptions, by altering brightness judgments, and by influencing chromatic discrimination [103]. There seems no a priori reason to assume that rod signals might not also influence color vision in other species in similar ways.

A dramatic impact of rod signals on human color vision is that, under the right viewing circumstances, the dimensionality of color vision can be altered; for instance, by allowing dichromatic individuals to make trichromatic discriminations [104] or a cone monochromat to perform as a dichromat [105]. It has been suggested that similar effects might have been seen in other species. For example, pinnipeds (seals, sea lions, walruses) have only a single class of cone photopigment [106] and it thus can be predicted that they lack color vision. However, an early experiment concluded that pinnipeds could make some color discriminations [107], a result that would imply that they were able to exploit a combination of rod and cone signals. A more recent experiment casts doubt on this conclusion, suggesting that the documented discriminations were based on brightness, not color, differences and thus there is no need to invoke joint rod and cone contributions to explain the earlier results [108]. Although at present there do not seem to be any clear cases where rod signals are known to clearly alter color dimensionality in nonhuman subjects, the topic is neither straightforward to investigate nor has it attracted much attention. Given the evidence from human color vision, an influence of rod signals on color vision cannot be ruled out in other species.

11.4 Evolution of Mammalian Color Vision

11.4.1 Opsin Evolution in Mammals

The first true mammals appeared during the early Mesozoic, about 250 mya [109]. Since that time, successive diversifications have yielded some 5400 contemporary species of mammal subdivided into three clades: Prototheria — egg-laying mammals, Metatheria — pouched mammals, and Eutheria — placental mammals (Fig. 11.6). As a group, mammals display a remarkably wide range of anatomical and physiological adaptations and that diversity has allowed them to successfully radiate into highly varied ecological niches.

As described above, early vertebrates acquired representation from four cone-opsin gene families (SWS1, SWS2, Rh2, LWS) at a time well before the emergence of mammals. The fate of these four gene families during the course of mammalian evolution is mostly one of loss. Although the timings of these various losses are only known by approximation, the overall pattern is summarized in Fig. 11.6. Included in this chronology are the following changes: (1) The Rh2 opsin gene family is not represented in any contemporary mammal, and thus it must have disappeared prior to major clade diversifications; (2) Functional SWS1 opsin genes are not present in the handful of species that comprises the extant prototheria — the monotremes — leaving them with cone-pigment representation drawn solely from the SWS2 and LWS gene families [110]; (3) Two cone-opsin gene families, SWS1 and LWS, have survived in contemporary marsupial and placental mammals [111, 112]. An additional change that occurred subsequent to the divergence of the prototheria from the metatheria/eutheria lines holds relevance for understanding the evolution of color vision. That change is the translocation of the LWS opsin genes from autosomal to X-chromosome sites [113].

Based on the current understanding of how structural changes in opsin genes link to the spectral positioning of photopigments, it is possible to use comparisons of gene sequences from contemporary mammals to infer what the spectral properties of the ancestral photopigments might have been. The inset in Fig. 11.6 (upper right) shows the predicted absorption spectra of the two pigments coded by the SWS1 and LWS gene families in ancestral mammals. These have respective λ_{max} values in the ultraviolet (~360 nm) and in the middle-to-long wavelengths (~555 nm) [114, 115]. Over time, gene-driven alterations in the structures of these opsins produced significant shifts in λ_{max}, of the photopigment absorption curves to the extent that contemporary mammals have cone pigments whose absorption peaks fall within the ranges (arrowhead lines in Fig. 11.6) of ~360–440 nm (SWS1 pigments) and ~500–560 nm (LWS pigments). With the exceptions of some primates

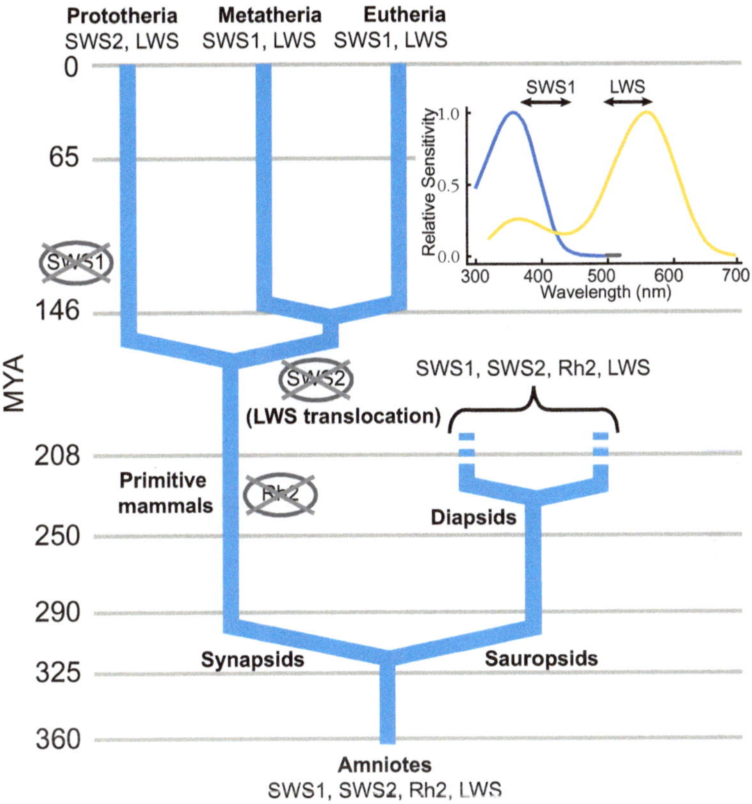

Fig. 11.6 Cladogram showing the fates of the cone-opsin gene families during the evolution of mammals. As described in the text, opsins derived from only two of the four families found in early vertebrates have survived in each of the three subclasses of contemporary mammals. The approximate locations of the losses of the original gene families are given by the symbols enclosed in the ovals. The timing of the translocation of the LWS genes from an autosome to the X-chromosome is also indicated. The *inset* to the upper right shows the estimated spectral sensitivities of the LWS (*yellow*) and SWS1 (blue) cone pigments present in ancestral mammals. Subsequent alterations in the structures of these genes in modern mammals has yielded cone pigments whose λ_{max} values fall within the spectral ranges indicated by the lines just above the curves

(see Sect. 11.4.4 below), the retinas of all mammals derive only a single cone pigment from each of these two gene families, thus limiting all these animals to having no better than dichromatic color vision.

11.4.2 Evolutionary Loss of Mammalian Opsin Gene Families

The standard explanation for the loss of cone-opsin gene families during mammalian evolution builds on an idea popularized by Gordon Walls nearly 75 years ago [116]. It assumes that early mammals, driven by predation pressures from diurnal

Archosaurs, adopted a nocturnal life style that extended over a time span encompassing some 150 million years [117]. The presence of numerous anatomical and physiological features in modern mammals, both in their visual systems and elsewhere, has been offered as evidence for this [118–120]. That period of mammalian evolution is often described as constituting a "nocturnal bottleneck." In line with that idea, several features that had been present in early vertebrates as specific adaptations for diurnal vision were lost, including multiple types of cone pigments and the color vision capacities they support.

Although the idea that a nocturnal bottleneck provides an explanation for the absence of whole opsin-gene families from contemporary mammals has been widely accepted (e.g., [119, 120]), some puzzles remain [121]. First, if color vision lacked utility for nocturnal mammals, why were only two of the opsin gene families eliminated rather than all four of them. And why, in particular, were those gene families, that were retained (SWS1 and LWS), the ones that code for the most spectrally separated cone pigments from among the four? Both of these facts would suggest that color vision continued to provide some functional utility during the mammalian nocturnal bottleneck. Second, questions have been raised as to whether the presumptive animals that preyed on mammals, the Mesozoic archosaurs, were themselves uniformly diurnal [122, 123]. If they were not, then perhaps not all early mammal lines were similarly driven to adopt a strictly nocturnal lifestyle. In addition, a recent study that examined eye morphology in non-mammalian synapsids suggests that nocturnality may not in fact have been an innovation unique to mammals, but rather it may have arisen much earlier [124]. Finally, and perhaps most relevant, there is abundant evidence from studies of contemporary mammals to show that the standard broad-scale categorizations of photic activity cycles (as being either nocturnal, diurnal, crepuscular, or arrhythmic) fail to capture the nuances of animal behavior [125]. Thus, species categorized as nocturnal (or diurnal, etc.) often show behavioral flexibility such that they readily alter the timing of their daily activity cycles in response to various environmental contingencies, e.g., changes in climate, food availability, or the presence of predators. All of these factors would suggest that although there seems little doubt that nocturnality was a central feature of early mammalian history, its impact on color vision, however, may have been more complex than the standard explanations now suggest.

11.4.3 Opsin Pseudogenes and Mammalian Color Vision

As a result of the evolutionary losses of two opsin-gene families most contemporary mammals have only two types of cone photopigments. Somewhat surprisingly, research conducted in recent years shows that a minority of mammals underwent further cone pigment loss. The first such instance was discovered when opsin antibody labeling of photoreceptors revealed a complete lack of S cones in two species of nocturnal primates [126]. The S-cone absence in these primates was traced subsequently to the presence of mutational changes in the SWS1 opsin genes that obviates expression of functional opsin proteins, i.e., the genes had transitioned to a pseudogene status [127]. Since then, application of genetic and functional measurements has revealed that numerous additional mammalian species, belonging to different mammalian orders, including some species of primates, cetaceans, chiropterans, rodents, and carnivores [58, 125], similarly lack S cones (reviewed by [125]). Correlative behavioral data are lacking for most of these species, but the loss of all S cones would be expected to render these animals completely colorblind. Although a large number of mammals have yet to be studied, it is estimated that as many as 10 % of all mammalian species may fall into this category [125]. We lack a clear understanding of why such a diverse collection of mammals have lost S cones in this manner. Many of the species so affected are categorized as being nocturnal, or at least their retinas show strong nocturnal specializations, and thus it is tempting to suggest that this cone loss has an explanation similar to that proffered for the loss of the two cone-opsin gene families in early mammalian evolution, i.e., that nocturnal animals are seldom active under illumination conditions necessary to support color vision and thus color vision yields no adaptive advantage under low-light conditions. If so, there might be little selective pressure against the loss of these cones. Complicating such an explanation are the facts that, (1) not all nocturnal mammals have been reduced to cone monochromacy, and (2) still other species that are classified as nocturnal, and have indeed lost their S cones, remain active under photic conditions that support cone-based vision [125]. It now seems that, although nocturnality may have set the stage for loss of function from the SWS1 opsin genes, it cannot be the sole circumstance promoting that change.

There is recent genetic evidence to suggest that a few mammals have suffered an even more severe loss of cone function. At least five whale species have been found to have inactivating mutations in *both* their SWS1 and LWS opsin genes, thus presumably rendering the retina completely without cone function [128]. In another, two extant species of Xenartha (the nine-banded armadillo and the two-toed sloth) have also apparently become rod monochromats as a result of both SWS1 and LWS opsin-gene mutations [129]. The whale species are all deep divers and the two species of Xenartha are believed to have evolved from fossorial ancestors. In both instances it has been suggested that the total loss of cone function may be traced to lives lived under ambient light conditions that allow for little cone function and thus their cone loss represents an adaptive response [128, 129].

In summary, although the majority of contemporary mammals have two types of cone pigment that can support a single dimension of color vision, a significant minority appears to lack color vision entirely, either as a result of the fact that through gene mutations they retain only a single cone class (cone monochromats) or they have no cones at all (rod monochromats).

11.4.4 Evolution of Primate Color Vision

There are two suborders of Primates: (1) *Strepsirrhini*—the galagos, lemurs, and lorises, and (2) *Haplorhini*—which includes the tarsiers, the Platyrrhines (New World monkeys), and the Catarrhines (Old World monkeys, Apes, and humans).

Depending on which of the various species definitions are employed, there are estimated to be some 350–450 extant species of primate. Over the past 30 years, there have been dramatic advances in our understanding of the evolution of primate color vision. That topic has been the subject of several detailed reviews (e.g., [33, 130–133], so only a brief summary is warranted.

There are four different cone-pigment patterns in primates, examples of which are sketched in Fig. 11.7. As in all mammals, these pigments are derived from opsin genes from the SWS1 and LWS gene families. Some primates maintain a cone pigment arrangement like that of many non-primate mammals having two types of cone pigment and potential dichromatic color vision (Fig. 11.7b). Another case (Fig 11.7a) involves species that are similar to those described above in that the SWS1 opsin gene has transitioned to pseudogene status, a change that renders these animals cone monochromats. A third arrangement, common to all catarrhine primates, features two LWS-derived cone pigments (Fig. 11.7d). All of these species have a total of three types of cone pigment and enjoy trichromatic color vision. The addition of a second cone pigment with maximum absorption in the middle-to-long wavelengths resulted from an X-chromosome opsin-gene duplication that occurred about 30-35 mya. Since that duplication event preceded the diversification of catarrhines, *all* of the members of this group share the same cone complements and, effectively, the same trichromatic color vision.

A significant insight into the evolution of primate color vision came from the discovery of cone-pigment polymorphisms in platyrrhine monkeys. In the most common case there are three allelic versions of the X-chromosome LWS opsin gene, yielding three possible middle-to-long wavelength cone pigments (dashed lines in Fig. 11.7c). Because these genes are located on the X-chromosome, the cone-pigment arrangements in male and female monkeys differ such that all male monkeys have only one of the three LWS-derived pigments and so are limited to having dichromatic color vision. Female monkeys that are homozygous at the X-chromosome opsin gene site also have a total of two types of cone pigments and dichromatic color vision. However, through the agency of early X-chromosome inactivation heterozygous females gain a second middle-to-long wavelength cone and they acquire trichromatic color vision. The remarkable result is that these polymorphic species contain individuals having any of one of six different cone pigment and color vision phenotypes: three dichromatic and three trichromatic. Beyond this basic variation, there are some additional species variations in the spectral positioning of the middle-to-long wavelength cone pigments. Furthermore, not all polymorphic species have a total of three alleles (for details see Ref. [33]).

Finally, unique among the platyrrhines, but similar to the catarrhines, howler monkeys (genus *Alouatta*) acquired a second LWS cone pigment and trichromacy as a result of an early X-chromosome opsin-gene duplication (see Chap. 1). Although the evolutionary pattern of this change is similar to that seen in the catarrhines, the two duplication events occurred independently. In the typical case, gene duplications are thought to follow allelic divergence [134]. It seems likely that is what occurred both in the case of the howler monkeys and of the catarrhines. If that

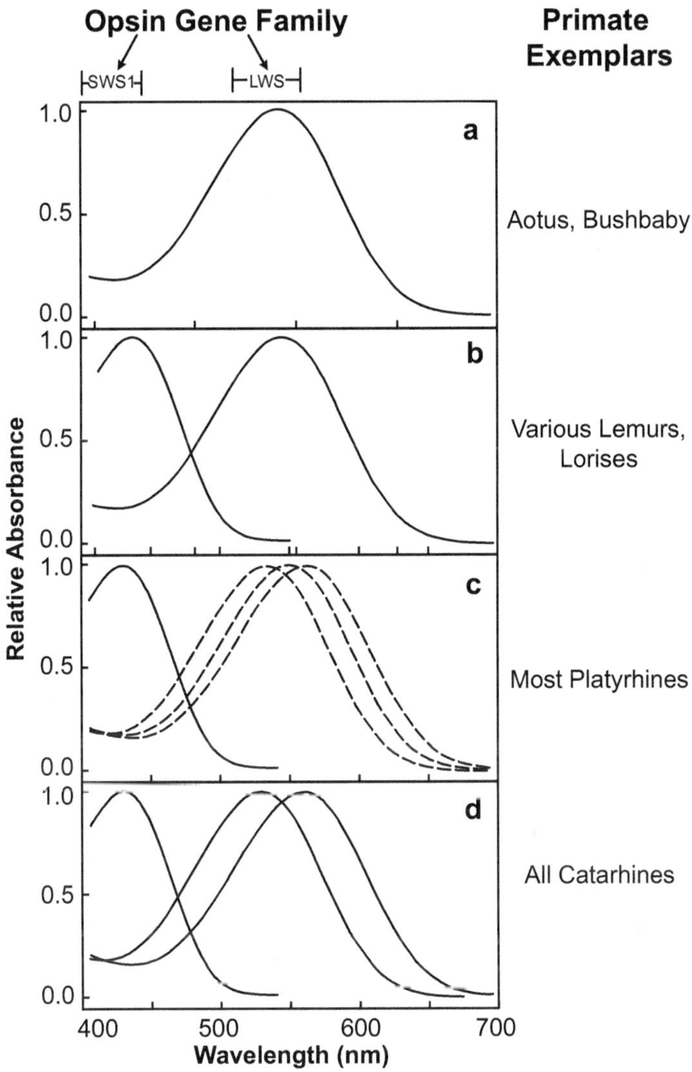

Fig. 11.7 Examples of the cone pigment complements detected in modern primates. As in all eutherian mammals, primate cone pigments are derived from opsin genes of the SWS1 and LWS families. SWS1 genes are autosomal while LWS genes are X-chromosome linked. Primate exemplars of each of the four patterns are noted to the right. (**a**) The retina contains a single cone pigment and thus these animals lack color vision. (**b**) Two cone pigments supporting dichromatic color vision. (**c**) In addition to an invariant SWS-derived pigment, individual animals have either any one of any two the three LWS-derived pigments. This arrangement allows individual to have either two or three cone photopigments; the former are dichromatic, the latter trichromatic. (**d**) Following a gene-duplication event, some primates (see text) acquired two separate genes coding for middle-to-long wavelength cone pigments. Such animals are routinely trichromatic

is so, both lines would have passed through an early period of polymorphic color vision characterized by a mixture of dichromatic and trichromatic individuals prior to the time they became routinely trichromatic.

11.4.5 Evolution of Color Vision in Catarrhine Primates

The key step in the evolution of catarrhine trichromacy was an X-chromosomal opsin-gene duplication. That event yielded separate genes (OPN1LW and OPN1MW) that code respectively for the L and M cone pigments. Reflective of this shared history, all contemporary catarrhines have L and M cone pigments with similar absorption spectra. Despite this overarching commonality, there is one striking difference between human color vision and that of the other catarrhines: the variations in opsin genes and color vision, that are such characteristic features of humans, are effectively absent in other catarrhine primates. The most dramatic human variation involves mutational changes in the M and L opsin genes that lead either to the loss of one of the corresponding cone pigment types, resulting in dichromatic color vision, or to shifts in the spectral absorption properties of one of the cone pigments, producing anomalous trichromatic color vision. Details of the relationships between opsin gene structure and human color vision defects are treated elsewhere in this book (Chap. 1). Although frequent among humans, in some populations affecting as many as 8 % of all males, similar color vision variations have seldom been detected in other catarrhine species [135, 136].

An early explanation offered for the high incidence of human color vision deficiencies was that there had been a relaxation of selection against such defects during recent human history [137]. Arguing against that idea, Kalmus [138] pointed out that it requires unrealistic assumptions about the stringency of selection in order to account for the high rates of human color vision defects. A second feature of the human color vision defects is that their incidence varies significantly across various racial groupings [139]. Some of this variation no doubt reflects differences in testing criteria, and there are instances where it seems clear that defective genes have spread throughout heavily inbred populations, but despite these factors there remain impressive overall differences. For example, a recent survey conducted in a large sample of California preschool children found that the incidence of defective color vision among boys varied from 1.4 % in African Americans to 5.6 % in Caucasians [140]. These racial variations have at times been attributed to differences in natural selection [141] or to reflect differences in photic environments [142], but there is at present no accepted explanation.

Variable selection pressure against color vision defects might provide a possible explanation for the frequent occurrence of defective color vision in humans and its near-absence in other catarrhines, but there are other potentially relevant differences. For instance, the genes coding for the catarrhine L and M opsins are adjacent to one another on the X-chromosome and are structurally very similar and these two features set the stage for unequal meiotic recombinations, events that can cause gene deletions and other structural rearrangements linked to color vision defects [143]. One characteristic feature of the human X-chromosome opsin gene array is that it typically consists of a single copy of the L opsin gene, but often has multiple (modal value = 2, with significant individual variation) copies of the M opsin gene [144]. Another feature is the very high-frequency polymorphism of the structure of the L opsin gene that results in small shifts in spectral sensitivity of the coded photopigment [145]. Parallel studies of the opsin genes of chimpanzees [146] and macaque monkeys [147] show that in nearly all cases these other catarrhines have only a single copy of the M opsin gene and that L opsin gene polymorphisms are nearly absent. The former fact seems particularly relevant because the presence of multiple copies of the M opsin gene is expected to greatly accelerate the number of rearrangements of the structures of the M and L opsin genes [145]. As Verrelli and colleagues [146] have pointed out, it seems likely that there have been different selective trajectories for the M and L opsin genes of humans and other catarrhines.

Besides the differences in the incidence of color vision defects, are there other behavioral consequences of these differences in opsin gene arrays among the catarrhines? As noted above, it is believed that in the human retina L cones outnumber M cones by ~2:1. Some—perhaps many—other catarrhines may differ from this norm. For example, estimates that were obtained from studies of large samples of both macaque monkeys [148] and chimpanzees [149] suggest their retinas contain about equal numbers of M and L cones. And there is additional evidence that there may be still further variation in L:M representation in other catarrhine species [150]. Presuming all these estimates of species differences in L:M representation are correct, they nevertheless seem to have little behavioral impact. Direct measurements do show that there are small but consistent differences in the spectral sensitivities measured on macaque monkeys and chimpanzees versus those obtained from human observers [149, 151]. These may reflect differences in lens density and/or the opsin gene variations of the kind just noted. At the same time, the performance of macaque monkeys and equivalently tested humans on direct tests of color vision are all but identical [152, 153].

11.4.6 Color Vision and Mammalian Behavior

A complete account of the evolution of color vision requires both (1) a documentation of the alterations in visual systems that accrue over time, and (2) an understanding of how color vision supports adaptive behaviors. Although considerable advances have been made in revealing the history of changes in mammalian visual systems that underlie color vision, progress toward meeting the second criterion—that of learning how various mammals employ color vision in those tasks required for survival—has been limited.

Because most primates are highly visual animals, much of the work that seeks to link mammalian color vision to behavior has been carried out in the context of primate color vision. It is a long-held idea that primate color vision evolved in the context of foraging for colorful fruits or edible foliage (e.g., [154]), but it has also been suggested alternatively that a keen color sense could provide an asset in various social signaling situations as, for example, in the evaluation of skin or pelt coloration, or in the detection of predators [130]. In recent years, various computational models, that combine information about photopigment properties and the reflectance spectra of natural objects with assumptions as to how receptor activation links to color vision, suggest that the trichromacy characteristic of many primates could aid in the detection and discrimination of fruits and leaves that appear (to humans) red or yellow from among the green foliage backgrounds in which they are often viewed [155–158]. Testing the predictions in the natural world has turned out to be challenging, because primates often live in large social groups that inhabit visually cluttered environments making it difficult for an observer to know what visual tasks are being performed by any given individual at any given moment.

Because individual animals with widely variant color vision capacities must face the common visual challenges, platyrrhine monkeys provide a natural laboratory for studying the utility of primate color vision. The variations in opsin genes that underlie the color vision polymorphisms in these monkeys that were described above have survived for 20 million years and so they must have been maintained by natural selection [159, 160]. Among the several possible mechanisms to explain the maintenance of this polymorphism [130] perhaps the most favored is heterosis (heterozygous advantage). Applied to this context, heterosis predicts that female monkeys that are heterozygous, and therefore trichromatic, should be advantaged over their homozygous (dichromatic) counterparts [161]. If that is so, trichromatic monkeys should outperform dichromats in those tasks where color vision is important and in doing so ultimately derive some fitness advantages.

To date, the most common approach to studying the utility of color vision in these polymorphic species is to compare the performances of trichromatic and dichromatic monkeys in foraging tasks, thus exploiting situations where color cues might be expected to provide useful information. Over the past several years there have been some thirty studies that have sought answers to questions like this. Unfortunately, the conclusions have been inconsistent: some have detected a positive linkage between color vision phenotype and visual success while others failed to find such a relationship (these studies are summarized in Ref. [131], and in Ref. [133]). Although the basis for this inconsistency is unclear, it could reflect, (1) the practical difficulties inherent in such studies, (2) that no single explanation is going to hold across species and highly varied visual environments, or (3) that investigators have thus far underestimated the subtleties of the uses of color vision in these animals. Whatever the case, thus far there is no simple story linking color vision capacity and natural behavior in the New World monkeys.

Direct fitness measurements could provide an alternative way to evaluate whether heterosis might account for the maintenance of the color vision polymorphisms in platyrrhine monkeys. Two recent studies have attempted to do just this by examining several measures that provide an index of fitness within groups of capuchin (*Cebus sp.*) monkeys resident at respective study sites in Costa Rica [162] and in

Argentina [163]. Heterozygous monkeys at the Costa Rica site did not differ from their homozygous counterparts in either birth rates or in infant survival [162], thus providing no evidence that heterosis might explain the maintenance of color vision polymorphisms in these monkeys. In contrast, in the other study heterozygous females were significantly superior to homozygous females in both birth rate and in weight measures [163]. Neither study detected differences between the life spans of the heterozygous and the homozygous females, although larger samples might well be required to detect a significant difference. These two initial attempts to link fitness differences and opsin gene heterozygosity thus reached opposite conclusions. A resolution of this issue, as well as that of trying to understand how color vision variations feed into visual success, remains a task for the future. That task is sure to be aided by continued study of the color vision of the polymorphic platyrrhine monkeys.

References

1. Mollon JD. The origins of modern color science. In: Shevell SK, editor. The science of color. 2nd ed. Amsterdam: Elsevier; 2003. p. 1–39.
2. Frisch K v. Weitere Untersuchungen über den Farbensinn der Fische. Zool Jb Zool Physiol Tiere. 1913;34:43–68.
3. Frisch KV. Der Farbensinn und Formensinn der Biene. Zool Jb Zool Physiol Tiere. 1914;35:1–188.
4. Walls GL. The evolution of color vision. Am J Phys. 1943;14:161–5.
5. Lamb TD. Evolution of phototransduction, vertebrate photoreceptors and retina. Prog Ret Eye Res. 2013;36:52–119. doi:10.1016/j.preteyeres.2013.06.001.
6. Hunt DM, Collin SP. The evolution of photoreceptors and visual photopigments in vertebrates. In: Hunt DM, Hankins MW, Collin SP, Marshall NJ, editors. Evolution of visual and non-visual pigments. New York: Springer; 2014. p. 163–218. doi:10.1007/978-1-4614-4355-1.
7. Henze MJ, Oakley TH. The dynamic evolutionary history of pancrustacean eyes and opsins. Int Comp Biol. 2013;55:830–42. doi:10.1093/icb/icv100.
8. Larhammar D, Nordström K, Larsson TA. Evolution of vertebrate rod and cone phototransduction genes. Philos Trans R Soc Lond B Biol Sci. 2009;364:2867–80. doi:10.1098/rstb.2009.0077.
9. Gärtner W, Ullrich D, Vogt K. Quantum yield of CHAPSO-solubilized rhodopsin and 3-hydroxyretinal containing bovine opsin. Photochem Photobiol. 1991;54:1047–55. doi:10.1111/j.1751-1097.1991.tb02128.x.
10. Arikawa K, Stavenga DG. Insect photopigments: photoreceptor spectral sensitivities and visual adaptations. In: Hunt DM, Hankins MW, Collin SP, Marshall NJ, editors. Evolution of visual and non-visual pigments. New York: Springer; 2014. p. 137–62. doi:10.1007/978-1-4614-4355-1.
11. Bok MJ, Porter KL, Place AR, Cronin TW. Biological sunscreens tune polychromatic ultraviolet vision in mantis shrimp. Curr Biol. 2014;24:1636–42. doi:10.1016/j.cub.2014.05.071.
12. Nilsson D-E. Eye evolution and its functional basis. Vis Neurosci. 2013;30:5–20. doi:10.1017/S0952523813000035.
13. Rister J, Desplan C. The retinal mosaics of opsin expression in invertebrates and vertebrates. Dev Neurobiol. 2011;71:1212–26. doi:10.1002/dneu.20905.
14. Jacobs GH, Williams GA, Cahill H, Nathans J. Emergence of novel color vision in mice engineered to express a human cone photopigment. Science. 2007;315:1723–5. doi:10.1126/science.1138838.

15. Mancuso K, Hauswirth WW, Li Q, Connor TB, Kuckenbecker JA, Mauck MC, et al. Gene therapy for red-green color blindness in adult primates. Nature. 2009;461:784–7. doi:10.1038/nature08401.
16. Osorio D, Vorobyev M. Photoreceptor spectral sensitivities in terrestrial animals: adaptations for luminance and colour vision. Proc R Soc Lond B Biol Sci. 2005;272:1745–52. doi:10.1098/rspb.2005.3156.
17. Warrant EJ, Kelber A, Frederiksen R. Ommatidial adaptations for spatial, spectral and polarisation vision in arthropods. In: North G, Greenspan R, editors. Invertebrate neurobiology. Woodbury: Cold Spring Harbor Laboratory Press; 2007. p. 123–54.
18. Nilsson D-E. The evolution of eyes and visually guided behaviour. Philos Trans R Soc Lond B Biol Sci. 2009;364:2833–47. doi:10.1098/rstb.2009.0083.
19. Kelber A, Osorio D. From spectral information to animal colour vision: concepts and terminology. Proc R Soc Lond B Biol Sci. 2010;277:1617–25. doi:10.1098/rspb.2009.2118.
20. Campenhausen CV. Photoreceptors, lightness constancy and color vision. Naturwissenschaften. 1986;73:674–5. doi:10.1007/BF00366692.
21. Johnsen S, Kelber A, Warrant EJ, Sweeney AM, Widder EA, Lee RL, et al. Twilight and nocturnal illumination and its effects on color perception by the nocturnal hawkmoth *Deilephila elpenor*. J Exp Biol. 2006;209:789–800. doi:10.1242/jeb.02053.
22. Maximov VV. Environmental factors which may have led to the appearance of colour vision. Philos Trans R Soc Lond B Biol Sci. 2000;355:1239–42. doi:10.1098/rstb.2000.0675.
23. Jacobs GH, Williams GA, Fenwick JA. Influence of cone pigment coexpression on spectral sensitivity and color vision in the mouse. Vision Res. 2004;44:1615–22. doi:10.1016/j.visres.2004.01.016.
24. Cronin TW. Invertebrate vision in water. In: Warrant EJ, Nilsson D-E, editors. Invertebrate vision. Cambridge: Cambridge University Press; 2006. p. 211–49.
25. Matsui S, Seidou M, Uchiyama I, Sekiya N, Hiraki K, Yoshihara K, et al. 4-Hydroxyretinal, a new visual pigment chromophore found in the bioluminescent squid *Watasenia scintillans*. Biochim Biophys Acta. 1988;966:370–4. doi:10.1016/0304-4165(88)90087-6.
26. Kelber A. Invertebrate colour vision. In: Warrant EJ, Nilsson DE, editors. Invertebrate vision. Cambridge: Cambridge University Press; 2006. p. 250–90.
27. Jacobs GH. The distribution and nature of colour vision among the mammals. Biol Rev. 1993;68:413–71. doi:10.1111/j.1469-185X.1993.tb00738.x.
28. Battelle BA, Kempler KE, Saraf SR, Marten CE, Dugger DR, Spiser DI, et al. Opsins in Limulus eyes: characterization of three visible light-sensitive opsins unique to and co-expressed in median eye photoreceptors and a peropsin/RGR that is expressed in all eyes. J Exp Biol. 2015;218:466–79. doi:10.1242/jeb.116087.
29. Wald G, Rayport S. Vision in annelid worms. Science. 1977;196:1434–9. doi:10.1126/science.196.4297.1434.
30. Osorio D, Vorobyev M. A review of the evolution of animal colour vision and visual communication signals. Vision Res. 2008;48:2042–51. doi:10.1016/j.visres.2008.06.018.
31. de Hempel Ibarra N, Vorobyev M, Menzel R. Mechanisms, functions and ecology of colour vision in the honeybee. J Comp Physiol A. 2014;200:411–33. 10.1007/s00359-014-0915-1.
32. Telles FJ, Lind O, Henze MJ, Rodriguez-Girones MA, Goyret J, Kelber A. Out of the blue: the spectral sensitivity of hummingbird hawkmoths. J Comp Physiol A. 2014;200:537–46. doi:10.1007/s00359-014-0888-0.
33. Jacobs GH. Primate color vision: a comparative perspective. Vis Neurosci. 2008;25:619–33. doi:10.1017/S0952523808080760.
34. Vorobyev M, Osorio D, Bennett ATM, Marshall NJ, Cuthill IC. Tetrachromacy, oil droplets and bird plumage colours. J Comp Physiol A. 1998;183:621–33. doi:10.1007/s003590050286.
35. Olsson P, Lind O, Kelber A. Bird colour vision: behavioural thresholds reveal underlying receptor noise. J Exp Biol. 2015;218:184–93. doi:10.1242/jeb.111187.
36. Munz FW, McFarland WN. Evolutionary adaptations of fishes to the photic environment. In: Crescitelli F, editor. Handbook of sensory physiology. The visual system in vertebrates. 7/5th ed. New York: Springer; 1977. p. 194–274. doi:10.1007/978-3-642-66468-7_4.

37. Futahashi R, Kawahara-Miki R, Kinoshita M, Yoshitake K, Yajima S, Arikawa K, et al. Extraordinary diversity of visual opsin genes in dragonflies. Proc Natl Acad Sci U S A. 2015;112:E1247–56. doi:10.1073/pnas.1424670112.
38. Koshitaka H, Kinoshita M, Vorobyev M, Arikawa K. Tetrachromacy in a butterfly that has eight varieties of spectral receptors. Proc R Soc Lond B Biol Sci. 2008;275:947–54. doi:10.1098/rspb.2007.1614.
39. Bybee SM, Yuan F, Ramstetter MD, Llorente-Bousquets J, Reed RD, Osorio D, et al. UV photoreceptors and UV-yellow wing pigments in *Heliconius* butterflies allow a color signal to serve both mimicry and intraspecific communication. Am Nat. 2012;179:38–51. doi:10.1086/663192.
40. Lythgoe JN. The ecology of vision. Oxford: Clarendon; 1979.
41. Kelber A, Vorobyev M, Osorio D. Colour vision in animals – behavioural tests and physiological concepts. Biol Rev Camb Philos Soc. 2003;78:81–118. doi:10.1017/S1464793102005985.
42. Prokopy RJ, Owens ED. Visual detection of plants by herbivorous insects. Ann Rev Entomol. 1983;28:337–64. doi:10.1146/annurev.en.28.010183.002005.
43. Kelber A. Receptor based models for spontaneous colour choices in flies and butterflies. Ent Exp Appl. 2001;99:231–44. doi:10.1046/j.1570-7458.2001.00822.x.
44. Taylor LA, Maier EB, Byrne KJ, Amin Z, Morehouse NI. Colour use by tiny predators: jumping spiders show colour biases during foraging. Anim Behav. 2014;90:149–57. doi:10.1016/j.anbehav.2014.01.025.
45. Goldsmith TH. The evolution of visual pigments and colour vision. In: Gouras P, editor. The perception of colour. London: Macmillan; 1991. p. 62–89.
46. Stevens M. Predator perception and the interrelation between different forms of protective coloration. Proc R Soc Lond B Biol Sci. 2007;274:1457–64. doi:10.1098/rspb.2007.0220.
47. Regier T, Kay P, Cook RS. Focal colors are universal after all. Proc Natl Acad Sci U S A. 2005;102:8386–91. doi:10.1073/pnas.0503281102.
48. Jacobs GH. Recent developments in comparative color vision. In: Werner JM, Chalupa M, editors. The new visual neurosciences. Cambridge: MIT Press; 2013. p. 557–68.
49. Osorio D. Colour generalization by birds. In: Tommasi L, Nadel L, Peterson MA, editors. Cognitive biology: evolutionary and developmental perspectives on mind, brain, and behavior. Cambridge, MS: MIT Press; 2009. p. 129–46.
50. Kram YA, Mantey S, Corbo JC. Avian cone photoreceptors tile the retina as five independent, self-organizing mosaics. PLoS ONE. 2010;5(2):e8992. doi:10.1371/journal.pone.0008992.
51. Manning JR, Brainard DH. Optimal design of photoreceptor mosaics: why we do not see color at night. Vis Neurosci. 2009;26:5–19. doi:10.1017/S095252380808084X.
52. Allison WT, Barthel LK, Skebo KM, Takechi M, Kawamura S, Raymond PA. Ontogeny of cone photoreceptor mosaics in zebrafish. J Comp Neurol. 2010;518:4182–95. doi:10.1002/cne.22447.
53. Loew ER, Govardovskii VI, Rohlich P, Szel A. Microspectrophotometric and immunocytochemical identification of ultraviolet photoreceptors in geckos. Vis Neurosci. 1996;13:247–56. doi:10.1017/S0952523800007483.
54. Field GD, Gauthier JL, Sher A, Greschner M, Machado TA, Jepson LH, et al. Functional connectivity in the retina at the resolution of photoreceptors. Nature. 2010;467:673–8. doi:10.1038/nature09424.
55. Jonas JB, Schneider U, Naumann GOH. Count and density of human retinal photoreceptors. Graefe's Arch Clin Exp Ophthalmol. 1992;230:505–10. doi:10.1007/BF00181769.
56. Nilsson D-E, Kelber A. A functional analysis of compound eye evolution. Arthr Struct Dev. 2007;36:373–85. doi:10.1016/j.asd.2007.07.003.
57. Kelber A, Henze MJ. Colour vision: fly colour pathway shares receptors with achromatic vision. Curr Biol. 2013;23:R1043–5. doi:10.1016/j.cub.2013.10.032.
58. Hunt DM, Peichl L. S cones: evolution, retinal distribution, development, and spectral sensitivity. Vis Neurosci. 2014;31:115–38. doi:10.1017/S0952523813000242.

59. Hart NS. The visual ecology of avian photoreceptors. Prog Ret Eye Res. 2001;20:675–703. doi:10.1016/S1350-9462(01)00009-X.
60. Ahnelt PK, Schubert C, Kübber-Heiss A, Schiviz A, Anger E. Independent variation of retinal S and M cone photoreceptor topographies: a survey of four families of mammals. Vis Neurosci. 2006;23:429–35. doi:10.1017/0S095252380623342X.
61. Temple SE. Why different regions of the retina have different spectral sensitivities: A review of mechanisms and functional significance of intraretinal variability in spectral sensitivity in vertebrates. Vis Neurosci. 2011;28:281–93. doi:10.1017/S0952523811000113.
62. Siebeck UE, Marshall NJ. Ocular media transmission in reef fish – can coral reef fish see ultraviolet light? Vision Res. 2001;41:133–49. doi:10.1016/S0042-6989(00)00240-6.
63. Lind O, Mitkus M, Olsson P, Kelber A. Ultraviolet vision in birds: the importance of transparent eye media. Proc R Soc B Biol Sci. 2014;281:20132209. doi:10.1098/rspb.2013.2209.
64. Douglas RH, Jeffery G. The spectral transmission of ocular media suggests ultraviolet sensitivity is widespread among mammals. Proc R Soc Lond B Biol Sci. 2014;281:20132995. doi:10.1098/rspb.2013.2995.
65. Roth LSV, Kelber A. Nocturnal colour vision in geckos. Proc R Soc Lond B Biol Sci. 2004;275(S6):S485–7. doi:10.1098/rsbl.2004.0227.
66. Lamb TD, Collin SP, Pugh Jr EN. Evolution of the vertebrate eye: opsins, photoreceptors, retina and eye cup. Nature Revs Neurosci. 2007;8:960–75. doi:10.1038/nrn2283.
67. Collin SP, Wayne L, Davies WL, Hart NS, Hunt DM. The evolution of early vertebrate photoreceptors. Phil Trans R Soc Lond B Biol Sci. 2009;364:2925–40. doi:10.1098/rstb.2009.0099.
68. Amemiya CT, Alföldi J, Lee AP, Fan S, Philippe H, MacCallum I, et al. The African coelacanth genome provides insights into tetrapod evolution. Nature. 2013;496:311–6. doi:10.1038/nature12027.
69. Nathans J, Thomas D, Hogness DS. Molecular genetics of human color vision: the genes encoding blue, green and red pigments. Science. 1986;233:193–202. doi:10.1126/science.2937147.
70. Wang Y, Macke JP, Zack DJ, Klaunberg B, Bennett J, Gerhart G, et al. A locus control region adjacent to the human red and green visual pigment genes. Neuron. 1992;9:429–40. doi:10.1016/0896-6273(92)90181-C.
71. Chiu MI, Nathans J. A sequence upstream of the mouse blue visual pigment gene directs blue cone-specific transgene expression in mouse retinas. Vis Neurosci. 1994;11:773–80. doi:10.1017/S0952523800003072.
72. Smallwood PM, Yang Y, Nathans J. Role of a locus control region in the mutually exclusive expression of human red and green cone pigment genes. Proc Natl Acad Sci U S A. 2002;99:1008–11. doi:10.1073/pnas.022629799.
73. Tsujimura T, Hosoya T, Kawamura S. A single enhancer regulating the differential expression of duplicated red-sensitive opsin genes in zebrafish. PLoS Genet. 2010;6(12), e1001245. doi:10.1371/journal.pgen.1001245.
74. Carleton KL. Visual photopigment evolution in speciation. In: Hunt DM, Hankins MW, Collin SP, Marshall NJ, editors. Evolution of visual and non-visual pigments. New York: Springer; 2014. p. 241–68. doi:10.1007/978-1-4614-4355-1.
75. Smith AR, D'Annunzio L, Smith AE, Sharma A, Hofmann CM, Marshall NJ, et al. Intraspecific cone opsin expression variation in the cichlids of Lake Malawi. Mol Ecol. 2011;20:299–310. doi:10.1111/j.1365-294X.2010.04935.x.
76. Sabbah S, Laria RL, Gray SM, Hawryshyn CW. Functional diversity in the color vision of cichlid fishes. BMC Biol. 2010;8:133. doi:10.1186/1741-7007-8-133.
77. O'Quinn KE, Smith D, Nasweer Z, Schulte J, Engel SD, Loh YH, et al. Divergence in cis-regulatory sequences surrounding the opsin gene arrays of African cichlid fishes. BMC Evol Biol. 2011;11:120. doi:10.1186/1471-2148-11-120.
78. Dalton BE, Cronin TW, Marshall NJ, Carleton KL. The fish eye view: are cichlids conspicuous? J Exp Biol. 2010;213:2243–55. doi:10.1242/jeb.037671.

79. Hofmann CM, Carleton KL. Gene duplication and differential gene expression play an important role in the diversification of visual pigments in fish. Integr Comp Biol. 2009;49:630–43. doi:10.1093/icb/icp079.
80. Bloch N. Evolution of opsin expression in birds driven by sexual selection and habitat. Proc Roy Soc B. 2015;282:20143321. doi:10.1098/rspb.2014.2321.
81. Röhlich P, van Veen T, Szel A. Two different visual pigments in one retinal cone cell. Neuron. 1994;13:459–66. doi:10.1016/0896-6273(94)90053-1.
82. Isayama T, Chen Y, Kono M, Fabre E, Slavsky M, DeGrip WJ, et al. Coexpression of three opsins in cone photoreceptors of the salamander *Ambystoma tigrinum*. J Comp Neurol. 2014;522:2249–65. doi:10.1002/cne.23531.
83. Saarinen P, Pahlberg J, Herczeg G, Viljanen M, Karjalainen M, Shikano T, et al. Spectral tuning by selective chromophore uptake in rods and cones of eight populations of nine-spined stickleback (*Pungitius pungitius*). J Exp Biol. 2012;215:2760–73. doi:10.1242/jeb.068122.
84. Stavenga DG, Wilts BD. Oil droplets of bird eyes: microlenses acting as spectral filters. Phil Trans R Soc Lond B Biol Sci. 2014;369:20130041. doi:10.1098/rstb.2013.0041.
85. Hess M. Triple cones in the retinae of three anchovy species: *Engraulis encrasicolus*, *Cetengraulis mysticetus* and *Anchovia macrolepidota* (Engraulididae, Teleostei). Vision Res. 2009;49:1569–82. doi:10.1016/j.visres.2009.03.016.
86. Li YN, Tsujimura T, Kawamura S, Kawamura S, Dowling JE. Bipolar cell-photoreceptor connectivity in the zebrafish (Danio rerio) retina. J Comp Neurol. 2012;520:3786–802. doi:10.1002/cne.23168.
87. Martin GR, Osorio D. Vision in birds. In: Masland D, Albright RM, editors. The senses: a comprehensive reference. 1st ed. San Diego: Academic; 2008. p. 25–52.
88. Pignatelli V, Champs C, Marshall NJ, Vorobyev M. Double cones are used for colour discrimination in the reef fish, *Rhinecanthus aculeatus*. Biol Lett. 2010;6:537–9. doi:10.1098/rsbl.2009.1010.
89. Lisney TJ, Theiss SM, Collin SP, Hart NS. Vision in elasmobranchs and their relatives: 21st century advances. J Fish Biol. 2012;80:2024–54. doi:10.1111/j.1095-8649.2012.03253.x.
90. Schluessel V, Rick IP, Plischke K. No rainbow for grey bamboo sharks: evidence for the absence of colour vision in sharks from behavioural discrimination experiments. J Comp Physiol A. 2014;200:939–47. doi:10.1007/s00359-014-0940-0.
91. Van-Eyck SM, Siebeck UE, Champ CM, Marshall N, Hart NS. Behavioural evidence for colour vision in an elasmobranch. J Exp Biol. 2011;214:4186–92. doi:10.1242/jeb.061853.
92. Spady TC, Parry JWL, Robinson PR, Hunt DM, Bowmaker JK, Carleton KL. Evolution of the cichlid visual palette through ontogenetic subfunctionalization of the opsin gene arrays. Mol Biol Evol. 2006;23:1538–47. doi:10.1093/molbev/msl014.
93. Partridge JC, Archer SN, Lythgoe JN. Visual pigments in the individual rods of deep-sea fishes. J Comp Physiol A. 1988;162:543–50. doi:10.1007/BF00612519.
94. Neumeyer C. Tetrachromatic colour vision in goldfish: evidence from color mixture experiments. J Comp Physiol A. 1992;171:639–49. doi:10.1007/BF00194111.
95. Bowmaker JK. Evolution of vertebrate visual pigments. Vision Res. 2008;48:2022–41. doi:10.1016/j.visres.2008.03.025.
96. Korenyak DA, Govardovskii VI. Photoreceptors and visual pigments in three species of newts. J Evol Biochem Physiol. 2013;49:399–407. doi:10.1134/S0022093013040038.
97. Mohun SM, Davies WL, Bowmaker JK, Pisani D, Himstedt W, Gower DJ, et al. Identification and characterization of visual pigments in caecilians (Amphibia: Gymnophiona), an order of limbless vertebrates with rudimentary eyes. J Exp Biol. 2010;213:3586–92. doi:10.1242/jeb.045914.
98. Rocha FAF, Saito CA, Silveira LCL, de Souza JM, Ventura DF. Twelve chromatically opponent ganglion cell types in turtle retina. Vis Neurosci. 2008;25:307–15. doi:10.1017/S0952523808080516.
99. Goldsmith TH, Butler BK. The roles of receptor noise and cone oil droplets in the photopic spectral sensitivity of the budgerigar *Melopsittacus undulatus*. J Comp Physiol A. 2003;189:135–42. doi:10.1007/s00359-002-0385-8.

100. Hart NS, Hunt DM. Avian visual pigments: characteristics, spectral tuning, and evolution. Am Nat. 2007;169 Suppl 1:S7–S26. doi:10.1086/510141.
101. Lind O, Delhey K. Visual modelling suggests a weak relationship between the evolution of ultraviolet vision and plumage coloration in birds. J Evol Biol. 2015;28:715–22. doi:10.1111/jeb.12595.
102. Stabell B, Stabell U. Duplicity theory of vision. Cambridge: Cambridge University Press; 2009.
103. Buck SL. The interaction of rod and cone signals: pathways and psychophysics. In: Werner JS, Chalupa LM, editors. The new visual neurosciences. Cambridge: MIT Press; 2014. p. 485–97.
104. Smith VC, Pokorny J. Large-field trichromacy in protanopes and deuteranopes. J Opt Soc Am. 1977;67:213–20. doi:10.1364/JOSA.67.000213.
105. Reitner A, Sharpe LT, Zrenner E. Is colour vision possible with only rods and blue-sensitive cones? Nature. 1991;352:798–800. doi:10.1038/352798a0.
106. Levenson DH, Ponganis PJ, Crognale MA, Deegan II JF, Dizon A, Jacobs GH. Visual pigments of marine carnivores: Pinnipeds, polar bear, and sea otter. J Comp Physiol A. 2006;192:833–43. doi:10.1007/s00359-006-0121-x.
107. Griebel U, Schmid A. Color vision in the california sea lion (*Zalophus californianus*). Vision Res. 1992;32:477–82. doi:10.1016/0042-6989(92)90239-F.
108. Scholtyssek C, Kelber A, Dehnhardt G. Why do seals have cones? Behavioural evidence for colour-blindness in harbor seals. Anim Cogn. 2015;18:184–93. doi:10.1007/s10071-014-0823-3.
109. Luo Z-X. Transformation and diversification in early mammal evolution. Nature. 2007;450:1011–9. doi:10.1038/nature06277.
110. Wakefield MJ, Anderson M, Chang E, Wei KJ, Kaul R, Graves JA, et al. Cone visual pigments of monotremes: filling the phylogenetic gap. Vis Neurosci. 2008;25:257–64. doi:10.1017/S0952523808080255.
111. Jacobs GH. Evolution of colour vision in mammals. Philos Trans R Soc Lond B Biol Sci. 2009;364:2957–67. doi:10.1098/rstb.2009.0039.
112. Davies WL, Collin SP, Hunt DM. Molecular ecology and adaptation of visual pigments in craniates. Mol Ecol. 2012;21:3121–58. doi:10.1111/j.1365-294X.2012.05617.x.
113. Deeb SS. Visual pigments and colour vision in marsupials and monotremes. In: Deakin JE, Waters PD, Marshall Graves JA, editors. Marsupial genetics and genomics. Dordrecht: Springer; 2010. p. 403–14.
114. Hunt DM, Wilkie SE, Bowmaker JK, Poopalasundaram S. Vision in the ultraviolet. Cell Mol Life Sci. 2001;58:1583–98. doi:10.1007/PL00000798.
115. Yokoyama S. Evolution of dim-light and color vision pigments. Annu Rev Genomics Hum Genet. 2008;9:259–82. doi:10.1146/annurev.genom.9.081307.164228.
116. Walls GL. The vertebrate eye and its adaptive radiation. Bloomfield Hills, MC: Cranbrook Institute of Science; 1942.
117. Kemp TS. The origin and evolution of mammals. Oxford: Oxford University Press; 2005.
118. Striedter GF. Principles of brain evolution. Sinauer: Sutherland; 2005.
119. Heesy CP, Hall MI. The nocturnal bottleneck and the evolution of mammalian vision. Brain Behav Evol. 2010;75:195–203. doi:10.1159/000314278.
120. Gerkema MP, Davies WIL, Foster RG, Menaker M, Hut RA. The nocturnal bottleneck and the evolution of activity patterns in mammals. Proc R Soc Lond B Biol Sci. 2013;280:20130508. doi:10.1098/rspb.2013.0508.
121. Jacobs GH. Evolution of color vision and its reflections in contemporary mammals. In: Elliott AJ, Fairchild MD, Franklin A, editors. Handbook of color psychology. Cambridge: Cambridge University Press; 2015. p. 110–30.
122. Kumar S, Hedges SB. A molecular timescale for vertebrate evolution. Nature. 1998;392:917–9220. doi:10.1038/31927.

123. Schmitz L, Motani R. Nocturnality in dinosaurs inferred from scleral ring and orbit morphology. Science. 2011;332:705–8. doi:10.1126/science.1200043.
124. Angielczyk KD, Schmitz L. Nocturnality in synapsids predates the origin of mammals by over 100 million years. Proc R Soc Lond B Biol Sci. 2014;281:20141642. doi:10.1098/rspb.2014.1642.
125. Jacobs GH. Losses of functional opsin genes, short-wavelength cone photopigments, and color vision – a significant trend in the evolution of mammalian vision. Vis Neurosci. 2013;30:39–53. doi:10.1017/S0952523812000429.
126. Wikler KC, Rakic P. Distribution of photoreceptor subtypes in the retina of diurnal and nocturnal primates. J Neurosci. 1990;10:3390–401.
127. Jacobs GH, Neitz M, Neitz J. Mutations in S-cone pigment genes and the absence of colour vision in two species of nocturnal primate. Proc R Soc Lond B Biol Sci. 1996;263:705–10. doi:10.1098/rspb.1996.0105.
128. Meredith RW, Gatesy J, Emerling CW, York VM, Springer MS. Rod monochromacy and the coevolution of cetacean retinal opsins. PLoS Genet. 2013;9(4):e1003432. doi:10.1371/journal.pgen.1003432.
129. Emerling CW, Springer MS. Genomic evidence for rod monochromacy in sloths and armadillos suggests early subterranean history for xenarthra. Proc R Soc Lond B Biol Sci. 2015;282:20142192. doi:10.1098/rspb.2014.2192.
130. Surridge AK, Osorio D, Mundy NI. Evolution and selection of trichromatic vision in primates. Trends Ecol Evol. 2003;18:198–206. doi:10.1016/S0169-5347(03)00012-0.
131. Jacobs GH. New world monkeys and color. Int J Primatol. 2007;28:729–59. doi:10.1007/s10764-007-9168-y.
132. Jacobs GH, Nathans J. The evolution of primate color vision. Sci Am. 2009;300:40–7. doi:10.1038/scientificamerican0409-56.
133. Kawamura S, Hiramatsu C, Melin AD, Schaffner CM, Aureli F, Fedigan LM. Polymorphic color vision in primates: evolutionary considerations. In: Hirai H, Imai H, Go Y, editors. Post-genome biology of primates. Tokyo: Springer; 2012. p. 93–120.
134. Proulx SR, Phillips PC. Allelic divergence precedes and promotes gene duplication. Evolution. 2006;60:881–92. doi:10.1554/05-507.1.
135. Onishi A, Koike S, Ida M, Imai H, Schichida Y, Osamu T, et al. Dichromatism in macaque monkeys. Nature. 1999;402:139–40. doi:10.1038/45966.
136. Jacobs GH, Williams GA. The prevalence of defective color vision in old world monkeys and apes. Color Res Appl. 2001;26:S123–7. doi:10.1002/1520-6378(2001)26:1+<::AID-COL27>3.0.CO;2-6.
137. Post RH. Population differences in red and green color vision deficiency: a review, and a query on selection relaxation. Eugen Q. 1962;9:131–46. doi:10.1080/19485565.1962.9987517.
138. Kalmus H. Kinds and intensity of selection at the colour vision loci. In: Mollon JD, Sharpe LT, editors. Colour vision physiology and psychophysics. London: Academic; 1983. p. 51–5.
139. Fletcher R, Voke J. Defective colour vision, fundamentals, diagnosis and management. Bristol: Adam Hilger; 1985.
140. Xie JZ, Tarczy-Hornoch K, Lin J, Cotter SA, Torres M, Varma R. Color vision deficiency in preschool children: the multi-ethnic pediatric eye disease study. Ophthalmology. 2014;121:1469–74. doi:10.1016/j.ophtha.2014.01.018.
141. Pickford RW. Natural selection and colour blindness. Eugen Rev. 1963;55:97–101.
142. Reimchen TE. Human color-vision deficiencies and atmospheric twilight. Soc Biol. 1987;34:1–11.
143. Nathans J, Piantanida TP, Eddy RL, Shows TB, Hogness DS. Molecular genetics of inherited variation in human color vision. Science. 1986;233:203–10. doi:10.1126/science.3485310.
144. Carroll J, Neitz J, Neitz M. 2002 Estimates of L:M cone ratio from ERG flicker phototometry and genetics. J Vis. 2002;2:531–42. doi:10.1167/2.8.1.
145. Neitz J, Neitz M. The genetics of normal and defective color vision. Vision Res. 2011;51:633–51. doi:10.1016/j.visres.2010.12.002.

146. Verrelli BC, Lewis Jr CM, Stone AC, Perry GH. Different selection pressures shape the molecular evolution of color vision in chimpanzee and human populations. Mol Biol Evol. 2008;25:2735–43. doi:10.1093/molbev/msn220.
147. Onishi A, Koike S, Ida-Hosonuma M, Imai H, Schichida Y, Takenaka O, et al. Variations in long- and middle-wavelength-sensitive opsin gene loci in crab-eating monkeys. Vision Res. 2002;42:281–92. doi:10.1016/S0042-6989(01)00293-0.
148. Jacobs GH, Deegan II JF. Spectral sensitivity of macaque monkeys measured with ERG flicker photometry. Vis Neurosci. 1997;14:921–8.
149. Jacobs GH, Deegan JF, Moran JL. ERG measurements of the spectral sensitivity of common chimpanzee (*Pan troglodytes*). Vis Res. 1996;16:2587–94. doi:10.1016/0042-6989(95)00335-5.
150. Jacobs GH, Deegan II JF. Uniformity of colour vision in Old World monkeys. Proc R Soc Lond B Biol Sci. 2001;266:2023–8.
151. Lindbloom-Brown Z, Tait LJ, Horwitz GD. Spectral sensitivity differences beween rhesus monkeys and humans: implications for neurophysiology. J Neurophysiol. 2014;112:3164–72. doi:10.1152/jn.00356.2014.
152. Gagin G, Bohon KS, Butensky A, Gates MA, Hu JY, Lafer-Sousa R, et al. Color-detection thresholds in rhesus macaque monkeys and humans. J Vis. 2014;14:1–15. doi:10.1167/14.8.12.
153. Horwitz GD. What studies of macaque monkeys have told us about human color vision. Neuroscience. 2015. doi:10.1016/j.neuroscience.2014.10.007.
154. Polyak S. The vertebrate visual system. Chicago: University of Chicago Press; 1957.
155. Vorobyev M, Osorio D. Colour vision as an adaptation to frugivory in primates. Proc R Soc Lond B Biol Sci. 1996;263:593–9. doi:10.1098/rspb.1996.0089.
156. Regan BC, Julliot C, Simmen B, Vienot F, Charles-Dominique P, Mollon JD. Fruits, foliage and the evolution of primate colour vision. Philos Trans R Soc Lond B Biol Sci. 2001;356:229–83. doi:10.1098/rstb.2000.0773.
157. Dominy NJ, Lucas PW. Ecological importance of trichromatic colour vision to primates. Nature. 2001;410:363–5. doi:10.1038/35066567.
158. Parraga CA, Troscianko T, Tolhurst DJ. Spatiochromatic properties of natural images and human vision. Curr Biol. 2002;12:483–587. doi:10.1016/S0960-9822(02)00718-2.
159. Boissinot S, Tan Y, Shyue SK, Schneider H, Sampaio I, Neiswanger K, et al. Origins and antiquity of x-linked triallelic color vision systems in new world monkeys. Proc Natl Acad Sci U S A. 1998;95:13749–54. doi:10.1073/pnas.95.23.13749.
160. Surridge AK, Mundy NI. Trans-specific evolution of opsin alleles and the maintenance of trichromatic colour vision in callitrichine primates. Mol Ecol. 2002;11:2157–69. doi:10.1046/j.1365-294X.2002.01597.x.
161. Mollon JD, Bowmaker JK, Jacobs GH. Variations of colour vision in a new world primate can be explained by polymorphism of retinal photopigments. Proc R Soc Lond B Biol Sci. 1984;222:373–99. doi:10.1098/rspb.1984.0071.
162. Fedigan LM, Melin AD, Addicott JF, Kawamura S. The heterozygote superiority hypothesis for polymorphic color vision is not supported by long-term fitness data from wild neotropical monkeys. PLoS ONE. 2014;9:e84872. doi:10.1371/journal.pone.0084872.
163. Green AT. Consequences of color vision variation on performance and fitness in capuchin monkeys. Doctoral Dissertation, University of Montana, Missoula. 2014.

Index

J. Kremers et al. (eds.), *Human Color Vision*, Springer Series in Vision Research, DOI 10.1007/978-3-319-44978-4

D

Kolibri 360 GmbH
Ferdinand-Jühlke-Straße 7
99095 Erfurt, Deutschland
produktsicherheit@kolibri360.de